—ADVANCED LEVEL—

P·U·R·E

MATHEMATICS

D.A. Bryars, Ph.D.

Deputy Headmaster
Wickersley Comprehensive School
Wickersley

Bell & Hyman

Published by BELL & HYMAN
an imprint of Unwin Hyman Ltd
Denmark House
37–39 Queen Elizabeth Street
London SE1 2QB

ISBN: 0 7135 2773 0

Typeset by MS Filmsetting Ltd, Frome, Somerset
Printed and bound in Great Britain by
Mackays of Chatham Ltd., Kent

Contents

Preface

This book has been written in the hope that it will be of value to students taking 'single' Mathematics at Advanced Level. It aims to cover almost all the pure mathematics in the various syllabuses available for this level of study.

In particular, it includes all the material representing a 'minimal core syllabus' recommended by the Standing Conference on University Entrance and the Council for National Academic Awards. I also trust that it will prove to be a suitable companion text for my *Advanced Level Statistics*.

I have endeavoured to produce a clear concise text, which is appropriately paced, taking account of the developing mathematical maturity of the student as the course progresses. There is ample cross-referencing and important statements and definitions are displayed for easy reference. Topics are developed in a logical sequence but can be re-ordered to suit personal preferences and the particular needs of students.

There are many worked examples and a substantial selection of questions. Some are routine, but many have been chosen to combine two or more topics to encourage the student to apply the techniques acquired in a variety of contexts.

I would like to thank the following examination authorities for permission to produce past examination questions: The Associated Examining Board; Cambridge University Local Examinations Syndicate; Joint Matriculation Board; University of London School Examinations Board; The North Regional Examinations Board; Oxford and Cambridge Schools Examination Board; University of Oxford Delegacy of Local Examinations; Southern Universities' Joint Board; Welsh Joint Education Committee.

I would also like to mention my indebtedness to Lois Hadfield, Gillian Haworth and R. I. Porter for supplying many of the solutions to the exercises, though I am, of course, responsible for the accuracy of these solutions. Finally, I would like to thank R. D. L. Artset Ltd for preparing the artwork and all the staff at Bell and Hyman for their assistance in the production of this book.

D. A. Bryars

List of symbols

Defined functions

$\sin x$	trig ratios, trigonometric functions	
$\cos x$		
$\tan x$		
$\sec x$	95	
$\operatorname{cosec} x$	95	
$\cot x$	96	
$\sin^{-1} x$	inverse trigonometric functions	234
$\cos^{-1} x$		
$\tan^{-1} x$		
a^x	an exponential function	9
e^x	the exponential function	250
e	e^1, an important number in mathematics	250
$\log_a x$	logarithm to the base a, "log to base a"	14
$\lg x$	log to base 10	18
$\ln x$	log to base e	251

Associated with calculus

$\rightarrow$	tends to	161
$\lim_{a \to b} f(x)$	the limit as a tends to b of $f(a)$	161
δx	a small change in x, a small increment	188
f'	the derived function, gradient function	165
f^n	the nth derivative, $f^2(x) \equiv f''(x)$	435
$\dfrac{dy}{dx}$	the derivative, rate of change of y w.r.t. x	168
$\displaystyle\int f(x)dx$	the (indefinite) integral of $f(x)$ w.r.t. x, sometimes denoted by $F(x)$	201
$\displaystyle\int_a^b f(x)dx$	the definite integral	319

Associated with summation

$\sum$	the sum of	288
AP	arithmetic progression	278
GP	geometric progression	282
$n!$	n factorial	301
$\dbinom{n}{r}$	$\equiv {}^nC_r \equiv \dfrac{n!}{(n-r)!r!}$	301

[in the context of vectors, $n\mathbf{i} + r\mathbf{j}$, page 461]

Vector notation

$\vec{AB}$	a vector	455
$\mathbf{a}, \mathbf{b}$	alternative notation for vectors	455

|a| or *a*	the modulus of the vector **a**, 'length'	460
â	a unit vector, length one unit	460
i, j, k	unit vectors, perpendicular to each other in specified directions	461, 462

$$\begin{pmatrix} a \\ b \\ c \end{pmatrix}$$ the vector $a\mathbf{i} + b\mathbf{j} + c\mathbf{k}$ 461, 462

r	often a variable vector	461, 462
a . b	the dot product	465

Complex number notation

i	usually $\sqrt{-1}$	492
$a + bi$	a complex number	493
z, w	usually variables ranging over the complex numbers	493
Re(z)	the real part of z	493
Im(z)	the imaginary part of z	493
z^{*}	the complex conjugate of z	494
$\|z\|$	the modulus of z	505
arg(z)	the argument of z	505

Chapter 1 Number

1.1 Number — sets — real numbers

Number

During the course of studying arithmetic one first meets examples of *natural numbers*. The natural numbers are 0, 1, 2, 3, 4, ..., where the '...' indicates that the sequence continues indefinitely. This can then be extended to include the negative numbers -1, -2, -3, -4, ... These are then referred to collectively as *integers*. Thus every natural number is an integer.

The next development involves the introduction of *rational numbers* or *rationals*. Any rational number can be expressed in the form m/n where m is an integer and n is a natural number, other than zero. In particular, if m is an integer then $m/1$ is a rational; and since $m = m/1$, every integer is a rational. Rationals are popularly called fractions.

An interesting property of rational numbers is that they can all be expressed as decimals which either *terminate* or *recur*. For example,

$$\tfrac{1}{2} = 0.5 \quad \text{and} \quad \tfrac{1}{8} = 0.125$$

where in each case the decimal terminates.

However,

$$\tfrac{1}{3} = 0.333\,333\,3\ldots$$
$$= 0.\dot{3}$$

and

$$\tfrac{1}{7} = 0.142\,857\,142\,857\,142\,85\ldots$$
$$= 0.\dot{1}42\,85\dot{7}$$

Now in this case the decimals are recurring.

In the first example one number recurs or is repeated indefinitely, and in the second a string of numbers is repeated indefinitely. Notice the use of the dots above the integers to indicate which numbers or string of numbers is repeated.

Conversely, *any terminating or recurring decimal is a rational*. The fact that a terminating decimal is a rational is obvious, for example

$$0.625 = \frac{625}{1000} = \tfrac{5}{8}$$

The proof that a recurring decimal is a rational is given on page 294.

Sets

A *set* is simply a collection of objects. If A is a set and x is a member of A, that is x belongs to A, then we write $x \in A$. If y does not belong to A we write $y \notin A$. We can represent this diagrammatically (see Fig. 1.1).

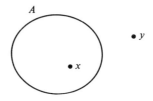

Fig. 1.1

Some important sets have special names.

Definition

> $\mathbb{N}$ is the set of natural numbers.
> $\mathbb{Z}$ is the set of integers.
> $\mathbb{Q}$ is the set of rationals.

To indicate that n is a natural number we can simply write $n \in \mathbb{N}$. Similarly $p \in \mathbb{Q}$ indicates that p is a rational.

Definition

> Two sets are equal if and only if they have the same members.

Definition

> A is called a subset of B, written $A \subseteq B$ if and only if every member of A is a member of B. Further, if A is a subset of B and $A \neq B$ then A is called a proper subset of B, written $A \subset B$.

Notice that if $A \subset B$ then $A \subseteq B$.

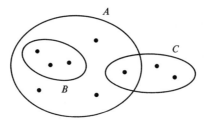

Fig. 1.2

In Fig. 1.2 it is evident that

$$B \subset A \quad \text{and} \quad C \nsubseteq A$$

It should be noted that $\mathbb{N} \subset \mathbb{Z} \subset \mathbb{Q}$.

A set is often defined using notation of the form:

$$\{x : x \text{ has some property}\}$$

Where the members are precisely those x with the property within the brackets $\{\ \}$. As an example, consider the set defined as

$$\{x : x \in N \quad \text{and} \quad x < 5\}$$

This set consists of those members of N which are less than 5, that is the natural numbers 0, 1, 2, 3, 4.

A *finite* set, such as this, may be defined simply by writing $\{0, 1, 2, 3, 4\}$. Using this notation we may define other sets related to N, Z and Q.

Definitions

$$N^+ = \{x : x \in N \quad \text{and} \quad x \neq 0\} = \{x : x \in N \quad \text{and} \quad x > 0\}$$

$$Z^+ = \{x : x \in Z \quad \text{and} \quad x > 0\}$$

$$Q^+ = \{x : x \in Q \quad \text{and} \quad x > 0\}$$

$$Z_0^+ = \{x : x \in Z \quad \text{and} \quad x \geqslant 0\}$$

$$Q_0^+ = \{x : x \in Q \quad \text{and} \quad x \geqslant 0\}$$

It may be convenient to write, for instance

$$q \in \{x : x \in Q \quad \text{and} \quad x < 3\}$$

rather than state in words that q is a rational number which is less than three.

Real numbers

The Ancient Greeks had thought that every number was a rational, and it came as quite a shock to them when they discovered that there were other numbers, which we shall call *irrationals*.

Consider the length of the diagonal of a square of side 1 unit (Fig. 1.3). By Pythagoras' theorem the length of the diagonal is $\sqrt{2}$, the square root of 2. We prove that $\sqrt{2}$ is not a rational.

The style of proof is almost as interesting as the result itself, and is called *proof by contradiction*.

We shall suppose that $\sqrt{2}$ is a rational, and from this statement alone deduce a contradiction — a statement which follows from the supposition but which cannot be true. Having done that we are obliged to accept that our supposition is false.

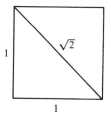

Fig. 1.3

Suppose then that $\sqrt{2}$ is rational. Then there are numbers m and n such that $m \in Z^+$, $n \in N^+$ and

$$\sqrt{2} = \frac{m}{n} \qquad \text{(since } \sqrt{2} \text{ is positive).}$$

Such a fraction could be cancelled if necessary, so we can agree that *m and n are not both even*. Squaring both sides we obtain

$$2 = \frac{m^2}{n^2} \qquad \Rightarrow 2n^2 = m^2$$

so m^2 is even. Since an odd number squared is odd, *m must be even*.

So

$$m = 2p, \qquad \text{for some } p \in \mathbb{N}$$

Substituting

$$2n^2 = (2p)^2$$

$$\Rightarrow \quad n^2 = 2p^2.$$

This time, since n^2 is even, *n must be even*. But this contradicts the fact, already agreed, that *m and n ar not both even*. So we have to accept that the supposition is false and $\sqrt{2}$ is *not* a rational!!

From what we have stated earlier, the decimal expansion of an irrational will be non-terminating and non-recurring.

An irrational number can be approximated as closely as we please but cannot be written down explicitly in decimal form. Now

$$\sqrt{2} \simeq 1.414 \qquad 3\,\text{d.p.}$$

and a better approximation is

$$\sqrt{2} \simeq 1.414\,214 \qquad 6\,\text{d.p.}$$

and we can continue as far as we please, but we can never attain equality.

The reader may be aware that π, the ratio of the circumference to the diameter of a circle is an irrational number. There are, in fact, an infinite number of irrational numbers, such as $\sqrt{2}, 2\sqrt{2}, \sqrt{3}, \sqrt{2} + \sqrt{5}, \pi^2, 3\pi$.

Definition

> A number which is either a rational or an irrational number is called a *real number*, or *real*. The set of real numbers is denoted by $\mathbb{R}$.

Notice that we have proved that $\mathbb{Q} \subset \mathbb{R}$ and so

$$\mathbb{N} \subset \mathbb{Z} \subset \mathbb{Q} \subset \mathbb{R}.$$

There is a further extension to the number system called the *complex numbers*, which is studied in Chapter 16.

Summary

$$\mathbb{N} = \{0, 1, 2, 3, \ldots\}$$
$$\mathbb{Z} = \{0, \pm 1, \pm 2, \pm 3, \ldots\}$$
$$\mathbb{Q} = \left\{\frac{m}{n} : m \in \mathbb{Z} \quad \text{and} \quad n \in \mathbb{N}^+\right\}$$
$$\mathbb{R} = \{x : \text{is a real number}\}$$
$$\mathbb{N} \subset \mathbb{Z} \subset \mathbb{Q} \subset \mathbb{R}$$

Exercise 1.1

1. Indicate which of the following statements are true.

    (a) $\mathbb{Q} = \mathbb{R}$ (b) $\mathbb{Q}^+ \subset \mathbb{R}$ (c) $\mathbb{Z} \subseteq \mathbb{R}$
    (d) $\mathbb{N} \subset \mathbb{R}^+$ (e) If $p \in \mathbb{R}$ then $p \in \mathbb{Q}$
    (f) If $p \in \mathbb{Z}^+$ then $p \in \mathbb{N}$.

2. Indicate which of the following statements are true and, for those which are not, give a counter example. (That is, an example which illustrates its falsity.)

    (a) If $x \in \mathbb{Q}$ and $y \in \mathbb{Q}$ then $x + y \in \mathbb{Q}$.
    (b) If $x \in \mathbb{N}$ and $y \in \mathbb{N}$ then $x + y \in \mathbb{N}$.
    (c) If $x \in \mathbb{N}$ and $y \in \mathbb{N}$ then $x - y \in \mathbb{N}$.
    (d) If $x \in \mathbb{N}$ and $y \in \mathbb{N}$ then $x - y \in \mathbb{Z}$.
    (e) If x and y are irrational then xy is irrational.
    (f) If $x \in \mathbb{R}$ and $y \in \mathbb{R}$ then $x + y \in \mathbb{R}$, $xy \in \mathbb{R}$ and $x/y \in \mathbb{R}$.
    (g) If $x \in \mathbb{Q}$ then $\sqrt{x}$ is irrational.
    (h) $x \in \mathbb{R}^+ \Leftrightarrow x$ is positive.

3. Indicate whether the following statements are true or false.

    (a) $3 \in \{x : x \in \mathbb{R} \text{ and } x < 3\}$.
    (b) $2 \in \{x : x \text{ is prime}\}$.
    (c) $1 \in \{x : x \text{ is prime}\}$.
    (d) $\sqrt{2} \in \{x : x \in \mathbb{Q} \text{ and } x^2 = 2\}$.

4. Prove that $\sqrt{3}$ is an irrational number.

5. Describe in words the following sets.

    (a) $\{x : x = 1 \text{ or } x = 2\}$ (b) $\{n : n \text{ a prime number}\}$
    (c) $\{n : \text{there is an } m \in \mathbb{N} \text{ and } n = 2m\}$ (d) $\{x : x < 0 \text{ and } x \in \mathbb{R}\}$
    (e) $\{x : x > 0 \text{ and } x \in \mathbb{R}\}$ referred to as $\mathbb{R}^+$
    (f) $\{x : x \geqslant 0 \text{ and } x \in \mathbb{R}\}$ referred to as $\mathbb{R}_0^+$

6. Use a proof by contradiction to prove that if m and n are integers such that mn is even, then at least one of m, n is even.

JMB 1982

7. Express

(a) $\frac{5}{8}$ as a decimal.

(b) (i) 0.25   (ii) 0.4375   (iii) $0.\dot{6}$

as fractions.

1.2   Surds

It should be noted that $\sqrt{2}$ is the *positive* real whose square is 2. The negative real whose square is 2 is written $-\sqrt{2}$.

The cube root of a number p, written $\sqrt[3]{p}$, is the real such that

$$\sqrt[3]{p} \times \sqrt[3]{p} \times \sqrt[3]{p} = p.$$

So

$$\sqrt[3]{27} = 3 \quad \text{and} \quad \sqrt[3]{-8} = -2.$$

Definition

> A surd is an irrational which can be expressed in terms of roots of rational numbers.

Examples of surds are

$$\sqrt{5}, \qquad \frac{\sqrt{2} + \sqrt{5}}{3}, \qquad (1 + \sqrt{3})^2, \qquad \frac{\sqrt[3]{10} + 1}{\sqrt{2}}.$$

However, $\sqrt{4}$, $\sqrt{16}$ and, interestingly, π, are not surds.

Now the square of $\sqrt{(a \cdot b)}$ is $a \cdot b$, and so also is the square of $\sqrt{a} \cdot \sqrt{b}$. So it follows that:

Rule of surds

1	$\sqrt{(a \cdot b)} = \sqrt{a} \cdot \sqrt{b}$	for $a \in \mathbb{R}$ and $b \in \mathbb{R}$
2	$\sqrt{\dfrac{a}{b}} = \dfrac{\sqrt{a}}{\sqrt{b}}$	for $a \in \mathbb{R}$ and $b \in \mathbb{R}$, where $b \neq 0$.

It should be remembered, however, that $\sqrt{(a + b)}$ cannot be simplified to $\sqrt{a} + \sqrt{b}$ (except when a or b is zero).

For example,

$$\sqrt{(9 + 16)} = \sqrt{25} = 5$$

but

$$\sqrt{9} + \sqrt{16} = 3 + 4 = 7.$$

It is not immediately apparent that

$$\sqrt{2} + 1 = \frac{1}{\sqrt{2} - 1}.$$

so our first aim is to develop techniques for simplifying or manipulating surds.

Example 1.1

Simplify the following surds.

(a) $\sqrt{72}$ (b) $\sqrt{2} + \sqrt{8} + \sqrt{32}$ (c) $\dfrac{\sqrt{14}}{\sqrt{63}}$ (d) $\dfrac{\sqrt{27}}{\sqrt{15}}$

(a) $\sqrt{72} = \sqrt{(36 . 2)}$

$\qquad = \sqrt{36} . \sqrt{2}$ by the first rule of surds

$\qquad = 6\sqrt{2}$

(b) $\sqrt{2} + \sqrt{8} + \sqrt{32} = \sqrt{2} + \sqrt{(4 . 2)} + \sqrt{(16 . 2)}$

$\qquad\qquad\qquad\qquad = \sqrt{2} + 2\sqrt{2} + 4\sqrt{2}$

$\qquad\qquad\qquad\qquad = 7\sqrt{2}$

(c) $\dfrac{\sqrt{14}}{\sqrt{63}} = \sqrt{\dfrac{14}{63}}$ *or* $\dfrac{\sqrt{14}}{\sqrt{63}} = \dfrac{\sqrt{2} . \sqrt{7}}{\sqrt{9} . \sqrt{7}}$

$\qquad\quad = \sqrt{\dfrac{2}{9}} \qquad\qquad\qquad\quad = \dfrac{\sqrt{2}}{3}$

$\qquad\quad = \dfrac{\sqrt{2}}{3}$

(d) $\dfrac{\sqrt{27}}{\sqrt{15}} = \sqrt{\dfrac{27}{15}}$

$\qquad\quad = \sqrt{\dfrac{9}{5}}$

$\qquad\quad = \dfrac{3}{\sqrt{5}}$

A surd is commonly written with a rational for the denominator when applicable. The process is called *rationalising the denominator*.

Thus

$$\frac{3}{\sqrt{5}} = \frac{3\sqrt{5}}{\sqrt{5}\sqrt{5}} = \frac{3\sqrt{5}}{5}$$

A useful general result to recall is

$$(a + b)(a - b) = a^2 - b^2.$$

So, in particular

$$(\sqrt{3} + \sqrt{2})(\sqrt{3} - \sqrt{2}) = 3 - 2$$
$$= 1.$$

Notice that in this example the product of two irrationals is rational.

Example 1.2

Rationalise the denominators of the following surds.

(a) $\dfrac{2\sqrt{3}}{3\sqrt{5}}$ (b) $\dfrac{1}{\sqrt{2} - 1}$ (c) $\dfrac{\sqrt{2} + 1}{\sqrt{5} - \sqrt{3}}$

(a) $\dfrac{2\sqrt{3}}{3\sqrt{5}} = \dfrac{2\sqrt{3}\cdot\sqrt{5}}{3\sqrt{5}\sqrt{5}}$ (The number is unchanged if the numerator and denominator are multiplied by the same number.)

$\qquad = \dfrac{2\sqrt{15}}{15}$

(b) $\dfrac{1}{\sqrt{2} - 1} = \dfrac{1(\sqrt{2} + 1)}{(\sqrt{2} - 1)(\sqrt{2} + 1)}$ (This is done to enable us to use the result $(a + b)(a - b) = a^2 - b^2$.)

$\qquad = \dfrac{\sqrt{2} + 1}{2 - 1}$

$\qquad = \sqrt{2} + 1$

(c) $\dfrac{\sqrt{2} + 1}{\sqrt{5} - \sqrt{3}} = \dfrac{(\sqrt{2} + 1)(\sqrt{5} + \sqrt{3})}{(\sqrt{5} - \sqrt{3})(\sqrt{5} + \sqrt{3})}$

$\qquad = \dfrac{(\sqrt{2} + 1)(\sqrt{5} + \sqrt{3})}{2}$

Exercise 1.2

1. Find

    (a) $(\sqrt{11})^2$ (b) $(\sqrt{15}\cdot\sqrt{20})^2$ (c) $\sqrt{3}(2\sqrt{27} - \sqrt{3})$

    (d) $\left(\dfrac{1}{2\sqrt{3}}\right)^2$ (e) $2\sqrt{2}(3\sqrt{6} - 5\sqrt{3})$ (f) $(2\sqrt{3} + 3\sqrt{2})^2$

2. Show that $\sqrt{80} = 4\sqrt{5}$. Similarly express each of the following in the form $a\sqrt{b}$ where b is the smallest possible integer.

    (a) $\sqrt{90}$ (b) $\sqrt{900}$ (c) $\sqrt{9000}$ (d) $\sqrt{18}$ (e) $\sqrt{128}$ (f) $\sqrt{1008}$

3. Express in the form $\sqrt{p}$ where $p \in \mathbb{Q}$.

    (a) $3\sqrt{2}$ (b) $5\sqrt{5}$ (c) $4\sqrt{2} + 2\sqrt{8}$ (d) $\dfrac{\sqrt{2}}{2}$ (e) $\dfrac{\sqrt{3}}{6}$ (f) $\dfrac{\sqrt{5}}{2\sqrt{3}}$

4. Simplify the following surds.

(a) $\sqrt{18} + \sqrt{8}$ (b) $\sqrt{50} + \sqrt{32} - \sqrt{72}$ (c) $\sqrt{27} + \sqrt{48} + \sqrt{75}$

(d) $\sqrt{75} + \sqrt{32} + \sqrt{200}$ (e) $(\sqrt{2} - \sqrt{3})^2$ (f) $(4\sqrt{3} - 3\sqrt{2})^2$

(g) $\dfrac{\sqrt{24} + \sqrt{40}}{2\sqrt{2}}$ (h) $\dfrac{\sqrt{15} + \sqrt{135}}{\sqrt{6}}$

5. Simplify the following:

(a) $(\sqrt{3} - \sqrt{5})(\sqrt{3} + \sqrt{5})$ (b) $(2 - \sqrt{3})(2 + \sqrt{3})$ (c) $(2\sqrt{2} + \sqrt{5})(2\sqrt{2} - \sqrt{5})$

6. Rationalise the denominators of the following:

(a) $\dfrac{1}{\sqrt{3}}$ (b) $\dfrac{5}{\sqrt{5}}$ (c) $\dfrac{\sqrt{2}}{\sqrt{5}}$ (d) $\dfrac{2}{\sqrt{2} + 2}$ (e) $\dfrac{5}{\sqrt{3} + \sqrt{2}}$

(f) $\dfrac{\sqrt{3}}{2\sqrt{5} - \sqrt{3}}$ (g) $\dfrac{1 + \sqrt{2}}{\sqrt{3} - 1}$ (h) $\dfrac{\sqrt{3} + 2\sqrt{2}}{\sqrt{5} - \sqrt{3}}$

(i) $\dfrac{\sqrt{2}}{1 - 1/\sqrt{2}}$ (j) $\dfrac{\sqrt{3}}{3/\sqrt{2} - 2/\sqrt{3}}$ (k) $\dfrac{1}{\sqrt{2} + 1} + \dfrac{1}{\sqrt{2} - 1}$

7. Use your calculator to evaluate the surds in Question 6 and also their rationalised form, correct to four significant figures.

8. Noting that $\cos 45° = \dfrac{1}{\sqrt{2}}$, $\sin 60° = \dfrac{\sqrt{3}}{2}$ and $\tan 30° = \dfrac{1}{\sqrt{3}}$, simplify:

(a) $\dfrac{1}{1 + \cos 45°}$ (b) $\dfrac{1 + \tan 30°}{1 - \tan 30°}$ (c) $\dfrac{\sin 60°}{\cos 45° - \sin 60°}$

9. (a) Find natural numbers a and b so that $7 + 2\sqrt{10} = (\sqrt{a} + \sqrt{b})^2$. Hence write down a simplification of $\sqrt{(7 + 2\sqrt{10})}$.
   (b) Similarly simplify $\sqrt{(19 - 8\sqrt{3})}$.

10. Prove that if $\sqrt{(a + b)} = \sqrt{a} + \sqrt{b}$ then one or both of a and b must be zero.

1.3 Indices

The product of n factors, where $n \in \mathbb{N}^+$ and each factor is equal to a is written a^n.
So

$$a^n = a \times a \times a \times \ldots \times a$$
$$\longleftarrow n \text{ factors} \longrightarrow$$

a^n is read as *a to the power n*, and n is called the *index* or *exponent*.
   Notice that

$$a^1 = a.$$

Now

$$a^2 \times a^3 = a \times a \ \times \ a \times a \times a = a^5$$

or, more generally,

$$a^n \times a^m = a^{n+m} \qquad \text{for} \quad n \in \mathbb{N}^+ \quad \text{and} \quad m \in \mathbb{N}^+.$$

Also

$$(a^5)^2 = a \times a \times a \times a \times a \quad \times \quad a \times a \times a \times a \times a$$

$$= a^{10} \qquad [a^{5 \times 2}]$$

or, again, more generally,

$$(a^m)^n = a^{mn} \qquad \text{for} \quad n \in \mathbb{N}^+ \quad \text{and} \quad m \in \mathbb{N}^+$$

It is similarly reasonable that

$$a^m \div a^n = a^{m-n} \qquad \text{for} \quad n \in \mathbb{N}^+ \quad \text{and} \quad m \in \mathbb{N}^+ \ [\text{at least when } m > n]$$

These ideas can be extended so that a^x has meaning for any real number x, providing a is a positive real.

Statement

> The definition of a^x can be extended to allow $x \in \mathbb{R}$ providing a is positive. The following rules can then be proved for all $x \in \mathbb{R}$ and $y \in \mathbb{R}$.
>
> *Rules of indices*
>
> (i) $a^x \times a^y = a^{x+y}$
> (ii) $a^x \div a^y = a^{x-y}$
> (iii) $(a^x)^y = a^{xy}$
>
> *Extended rules of surds*
>
> (iv) $(a.b)^x = a^x.b^x$ ⎫
> (v) $a^x/b^x = (a/b)^x$ ⎭   where a and b are both positive.

Interpretation of a^x

Using rule (i) when $x = 0$ and any choice for y

$$a^0 \times a^y = a^{0+y} = a^y. \qquad \text{Thus } a^0 = 1 \qquad !!$$

Now using rule (ii) when $x = 0$

$$a^0 \div a^y = a^{0-y}$$

$$\Rightarrow \qquad \frac{1}{a^y} = a^{-y} \qquad \text{since } a^0 = 1.$$

So, for example,

$$a^{-3} = \frac{1}{a^3}.$$

Using rule (i) when $x = \frac{1}{2}$ and $y = \frac{1}{2}$

$$a^{1/2} \cdot a^{1/2} = a^{1/2 + 1/2} = a^1 = a$$

So

$$a^{1/2} = \sqrt{a} \qquad \text{the square root of } a \quad \text{!!!}$$

More generally, when $n \in \mathbb{N}^+$

$$\underbrace{a^{1/n} \times a^{1/n} \times \ldots \times a^{1/n}}_{n \text{ factors}} = a^1 = a \qquad \text{by repeated application of rule (i).}$$

So $a^{1/n}$ is the nth root of a, written $\sqrt[n]{a}$ for $n \in \mathbb{N}^+$ [$\sqrt[2]{a}$ is written $\sqrt{a}$ by convention].

Example 1.3

Evaluate (a) $8^{2/3}$ (b) $16^{3/4}$

(a) $8^{2/3} = (8^{1/3})^2$ by rule (iii)

$\qquad = 2^2$ since $8^{1/3} = \sqrt[3]{8}$

$\qquad = 4$

(b) $16^{3/4} = (16^{1/4})^3$ by rule (iii)

$\qquad = 2^3$

$\qquad = 8$

In fact, using rule (iii) we have

$$a^{m/n} = a^{(1/n)m}$$

$$= (a^{1/n})^m$$

So when $n \in \mathbb{N}^+$, $a^{m/n}$ is the nth root of a, raised to the power m.
    Or alternatively

$$a^{m/n} = a^{m(1/n)}$$

$$= (a^m)^{1/n}$$

Which is the same as the nth root of a raised to the power m.

Advanced Pure Mathematics

Summary

> Whenever $x \in \mathbb{R}$, $y \in \mathbb{R}^+$ and $a \in \mathbb{R}^+$ then
>
> $$a^{-x} = 1/a^x$$
>
> $$a^0 = 1$$
>
> $$a^{x/y} = (a^{1/y})^x = (a^x)^{1/y}$$
>
> and, in particular
>
> $$a^{1/n} = \sqrt[n]{a} \qquad \text{for} \quad n \in \mathbb{N}^+$$

We cannot use the above rules to establish the value of say $3^{\sqrt{2}}$, indeed since it is irrational we cannot write down its exact value using decimal notation.

Those with an 'x^y' button on their calculator, could use this to good effect, to find at least an approximation of a^x. For example

$$3^{\sqrt{2}} \simeq 3^{1.414\,21}$$

$$\simeq 4.7288 \qquad \text{4 d.p.} \qquad \boxed{\text{C}} \dagger$$

When a is zero or negative, the situation is more complicated. For example 0^0 has no meaning. However $(-1)^{1/3}$ can consistently taken to be -1, since $(-1).(-1).(-1) = -1$. However, for an interpretation of $(-1)^{1/2}$ we shall have to wait until Complex Numbers are dealt with. Certainly there is no real number whose square is -1.

Example 1.4

Evaluate the following, without using a calculator.

(a) $8^{-1/3}$ (b) 3^{-2} (c) $27^{2/3}$ (d) 9^0 (e) $\left(\dfrac{16}{81}\right)^{-3/4}$

(a) $8^{-1/3} = \dfrac{1}{8^{1/3}} = \tfrac{1}{2}$

(b) $3^{-2} = \dfrac{1}{3^2} = \dfrac{1}{9}$

(c) $27^{2/3} = (27^{1/3})^2 = 3^2 = 9$

(d) $9^0 = 1$ [any positive number to the power zero is 1]

(e) $(16/81)^{-3/4} = \dfrac{1}{(16/81)^{3/4}} = \dfrac{1}{((16/81)^{1/4})^3} = \dfrac{1}{(2/3)^3} = \dfrac{1}{8/27} = \dfrac{27}{8}$

Example 1.5

Simplify

(a) $x^{3/2} \times x^6$ (b) $x^{1/2} \times x^{3/2}$ (c) $(2x)^{1/2} \times (6x)^{-1/2}$

$\dagger$ Notice the use of the symbol $\boxed{\text{C}}$, to indicate that a calculation has been performed by the use of a calculator.

(a) $x^{3/2} \times x^6 = x^{3/2+6} = x^{15/2}$

(b) $x^{1/2} \times x^{3/2} = x^{1/2+3/2} = x^2$

(c) $(2x)^{1/2} \times (6x)^{-1/2} = \dfrac{2^{1/2} \times x^{1/2}}{6^{1/2} \times x^{1/2}} = \dfrac{1}{3^{1/2}} = \dfrac{3^{1/2}}{3}$

Exercise 1.3

1. Evaluate each of the following:

    (a) 15^0 (b) $16^{1/2}$ (c) $36^{-1/2}$ (d) $81^{3/4}$ (e) $81^{-3/4}$
    (f) $64^{-1/4}$ (g) 16^0 (h) $100^{3/2}$ (i) $(-8)^{1/3}$ (j) $(-27)^{-1/3}$
    (k) $(1/2)^{-3}$ (l) $1/(2^{-2})$ (m) $(1/4)^{3/2}$
    (n) $(125/27)^{-1/3}$ (o) $(0.25)^{1/2}$ (p) $(0.0625)^{-1/4}$

2. Solve by inspection the following equations.

    (a) $2^x = 8$ (b) $3^x = 27^{2/3}$ (c) $3^x = 27^{1/4}$
    (d) $10^x = 0.01$ (e) $(2^x)^3 = 16$ (f) $(2^x)^2 = 2^5$

3. Express as powers of 3:

    (a) $(3^4)^5$ (b) $3^5 \times 3^4$ (c) 9 (d) 9^2 (e) $9^{-1/2}$ (f) $81^{1/3} \div 27^{1/4}$

4. Simplify the following expressions.

    (a) $\sqrt{(64 . x^4 . y^2)}$ (b) $\left(\dfrac{a^3 b^6}{c^9}\right)^{1/3}$

    (c) $(8x^2 . y^3)^{1/2}$ (d) $\left(\dfrac{9x^2}{16(x+1)^4}\right)^{1/2}$

    (e) $\dfrac{4x^{1/4}}{2x^{1/2}}$

5. Simplify the following expressions.

    (a) $x^{1/2} . x^{3/2}$ (b) $2x^2 . 3x^3$
    (c) $2x^3 \div 6x^5$ (d) $2x^{3/2} . 5x^{-3/4}$
    (e) $x^{-3} . 4x^5$ (f) $4x^{3/4} \div (2x^{1/2})$
    (g) $(x^{1/3})^4 \div (x^2)^{1/3}$ (h) $(1+x)^{1/2} . (1+x)^{3/2}$
    (i) $3x^{-2} . (3x)^{-2}$ (j) $(1-x^2) . (1+x)^{-1/2}$

6. Expand and simplify the following expressions

    (a) $x^{1/2}(x^{3/2} + x^{-1/2})$ (b) $x^{2/3}(x^{1/3} - x^{-1/3})$
    (c) $(x^{1/2} + 1)(x^{-1/2} - 1)$ (d) $(2x^{1/2} - 3)(4x^{3/2} + 1)$

    (e) $\dfrac{x^{1/2}(x-1) + x^{-1/2}(x-1)^{-1}}{x^{1/2}}$

1.4 Logarithms

In the previous section it was stated that for any positive a, a^x could be evaluated for any $x \in \mathbb{R}$ (at least approximately when a^x is irrational).

A sketch of the graph of $y = a^x$ is shown in Fig. 1.4, where it has been assumed that $a > 1$. When $a > 1$, the value of a^x increases rapidly for increasing x. If $0 < a < 1$ then the value of a^x decreases rapidly for increasing x, and the graph of $y = a^x$ is indicated by the dotted curve.

Of course, if $a = 1$, then the graph of $y = a^x$ is simply that of $y = 1$; which is a straight line parallel to the x axis. For any $a \neq 0$, the graph of $y = a^x$ will pass through the point $(0, 1)$ since $a^0 = 1$.

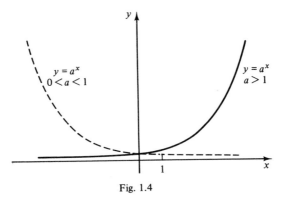

Fig. 1.4

It can be seen from the graph that for *any* $y > 0$, there is a *unique* x, such that $y = a^x$. This value of x is called the logarithm (or log for short) to the base a of y and is written $x = \log_a y$, so $\log_a y$ is the power that a has to be raised by to equal y.

Thus

$$\boxed{y = a^x \Leftrightarrow x = \log_a y}$$

It follows that

$$4 = 16^{1/2} \qquad \Rightarrow \log_{16} 4 = \tfrac{1}{2}$$
$$8 = 2^3 \qquad \Rightarrow \log_2 8 = 3$$

Similarly

$$\log_5 125 = 3 \qquad \text{since } 125 = 5^3$$

and

$$\log_9 \left(\tfrac{1}{81}\right) = -2 \qquad \text{since } \tfrac{1}{81} = 9^{-2}$$

Suppose

$$y = a^x$$

then since $x = \log_a y$, we have

$$\boxed{y = a^{\log_a y}} \qquad \text{substituting for } x.$$

For example

$$4^{\log_4 8} = 8$$

Similarly if

$$\log_a y = x$$

then $y = a^x$, so

$$\boxed{\cdot\log_a a^x = x}$$ substituting for y.

For example

$$\log_3 3^5 = 5$$

and

$$\boxed{\log_a a = 1}$$ since $a = a^1$

Assuming a, b and c are positive, it is obvious that

$$\text{if} \qquad b = c$$
$$\text{then} \quad \log_a b = \log_a c.$$

This is called 'taking logs to base a'.
   For example,

$$2^x = 5$$
$$\Rightarrow \log_2 2^x = \log_2 5$$
$$\Rightarrow \qquad x = \log_2 5 \qquad \text{since } \log_2 2^x = x.$$

Statement

The following *laws of logarithms* can be derived using the comparable laws of indices.

$$\text{(i) } \log_a(M \cdot N) = \log_a M + \log_a N \quad\left.\begin{array}{l} \text{for } M > 0 \\ N > 0 \\ a > 0 \\ \text{and } n \in \mathbb{R} \end{array}\right.$$
$$\text{(ii) } \log_a(M/N) = \log_a M - \log_a N$$
$$\text{(iii) } \log_a(M^n) = n \log_a M$$

Proof

Suppose that $M = a^x$ and $N = a^y$ so $\log_a M = x$ and $\log_a N = y$. [x and y will exist since M and N are positive.]

(i) $a^x \times a^y = a^{x+y}$

$$\Leftrightarrow \log_a(a^x \times a^y) = \log_a a^{x+y} \qquad \text{(taking logs to base } a)$$

$\Leftrightarrow \log_a(a^x \times a^y) = x + y$

$\Leftrightarrow \log_a(M \cdot N) = \log_a M + \log_a N$ (substituting)

(ii) $a^x \div a^y = a^{x-y}$

$\Leftrightarrow \log_a(a^x \div a^y) = x - y$ (taking logs to base a)

$\Leftrightarrow \log_a(M/N) = \log_a M - \log_a N$ (substituting)

(iii) $(a^x)^n = a^{nx}$

$\Leftrightarrow \log_a((a^x)^n) = nx$ (taking logs to base a)

$\Leftrightarrow \log_a(M^n) = n \log_a M$ (substituting)

Example 1.6

Evaluate

(a) $\log_3 243$ (b) $\log_{243} 3$ (c) $\log_5(1/125)$

(a) $\log_3 243 = \log_3 3^5 = 5$

(b) $\log_{243} 3 = \log_{243} 243^{1/5} = \frac{1}{5}$

(c) $\log_5(1/125) = \log_5(1/5^3) = \log_5 5^{-3} = -3$

Example 1.7

Express as a simple logarithm

(a) $\log_3 5 + \log_3 6$ (b) $5 \log_4 3$ (c) $2 \log_4 3 + 3 \log_4 2$
(d) $\log_5 10 - 3 \log_5 4$ (e) $- \log_2 5$

(a) $\log_3 5 + \log_3 6 = \log_3(5 \times 6) = \log_3 30$

(b) $5 \log_4 3 = \log_4 3^5 = \log_4 243$

(c) $2 \log_4 3 + 3 \log_4 2 = \log_4 3^2 + \log_4 2^3$

$$= \log_4(3^2 \times 2^3)$$

$$= \log_4 72$$

(d) $\log_5 10 - 3 \log_5 4 = \log_5 10 - \log_5 4^3$

$$= \log_5(10/64)$$

$$= \log_5(5/32)$$

(e) $- \log_2 5 = \log_2 5^{-1}$. Note that $- \log_2 5 = (-1) \log_2 5$.

Example 1.8

(a) Expand $\log_a(a^3 \cdot b^{-2} \cdot c)$.
(b) Express as a single logarithm $\log_a x^2 - 2 \log_a y + 3$.

(a) $\log_a(a^3 . b^{-2} . c) = \log_a a^3 + \log_a b^{-2} + \log_a c$

$$= 3\log_a a - 2\log_a b + \log_a c$$

$$= 3 - 2\log_a b + \log_a c \qquad \text{since } \log_a a = 1$$

(b) $\log_a x^2 - 2\log_a y + 3 = \log_a x^2 + \log_a y^{-2} + \log_a a^3$

$$= \log_a x^2 . y^{-2} . a^3$$

Exercise 1.4

1. Express in logarithmic notation:

   (a) $3^3 = 27$ (b) $4^2 = 16$ (c) $5^0 = 1$ (d) $4^{1/2} = 2$
   (e) $3^{-1} = \frac{1}{3}$ (f) $4^{-2} = \frac{1}{16}$ (g) $a^b = C$ (h) $4^{3/2} = 8$

2. Simplify:

   (a) $\log_3 3^7$ (b) $\log_3 27$ (c) $\log_4 16$ (d) $\log_8 \frac{1}{8}$
   (e) $\log_2 \frac{1}{4}$ (f) $\log_6 36^{1/3}$ (g) $\log_{10} 10$ (h) $\log_{10}(0.01)$
   (i) $\log_{\sqrt{2}} 2$

3. Expand the following expressions.

   (a) $\log_3(8 \times 4)$ (b) $\log_4 \frac{3}{5}$ (c) $\log_2 5^3$ (d) $\log_a(b . c)$
   (e) $\log_a(b^2 . c)$ (f) $\log_a(b^3/c)$ (g) $\log_a(b/c)^2$ (h) $\log_a(b/a)$

   (i) $\log_a(a^3/b^2)$ (j) $\log_a\left(\dfrac{\sqrt{b} \times c}{a}\right)$ (k) $\log_a\sqrt{\left(\dfrac{x+1}{x-1}\right)}$

   (l) $\log_a\left(\dfrac{x^2+1}{(x^2-1)^{1/2}}\right)$

4. Express as a single logarithm:

   (a) $\log_2 3 + \log_2 5$ (b) $\log_3 8 - \log_3 6$ (c) $3\log_2 5$
   (d) $-2\log_3 4$ (e) $-\log_5 4$ (f) $2\log_3 5 + \log_3 2$
   (g) $4\log_2 3 - \log_2 9$ (h) $2\log_2 5 + \log_2 3 - 2\log_2 15$
   (i) $(\log_3 2) + 1$ (j) $(\log_2 5) - 3$

5. Express in the form of a single logarithm

   (a) $\log_a b + \log_a c$ (b) $\log_c c + \log_c b$
   (c) $1 + \log_c b$ (d) $2 + \log_a b$
   (e) $2\log_a b$ (f) $\frac{1}{2}\log_a b^3$
   (g) $2\log_a b^3 - 3\log_a b^2$ (h) $\log_a b^2 + 2\log_a c - 2$
   (i) $\frac{1}{2}\log_a(x-1) + \log_a(x-3)$ (j) $3\log_2 x^2 - \log_2 y^3$

1.5   Evaluation of logarithms

So far we have used the definition and rules of logarithms to simplify, and in certain cases to evaluate, expressions containing logarithms. We now consider how to evaluate logarithms in general.

Logarithms were invented by John Napier in 1614 and further developed by Henry Briggs, who first computed tables for logarithms to base 10. These are called common logarithms and their values are readily available on calculators and in books of tables.

> $\log_{10} x$ is often written as $\log x$ or $\lg x$

Some calculators possess both an 'lg' button and an 'ln' button. The meaning of $\ln x$ will be made clear in Chapter 8.

Now for example

$$\lg 5 = 0.699 \qquad 3\,\text{d.p.} \qquad \boxed{\text{C}}$$

and

$$\lg 2 = 0.301 \qquad 3\,\text{d.p.} \qquad \boxed{\text{C}}$$

To evaluate logarithms to other bases we can use the following result to 'change the base'.

Statement

> For positive a, b and x $\qquad \log_a x = \dfrac{\log_b x}{\log_b a}$
>
> In particular, when $x = b$ $\qquad \log_a b = \dfrac{\log_b b}{\log_b a} = \dfrac{1}{\log_b a}$

Proof

Suppose $\log_a x = M$. So

$$a^M = x$$

$$\Leftrightarrow \qquad \log_b a^M = \log_b x \qquad \text{taking logs to base } b$$

$$\Leftrightarrow \qquad M \log_b a = \log_b x$$

$$\Leftrightarrow \log_a x \log_b a = \log_b x \qquad \text{since } M = \log_a x$$

$$\Leftrightarrow \qquad \log_a x = \frac{\log_b x}{\log_b a}$$

Example 1.9

Evaluate   (a) $\log_2 5$ (b) $\log_5 2$

(a) $\log_2 5 = \dfrac{\log_{10} 5}{\log_{10} 2} = \dfrac{\lg 5}{\lg 2}$ $\left[\simeq \dfrac{0.699}{0.301} \right]$

$\qquad\qquad\qquad\qquad = 2.322 \qquad 3\,\text{d.p.} \qquad \boxed{C}$

(b) $\log_5 2 = \dfrac{1}{\log_2 5}$

$\qquad\qquad = 0.431 \qquad 3\,\text{d.p.} \qquad \boxed{C}$

Example 1.10

Solve the following equations.

(a) $2^x = 5$ (b) $6^x = \frac{1}{3}$ (c) $5^{x-1} = 3^{2x-1}$

(a) $\qquad 2^x = 5$

$\qquad \Rightarrow \lg 2^x = \lg 5 \qquad\qquad$ [taking logs to base 10]

$\qquad \Rightarrow x \lg 2 = \lg 5$

$\qquad \Rightarrow x = \dfrac{\lg 5}{\lg 2}$

$\qquad \Rightarrow x \simeq 2.322 \qquad 3\,\text{d.p.} \qquad$ from the earlier example.

Alternatively

$\qquad\qquad 2^x = 5$

$\qquad \Rightarrow x = \log_2 5 \qquad$ by definition

$\qquad \Rightarrow x = 2.322 \qquad$ as before

(b) $\qquad 6^x = \frac{1}{3}$

$\qquad \Rightarrow \lg 6^x = \lg \frac{1}{3}$

$\qquad \Rightarrow x \lg 6 = \lg \frac{1}{3}$

$\qquad \Rightarrow x = \dfrac{\lg \frac{1}{3}}{\lg 6} \quad \left[\simeq \dfrac{-0.477}{0.778} \right]$

$\qquad\qquad = -0.613 \qquad 3\,\text{d.p.} \qquad \boxed{C}$

(c) $\qquad 5^{x-1} = 3^{2x-1}$

$\qquad \Rightarrow \lg 5^{x-1} = \lg 3^{2x-1}$

$\qquad \Rightarrow (x-1)\lg 5 = (2x-1)\lg 3$

$\qquad \Rightarrow x \lg 5 - \lg 5 = 2x \lg 3 - \lg 3$

$\qquad \Rightarrow 2x \lg 3 - x \lg 5 = -\lg 5 + \lg 3$

$\qquad \Rightarrow x \lg 9 - x \lg 5 = \lg 3 - \lg 5$

$\qquad \Rightarrow x (\lg \tfrac{9}{5}) = \lg \tfrac{3}{5}$

$$\Rightarrow x = \frac{\lg 0.6}{\lg 1.8}$$

$\Rightarrow x \simeq -0.869$ 3 d.p. $\boxed{C}$

Example 1.11

Find x if   (a) $\lg x = 1.5$ (b) $\log_3 x = 6.2$

(a) $\lg x = 1.5$

   $\Rightarrow x = 10^{1.5}$

   $\Rightarrow x \simeq 31.623$ 3 d.p. $\boxed{C}$

(b) $\log_3 x = 6.2$

   $\Rightarrow x = 3^{6.2}$

   $\Rightarrow x \simeq 908.138$ 3 d.p. $\boxed{C}$

Exercise 1.5

1. Evaluate:

    (a) $\log_2 3$ (b) $\log_3 2$ (c) $\log_5 6$ (d) $\log_2 0.1$ (e) $\log_3 0.5$

2. Solve the following equations.

    (a) $3^x = 8$ (b) $5^x = 15$ (c) $2^{2x+1} = 0.01$
    (d) $10^x = 0.3$ (e) $8(2^{2x}) = 5$ (f) $3^{x-2} = 5^{2x+1}$

3. Find x in each of the following equations.

    (a) $\log_8 x = 1.5$ (b) $\log_3 x = 0.7$ (c) $\log_3 x = 2.2$ (d) $\log_2 x = 0.05$

4. Find x in each of the following equations.

    (a) $\log_x 65 = 6$ (b) $\log_x 2 = 3$ (c) $\log_x 5 = 1.2$

5. Simplify:

    (a) $\frac{1}{2}\lg(x + 1) - 2\lg(x - 1)$ (b) $\lg(x - 1) - \frac{1}{2}\lg(x^2 - 1)$
    (c) $\log_a b^2 \times \log_b a^2$

6. Solve the equation $2\log_{25} 2x = \log_5(1 - x)$

7. Solve the equation $3^{x+1} = 4^{x-1}$, giving the value of x correct to three significant figures.

8. Given that $y = 6^x$, find, to 2 decimal places, the value of x when $y = 0.5$.

9. Evaluate $(\log_a 27)(\log_3 a)$.

10. Prove that $\log_a b \cdot \log_b c \cdot \log_c a = 1$ and hence find $\log_2 7 \cdot \log_7 25 \cdot \log_5 8$.

11. Prove that $\log_x y + \log_{1/x} y = 0$.

12. Solve the equation $\log_3 (2x + 1) = \log_9 (2(x + 1)^2)$.

13. Without using tables or a calculator find

    (a) $\log_4 8\sqrt{2}$ (b) $(\log_3 25)(\log_5 27)$.

14. Given that $\log_4 (x + 2) = \log_2 x$, find the value of $\log_4 x$.

Chapter 2   Polynomials

2.1   Identity

In this chapter we shall be concerned with *algebraic expressions* rather than numerical ones. Algebraic expressions such as $2x + x^2$ or $a^2 - b^2$ are built up from numbers, normal mathematical symbols and letters, often called *variables*, which stand for as yet unstated numbers.

It is characteristic of an algebraic expression that it can be evaluated when the values of the variables within the expression are known. Thus $2x^2$ assumes the value 18 when $x = 3$ and the values 32 when $x = 4$ or $x = -4$.

Definition

> Two algebraic expressions are said to be identically equal, $\equiv$, if they each assume the same numerical value for each choice of values of the variables.

So we would state that

$$2(x + x^2) \equiv 2x + 2x^2$$

and that

$$a^2 - b^2 \equiv (a - b)(a + b)$$

since in each case the l.h.s. = r.h.s. for each choice of value for x or a and b.

It is useful to adopt notation such as $P(x)$ to represent a particular expression which may be defined using the concept of identical equality. For example, defining

$$P(x) \equiv 2x - x^2$$

then

$$P(1) = 2 - 1 = 1$$
$$P(2) = 4 - 4 = 0$$
$$P(3) = 6 - 9 = -3$$

and generally $P(n)$ represents the value of the expression when $x = n$.

An expression such as $P(x)$, which varies for different values of x, is often called a *function*. This idea is further refined in Chapter 5.

An algebraic expression is not necessarily defined for all possible choice of values of a variable. The expression $x/(1 - x)$ is not defined for $x = 1$ as the following table suggests

x	0.9	0.99	0.999	0.9999...	...1.0001	1.001	1.01	1.1
$\dfrac{x}{1-x}$	9	99	999	9999	-10001	-1001	-101	-11

A graph to indicate the value of $x/(1 - x)$ as x approaches 1 is given in Fig. 2.1.

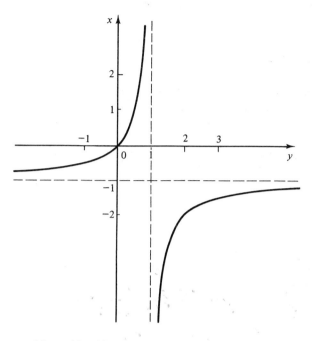

Advanced Level Pure Mathematics **Fig. 2.1**

In order to use the notion of identical equality such values are excluded. So, for example

$$x \equiv \frac{1}{1/x} \qquad x \neq 0$$

and

$$\frac{a^2 - b^2}{a - b} \equiv a + b \qquad a \neq b$$

2.2   Polynomials

Definition

An expression of the form

$$a_0 + a_1 x + a_2 x^2 + \ldots + a_s x^s + \ldots + a_n x^n$$

where each $a_s \in \mathbb{R}$ and $a_n \neq 0$ is called a *polynomial of degree n*. In this case the polynomial has been written in ascending powers of x. When written the other way round, with the highest power of x first, then it is said to be expressed in descending powers of x. The number a_s is called the *coefficient* of x^s, and a_0 is called the *constant term*.

Polynomials can be added, subtracted or multiplied to obtain new polynomials. Generally, the division of two polynomials does not result in a new polynomial. Notice that a polynomial is defined for all values of x.

Example 2.1

Suppose that $P(x) \equiv x^2 + 2$ and $Q(x) \equiv x^2 + x - 1$. Simplify

(a) $P(x) + Q(x)$ (b) $P(x) - Q(x)$ (c) $P(x) \cdot Q(x)$.

(a) $P(x) + Q(x) \equiv x^2 + 2 + x^2 + x - 1$

$\qquad\qquad\quad \equiv 2x^2 + x + 1$

(b) $P(x) - Q(x) \equiv x^2 + 2 - (x^2 + x - 1)$

$\qquad\qquad\quad \equiv x^2 + 2 - x^2 - x + 1$

$\qquad\qquad\quad \equiv -x + 3$

(c) $P(x) \cdot Q(x) \equiv (x^2 + 2) \cdot (x^2 + x + 1)$

$\qquad\qquad\quad \equiv x^2(x^2 + x + 1) + 2(x^2 + x + 1)$

$\qquad\qquad\quad \equiv x^4 + x^3 + x^2 + 2x^2 + 2x + 2$

$\qquad\qquad\quad \equiv x^4 + x^3 + 3x^2 + 2x + 2$

Some polynomials have special names:

Degree of polynomial	Special name	General form	Example
1	Linear	$ax + b$	$3x + 4$
2	Quadratic	$ax^2 + bx + c$	$2x^2 - x + 6$
3	Cubic	$ax^3 + bx^2 + cx + d$	$x^3 + 4x^2 - 6x + 2$
4	Quartic	$ax^4 + bx^3 + cx^2 + dx + c$	$x^4 - 5x + 2$

An expression which is the sum of two terms is called a *binomial*, so, for example, $3x^2 + 6$ is a quadratic binomial.

Suppose $x^2 + 2x + 3 \equiv ax^2 + bx + c$, then the l.h.s. will equal the r.h.s. for any value of x. So, when $x = 0$, $3 = c$.

Thus

$$x^2 + 2x + 3 \equiv ax^2 + bx + 3$$

Now when $x = 1$,

$$6 = a + b + 3$$

$$\Rightarrow a + b = 3 \qquad \text{①}$$

and when $x = -1$,

$$2 = a - b + 3$$

$$\Rightarrow a - b = -1 \qquad ②$$

Adding equations ① and ② (written simply as ① + ②) we obtain

$$2a = 2$$

$$\Rightarrow \quad a = 1$$

and ① − ②,

$$\Rightarrow 2b = 4$$

$$\Rightarrow \quad b = 2.$$

Thus if

$$x^2 + 2x + 3 \equiv ax^2 + bx + c$$

then

$$a = 1, \quad b = 2 \quad \text{and} \quad c = 3.$$

The two quadratics have the same coefficients for each power of x. This is an example of a far more general result.

Statement

> For any two polynomials $P(x)$ and $Q(x)$,
>
> $$P(x) \equiv Q(x)$$
>
> if and only if for each power of x, the coefficient in $P(x)$ is the same as that in $Q(x)$.

Of course, if the corresponding coefficients in $P(x)$ and $Q(x)$ are the same, then it is obvious that the two polynomials are identically equal. However, the converse is not easy to prove, and we shall simply accept it as a fact.

Example 2.2

Find the value of a and b if

$$1 + 3x + 5x^2 + 3x^3 \equiv (1 + x)(1 + ax + bx^2)$$

Now

$$(1 + x)(1 + ax + bx^2) \equiv 1 + ax + bx^2 + x(1 + ax + bx^2)$$

$$\equiv 1 + ax + bx^2 + x + ax^2 + bx^3$$

$$\equiv 1 + (a + 1)x + (a + b)x^2 + bx^3$$

Thus

$$1 + 3x + 5x^2 + 3x^3 \equiv 1 + (a + 1)x + (a + b)x^2 + bx^3$$

Comparing coefficients

$$3 = a + 1$$
$$5 = a + b$$
$$3 = b$$

It follows that

$$a = 2 \quad \text{and} \quad b = 3$$

Example 2.3

Find the values of the constants A, B and C so that

$$x^2 - 5x + 12 \equiv A(x - 1)(x - 2) + B(x + 1)(x - 1) + C(x - 2)(x + 3)$$

We could simplify the r.h.s. and compare coefficients. However, since the two polymonials are identically equal they will be equal for any chosen value of x—say $x = 1$, $x = 2$ and $x = 0$.

For $x = 1$

$$1^2 - 5.1 + 12 = A.0 + B.0 - 4.C$$
$$\Rightarrow 8 = -4C$$
$$\Rightarrow C = -2$$

For $x = 2$

$$2^2 - 5.2 + 12 = A.0 + 3.B + C.0$$
$$\Rightarrow 6 = 3B$$
$$\Rightarrow B = 2$$

For $x = 0$

$$12 = 2.A - B - 6.C$$

and using the fact that $C = -2$, $B = 2$ we have

$$12 = 2A - 2 + 12$$
$$\Rightarrow A = 1$$

So $A = 1$, $B = 2$ and $C = -2$.

Notice that the choice of $x = 1$ and $x = 2$ ensured that in each case two of the constants were eliminated in the resulting equations. This made the equations easier to solve.

The choice of $x = 0$ is, in fact, equivalent to comparing the constant term on each side of the equation. In problems of this kind it is often convenient to mix both the method of comparing coefficients and the method of selecting particular values of x.

2.3 Completing the square

Any quadratic $ax^2 + bx + c$ can be expressed in the form $a(x + A)^2 + B$ for some A and B. The process is called completing the square, and has many applications.

Example 2.4

Find A and B so that

$$x^2 + 4x + 1 \equiv (x + A)^2 + B$$

Now, on expanding the r.h.s. we have

$$x^2 + 4x + 1 \equiv x^2 + 2Ax + A^2 + B$$

Comparing coefficients

$$2A = 4 \qquad \Rightarrow A = 2$$

and

$$A^2 + B = 1$$
$$\Rightarrow 4 + B = 1 \qquad \Rightarrow B = -3$$

Thus

$$x^2 + 4x + 1 \equiv (x + 2)^2 - 3$$

One immediate application is to note that no matter which value for x is chosen, $(x + 2)^2$ will be positive or zero. So the expression $(x + 2)^2 - 3$ will equal -3 when $x = -2$, and more than -3 for any other value of x. It follows that the least or minimum value of the quadratic $x^2 + 4x + 1$ is -3, which occurs when $x = -2$.

Example 2.5

Find the *maximum* value of the quadratic $1 + x - 2x^2$.

We first complete the square by supposing

$$-2x^2 + x + 1 \equiv -2(x + A)^2 + B \qquad \text{as suggested above}$$
$$\equiv -2(x^2 + 2Ax + A^2) + B$$
$$\equiv -2x^2 - 4Ax - 2A^2 + B$$

Comparing coefficients

$$-4A = 1 \qquad \Rightarrow A = -\tfrac{1}{4}$$
$$-2A^2 + B = 1$$
$$\Rightarrow \qquad -\tfrac{1}{8} + B = 1 \qquad \Rightarrow B = \tfrac{9}{8}.$$

Thus

$$-2x^2 + x + 1 \equiv -2(x - \tfrac{1}{4})^2 + \tfrac{9}{8}.$$

Now again $(x - \tfrac{1}{4})^2$ will be positive or zero, so $-2(x - \tfrac{1}{4})^2$ will always be negative or zero (when $x = \tfrac{1}{4}$). Thus the *maximum* value of $-2x^2 + x + 1$ is $\tfrac{9}{8}$, occurring when $x = \tfrac{1}{4}$.

In general

> The quadratic $ax^2 + bx + c$ will have a maximum value if $a < 0$ and a minimum value if $a > 0$.

Exercise 2.1

1. If $P(x) \equiv 2x^2 - 3x + 4$, find

    (a) $P(1)$ (b) $P(\tfrac{1}{2})$ (c) $P(0)$.

2. If $G(x) \equiv x^3 - 4x^2 + 2x - 1$, find

    (a) $G(-1)$ (b) $G(-2)$ (c) $G(3/2)$.

3. In each case find the value of the constants A, B and C, assuming:

    (a) $3x + 3 \equiv A(x - 1) + B(2 + x)$,
    (b) $7x + 6 \equiv A(x - 2) + B(x + 3)$,
    (c) $2x + 5 \equiv A(x + 1) + B(x - 2)$,
    (d) $2x^2 - 5x + 7 \equiv A(x + 1)(x - 2) + B(x + 1)(x - 1) + C(x - 2)(x - 1)$,
    (e) $x^2 - 6x - 19 \equiv A(x + 5)(x - 1) + B(x^2 - 1) + C(x + 5)(x + 1)$.

4. State the values for which the following expressions are undefined.

    (a) $\dfrac{2}{2 - x}$ (b) $\dfrac{1}{x + 3}$ (c) $\dfrac{x - 1}{2x + 1}$

    (d) $\dfrac{x^2 + 1}{(x + 2)(x + 5)}$ (e) $\dfrac{x}{x^2 - 1}$

5. If $P(x) \equiv x^2 + x - 1$ and $Q(x) \equiv 1 + 2x$, simplify

    (a) $P(x) + 2 . Q(x)$ (b) $P(x) - Q(x)$ (c) $3 . P(x) + x . Q(x)$
    (d) $(1 + x) . P(x)$ (e) $Q(x) . P(x)$ (f) $P(x) + Q(x) . Q(x)$.

6. Find values for A and B so that

    $$x^3 + 4x^2 + 4x + 1 \equiv (x + 1)(x^2 + Ax + B).$$

7. Find $P(x)$ if $x^3 + 3x^2 + x - 2 \equiv (x + 2) . P(x)$.

8. Find values for A and B if

   (a) $x^2 + 6x + 2 \equiv (x + A)^2 + B$ (b) $x^2 + 2x - 1 \equiv (x + A)^2 + B$
   (c) $x^2 + 3x - 1 \equiv (x + A)^2 + B$ (d) $x^2 - 5x + 6 \equiv (x + A)^2 + B$

9. Find values for A and B if

   (a) $2x^2 + 4x + 1 \equiv 2(x + A)^2 + B$ (b) $3x^2 + 2x + 6 \equiv 3(x + A)^2 + B$
   (c) $-x^2 + x - 2 \equiv -(x + A)^2 + B$ (d) $-2x^2 + 3x + 4 \equiv -2(x + A)^2 + B$

10. Complete the square for each of the following quadratics and hence find the minimum value of the quadratic, stating the value of x for which this occurs.

    (a) $x^2 + 3x + 2$ (b) $2x^2 - x + 4$

11. (a) State whether each of the quadratics given below has a maximum or a minimum value.

    (i) $2x^2 - x + 1$ (ii) $-1 - 2x + 2x^2$ (iii) $1 - 3x - x^2$
    (iv) $-3x^2 + 1 - 2x$ (v) $(x - 2)(x + 3)$ (vi) $(1 - x)(2 - x)$
    (vii) $(1 + x)(2 - x)$ (viii) $(1 + 2x)(1 - x)$

    (b) Find the maximum values of the above quadratics when they exist, indicating the value of x for which this maximum is achieved.

12. Show that the expression $x^2 + 2x + 2$ is always positive.
    (Hint: show that the minimum value is positive.)

13. Show that the expression $-2x^2 + 8x - 9$ is always negative.

2.4 Factorising polynomials

Statement

> The process by which a polynomial is expressed as the product of polynomials of lower degree is called factorising.

For example

$$1 + 3x + 2x^2 \equiv (1 + 2x)(1 + x)$$

as can be checked by expanding the r.h.s. Thus the *factors* of $1 + 3x + 2x^2$ are $1 + 2x$ and $1 + x$. Some polynomials, for example $1 + x$ and $1 + x^2$, cannot be factorised using real coefficients.

Example 2.6

Factorise the following polynomials

(a) $4 + 5x + x^2$ (b) $9 - 4x^2$ (c) $1 + 6x + 11x^2 + 6x^3$

(a) $4 + 5x + x^2 = (4 + x)(1 + x)$

(b) $9 - 4x^2 = (3 - 2x)(3 + 2x)$

(c) $1 + 6x + 11x^2 + 6x^3 = (1 + x)(1 + 2x)(1 + 3x)$

These results can be checked by expanding the r.h.s. in each equation. It is assumed that the reader is familiar with the methods of factorising (at least simple) quadratics. As an aid in developing techniques for factorising polynomials of higher degree we next consider the division of polynomials.

Division of polynomials

The method will be illustrated by way of some examples.

Example 2.7

Divide $(1 + 6x + 11x^2 + 6x^3)$ *by* $(1 + 2x)$

$\underbrace{\qquad\qquad}_{\text{Dividend}} \quad \underbrace{\quad}_{\text{Divisor}}$

$$\overset{\overbrace{\quad\text{Quotient}\quad}}{3x^2 + 4x + 1}$$

$2x + 1 \overline{)\, 6x^3 + 11x^2 + 6x + 1}$

$3x^2(2x + 1) \qquad\qquad \begin{cases} 6x^3 + 3x^2 \end{cases}$

$\qquad\qquad\qquad\qquad\qquad \begin{cases} 8x^2 + 6x \\ 8x^2 + 4x \end{cases}$ subtracting

$4x(2x + 1)$

$\qquad\qquad\qquad\qquad\qquad \begin{cases} 2x + 1 \\ 2x + 1 \end{cases}$ subtracting

$1.(2x + 1)$

$\qquad\qquad\qquad\qquad\qquad\qquad\quad 0$ no remainder

The quotient is chosen term by term to ensure that these leading terms are equal.

Notice that both the divisor and dividend have been expressed in decending powers of x.

It is perhaps no surprise that there is no remainder as this simply means that $2x + 1$ is a factor of $1 + 6x + 11x^2 + 6x^3$, a fact which was noted in the previous example.

Of course just as with the division of integers there may be a remainder.

Example 2.8

Find the quotient and remainder when $3x^4 + 6x^3 + 2x + 6$ is divided by $(x - 1)$.

$$
\begin{array}{r}
3x^3 + 9x^2 + 9x + 11 \\
x - 1 \overline{)3x^4 + 6x^3 + 0x^2 + 2x + 6} \\
3x^4 - 3x^3 \\
\hline
9x^3 + 0x^2 \\
9x^3 - 9x^2 \\
\hline
9x^2 + 2x \\
9x^2 - 9x \\
\hline
11x + 6 \\
11x - 11 \\
\hline
17 \qquad \text{Remainder}
\end{array}
$$

So in this case the quotient is $3x^3 + 9x^2 + 9x + 11$ and the remainder is 17.

To avoid confusion it is important to enter terms with zero coefficients so that there are no 'gaps'. This can be seen again in the following example.

Example 2.9

Find the quotient and remainder by dividing out $\dfrac{2x^3 + 3x^2 - 5}{x^2 + 2}$.

$$
\begin{array}{r}
2x + 3 \\
x^2 + 0x + 2 \overline{)2x^3 + 3x^2 + 0x - 5} \\
2x^3 + 0x^2 + 4x \\
\hline
3x^2 - 4x - 5 \\
3x^2 + 0x + 6 \\
\hline
-4x - 11 \qquad \text{Remainder}
\end{array}
$$

So the quotient is $2x + 3$ and the remainder is $-4x - 11$.

Statement

> The degree of the remainder will always be less than that of the divisor — otherwise the process of division would continue.

2.5   The remainder theorem

When dividing integers we obtain for example $17/6 = 2$ remainder 5 and this can be written in the form

$$
\frac{17}{6} = 2 + \frac{5}{6} \qquad \text{or} \qquad 17 = 2 \times 6 + 5
$$

Statement

> When dividing polynomials we can write
>
> $$\frac{F(x)}{G(x)} \equiv Q(x) + \frac{R(x)}{G(x)} \qquad \text{when } G(x) \neq 0$$
>
> or
>
> $$F(x) \equiv Q(x) . G(x) + R(x)$$
>
> where $Q(x)$ is the quotient
> $\qquad\qquad R(x)$ is the remainder
> and $G(x)$ is the divisor.

Since the degree of the remainder will be lower than the degree of the divisor, the remainder will be a constant — say R, when the divisor is linear — say $ax + b$.

That is, for example

$$\frac{F(x)}{x - a} \equiv Q(x) + \frac{R}{x - a} \qquad \text{when } x \neq a$$

in the case when the divisor is $x - a$.

Now, multiplying both sides by $x - a$

$$F(x) \equiv Q(x)(x - a) + R$$

and substituting $x = a$ so that $x - a = 0$ we obtain $F(a) = R$. Thus

Statement

> **The remainder theorem**
> When a polynomial $F(x)$ is divided by $x - a$ the remainder is $F(a)$.

Repeating the argument with the divisor $ax + b$ rather than $x - a$ we can similarly obtain

Statement

> When a polynomial $F(x)$ is divided by $ax + b$ the remainder is $F(-b/a)$.

Notice that the choice of $x = -b/a$ is to ensure that $ax + b$ is zero.

Example 2.10

Find the remainder when $2x^3 + 6x - 8$ is divided by

(a) $(x - 1)$ (b) $(x + 1)$ (c) $(2x + 1)$.

Let $F(x) \equiv 2x^3 + 6x - 8$

(a) The remainder on division by $x - 1$ is $F(1)$ where

$$F(1) = 2.1 + 6.1 - 8$$
$$= 2 + 6 - 8$$
$$= 0$$

So $x - 1$ is a factor of $2x^3 + 6x - 8$, since there is no remainder.

(b) The remainder on division by $x + 1$ is $F(-1)$ $[(x + 1) \equiv (x - (-1))]$ where

$$F(-1) = 2(-1)^3 + 6(-1) - 8$$
$$= -2 - 6 - 8$$
$$= -16$$

(c) The remainder on division by $2x + 1$ is $F(-\frac{1}{2})$ where

$$F(-\tfrac{1}{2}) = 2(-\tfrac{1}{2})^3 + 6(-\tfrac{1}{2}) - 8$$
$$= -\tfrac{1}{4} - 3 - 8$$
$$= -11\tfrac{1}{4}$$

Notice that the choice of $x = -\frac{1}{2}$ ensures that $2x + 1 = 0$.

Statement

The factor theorem

If $x - a$ is a factor of $F(x)$ then there is no remainder when $F(x)$ is divided by $(x - a)$, so by the remainder theorem $F(a) = 0$. In other words $x - a$ is a factor of $F(x)$ if and only if $F(a) = 0$.

The factor theorem can be extended to:

Statement

$ax + b$ is a factor of $F(x)$ if and only if $F(-b/a) = 0$

We now return to the problem of factorising polynomials.

Example 2.11

Factorise $P(x) \equiv x^3 - 7x - 6$

$$P(1) = 1^3 - 7.1 - 6$$
$$= -12 \quad (\neq 0 \quad \text{so } x - 1 \text{ is not a factor})$$
$$P(2) = 2^3 - 7.2 - 6$$
$$= 8 - 14 - 6$$
$$= -12 \quad (\neq 0 \quad \text{so } x - 2 \text{ is not a factor})$$

$$P(-1) = (-1)^3 - 7(-1) - 6$$
$$= -1 + 7 - 6$$
$$= 0 \qquad \text{so } (x - (-1)) \equiv x + 1 \text{ is a factor}$$

Notice that the first factor has been found by trial and error. Common sense is required, but obvious first choices are factors of the constant term -6.

Having found a factor we can proceed in one of three ways.

(a) We can continue looking for additional factors using the factor theorem. The factors of the constant term are $\pm 1 \pm 2 \pm 3 \pm 6$ and these are the ones to begin with.

(b) By division

$$
\begin{array}{r}
x^2 - x - 6 \\
x + 1 \overline{\smash{\big)}\, x^3 + 0x^2 - 7x - 6} \\
\underline{x^3 + x^2} \\
-x^2 - 7x \\
\underline{-x^2 - x} \\
-6x - 6 \\
\underline{-6x - 6} \\
0 \qquad \text{as expected}
\end{array}
$$

So $P(x) \equiv (x + 1)(x^2 - x - 6)$

$\qquad \equiv (x + 1)(x + 2)(x - 3)$ \qquad by inspection

(c) Since $P(x)$ is a cubic there will be a quadratic $ax^2 + bx + c$ so that

$$P(x) \equiv x^3 - 7x - 6 \equiv (x + 1)(ax^2 + bx + c)$$

but

$$(x + 1)(ax^2 + bx + c) \equiv ax^3 + bx^2 + cx + ax^2 + bx + c$$
$$\equiv ax^3 + (a + b)x^2 + (b + c)x + c$$

and comparing coefficients

$$a = 1$$
$$a + b = 0$$
$$b + c = -7$$
$$c = -6$$

So $a = 1$, $b = -1$ and $c = -6$ and we can continue as in part (b). This latter method which avoids division is particularly useful when dealing with a cubic, as we are in this case. Much of the working can be done mentally.

Statement

> It can readily be checked that
> $$a^3 + b^3 \equiv (a + b)(a^2 - ab + b^2)$$
> and
> $$a^3 - b^3 \equiv (a + b)(a^2 + ab + b^2)$$

These relationships are worth remembering as they can be used to factorise a certain group of polynomials directly.

Example 2.12

Factorise

(a) $1 + x^3$　　(b) $8 - x^3$　　(c) $27 - 8x^3$

(a) $1 + x^3 \equiv (1 + x)(1 - x + x^2)$

(b) $8 - x^3 \equiv (2 - x)(4 + 2x + x^2)$

(c) $27 - 8x^3 \equiv (3^3 - (2x)^3)$
$$\equiv (3 - 2x)(9 - 6x + 4x^2)$$

Exercise 2.2

1. $P(x) \equiv x^3 + x^2 + x + 1$ is divided by each of the polynomials

   (a) $x + 2$　　(b) $x^2 - 2x$　　(c) $x^2 + 1$　　(d) $x^3 + 1$.

   Find the quotient and remainder.

2. Show by direct division that $x + 2$ is a factor of

   (a) $x^3 + 5x^2 + 7x + 2$　　(b) $x^3 + x^2 - 3x - 2$.

3. Using the remainder theorem find the remainder when the following polynomials are divided by the linear polynomials indicated.

   (a) $x^2 + 4x^2 - 3x + 2$,　$x - 2$　　　　(b) $x^4 - x^2 + 1$,　$x - 1$
   (c) $x^3 + 2x - 3$,　$x + 1$　　　　　　　(d) $x^4 - 3x^3 + 2x - 1$,　$x + 2$
   (e) $x^5 - x^3 + 6$,　$x + 1$　　　　　　　(f) $x^3 + 2x + 1$,　$2x - 1$
   (g) $x^3 - 2x^2 + x + 1$,　$2x - 3$　　　　(h) $x^3 + 15x - 1$,　$3x + 1$

4. If you need more practise at division you should check the results of questions 3 and 5 by direct division.

5. Determine whether the linear polynomials indicated are factors of the associated polynomials.

(a) $x + 2$, $x^4 + 4x^3 + 4x^2$ (b) $x - 1$, $x^5 + 3x^2 - 6x + 3$
(c) $x + 1$, $x^3 - 2x^2 + 6x + 9$ (d) $x - 2$, $x^2 + 2x - 7$
(e) $x - 2$, $x^3 - 4x^2 + 3x + 2$ (f) $2x + 1$, $2x^3 - 3x^2 + 2x + 2$
(g) $2x - 1$, $x^3 + 4x^2 - 3x - 1$ (h) $x - 1$, $x^5 - x^4 + x^3 - x^2 + x - 1$

6. Factorise the following polynomials as far as possible.

(a) $x^3 - 6x^2 + 11x - 6$ (b) $x^3 + 3x^2 + 3x + 1$
(c) $x^3 - x^2 + x - 1$ (d) $x^4 + x^3 - 3x^2 - x + 2$
(e) $1 + x^3$ (f) $27 - x^3$
(g) $64 + 27x^3$

7. If $x - 2$ is a factor of $ax^3 + 3x^2 - 2x + a$ find the value of a.

8. Given that $x + 1$ is a factor of $ax^3 + 2x - 6$ find the value of a.

9. Show that $x - a$ is a factor of $x^3 + (1 - a)x^2 + (3 - a)x - 3a$.

10. Factorise $4x^3 - 9x^2 - 16x + 36$ and hence solve the equation
$$4x^3 - 9x^2 - 16x + 36 = 0.$$

11. Factorise $2x^3 - x^2 - 8x + 4$ and hence solve the equation
$$2 \cdot 10^{3x} - 10^{2x} - 8 \cdot 10^x + 4 = 0.$$

12. Given that $2x - 1$ is a factor of the polynomial
$$8x^3 - 12x^2 + ax - 1,$$

find the value of the constant a.

JMB 1982

2.6 The binomial expansion

[This section is further developed in Chapter 9 on page 300.]

It is useful to have an efficient method of expanding an expression of the form $(a + b)^n$. For example

$$(2 + 3x)^3 \equiv (2 + 3x)(2 + 3x)^2$$
$$\equiv (2 + 3x)(4 + 12x + 9x^2)$$
$$\equiv 2(4 + 12x + 9x^2) + 3x(4 + 12x + 9x^2)$$
$$\equiv 8 + 24x + 18x^2 + 12x + 36x^2 + 27x^3$$
$$\equiv 8 + 36x + 54x^2 + 27x^3$$

We could obtain the expansion of $(2 + 3x)^5$ or even $(2 + 3x)^9$ in like manner, but the procedure is tedious.

A natural first step in trying to find a more effective way to expand such expressions is to be systematic and look for a pattern. Now

$$(a + b)^0 \equiv 1$$
$$(a + b)^1 \equiv a + b$$
$$(a + b)^2 \equiv a^2 + 2ab + b^2$$
$$(a + b)^3 \equiv a^3 + 3a^2b + 3ab^2 + b^3$$
$$(a + b)^4 \equiv a^4 + 4a^3b + 6a^2b^2 + 4ab^3 + b^4$$
$$\vdots$$

It is reasonable to suppose (and in fact correct) that

$$(a + b)^5 \equiv a^5 + 5a^4b + ?a^3b + ?a^2b^3 + 5ab^4 + b^5$$

The problem reduces to finding the numerical parts of the terms.

We have so far

when $n = 0$					1			
$n = 1$				1		1		
$n = 2$			1		2		1	
$n = 3$		1		3		3		1
$n = 4$	1		4		6		4		1

The pattern emerging was first studied by Blaise Pascal and is known as Pascal's triangle. Each number is obtained by summing the two numbers above and immediately to the left and right. (Apart from the 1 at the end of each row.)

We may continue as far as we please using this method.

For $n = 5$	1	5	10	10	5	1
$n = 6$	1	6	15	20	15	6	1
			$\vdots$			

Example 2.13

Expand each of the following

(a) $(a + b)^5$ (b) $(2 + 3x)^4$ (c) $(1 - x)^6$.

(a) $(a + b)^5 \equiv a^5 + 5a^4b + 10a^3b^2 + 10a^2b^3 + 5ab^4 + b^5$

(b) $(2 + 3x)^4 \equiv 2^4 + 4 \cdot 2^3(3x) + 6 \cdot 2^2(3x)^2 + 4 \cdot 2(3x)^3 + (3x)^4$

$$\equiv 16 + 96x + 216x^2 + 216x^3 + 81x^4$$

(c) $(1 - x)^6 \equiv 1^6 + 6(1^5)(-x) + 15(1^4)(-x)^2 + 20(1^3)(-x)^3 + 15(1^2)(-x)^4$

$$+ 6(1)(-x)^5 + (-x)^6$$

$$\equiv 1 - 6x + 15x^2 - 20x^3 + 15x^4 - 6x^5 + x^6$$

Statement

> Each row in Pascal's triangle can be obtained directly as
> $$1, \frac{n}{1}, \frac{n(n-1)}{2 \times 1}, \frac{n(n-1)(n-2)}{3 \times 2 \times 1}, \ldots, \frac{n}{1}, 1$$

This result cannot be proved at this stage. However, for $n = 3$ we obtain

$$1, \frac{3}{1}, \frac{3.2}{2.1}, \frac{3.2.1}{3.2.1} \qquad \text{giving 1, 3, 3, 1}$$

and for $n = 4$

$$1, \frac{4}{1}, \frac{4.3}{2.1}, \frac{4.3.2}{3.2.1}, \frac{4.3.2.1}{4.3.2.1} \qquad \text{giving 1, 4, 6, 4, 1}$$

The binomial theorem

> $$(a + b)^n \equiv a^n + na^{n-1}b + \frac{n(n-1)}{2.1}a^{n-2}b^2 + \frac{n(n-1)(n-2)}{3.2.1}a^{n-3}b^3 + \ldots$$
>
> $$\ldots + b^n \qquad \text{for } n \in \mathbb{N}$$

This important formula should be learned.

Example 2.14

Find the first four terms in the expansion of the following in ascending powers of x:

(a) $(x + 1)^8$ (b) $(3 - 2x)^8$.

(a) $(x + 1)^8 \equiv (1 + x)^8$ (this ensures an expansion in ascending powers of x)

$$\equiv 1 + \frac{8x}{1} + \frac{8.7x^2}{2.1} + \frac{8.7.6x^3}{3.2.1} + \ldots$$

$$\equiv 1 + 8x + 28x^2 + 56x^3 + \ldots$$

(b) $(3 - 2x)^8 \equiv 3^8 + \frac{8}{1}3^7(-2x) + \frac{8.7}{2.1}3^6(-2x)^2 + \frac{8.7.6}{3.2.1}3^5(-2x)^3 + \ldots$

$$\equiv 3^8 - 16.3^7x + 112.3^6x^2 - 448.3^5x^3 + \ldots$$

Example 2.15

Expand $(3 + 2x)^2$.

It is tempting to apply the binomial theorem at this stage, and it would of course give an accurate result. However, recall that

$$(a + b)^2 \equiv a^2 + 2ab + b^2$$

so we can write directly

$$(3 + 2x)^2 \equiv 9 + 12x + 4x^2$$

Example 2.16
Find the term in x^2 in the expansion $(1 + x^2)(1 - x)^4$.

Now

$$(1 + x^2)(1 - x)^4 \equiv (1 + x^2)[1 + 4(-x) + 6(-x)^2 + \text{higher terms}]$$
$$\equiv 1 - 4x + 6x^2 + \ldots + x^2 - 4x^3 + \ldots$$

It follows that the term required is $7x^2$.

Example 2.17
Assuming that x is small enough for powers of x^3 and above to be neglected, find a quadratic approximation to $(1 + 2x)(1 + x)^6$

$$(1 + 2x)(1 + x)^6 \equiv (1 + 2x)[1 + 6x + 15x^2 + \ldots]$$
$$\equiv 1 + 6x + 15x^2 + \ldots$$
$$+ 2x + 12x^2 + \ldots \quad \text{neglecting powers of } x^3 \text{ and above}$$
$$\equiv 1 + 8x + 27x^2 + \ldots$$

Example 2.18
Without using a calculator, find the value of $(1.01)^6$ correct to 3 d.p.

Now

$$(1 + x)^6 = 1 + 6x + 15x^2 + 20x^3 + 15x^4 + 6x^5 + x^6$$

Setting $x = 0.01$

$$(1.01)^6 = 1 + 6.(0.01) + 15.(0.01)^2 + 20.(0.01)^3 + \ldots$$
$$= 1 + 0.06 + 0.0015 + 0.000\,020 + \ldots$$
$$\simeq 1.061\,520\ldots$$
$$\simeq 1.062 \qquad 3\,\text{d.p.}$$

Notice that only the first few terms contributed to the final answer.

Exercise 2.3

1. Simplify the following polynomials

   (a) $(3x^2 + 5x - 6) + (9x^3 - 2x^2 + 6x - 5)$
   (b) $4x(2x^3 - 3x^2 + 1)$ (c) $(3x + 4)(3x^2 + 2x - 6)$
   (d) $(1 + 2x - 3x^2)(4 - 2x + x^2)$ (e) $(1 + 2x)(2 - x)^2$

2. Write down the expansion of each of the following powers of binomials:

(a) $(1 + 3x)^2$ (b) $(5 - 2x)^2$ (c) $(2 + 3x)^2$
(d) $(1 + 3x)^3$ (e) $(2 - x)^3$ (f) $(2x - 1)^3$.

3. Use the binomial theorem to expand the following polynomials in ascending powers of x.

(a) $(1 + 2x)^4$ (b) $(1 - x)^4$ (c) $(2x + 1)^5$

(d) $(x - 1)^7$ (e) $\left(1 + \dfrac{x}{2}\right)^4$ (f) $\left(1 - \dfrac{2x}{3}\right)^5$

(g) $(2 - x)^5$ (h) $(x + 3)^4$ (i) $\left(2 - \dfrac{x}{2}\right)^4$

(j) $\left(1 + \dfrac{x}{2}\right)^4 - \left(1 - \dfrac{x}{2}\right)^4$ (k) $(2 + x^2)^5$

4. Find the coefficient of x^3 in the expansion of:

(a) $(1 - 3x)(1 - x)^4$ (b) $(1 + x - x^2)(1 + 2x)^5$
(c) $(1 + 2x - x^2)(1 - x)^6$.

5. Expand the following and simplify the resulting expressions.

(a) $(1 + 2\sqrt{2})^5$ (b) $(3 + \sqrt{3})^4$ (c) $(\sqrt{3} + \sqrt{2})^4$

6. By expanding $(1 + x)^4$ and setting $x = 0.01$ find $(1.01)^4$ without using a calculator.

7. Expand $(a + b)^6$. By setting $a = x$ and $b = \dfrac{1}{x}$ find an expansion for

$$\left(x + \frac{1}{x}\right)^6.$$

8. Expand $\left(x^2 - \dfrac{1}{x}\right)^5$ in ascending powers of x.

9. Find, without using a calculator, an approximation to $(1.01)^8$ correct to 3 d.p.

10. Assuming that x is small enough for powers of x^3 and above to be neglected, find a quadratic approximation to $(2 + x)(1 - x)^6$.

11. Expand $(1 + a)^3$ and by setting $a = x(1 + 2x)$ find the expansion of $(1 + x + 2x^2)^3$ in ascending powers of x.

12. Expand $(2 + x - x^2)^3$

2.7 Partial fractions

Statement

> If $F(x)$ and $G(x)$ are polynomials, then an expression of the form $F(x)/G(x)$ is called an algebraic rational or rational function.

Algebraic rationals are in many respects similar to rationals (fractions). They can be added, subtracted, multiplied or divided to obtain new algebraic rationals.

An algebraic rational is said to be *proper* if the degree of the numerator is less than the degree of the denominator, otherwise it is *improper*.

Any improper algebraic rational $F(x)/G(x)$ can be divided to obtain a polynomial and a proper algebraic rational. If, however, $G(x)$ is a factor of $F(x)$ then the algebraic rational reduces to a polynomial.

Example 2.19

Simplify the following:

(a) $\dfrac{x^2 - 1}{x^3 - 1}$ (b) $\dfrac{2x}{(1 + x)} \cdot \dfrac{3}{(2 + x)}$

(c) $\dfrac{x - 1}{x^2 + 3x + 2} \div \dfrac{x^2 - 1}{x + 2}$ (d) $\dfrac{2}{3 + x} - \dfrac{1}{1 + x}$

(a) $\dfrac{x^2 - 1}{x^3 - 1} \equiv \dfrac{(x + 1)(x - 1)}{(x - 1)(x^2 + x + 1)}$ factorising

$\qquad\quad \equiv \dfrac{x + 1}{x^2 + x + 1}$ cancelling common factors

(b) $\dfrac{2x}{(1 + x)} \cdot \dfrac{3}{(2 + x)} \equiv \dfrac{6x}{(1 + x)(2 + x)}$

(c) $\dfrac{x - 1}{x^2 + 3x + 2} \div \dfrac{x^2 - 1}{x + 2} \equiv \dfrac{x - 1}{(x + 2)(x + 1)} \cdot \dfrac{x + 2}{(x - 1)(x + 1)}$

$\qquad\qquad\qquad\qquad\qquad\quad \equiv \dfrac{1}{(x + 1)^2}$

(d) $\dfrac{2}{3 + x} - \dfrac{1}{1 + x} \equiv \dfrac{2(1 + x) - 1(3 + x)}{(3 + x)(1 + x)}$

$\qquad\qquad\qquad\quad \equiv \dfrac{-(1 - x)}{(3 + x)(1 + x)}$

Example 2.20

Find constants A and B so that

$$\frac{x-8}{(x+1)(x-2)} \equiv \frac{A}{x+1} + \frac{B}{x-2} \qquad x \neq 2, -1.$$

Multiply both sides by $(x+1)(x-2)$ to obtain

$$x - 8 \equiv A(x-2) + B(x+1)$$

Since the two polynomials are identically equal, they will be equal for all values of x. In particular:

When $x = 2$,

$$-6 = 0 + 3B \qquad \Rightarrow B = -2.$$

When $x = -1$,

$$-9 = -3A + 0 \qquad \Rightarrow A = 3.$$

Any two values of x could have been selected, the choice of $x = 2$ and $x = -1$ ensured that the resulting equations contained just one unknown, which simplified the working. So

$$\frac{x-8}{(x+1)(x-2)} \equiv \frac{3}{x+1} - \frac{2}{x-2} \qquad x \neq 2, -1$$

This relationship holds for all values of x for which the expressions can be evaluated. [We must avoid $x = -1$ and $x = 2$ since we cannot divide by zero.]

In order to ensure that no error is made, it is worthwhile selecting a new value for x, say $x = 0$ and checking that the above relationship holds.

Statement

> The process of splitting a proper algebraic rational into the sum or difference of other proper algebraic rationals is called 'expressing the rational function in *partial fractions*'. An *improper algebraic rational is first divided.*

Example 2.21

Express the following in partial fractions

(a) $\dfrac{6}{(x+1)(x-2)}$ (b) $\dfrac{x}{(x-2)^2}$

(a) We assume there are constants A and B so that

$$\frac{6}{(x+1)(x-2)} \equiv \frac{A}{x+1} + \frac{B}{x-2} \qquad x \neq -1, 2$$

Multiplying both sides by $(x + 1)(x - 2)$

$$6 \equiv A(x - 2) + B(x + 1)$$

For $x = 2$

$$6 = 0 + 3B \qquad \Rightarrow B = 2$$

For $x = -1$

$$6 = -3A + 0 \qquad \Rightarrow A = -2$$

Thus

$$\frac{6}{(x + 1)(x - 2)} \equiv \frac{-2}{x + 1} + \frac{2}{x - 2} \qquad x \neq -1, 2$$

Checking for $x = 0$ (say)

$$\text{l.h.s.} = \frac{6}{-2} = -3 \qquad \text{r.h.s.} = \frac{-2}{1} + \frac{2}{-1} = -3$$

and the check is complete.

(b) If we assume there are constants A and B so that

$$\frac{x}{(x - 2)(x - 2)} \equiv \frac{A}{x - 2} + \frac{B}{x - 2} \qquad x \neq 2$$

then multiplying both sides by $(x - 2)^2$ we find

$$x \equiv A(x - 2) + B(x - 2)$$

Setting $x = 2$ we obtain

$$2 = 0 + 0 \qquad \text{which is inconsistent!}$$

Thus our assumption is unacceptable.
    On reflection

$$\frac{A}{x - 2} + \frac{B}{x - 2} = \frac{A + B}{x - 2}$$

which does not have the same form as

$$\frac{x}{(x - 2)(x - 2)}.$$

Note that

$$\frac{5}{9} = \frac{3}{9} + \frac{2}{9},$$

that is

$$\frac{5}{3^2} = \frac{1}{3} + \frac{2}{3^2}.$$

By analogy we assume there are constants A and B so that

$$\frac{x}{(x-2)^2} \equiv \frac{A}{(x-2)} + \frac{B}{(x-2)^2} \qquad x \neq 2.$$

Again, multiplying by $(x-2)^2$ we require

$$x \equiv A(x-2) + B$$

When $x = 2$,

$$2 = 0 + B \qquad \Rightarrow B = 2.$$

When $x = 0$,

$$0 = -2A + B \qquad \text{but} \quad B = 2,$$

$$\Rightarrow 0 = -2A + 2$$

$$\Rightarrow A = 1.$$

Thus

$$\frac{x}{(x-2)^2} \equiv \frac{1}{x-2} + \frac{2}{(x-2)^2}.$$

We still need to check our assumption and cannot use $x = 0$ since we have already used this value to establish the value of one of the constants. Of course we could simply add the terms on the r.h.s. and show directly that l.h.s. $\equiv$ r.h.s.
Alternatively we try another value of x, say $x = 3$.

$$\text{l.h.s.} = \frac{3}{1^2} = 3 \qquad \text{r.h.s.} = \frac{1}{1} + \frac{2}{1^2} = 3$$

Statement

The following rules will be found helpful when trying to express a rational function in partial fractions.

(i) Divide out an improper algebraic rational to obtain a polynomial and a proper rational function.

(ii) For each linear factor $ax + b$ in the denominator of a proper rational function, introduce a partial fraction of the form

$$\frac{A}{ax+b}.$$

(iii) (a) If the linear factor is repeated, such as $(ax + b)^2$, introduce partial fractions of the form

$$\frac{A}{ax+b} + \frac{B}{(ax+b)^2}$$

(b) Similarly for $(ax + b)^3$, introduce

$$\frac{A}{ax + b} + \frac{B}{(ax + b)^2} + \frac{C}{(ax + b)^3} \qquad \text{and so on.}$$

(iv) If a quadratic factor $ax^2 + bx + c$ appears in the denominator which does not factorise then a partial fraction of the form

$$\frac{Ax + B}{ax^2 + bx + c}$$

should be introduced

Example 2.22

Express the following in partial fractions.

(a) $\dfrac{x^2 + 2x + 1}{x^2 - 1}$ $\quad x \neq \pm 1$ $\qquad$ (b) $\dfrac{x^2 - 2x - 1}{2x(x^2 + 1)}$ $\quad x \neq 0$

(a) The rational function is improper so we first divide

$$x^2 + 0x - 1 \overline{\smash{\big)}\, x^3 + 0x^2 + 2x + 1} \quad \overset{\displaystyle x}{}$$
$$\underline{x^3 + 0x^2 - x}$$
$$3x + 1$$

So

$$\frac{x^3 + 2x + 1}{x^2 - 1} \equiv x + \frac{3x + 1}{x^2 - 1} \qquad x \neq \pm 1$$

Now suppose

$$\frac{3x + 1}{x^2 - 1} \equiv \frac{A}{x - 1} + \frac{B}{x + 1} \qquad \left(\begin{array}{l}\text{noting that}\\ x^2 - 1 \equiv (x - 1)(x + 1).\end{array}\right)$$

Multiply both sides by $(x + 1)(x - 1)$. We require

$$3x + 1 \equiv A(x + 1) + B(x - 1).$$

When $x = 1$,

$$4 = 2A + 0 \qquad \Rightarrow A = 2.$$

When $x = -1$,

$$-2 = 0 - 2B \qquad \Rightarrow B = 1.$$

Thus

$$\frac{x^3 + 2x + 1}{x^2 - 1} \equiv x + \frac{2}{x - 1} + \frac{1}{x + 1} \qquad x \neq \pm 1$$

Checking for $x = 0$

$$\text{l.h.s.} = -1 \qquad \text{r.h.s.} = 0 - 2 + 1 = -1$$

(b) Assume constants A, B and C exist so that

$$\frac{x^2 - 2x - 1}{2x(x^2 + 1)} \equiv \frac{A}{2x} + \frac{Bx + C}{x^2 + 1} \qquad x \neq 0$$

Multiply both sides by $2x(x^2 + 1)$. We require

$$x^2 - 2x - 1 \equiv A(x^2 + 1) + (Bx + C)2x$$

When $x = 0$

$$-1 = A + 0 \qquad \Rightarrow A = -1$$

Compare the coefficients of x^2

$$1 = A + 2B \qquad \text{and} \qquad A = -1$$
$$\Rightarrow B = 1$$

Set $x = 1$

$$-2 = 2A + 2(B + C)$$
$$\Rightarrow -1 = A + B + C \qquad \text{but } A = -1 \text{ and } B = 1$$
$$\Rightarrow -1 = -1 + 1 + C$$
$$\Rightarrow C = -1$$

Thus

$$\frac{x^2 - 2x - 1}{2x(x^2 + 1)} \equiv \frac{-1}{2x} + \frac{x - 1}{x^2 + 1} \qquad x \neq 0$$

Checking for $x = -1$

$$\text{l.h.s.} = \frac{2}{-4} = \frac{-1}{2} \qquad \text{r.h.s.} = \frac{-1}{-2} + \frac{-2}{2} = \frac{-1}{2}$$

Exercise 2.4

1. Simplify the following algebraic rationals.

(a) $\dfrac{x(x + 2)}{x + 3} \cdot \dfrac{(x + 3)}{x^2 + 2x}$ (b) $\dfrac{x^2 - 1}{x + 2} \div \dfrac{x + 1}{x - 1}$

(c) $\dfrac{x - 2}{x + 3} \div \dfrac{x - 2}{x + 2}$ (d) $\dfrac{x^2 + 2x + 1}{x^2 + 3x + 2} \div \dfrac{x^2 + 4x + 3}{x^2 + 5x + 6}$

2. Simplify the following.

(a) $\dfrac{1}{1+x} + \dfrac{1}{1-x}$ (b) $\dfrac{1}{1+x} - \dfrac{1}{2+x}$

(c) $\dfrac{2x}{1-x} + \dfrac{1}{1+x}$ (d) $\dfrac{1}{2+x} + \dfrac{3}{3+x} - \dfrac{4}{4+x}$

3. Express the following in partial fractions.

(a) $\dfrac{2x+4}{(x+1)(x+2)}$ (b) $\dfrac{x-4}{(x+2)(x-1)}$ (c) $\dfrac{3}{x^2-9}$

(d) $\dfrac{x}{x^2-1}$ (e) $\dfrac{x+1}{x^2-4}$ (f) $\dfrac{x+3}{x^2+5x+4}$

4. Express the following in partial fractions.

(a) $\dfrac{4x+10}{(2+x)(3+x)(4+x)}$ (b) $\dfrac{4x+2}{(x+2)(x^2-1)}$

(c) $\dfrac{4x}{(2-x)^2}$ (d) $\dfrac{x}{(1+x)^2}$

5. Divide the following and hence express in partial fractions.

(a) $\dfrac{x^2+1}{x^2-1}$ (b) $\dfrac{x^2+x-1}{(x+1)(x+2)}$ (c) $\dfrac{2x^2+3x+2}{x^2+3x+2}$ (d) $\dfrac{x^3+x+1}{(x+1)^2}$

6. Express the following in partial fractions.

(a) $\dfrac{x(x-1)}{(x+1)(x^2+1)}$ (b) $\dfrac{1}{(x^2+1)(x^2+2)}$

(c) $\dfrac{2x^2+x+2}{(x^2+1)(x^2+2)}$ (d) $\dfrac{2x-12}{(x^2+9)(x+3)}$

7. Express the following in partial fractions.

(a) $\dfrac{1}{x(x-1)^2}$ (b) $\dfrac{-2x}{(x^2+2)(x+1)}$

(c) $\dfrac{x^2}{x^2+3x+2}$ (d) $\dfrac{4x^3+2x^2+x+2}{x^2(x^2+1)}$

8. Factorise the cubic $x^3 - 6x^2 + 11x - 6$ and hence express in partial fractions

$$\dfrac{2}{x^3 - 6x^2 + 11x - 6}$$

9. Express in partial fractions $\dfrac{x^2+2}{x(x^2+1)}$

Chapter 3 Equations and inequalities

3.1 Introduction

Many problems in mathematics can be reduced to finding the solution or *roots* of an equation. So it is important that one should be familiar with the methods of solving standard equations.

Example 3.1

Solve the equation $(x - 1)(x - 2) = 0$.

The fundamental point to note is that if the product of two (or more) numbers is zero then one of them must be zero. Thus

$$(x - 1)(x - 2) = 0$$
$$\Rightarrow x - 1 = 0 \quad \text{or} \quad x - 2 = 0$$
$$\Rightarrow \quad x = 1 \quad \text{or} \quad x = 2.$$

We have found two roots or solutions in this case. If either value is substituted into the original equation then the resulting statement will be found to be true. Moreover, no other value for x has this property.

If the graph of the equation $y = (x - 1)(x - 2)$ is drawn as in Fig. 3.1 then the points where the graph cuts the x axis correspond to the solutions of the equation $0 = (x - 1)(x - 2)$.

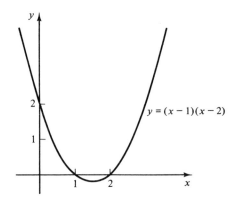

Fig. 3.1

This observation applies quite generally and so graphs are sometimes useful in solving equations. (See also Chapter 17.)

A clear distinction should be maintained between an equation and an identity. If

$$ax + b = 1 \qquad a \neq 0$$

then

$$x = \frac{1 - b}{a} \qquad \text{is the solution.}$$

However if

$$ax + b \equiv 1$$

then

$$a = 0 \text{ and } b = 1 \qquad ! \quad \text{by comparing coefficients.}$$

An attempt to 'solve' an identity will result in the statement $0 = 0$.
For example

$$(x + 1)^2 \equiv x^2 + 2x + 1$$
$$\Rightarrow \quad x^2 + 2x + 1 \equiv x^2 + 2x + 1$$
$$\Rightarrow \qquad\qquad 0 = 0 \qquad\qquad !!$$

3.2   The quadratic equation

A quadratic equation can be expressed in the form

$$ax^2 + bx + c = 0 \qquad \text{where } a \neq 0$$

[It would not be a quadratic equation if $a = 0$.] If b or c is zero, the solutions are easy
to obtain.

Example 3.2

Solve the equations

(a) $x^2 + 2x = 0$ (b) $x^2 - 2 = 0$ (c) $x^2 + 2 = 0$ (d) $x^2 + 4x + 2 = 0$

(a) $x^2 + 2x = 0 \qquad \Leftrightarrow x(x + 2) = 0$
$$\Leftrightarrow x = 0 \quad \text{or} \quad x + 2 = 0$$
$$\Leftrightarrow x = 0 \quad \text{or} \quad x = -2$$

One should avoid the temptation to divide both sides by x, for in that case the
solution $x = 0$ is lost. [Division by x carries with it the assumption that $x \neq 0$.]

(b) $x^2 - 2 = 0 \qquad \Leftrightarrow x^2 = 2$
$$\Leftrightarrow x = \sqrt{2} \quad \text{or} \quad x = -\sqrt{2} \qquad \text{usually written } x = \pm\sqrt{2}$$

(c) $x^2 + 2 = 0 \qquad \Leftrightarrow x^2 = -2$

There are no real solutions in this case. The phrase 'no real solutions' means that there are no solutions which are real numbers. This is in anticipation of the fact that there are solutions amongst the *complex numbers* [dealt with in Chapter 16].

(d) $x^2 + 4x + 2 = 0$.

The quadratic $x^2 + 4x + 2$ is difficult to factorise, without using the technique called completing the square. As usual, assume there are constants A and B so that

$$x^2 + 4x + 2 \equiv (x + A)^2 + B$$
$$\equiv x^2 + 2Ax + A^2 + B$$

and by comparing coefficients we obtain

$$A = 2 \quad \text{and} \quad B = -2$$

So

$$x^2 + 4x + 2 \equiv (x + 2)^2 - 2$$

Return to the original equation

$$x^2 + 4x + 2 = 0$$
$$\Leftrightarrow (x + 2)^2 - 2 = 0$$
$$\Leftrightarrow (x + 2)^2 = 2$$
$$\Leftrightarrow x + 2 = \sqrt{2} \quad \text{or} \quad x + 2 = -\sqrt{2}$$
$$\Leftrightarrow x = -2 + \sqrt{2} \quad \text{or} \quad x = -2 - \sqrt{2}$$
$$\Leftrightarrow x = -0.586 \quad \text{or} \quad x = -3.414 \qquad \text{3d.p.} \qquad \boxed{C}$$

This means that the factors of $x^2 + 4x + 2$ are $(x + 2 - \sqrt{2})$ and $(x + 2 + \sqrt{2})$ by the factor theorem.

Example 3.3

Factorise the quadratic

$$P(x) \equiv x^2 + 3x + 1.$$

Suppose

$$x^2 + 3x + 1 \equiv (x + A)^2 + 6$$
$$\equiv x^2 + 2Ax + A^2 + B.$$

Now

$$2A = 3 \qquad \Rightarrow A = \frac{3}{2}$$

and

$$A^2 + B = 1 \qquad \Rightarrow \frac{9}{4} + B = 1$$
$$\Rightarrow B = -\frac{5}{4}.$$

So
$$P(x) \equiv x^2 + 3x + 1 \equiv \left(x + \frac{3}{2}\right)^2 - \frac{5}{4}.$$

Recalling that
$$a^2 - b^2 \equiv (a + b)(a - b)$$

and setting $a = x + \dfrac{3}{2}$ and $b = \dfrac{\sqrt{5}}{2}$, we have

$$P(x) \equiv \left(x + \frac{3}{2} + \frac{\sqrt{5}}{2}\right)\left(x + \frac{3}{2} - \frac{\sqrt{5}}{2}\right).$$

Statement

> The process of finding constants A and B so that
> $$ax^2 + bx + c \equiv a(x + A)^2 + B$$
> is called completing the square.
>   When $B = 0$ and a is one we say that the quadratic is a perfect square.

Rather than complete the square each time we find a quadratic difficult to factorise we can consider the general situation.
  Suppose
$$ax^2 + bx + c = 0 \qquad a \neq 0$$

Assume there are constants A and B so that
$$ax^2 + bx + c \equiv a(x + A)^2 + B$$
$$\equiv ax^2 + 2aAx + aA^2 + B.$$

Comparing coefficients we have
$$2aA = b \qquad \Rightarrow A = \frac{b}{2a} \qquad [\text{recall that } a \neq 0]$$

and
$$aA^2 + B = c$$
$$\Rightarrow \frac{b^2}{4a} + B = c$$
$$\Rightarrow B = \frac{4ac - b^2}{4a}.$$

So
$$ax^2 + bx + c \equiv a\left(x + \frac{b}{2a}\right)^2 + \frac{4ac - b^2}{4a}.$$

Returning to the original equation

$$ax^2 + bx + c = 0$$

$$\Rightarrow a\left(x + \frac{b}{2a}\right)^2 + \frac{4ac - b^2}{4a} = 0$$

$$\Rightarrow \left(x + \frac{b}{2a}\right)^2 = \frac{b^2 - 4ac}{4a^2}$$

$$\Rightarrow x + \frac{b}{2a} = \frac{\sqrt{(b^2 - 4ac)}}{2a} \quad \text{or} \quad x + \frac{b}{2a} = -\frac{\sqrt{(b^2 - 4ac)}}{2a}$$

and so finally

$$x = \frac{-b + \sqrt{(b^2 - 4ac)}}{2a} \quad \text{or} \quad x = \frac{-b - \sqrt{(b^2 - 4ac)}}{2a}.$$

These two solutions are conventionally combined to give the familiar formula.

Statement

> The solutions of the quadratic equation
>
> $$ax^2 + bx + c = 0$$
>
> are
>
> $$x = \frac{-b \pm \sqrt{(b^2 - 4ac)}}{2a}$$

Example 3.4

Solve the equation

$$2x^2 + 5x + 1 = 0$$

Using the formula where $a = 2$, $b = 5$ and $c = 1$,

$$x = \frac{-5 \pm \sqrt{(25 - 8)}}{4} = \frac{-5 \pm \sqrt{17}}{4}$$

$$\Rightarrow x = -0.219 \quad \text{or} \quad x = -2.281. \qquad 3\text{d.p.} \qquad \boxed{C}$$

The nature of the roots of a quadratic equation

Statement

> The expression $b^2 - 4ac$ is called the discriminant of the quadratic equation $ax^2 + bx + c = 0$.

If $b - 4ac > 0$ then the square root of the discriminant exists and there will be two distinct roots. So the graph of $y = ax^2 + bx + c$ will cut the x axis at two distinct points.

If $b^2 - 4ac = 0$ then there will be only the value $x = -b/2a$ which satisfies the equation. We saw that there are *two* equal or repeated roots. In this case, it will be found that the graph of $y = ax^2 + bx + c$ just touches the x axis.

If $b^2 - 4ac < 0$ then there is no (real) square root and so no real solutions in this case. The equation has no roots. The graph of $y = ax^2 + bx + c$ fails to cross the x axis.

Example 3.5

Without solving the equations directly discuss the nature of the solutions to the following equations.

(a) $3x^2 + 4x + 1 = 0$ (b) $4x^2 + 12x + 9 = 0$ (c) $x^2 + x + 1 = 0$

In each case we calculate the value of the discriminant.

(a) $a = 3$, $b = 4$, $c = 1$ so the discriminant is $4^2 - 4 . 3 . 1 = 4 [>0]$. So there are two solutions. Since the square root of the discriminant is the square of an integer we could have solved the equation by factorising.

(b) $a = 4$, $b = 12$, $c = 9$ so the discriminant is $12^2 - 4 . 4 . 9 = 0$. There are repeated roots and the quadratic is a perfect square — in fact $(2x + 3)^2$.

(c) $a = 1$, $b = 1$, $c = 1$ so the discriminant is $1^2 - 4 . 1 . 1 = -3 [<0]$ and so there are no real roots.

The graphs in Fig. 3.2 indicate geometrically the nature of the roots. These graphs are smooth and symmetrical.

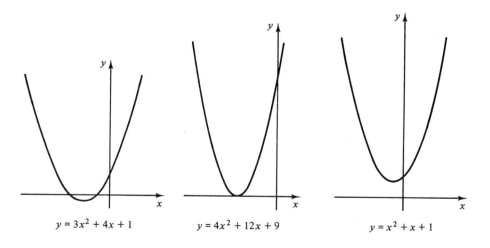

$y = 3x^2 + 4x + 1$ $y = 4x^2 + 12x + 9$ $y = x^2 + x + 1$

Fig. 3.2

Example 3.6

Find the relationship between b and c if the equation $x^2 + bx + c = 0$ has equal roots.

Assuming equal roots the discriminant must be zero. That is

$$b^2 - 4c = 0.$$

Cubic equations

Example 3.7

Solve the cubic

$$(x + 1)(x - 2)(2x + 1) = 0$$

As stated before, if the product of two or more terms is zero, then at least one of the terms must be zero. So

$$x + 1 = 0 \quad \text{or} \quad x - 2 = 0 \quad \text{or} \quad 2x + 1 = 0$$
$$\Rightarrow x = -1 \quad \text{or} \quad x = 2 \quad \text{or} \quad x = -\tfrac{1}{2}$$

Example 3.8

Solve the equation

$$x^3 + 6x^2 + 11x + 6 = 0.$$

We first attempt to factorise the cubic polynomial

$$P(x) \equiv x^3 + 6x^2 + 11x + 6$$

using the factor theorem.

When

$$x = 1 \qquad P(1) = 1 + 6 + 11 + 6 = 24 \qquad [\neq 0]$$
$$x = -1 \qquad P(-1) = -1 + 6 - 11 + 6 = 0$$

so $x + 1$ is a factor of $P(x)$.

It follows that there will be constants a, b and c so that

$$P(x) \equiv (x + 1)(ax^2 + bx + c)$$
$$\equiv ax^3 + (b + a)x^2 + (b + c)x + c$$

and comparing coefficients

$$a = 1, \quad b = 5 \quad \text{and} \quad c = 6.$$

Thus

$$P(x) \equiv (x + 1)(x^2 + 5x + 6)$$
$$\equiv (x + 1)(x + 2)(x + 3).$$

[We could have reached this point by dividing $P(x)$ by the factor $(x + 1)$.] Returning to the original equation $P(x) = 0$,

$$\Leftrightarrow (x + 1)(x + 2)(x + 3) = 0$$
$$\Leftrightarrow (x + 1) = 0 \quad \text{or} \quad (x + 2) = 0 \quad \text{or} \quad (x + 3) = 0$$
$$\Leftrightarrow x = -1 \quad \text{or} \quad x = -2 \quad \text{or} \quad x = -3.$$

There are 3 solutions in this case.

Some polynomials are difficult to factorise (indeed some are impossible!) and particular techniques need to be adopted.

Exercise 3.1

1. Solve the following quadratic equations by completing the square.

    (a) $x^2 + 6x + 5 = 0$ (b) $x^2 + 5x + 3 = 0$ (c) $2x^2 - 3x - 2 = 0$
    (d) $2x^2 + 5x - 1 = 0$

2. Use the appropriate method to solve these quadratic equations.

    (a) $x^2 + 3x + 1 = 0$ (b) $2x^2 - 6x + 3 = 0$ (c) $x^2 - x - 1 = 0$
    (d) $x(x - 1) = 2(1 - x)$ (e) $2x^2 + 3x - 2 = 0$ (f) $x^2 - 5 = 0$
    (g) $x^2 + 4x = 0$ (h) $3x^2 + 2x = 0$ (i) $5x^2 - 3 = 0$
    (j) $2x^2 = 9$

3. Discuss the existence of solutions to the following equations without solving them.

    (a) $x^2 + x + 1 = 0$ (b) $2x^2 + x - 1 = 0$
    (c) $4x^2 + 12x + 9 = 0$ (d) $x^2 - 5x + 4 = 0$
    (e) $x^2 + x - a = 0$

4. Solve the following equations.

    (a) $(x + 1)(x - 3) = 0$ (b) $(x - 2)(x + 1)(x + 6) = 0$
    (c) $(2x + 1)(3x - 5) = 0$ (d) $(2x - 1)(x + 2)(3x - 7) = 0$
    (e) $x^2 + 7x + 12 = 0$ (f) $2x^2 + 3x - 2 = 0$

5. Factorise the polynomials and so solve the following equations.

    (a) $2x^3 - x^2 - 13x - 6 = 0$ (b) $3x^2 + 2x + 2 = x^2 + 3x + 3$
    (c) $(3x - 1)(x + 2) = 3x(2 - x)$ (d) $x^4 - 1 = 0$

6. State which of the following are identities and solve the other equations.

    (a) $x^2 + 3x + 6 = 2(x + 3) + x(x + 1)$
    (b) $x(x - 1) + 3(x - 2) = 3(x - 3) + x + 6$

(c) $\dfrac{1}{1-x} + \dfrac{2}{x+1} = \dfrac{x-3}{x^2-1}$

(d) $(x-1)(x-2) + (x+3)(x-2) = 2(x-2)(x+1)$

7. Solve the following cubic equations

   (a) $x^3 + x^2 + x + 1 = 0$ (b) $4x^3 - 4x^2 - x + 1 = 0$
   (c) $4x^3 + 6x^2 + 3x + 1 = 0$ (d) $2x^3 + 7x^2 + 8x + 4 = 0$

8. For what value of b is $4x^2 + bx + 9$ a perfect square?

9. Find the relationship between a and b if the equation

$$ax^2 + bx + 9 = 0$$

   has equal roots.

10. Prove that the equation of the form $x^2 + x - b^2 = 0$ has real roots regardless of the value of b.

11. Show that the roots of

$$px^2 + (p+q)x + q = 0$$

   are real, regardless of the value of p and q.

12. Find the value of k for which the equation

$$k(x+2) - (x-1)(x-2) = 0$$

   has just one solution.

3.3 Inequalities

The following symbols may already have been met:

    $<$: less than $>$: greater than
    $\leqslant$: less than or equal to $\geqslant$: greater than or equal to

The statement $a < b$ means the same as $b > a$, and $a \nless b$ means $a \geqslant b$ ($a \nless b$ is read as 'a is not less than b'.)

If $a < b$ then a is to the left of b when plotted on a number line.

If any number c is added to (or subtracted from) both a and b then $a + c$ will remain to the left of $b + c$ and thus

$$a < b \Leftrightarrow a + c < b + c$$

Similarly if $a < b$ and both a and b are multiplied (or divided) by a positive number c

then again ac will remain to the left of bc thus

$$a < b \Leftrightarrow ac < bc \qquad \text{for } c > 0$$

However, if c is negative, then ac will be to the *right* of bc, that is

$$a < b \Leftrightarrow ac > bc \qquad \text{for } c < 0$$

For example

$$2 < 3 \quad \text{but} \quad -2 > -3$$

Similarly

$$-3 < 1 \quad \text{but} \quad 3 > -1$$

Statement

> If both sides of an inequality are multiplied (or divided) by a negative number then the inequality sign is reversed.
> So $>$ is replaced by $<$ and $\leqslant$ is replaced by $\geqslant$.

Example 3.9

Solve the inequalities

(a) $2x + 1 \leqslant x - 5$ (b) $3x + 2 < 5x - 8$
(c) $3x + 1 > x - 2$ and $4x - 1 < x + 2$

(a) $2x + 1 \leqslant x - 5$

     $\Leftrightarrow x + 1 \leqslant -5$ subtracting x from both sides

     $\Leftrightarrow x \leqslant -6$ subtracting 1 from both sides.

In other words the original inequality is satisfied by any (and all) value of x which is less than or equal to -6.

(b) $3x + 2 < 5x - 8$

     $\Leftrightarrow -2x \quad < -10$

     $\Leftrightarrow \quad x \quad > 5$ reversing the sign upon division by -2.

We could avoid this division as follows

$$3x + 2 < 5x - 8$$

$$\Leftrightarrow \quad 10 < 2x$$

$$\Leftrightarrow \quad 5 < x$$

which amounts to the same thing!

(c) For this example we have to find those values of x for which both

$$3x + 1 > x - 2 \quad \text{and} \quad 4x - 1 < x + 2$$

For the first part we find

$$3x + 1 > x - 2$$

$$\Leftrightarrow \quad 2x > -3$$

$$\Leftrightarrow \quad x > -\tfrac{3}{2}$$

and for the second we obtain

$$4x - 1 < x + 2$$

$$\Leftrightarrow \quad 3x < 3$$

$$\Leftrightarrow \quad x < 1$$

Now we can find solutions which satisfy both inequalities providing

$$x > -\tfrac{3}{2} \quad \text{and} \quad x < 1$$

that is

$$-\tfrac{3}{2} < x \quad \text{and} \quad x < 1$$

A convenient shorthand for such a condition is

$$-\tfrac{3}{2} < x < 1$$

It is sometimes convenient to display the set of solutions of an inequality as an interval on the number line. So the condition $x \leqslant 2$ could be displayed as

$$\begin{array}{c|c|c|c|c} \hline & & & &] \\ \hline -1 & 0 & 1 & 2 & 3 \end{array}$$

In order to display $x > 1$ we have

$$\begin{array}{c|c|c|c|c} \hline & & (& & \\ \hline -1 & 0 & 1 & 2 & 3 \end{array}$$

A (is used at the end point rather than [, as in the first example, to indicate that this end point is not included. We need this notation since $2 \leqslant 2$ but $1 \not> 1$.

Notice that the condition

$$x > 1 \quad \text{and} \quad x \leqslant 2$$

that is

$$1 < x \quad \text{and} \quad x \leqslant 2$$

which is conveniently written

$$1 < x \leqslant 2$$

is displayed as

$$\begin{array}{c|c|c|c|c} \hline & & (\text{---}] & & \\ \hline -1 & 0 & 1 & 2 & 3 \end{array}$$

and sometimes described as the interval $(1, 2]$.

Similarly we could write

$$x \in (1, 4) \Leftrightarrow 1 < x < 4$$

Now if $x \leqslant 2$ we would write

$$x \in (-\infty, 2]$$

and similarly

$$x \in (3, \infty) \Leftrightarrow 3 < x$$

If we know that

$$1 < x < 5 \quad and \quad 0 \leqslant x \leqslant 3$$

then we deduce $1 < x \leqslant 3$ as can be seen from the following diagram.

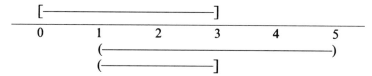

However, if

$$1 < x < 5 \quad or \quad 0 \leqslant x \leqslant 3$$

then we simply know that

$$0 \leqslant x < 5$$

The modulus of an expression

Statement

> The modulus of x, written $|x|$, is the positive number which has the same magnitude as x.

So

$$|2| = 2 \quad and \quad |-2| = 2$$

Generally

$$x \geqslant 0 \Rightarrow |x| = x \quad and \quad x < 0 \Rightarrow |x| = -x$$

Example 3.10

Find x if (a) $|x + 3| = 2$ (b) $|x + 1| > 5$ (c) $|x + 1| \leqslant 5$

(a) If $|x + 3| = 2$, then

$$x + 3 = 2 \quad or \quad x + 3 = -2$$
$$\Rightarrow \quad x = -1 \quad or \quad x = -5.$$

(b) If $|x + 1| > 5$, then

$$x + 1 > 5 \quad \text{or} \quad x + 1 < -5$$
$$\Rightarrow \quad x > 4 \quad \text{or} \quad x < -6.$$

The solutions may be displayed on a number line

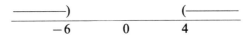

(c) By inspection of the solution to part (b), if

$$|x + 1| \leqslant 5$$

then

$$-6 \leqslant x \leqslant 4.$$

More directly we may argue as follows:

$$|x + 1| \leqslant 5$$
$$\Rightarrow x + 1 \leqslant 5 \quad \text{and} \quad x + 1 \geqslant -5$$
$$\Rightarrow \quad x \leqslant 4 \quad \text{and} \quad x \geqslant -6$$
$$\Rightarrow \quad -6 \leqslant x \leqslant 4.$$

A geometrical interpretation of the constraint $|x + 1| \leqslant 5$ is to note that x must lie within 5 units of the number -1 on the number line.

Similarly if $|x - 2| < 3$ then x lies within 3 units of the number 2 that is

$$-1 < x < 5$$

To check this, notice that

$$|x - 2| < 3$$
$$\Rightarrow x - 2 < 3 \quad \text{and} \quad x - 2 > -3$$
$$\Rightarrow x < 5 \quad \text{and} \quad x > -1$$
$$\Rightarrow -1 < x < 5.$$

In general

$$|x| < N \Leftrightarrow -N < x < N$$

and

$$|x - a| < N \Leftrightarrow a - N < x < a + N$$

Exercise 3.2

1. Solve the following inequalities.

   (a) $3x + 7 < 2x + 9$

   (b) $2x - 5 \geqslant 3x + 4$

   (c) $2(x + 1) + 5 \leqslant 4(x - 6)$

   (d) $\left(\dfrac{x + 1}{3}\right) \leqslant \dfrac{2x - 5}{4}$

   (e) $3x + 7 > \frac{5}{3}(x - 9)$

2. Simplify the following statements.

   (a) $1 \leqslant x < 3$ and $0 < x < 2$

   (b) $0 \leqslant x < 4$ and $3 < x \leqslant 4$

   (c) $2 < x < 4$ and $0 < x < 3$

   (d) $0 < x < 2$ or $-1 < x \leqslant 2$

3. Rewrite the following using the appropriate interval notation:

   (a) $1 < x \leqslant 3$

   (b) $0 \leqslant x < 5$

   (c) $3 > x$ and $x \geqslant 2$

   (d) $x < 4$

   (e) $x \geqslant -1$

   (f) $x < -1$ and $x \leqslant 2$

   (g) $x < -1$ or $x \geqslant -2$

   (h) $x > 3$ and $x < 0$

4. Write down the value of each of the following:

   (a) $|-9|$ (b) $|10|$ (c) $|-3| + 1$ (d) $-1 + |-3|$ (e) $|-1 + |-3||$

5. Solve each of the following equations.

   (a) $|x + 2| = 1$ (b) $|x - 2| = 2$ (c) $|x| = -1$ (d) $2|x - 3| = 5$

6. Explain why $|x + a|^2 = (x + a)^2$ for all x.

7. Solve the following inequalities.

   (a) $|x - 2| > 4$ (b) $|x + 3| \geqslant 1$ (c) $|x - a| \geqslant N$

8. Solve the following inequalities.

   (a) $|x + 2| < 3$ (b) $|x| \leqslant 1$ (c) $|x| < 3$ (d) $|x - 3| \leqslant 4$

3.4   Equations which reduce to the solution of a quadratic

Example 3.11

Solve the equation

$$2^{2x} - 3.2^x - 4 = 0.$$

Set $y = 2^x$, then we have

$$y^2 - 3y - 4 = 0 \qquad \text{since } 2^{2x} = (2^x)^2$$

$$\Rightarrow (y - 4)(y + 1) = 0$$
$$\Rightarrow y = 4 \quad \text{or} \quad y = -1$$

That is

$$2^x = 4 \quad \text{or} \quad 2^x = -1$$

By inspection $x = 2$ since there are no values of x for which $2^x < 0$.

Example 3.12

Solve the equation

$$16 \log_x 2 + \log_2 x = 10.$$

In this type of equation it is best to ensure a common base — say base 2. Now

$$\log_x 2 = \frac{\log_2 2}{\log_2 x} = \frac{1}{\log_2 x}$$

Thus

$$\frac{16}{\log_2 x} + \log_2 x = 10$$

Set

$$y = \log_2 x$$

$$\frac{16}{y} + y = 10$$

$$\Rightarrow y^2 - 10y + 16 = 0 \qquad \text{following multiplication of both sides by } y$$
$$\Rightarrow (y - 8)(y - 2) = 0$$
$$\Rightarrow y = 8 \quad \text{or} \quad y = 2$$
$$\Rightarrow \log_2 x = 8 \quad \text{or} \quad \log_2 x = 2$$
$$\Rightarrow x = 2^8 \quad \text{or} \quad x = 2^2$$
$$\Rightarrow x = 256 \quad \text{or} \quad x = 4$$

Example 3.13

Solve the equation

$$\sqrt{3(x - 1)} - \sqrt{x} = 1.$$

Squaring *both* sides

$$3(x - 1) - 2\sqrt{x} \cdot \sqrt{3(x - 1)} + x = 1$$
$$\Rightarrow 2\sqrt{3x(x - 1)} = 4x - 4$$
$$\Rightarrow \sqrt{3x(x - 1)} = 2x - 2$$

Squaring again

$$3x(x-1) = 4x^2 - 8x + 4$$

$$\Rightarrow x^2 - 5x + 4 = 0$$

$$\Rightarrow (x-4)(x-1) = 0$$

$$\Rightarrow x = 4 \quad \text{or} \quad x = 1$$

Now the process of squaring both sides may introduce additional solutions. For consider

$$x = 2$$

$$\Rightarrow x^2 = 4$$

$$\Rightarrow x = 2 \quad \text{or} \quad x = -2 \quad !!$$

So it is necessary to check each possible solution in the original equation.

When $x = 4$,

l.h.s. $= \sqrt{3(4-1)} - \sqrt{4} = 1$

r.h.s. $= 1$

so $x = 4$ is a solution.

When $x = 1$,

l.h.s. $= \sqrt{3(1-1)} - \sqrt{1} = -1$

r.h.s. $= 1$

so $x = 1$ is *not* a solution of the original equation.

3.5   Simultaneous equations

The equations we have met so far have contained just one variable — usually x, but there is no reason why this should always be so. Consider the following equation, linear in x and y:

$$2x + 3y = 7$$

A feature of such an equation is that we can substitute any value for x and then attempt to solve the resulting equation of y. For example, when $x = 5$ we have

$$10 + 3y = 7$$

$$\Rightarrow \quad y = 1$$

When $x = -4$ we have

$$-8 + 3y = 7$$

$$\Rightarrow \quad y = 5$$

More generally we can make y the *subject of the equation*

$$y = \tfrac{1}{3}(7 - 2x)$$

We could equally well, of course, substitute for y and solve the resulting equation for x.

We are often faced with two such equations, say,

$$2x + 3y = 7 \qquad ①$$

$$3x - 2y = 4 \qquad ②$$

and are asked to solve them *simultaneously*. That is, we find a value for x and y which satisfy *both* equations. In this example the solution is $x = 2$ and $y = 1$. It is assumed that the reader is familiar with the methods used to solve such linear equations.

Notice from Fig. 3.3 that the point of intersection of the two graphs corresponds to the simultaneous solution of the two equations.

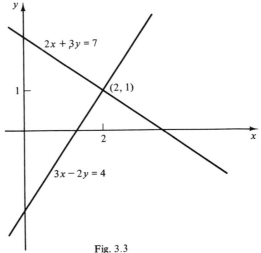

Fig. 3.3

Example 3.14

Solve simultaneously

$$3x - 4y = 5 \qquad ①$$

$$6x + 3y = -1 \qquad ②$$

$2 \times ① - ②$

$$\Rightarrow -11y = 11$$

$$\Rightarrow \quad y = -1$$

Substitute into ①

$$3x - -4 = 5$$

$$\Rightarrow \qquad 3x = 1$$

$$\Rightarrow \qquad x = \tfrac{1}{3}$$

[Check in ②: $2 + -3 = -1$]

Thus

$$x = \tfrac{1}{3}, \quad y = -1$$

We now consider the situation where we are asked to solve two equations simultaneously, when one of the equations is linear and the other quadratic (at least in one of the variables).

Example 3.15

Solve simultaneously

$$2x + y = 1 \qquad\qquad ①$$

$$y^2 + 3xy + 2x = 0 \qquad ②$$

A typical method is to express y in terms of x using the first equation to obtain

$$y = 1 - 2x$$

and then to substitute into the second equation.

$$(1 - 2x)^2 + 3x(1 - 2x) + 2x = 0$$

$$\Rightarrow 1 - 4x + 4x^2 + 3x - 6x^2 + 2x = 0$$

$$\Rightarrow 2x^2 - x - 1 = 0$$

$$\Rightarrow (2x + 1)(x - 1) = 0$$

$$\Rightarrow x = -\tfrac{1}{2} \quad \text{or} \quad x = 1$$

We now substitute into the first equation

$$y = 1 - 2x$$

to obtain the corresponding values for y.

$$x = -\tfrac{1}{2} \qquad \Rightarrow y = 2$$

$$x = 1 \qquad \Rightarrow y = -1$$

In this case there are two pairs of values which satisfy both equations simultaneously. Notice that we express y in terms of x for convenience only, we could equally well have expressed x in terms of y.

Example 3.16

Solve simultaneously the equations

$$\text{(i)}\ 2\log_y x + 2\log_x y = 5 \qquad \text{(ii)}\ xy = 27$$

We can convert the logarithms in equation (i) to a common base by noting that

$$\log_y x = \frac{1}{\log_x y}.$$

So substituting into equation (i) we obtain

$$\frac{2}{\log_x y} + 2\log_x y = 5.$$

Setting $t = \log_x y$ we have

$$\frac{2}{t} + 2t = 5$$

$$\Rightarrow 2t^2 - 5t + 2 = 0$$

$$\Rightarrow (2t - 1)(t - 2) = 0$$

$$\Rightarrow t = \tfrac{1}{2} \quad \text{or} \quad t = 2$$

$$\Rightarrow \log_x y = \tfrac{1}{2} \quad \text{or} \quad \log_x y = 2$$

$$\Rightarrow y = x^{1/2} \quad \text{or} \quad y = x^2.$$

Consider now the second equation.

When $y = x^{1/2}$

$$x \cdot x^{1/2} = 27$$

$$\Rightarrow \quad x^{3/2} = 27$$

$$\Rightarrow \quad x = 27^{2/3} \qquad \text{taking the power of } \tfrac{2}{3} \text{ of both sides}$$

$$\Rightarrow \quad x = 9$$

and when $y = x^2$

$$x \cdot x^2 = 27$$

$$\Rightarrow \quad x^3 = 27$$

$$\Rightarrow \quad x = 3$$

Using equation (ii) we can find the corresponding values of y to obtain the pairs $x = 9$ and $y = 3$ or $x = 3$ and $y = 9$ as solutions.

Exercise 3.3

Solve equations 1 to 12.

1. $2^{2x} - 5.2^x + 4 = 0$

2. $3^{2x} - 12.3^x + 27 = 0$

3. $5^{2x} - 25 = 0$

4. $6\log_x 5 + \log_5 x = 5$

5. $\log_3 x = 1 + 2\log_x 3$

6. $\log_x 4 = \log_2 x^2 + 3$

7. $\dfrac{14}{x + 1} = x + 6$

8. $\dfrac{1}{x + 1} + \dfrac{2}{x + 2} = 1$

9. $\tfrac{3}{4} + \dfrac{2}{x} = \dfrac{5}{x - 3}$

10. $\sqrt{(2x + 5)} - \sqrt{(3x - 5)} = 2$

11. $\sqrt{(2x - 3)} - \sqrt{(x - 2)} = 1$

12. $\sqrt{(2x + 1)} + \sqrt{(4x + 3)} = 5$

13. Consider the equation $3n^2 - 4tn - 4t^2 = 0$.
    It can be viewed either as a quadratic in n or a quadratic in t. In each case solve the equation treating the other variable as a constant.

14. Solve the equation $2^{2x+1} = 3.2^x - 1$.

15. Solve the equation $2n(t + n) + t(n - 2t) = 0$

    (a) as a quadratic in n (b) as a quadratic in t

16. Solve the following pairs of equations.

    (a) $x + y = 3$

    $y^2 + 2xy + x - 9 = 0$

    (b) $2x - y = 4$

    $y^2 + 4xy + 2(x + 1) = 0$

17. Find rational values of a and b so that
    $$\sqrt{52 - 30\sqrt{3}} = a + b\sqrt{3}.$$

18. Solve the equation $2^{x/2} - 2^{-x/2} = 1$.

19. Solve simultaneously $\dfrac{1}{x} + \dfrac{1}{y} = \dfrac{1}{3}$

    $2x^2 - xy - y^2 = 0$

20. In the equation $y = a \cdot 2^{bx}$, a and b are constants.
    When $x = 2$, $y = 3$ and when $x = -2$, $y = \tfrac{3}{2}$. Find the value of the constants.

21. Solve simultaneously $3^{x+3} = 9^{2-y}$

    $3^{-y} = 9^{x-4}$

22. Solve the equation $\sqrt{3(1-x)} - \sqrt{2-x} = 1$.

23. If x and y are such that

$$x^2 = 3x + 2y \qquad ①$$
$$\text{and} \quad y^2 = 2x + 3y \qquad ②$$

show that

$$(x - y)(x + y) = x - y.$$

Hence solve the given simultaneous equations.

24. Solve the following equations simultaneously.

(a) $x - y = 2$ (b) $x - 2y = 11$ (c) $x + y = 3$

     $y^2 + 2xy + x^2 = 0$ $x^2 + y^2 = 25$ $x^2 - y^2 = 12$

25. Solve the following equations simultaneously.

$$xy = 1$$
$$2xy + x - 2y = 3.$$

26. Solve the following three equations simultaneously.

(i) $2x + y - z = 1$

(ii) $3x + y + z = 8$

(iii) $\quad x - y + z = 5$

Hint: Eliminate z between equations (i) and (ii) and between equations (i) and (iii), to obtain two equations involving x and y only.

3.6 Relationships between the roots and the coefficients of quadratic equations

Suppose α and β are the roots (solutions) of the quadratic equation

$$x^2 + bx + c = 0$$

then $x - \alpha$ and $x - \beta$ are factors of the quadratic.
    It is evident that

$$(x - \alpha)(x - \beta) \equiv x^2 + bx + c.$$

On the expansion we find that

$$x^2 - (\alpha + \beta)x + \alpha\beta \equiv x^2 + bx + c.$$

So comparing coefficients

$$-(\alpha + \beta) = b \quad \text{and} \quad \alpha\beta = c.$$

Thus

$$\text{(i) } \alpha + \beta = -b,$$

$$\text{(ii)} \quad \alpha\beta = c.$$

For the general quadratic equation

$$ax^2 + bx + c = 0$$

$$\Rightarrow x^2 + \frac{bx}{a} + \frac{c}{a} = 0$$

and so:

Statement

> If α and β are the roots of the quadratic equation $ax^2 + bx + c = 0$ then the sum of the roots $\alpha + \beta = -b/a$ and the product of the roots $\alpha\beta = c/a$

It is important to realise that the sum and product of the roots of a quadratic equation can be found by inspection even if the roots themselves had to be obtained by using the formula.

Example 3.17

(i) Write down the sum and product of the roots of the quadratic equation

$$2x^2 + 5x - 6 = 0$$

Assuming the roots are α and β and noting that $a = 2$, $b = 5$ and $c = -6$ we have

$$\alpha + \beta = -5/2$$

$$\alpha\beta = -3$$

An immediate consequence of this is that one can establish the values of certain expressions involving the roots α and β.

(ii) Using the sum and product of the roots found in part (i) establish the values of

(a) $\alpha^2 + \beta^2$ (b) $\dfrac{1}{\alpha} + \dfrac{1}{\beta}$ (c) $\alpha^3 + \beta^3$

The object in each case is to transform the expression into a form consisting of the sum and product of the roots — since these are known.

(a) $\alpha^2 + \beta^2 \equiv (\alpha + \beta)^2 - 2\alpha\beta = \left(\dfrac{-5}{2}\right)^2 - 2(-3) = 12\frac{1}{4}$

(b) $\dfrac{1}{\alpha} + \dfrac{1}{\beta} = \dfrac{\alpha + \beta}{\alpha\beta} = \dfrac{-5/2}{-3} = \dfrac{5}{6}$

(c) $\alpha^3 + \beta^3 \equiv (\alpha + \beta)(\alpha^2 - \alpha\beta + \beta^2)$

$\equiv (\alpha + \beta)((\alpha + \beta)^2 - 3\alpha\beta)$ \qquad using (a)

$= \left(\dfrac{-5}{2}\right)\left(\left(\dfrac{-5}{2}\right)^2 - 3(-3)\right)$

$= \dfrac{-5}{2}\left(\dfrac{25}{4} + 9\right)$

$= \dfrac{-305}{8}$

It is interesting to note that any expression symmetrical in α and β can be evaluated simply from a knowledge of $\alpha + \beta$ and $\alpha\beta$.

A symmetric expression in α and β is one which remains unchanged when α and β are interchanged throughout the expression.

Example 3.18

Write down the quadratic equation whose roots are such that:

(a) their sum is $\frac{3}{2}$ \quad and \quad (b) their product is $\frac{1}{3}$

The quadratic equation has the form

$$x^2 - (\text{sum of roots})\, x + \text{product of roots} = 0$$

So the required equation is

$$x^2 - \tfrac{3}{2}x + \tfrac{1}{3} = 0$$
$$\Rightarrow 6x^2 - 9x + 2 = 0$$

Combining these ideas we have:

Example 3.19

Assuming the roots of the quadratic equation

$$3x^2 - 2x - 1 = 0$$

are α and β, find the quadratic equation whose roots are α/β and β/α.

We first observe that

$$\alpha + \beta = -(-\tfrac{2}{3}) = \tfrac{2}{3}$$
$$\alpha\beta = -\tfrac{1}{3}$$

Now we shall require the sum and product of the roots of the new equation.

That is

$$\frac{\alpha}{\beta} + \frac{\beta}{\alpha} = \frac{\alpha^2 + \beta^2}{\alpha\beta}$$

$$= \frac{(\alpha + \beta)^2 - 2\alpha\beta}{\alpha\beta}$$

$$= \frac{(\frac{2}{3})^2 - 2(-\frac{1}{3})}{-\frac{1}{3}}$$

$$= \frac{-10}{3}$$

and

$$\frac{\alpha}{\beta} \cdot \frac{\beta}{\alpha} = 1$$

So the required equation is

$$x^2 - \left(\frac{-10}{3}\right)x + 1 = 0$$

$$\Rightarrow 3x^2 + 10x + 3 = 0$$

Example 3.20

Find the relationship between b and c if the equation $x^2 + bx + c = 0$ has equal roots.

Assume the roots are α and α so

(i) $\alpha + \alpha = -b$

(ii) $\alpha^2 = c$

Squaring equation (i) we have $4\alpha^2 = b^2$ and substituting from equation (ii),

$$4c = b^2$$

Notice that this question was attempted previously by considering the discriminant.

Exercise 3.4

1. Write down the sum and products of the roots of the following equations.

   (a) $x^2 + 5x - 6 = 0$ (b) $x^2 + 6x + 3 = 0$
   (c) $x(x + 4) = -1$ (d) $2x^2 - 9x + 5 = 0$

2. Write down the quadratics whose sum and products are:

   (a) sum $= 9$ product $= 4$ (b) sum $= \frac{2}{3}$ product $= -1$

3. If α and β are the roots of the quadratic $2x^2 - 5x - 1 = 0$, find the values of

    (a) $\alpha^2 + \beta^2$ (b) $\dfrac{\alpha}{\beta} + \dfrac{\beta}{\alpha}$ (c) $\dfrac{1}{\alpha} + \dfrac{1}{\beta}$

4. If α and β are the roots of the quadratic $x^2 + 2x - 1 = 0$, find the values of

    (a) $\alpha^3 + \beta^3$ (b) $(\alpha + \beta)^2$ (c) $(\alpha - \beta)^2$

5. The roots of the equation $2x^2 + 6x - 5 = 0$ are α and β. Find the equation whose roots are

    (a) $\alpha + 1, \beta + 1$ (b) $\dfrac{1}{\alpha}, \dfrac{1}{\beta}$ (c) α^2, β^2

6. The roots of the equation $x^2 + 3x + 1 = 0$ are α and β. Find the equation whose roots are

    (a) $2\alpha, 2\beta$ (b) $\dfrac{\alpha}{\beta}, \dfrac{\beta}{\alpha}$ (c) $2\alpha - \beta, 2\beta - \alpha$

7. Write down the sum and product of the roots of the equation $x^2 + 2ax - c^2 = 0$ and hence obtain the quadratic equation whose roots are the reciprocal of the roots of the above equation.

8. Find the quadratic equation which has the sum of its roots equal to ab and product equal to $a + b$. Hence write down the quadratic equation whose roots are double the roots of this quadratic.

9. By noting that $\alpha - \beta = \sqrt{((\alpha - \beta)^2)}$ when $\alpha \geqslant \beta$, show that
    $$\alpha - \beta = \sqrt{((\alpha + \beta)^2 - 4\alpha\beta)}.$$
    Hence find the positive difference between the roots of the quadratic
    $$64x^2 - 80x - 11 = 0$$

10. If one root of the equation $mx^2 + nx + p = 0$ is twice the other find the relationship connecting m, n and p. [Hint: Assume the roots are α and 2α respectively.]

11. If α and β are the roots of the equation $mx^2 + mx + p = 0$ find the quadratic equation whose roots are $\dfrac{1}{\alpha}, \dfrac{1}{\beta}$ respectively.

12. If one root of the quadratic equation $x^2 - 4px + 27 = 0$ is the square of the other, find the value of p.

13. If $x^2 + 6x + c = 0$ has roots which differ by $2a$ show that $a^2 = 9 - c$.

14. Given that the equation $ax^2 + bx + c = 0$ has roots α and β, find an expression for

$$\alpha^2 + \frac{1}{\alpha} + \beta^2 + \frac{1}{\beta}$$

W 1982

15. If α and β are the roots of the equation $x^2 - x + 2 = 0$, show that

$$\alpha^2 + \beta^2 = -3.$$

Hence, or otherwise, find the quadratic equation whose roots are

$$\frac{1}{\alpha^3 \beta} \quad \text{and} \quad \frac{1}{\alpha \beta^3}$$

SU 1981

16. Express $\alpha^3 + \beta^3$ as the product of a linear and a quadratic factor.

Given that

$$\alpha + \beta = 2 \quad \text{and} \quad a^3 + \beta^3 = 32,$$

find the value of $\alpha\beta$.

Write down a quadratic equation with numerical coefficients and roots α, β. Hence calculate the exact values of α and β.

JMB 1982

Revision exercise A

1. Solve the equations
$$xy = 6$$
$$x - 2y + y^2 = 5$$

W 1981

2. Given that one of the roots of the equation $x^3 + x^2 - 2x - 2 = 0$ is an integer, find all of the roots.

SU 1983

3. Find the value of x for which $2^{3x+1} = 3^{x+2}$, giving three significant figures in your answer.

CAMB 1982

4. Prove that $\log_b a = \dfrac{1}{\log_a b}$.

By using the substitution $t = \log_x 3$, or otherwise, solve the equation

$$2\log_9 x + 1 = 2\log_x 3.$$

SU 1981

5. Given that $\log_{10} 2 = p$ and $\log_{10} 3 = q$, solve for x in terms of p and q

    (a) $\log_3 x = \log_x 2$

    (b) $(6^{3-4x})(4^{x+4}) = 2$

AEB 1982

6. Given that $x + 1$ and $x - 2$ are factors of the expression

    $$x^5 + ax^3 + bx^2 + 6x - 4,$$

    prove (do not merely verify) that $a = -7$ and $b = 4$.
    Solve the equation $x^5 - 7x^3 + 4x^2 + 6x - 4 = 0$.

SU 1981

7. Solve the equation $\qquad x + 5 = \sqrt{(7 - x)}$.

    [The expression $\sqrt{u}$ denotes the positive square root of u.]

CAMB 1983

8. Find the values of x and y, such that $x = 2y$ and $\log_2 x + \log_2 y = 3$.

SU 1980

9. The variables x and y are connected by the relationship $y = ax^n$, where a and n are constants.
    When $x = 16$, $y = 27$. When $x = 81$, $y = 8$. Find the values of a and n.

SU 1983

10. Solve $\qquad x^2 + xy + y^2 = 28$

    $$x^2 - xy + y^2 = 12$$

W 1982

11. The quadratic equation $(Ax + B)^2 = x$ has roots 1 and 4. Find all the possible pairs of values of the constants A and B.

SU 1982

12. Given that the equation $x^2 + 2x + 4 = k(x^2 + 4)$ has equal roots, find the possible values of k.

SU 1983

13. Given that α and β are the roots of the equation $3x^2 + x + 2 = 0$,

    (a) evaluate $\dfrac{1}{\alpha^2} + \dfrac{1}{\beta^2}$

    (b) find an equation whose roots are $\dfrac{1}{\alpha^2}$ and $\dfrac{1}{\beta^2}$

    (c) show that $27\alpha^4 = 11\alpha + 10$.

AEB 1982

14. The quadratic equation $x^2 - kx - 1 = 0$ has roots α and β. Find the quadratic equation whose roots are $\alpha^2 + \beta^2$ and $\alpha\beta$.

W 1981

15. Given that a_1 and a_2 are the roots of the quadratic equation

$$(x - b_1)(x - b_2) = c$$

write down expressions for $a_1 + a_2$ and $a_1 a_2$ in terms of the constants b_1, b_2 and c.

Show that $-a_1$ and $-a_2$ are the roots of the equation

$$(x + b_1)(z + b_2) = c$$

LOND 1983

16. Solve for x and y the simultaneous equations

$$2x + y = 4$$
$$x^2 + 2y^2 = 17$$

CAMB 1983

17. Solve the equation

$$\log_x a + 2\log_a x = 3$$

where a is a positive constant not equal to 1, giving the values of x in terms of a.

[The result $\log_c b = \dfrac{1}{\log_b c}$ may be used without proof.]

CAMB 1983

18. Find, in surd form, the value of x, given that

$$\log_3(x^2) + \log_3 x = \log_9 27$$

CAMB 1982

19. Find the values of the constants p and q so that the polynomial

$$px^3 - 11x^2 + qx + 4$$

is divisible by $(x - 1)$, and has remainder 70 when divided by $(x - 3)$.

Express the polynomial as a product of 3 linear factors.

LOND 1984

Chapter 4   Trigonometry

4.1   The sine and cosine rules

It is assumed that the reader has studied trigonometry related to the *right-angled triangle*. The use of a calculator or tables to evaluate the trigonometric ratios enables one to 'solve' the triangle completely. That is, if one is given sufficient information to determine the triangle uniquely, then the unknown sides and angles can readily be established. It is reasonable to suppose that if we are given sufficient information to determine *any* triangle uniquely, whether right-angled or not, we should be able to solve the triangle completely. Indeed we could draw the triangle to scale and then measure the unknown sides and angles. However, this is unsatisfactory, since it is time consuming and no great accuracy is achieved without going to a great deal of trouble. It would be better if we could *calculate* the unknowns.

The following rules are useful for this purpose when dealing with the general triangle.

Statement

By convention, if the angles of a triangle are denoted by the letters A, B and C, then the sides opposite these angles are respectively denoted by a, b and c.

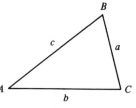

Fig. 4.1

The sine rule

For any triangle, $\triangle ABC$, the ratio of one side to the sine of the opposite angle is a constant. That is:

$$\frac{a}{\sin A} = \frac{b}{\sin B} = \frac{c}{\sin C}$$

The cosine rule

This rule has two equivalent forms.

$$a^2 = b^2 + c^2 - 2bc \cos A$$

$$\cos A = \frac{b^2 + c^2 - a^2}{2bc}$$

With corresponding results for b^2 and c^2.

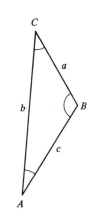

Fig. 4.2

Before we proceed to the proofs of these statements we need to note some facts which enable us to deal with obtuse angles.

Statement

> For any angle A
>
> (i) $\sin A = \sin(180° - A)$
>
> (ii) $\cos A = -\cos(180° - A)$

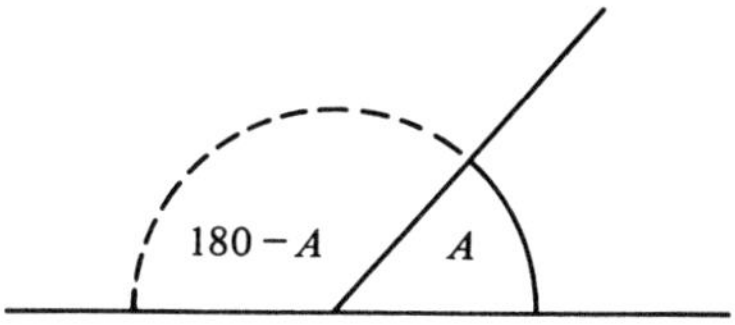

Fig. 4.3

So, for example

$$\sin 160° = \sin(180° - 160°)$$
$$= \sin 20°$$
$$= 0.3420 \quad 4\,\text{d.p.} \qquad \boxed{\text{C}}$$

and

$$\cos 160° = -\cos(180° - 160°)$$
$$= -\cos 20°$$
$$= -0.9397 \quad 4\,\text{d.p.} \qquad \boxed{\text{C}}$$

Most modern calculators can compute the trigonometric ratios for obtuse angles directly. The values obtained will agree with the results noted above, the results will be *proved* later in the chapter.

When we are trying to find an unknown angle, we may meet an equation of the form

$$\cos A = -0.602$$

In such a case we can assume that A is obtuse, since the cosine of an acute angle is positive.

In fact

$$A = 127° \qquad \boxed{\text{C}}$$

On the other hand if we have

$$\sin A = 0.743$$

we find

$$A = 48.0° \qquad \boxed{\text{C}}$$

but since

$$\sin A = \sin(180° - A)$$

it follows that $180° - 48.0° = 132°$ is also a solution!

Proof of the sine rule

Consider the triangle, $\triangle ABC$, where we first assume that $\angle A$ is acute (Fig. 4.4 (i)). Let the perpendicular from C meet AB (or AB produced if B is obtuse) at D, and suppose that $CD = h$.

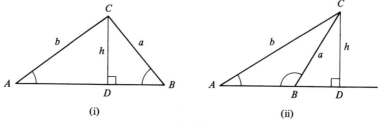

(i)   (ii)

Fig. 4.4

By considering the right-angled triangles $\triangle ADC$ and $\triangle DBC$ we obtain

$$\sin A = \frac{h}{b} \quad \text{and} \quad \sin B = \frac{h}{a}$$

[If B is obtuse, as in Fig. 4.4 (ii), then we obtain $\sin(180° - B) = h/a$ but recalling that $\sin(180° - B) = \sin B$ we have the same result!]

Now eliminating h note that

$$b \sin A = a \sin B$$

$$\Rightarrow \frac{a}{\sin A} = \frac{b}{\sin B}$$

If A is obtuse, then B is acute and we obtain

$$\sin(180° - A) = \frac{h}{b} \quad \text{and} \quad \sin B = \frac{h}{a}$$

However, noting again that $\sin A = \sin(180° - A)$ the previous result follows.
We could similarly show that

$$\frac{b}{\sin B} = \frac{c}{\sin C}$$

and so combining these results, we obtain

$$\frac{a}{\sin A} = \frac{b}{\sin B} = \frac{c}{\sin C}$$

Statement

> The sine rule can be used when
>
> (i) two angles and one side are known
> (ii) two sides and the non-included angle are known.

Example 4.1

In the triangle, $\triangle ABC$, $\angle A = 62.4°$, $\angle B = 40.1°$ and $c = 21$ cm (Fig. 4.5). Find b correct to 1 d.p.

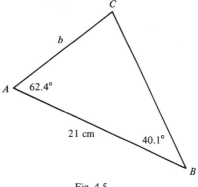

Fig. 4.5

The sum of the three angles of a triangle is 180°, therefore $\angle C = 77.5°$.
Applying the sine rule in the form

$$\frac{b}{\sin B} = \frac{c}{\sin C}$$

we have

$$\frac{b}{\sin 40.1} = \frac{21}{\sin 77.5}$$

$$\Rightarrow \qquad b = \frac{21 \sin 40.1}{\sin 77.5}$$

$$= 13.855 \qquad \boxed{C}$$

$$\Rightarrow \qquad b = 13.9 \qquad 1\,\text{d.p.}$$

Example 4.2

In the triangle, $\triangle ABC$, $\angle B = 35°$, $c = 25$ cm and $b = 15$ cm (Fig. 4.6). Find $\angle C$.

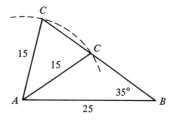

Fig. 4.6

In attempting to construct the triangle with the above lengths and angles notice that there are two possible triangles. One where $\angle C$ is acute and the other where $\angle C$ is obtuse. This is called the *ambiguous case*, since there are two triangles satisfying the given data. It will be interesting to see how the sine rule copes with this case!

Using the form

$$\frac{c}{\sin C} = \frac{b}{\sin B}$$

we have

$$\frac{25}{\sin C} = \frac{15}{\sin 35°}$$

$$\Rightarrow \sin C = \frac{25 \sin 35°}{15}$$

$$\Rightarrow \sin C = 0.956 \qquad \boxed{C}$$

$$\Rightarrow \quad C = 72.9° \qquad \boxed{C}$$

$$\text{or} \quad C = 180° - 72.9°$$

$$= 107.1° \qquad !!$$

As expected we obtain two possible angles.

Whenever we are given two sides and a non-included angle, we run the risk of being in the ambiguous case. However, it should not be assumed that there will always be *two* triangles satisfying the conditions.

Example 4.3

In the triangle, $\triangle ABC$, find $\angle B$ if $c = 25\,$cm, $b = 15\,$cm and $C = 35°$ (Fig. 4.7).

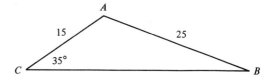

Fig. 4.7

Again, using the form

$$\frac{b}{\sin B} = \frac{c}{\sin C}$$

$$\frac{15}{\sin B} = \frac{25}{\sin 35°}$$

$$\Rightarrow \sin B = \frac{15 \sin 35°}{25}$$

$$\Rightarrow \sin B = 0.344 \qquad \boxed{C}$$

$$\Rightarrow B = 20.1° \quad \text{or} \quad B = 159.9°. \qquad \boxed{C}$$

But this time the solution $\angle B = 159.9°$ is inadmissible. [It is too large since we know that $C = 35°$ and the sum of the angles of a triangle is $180°$!]

Again an attempt at constructing the triangle is helpful.

Proof of the cosine rule

Consider the triangle, $\triangle ABC$, where we first assume that $\angle A$ is acute (Fig. 4.8(i)). Let the perpendicular from C meet AB (or AB produced if B is obtuse) at D and suppose that $CD = h$ and $AD = x$.

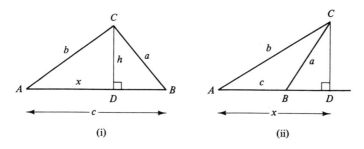

(i)   (ii)

Fig. 4.8

Using Pythagoras' theorem for each of the right-angled triangles, $\triangle ADC$ and $\triangle DBC$, we have

$$b^2 = h^2 + x^2 \quad \text{and} \quad a^2 = h^2 + (c - x)^2 \quad \text{(since } DB = c - x)$$

[In the event that $\angle B$ is obtuse, as in Fig. 48 (ii), we obtain $a^2 = h^2 + (x - c)^2$.] In either case, eliminating h^2 we obtain

$$a^2 = b^2 - x^2 + c^2 - 2cx + x^2$$
$$a^2 = b^2 + c^2 - 2cx.$$

Now considering $\triangle ADC$

$$\cos A = \frac{x}{b} \quad \Rightarrow x = b \cos A$$

and so finally we have, substituting for x,

$$a^2 = b^2 + c^2 - 2bc \cos A.$$

If we now suppose that $\angle A$ is obtuse (Fig. 4.9), so the perpendicular CD meets BA produced. Again

$$b^2 = h^2 + x^2$$

but this time

$$a^2 = h^2 + (c + x)^2$$

So

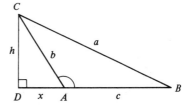

Fig. 4.9

$$a^2 = b^2 - x^2 + c^2 + 2cx + x^2$$
$$\Rightarrow a^2 = b^2 + c^2 + 2cx.$$

Now

$$\cos(180° - A) = \frac{x}{b} \quad \text{but} \quad \cos(180° - A) = -\cos A$$

$$x = -b\cos A$$

and so yet again, substituting for x

$$a^2 = b^2 + c^2 - 2bc\cos A.$$

Naturally we could obtain the corresponding results

$$b^2 = a^2 + c^2 - 2ac\cos B$$

and

$$c^2 = a^2 + b^2 - 2ab\cos C$$

by considering the angles $\angle B$ and $\angle C$ respectively.

The related result

$$\cos A = \frac{b^2 + c^2 - a^2}{2bc}$$

follows by algebraic manipulation of the above results.

Statement

> The cosine rule can be used whenever
>
> (i) two sides and the included angle are known
> (ii) three sides are known.

Example 4.4

Find the largest angle in the triangle, $\triangle ABC$, where $a = 8\,\text{cm}$, $b = 7\,\text{cm}$ and $c = 14\,\text{cm}$.

The largest angle will be opposite the longest side, that is $\angle C$. Using the cosine rule in the form

$$\cos C = \frac{a^2 + b^2 - c^2}{2ab}$$

$$= \frac{8^2 + 7^2 - 14^2}{2.8.7}$$

$$= -0.741 \qquad \boxed{C}$$

$$\Rightarrow \angle C = 137.8° \qquad \boxed{C}$$

Example 4.5

In the triangle, $\triangle ABC$, $\angle A = 38°$, $b = 10.1\,\text{cm}$ and $c = 14.3\,\text{cm}$ (Fig. 4.10). Find $\angle A$.

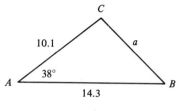

Fig. 4.10

We know two sides and the included angle and so can apply the cosine rule

$$a^2 = b^2 + c^2 - 2ab \cos A$$

to obtain a. Then we shall apply the sine rule to obtain $\angle B$.

Now

$$a^2 = 10.1^2 + 14.3^2 - 2 \times 10.1 \times 14.3 \times \cos 38°$$

$$= 78.88 \qquad \boxed{C}$$

$$\Rightarrow a = 8.88\,\text{cm} \qquad \boxed{C}$$

Now

$$\frac{10.1}{\sin B} = \frac{8.88}{\sin 38°}$$

$$\Rightarrow \sin B = \frac{10.1 \times \sin 38°}{8.88}$$

$$= 0.700 \qquad \boxed{C}$$

$$\Rightarrow \angle B = 44.4° \quad \text{or} \quad \angle B = 135.6°$$

It should seem odd that we have two solutions in this case since a triangle is *uniquely* determined if we are given two sides and the *included* angle.

However, we can exclude the solution $\angle B = 135.6°$ since $\angle B$ is not the largest angle. [It is not opposite the longest side.] Thus $\angle B = 44.4°$.

Area of a triangle

Consider the triangle, $\triangle ABC$, where the perpendicular from C meets AB at D (or BA produced if A is obtuse) (Figs. 4.11 (i), (ii)—see over). Suppose that $CD = h$.

Regardless of whether $\angle A$ is acute or obtuse

$$\sin A = \frac{h}{b} \qquad \Rightarrow h = b \sin A$$

[In the event $\angle A$ is obtuse, $\sin(180° - A) = \dfrac{h}{b}$, but $\sin(180° - A) = \sin A$.]

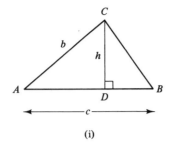
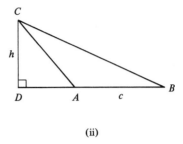

(i)   (ii)

Fig. 4.11

Now the area of the triangle is

$$\tfrac{1}{2}ch = \tfrac{1}{2}bc \sin A \qquad \text{substituting for } h$$

We can similarly show that the area is given by

$$\tfrac{1}{2} ab \sin C \quad \text{or} \quad \tfrac{1}{2} ac \sin B.$$

Statement

> The area of a triangle is given by the formula
>
> $$\text{Area} = \tfrac{1}{2} bc \sin A$$
>
> or the corresponding formula for $\angle B$ and $\angle C$.

Exercise 4.1

1. Find the unknown sides and angles of $\triangle ABC$ given

    (a) $\angle A = 95°$ $\angle B = 35°$ and $a = 42\,\text{m}$
    (b) $\angle B = 44°$ $\angle C = 63°$ and $b = 3.8\,\text{m}$
    (c) $\angle C = 35°$ $b = 4\,\text{m}$ and $c = 2.5\,\text{m}$
    (d) $\angle A = 50°$ $c = 2\,\text{m}$ and $a = 5\,\text{m}$

2. Study the information given in
    Fig. 4.12 and find length CD.

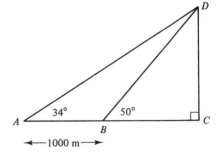

Fig. 4.12

3. Find the unknown sides and angles in $\triangle ABC$ given

   (a) $\angle A = 35°$ $b = 15\,\text{cm}$ and $c = 13\,\text{cm}$
   (b) $\angle B = 14°$ $a = 2\,\text{m}$ and $c = 3.2\,\text{m}$
   (c) $a = 5\text{m}$ $b = 4\,\text{m}$ and $c = 7\,\text{m}$
   (d) $a = 12\,\text{cm}$ $b = 28\,\text{cm}$ and $c = 17\text{cm}$

4. Four rods are joined to form a quadrilateral $ABCD$. The lengths AB, BC, CD and DA are respectively $10\,\text{m}$, $9\,\text{m}$, $8\,\text{m}$ and $7\,\text{m}$. If $\angle ABC = 100°$ find the distance AC and $\angle ADC$.

5. Find the area of $\triangle ABC$ given

   (a) $\angle A = 60°$ $b = 8.4\,\text{cm}$ and $c = 7.6\,\text{cm}$
   (b) $\angle B = 53°$ $a = 14\,\text{cm}$ and $c = 8.5\,\text{cm}$
   (c) $\angle C = 48°$ $a = 6\,\text{cm}$ and $c = 6\,\text{cm}$

6. The triangle, $\triangle ABC$, is cut from a disc. If $\angle A = 50°$ and $b = c = 8\,\text{cm}$ find

   (a) the smallest possible radius of the disc,
   (b) the smallest possible area discarded.

7. The triangle, $\triangle ABC$, is right-angled at A. The bisector of $\angle A$ meets BC at D. Prove that

   (a) $\dfrac{c}{b} = \dfrac{BD}{DC}$,   (b) $AD = \dfrac{\sqrt{2}\,.\,a\sin B}{1 + \tan B}$.

8. $\triangle ABC$ is a plane triangle with $AB = 4\,\text{cm}$, $AC = 5\,\text{cm}$ and angle $ACB = 30°$. Find, by calculation, the two possible lengths of BC, each correct to 3 significant figures, and the corresponding values of angle ABC, correct to the nearest tenth of a degree.

   SU 1980

9. Fig. 4.13 shows the points A, B and C on a horizontal plane, where $AB = 5a$, $AC = 3a$ and the angle $CAB = 120°$. The bisector of angle CAB meets BC at the point P.

   (a) Prove that $BC = 7a$.

   (b) Prove that $CP = \dfrac{21a}{8}$.

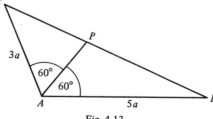

Fig. 4.13

A vertical pole CV is placed at C. The angle of elevation of V from A is $30°$.

   (c) Calculate the angle of elevation of V from P.

   AEB 1983

10. The point D divides the side AB of $\triangle ABC$ internally in the ratio $AD:DB$ $= m:n$. Given that $\angle ACD = \alpha$, $\angle BCD = \beta$ and $\angle BDC = \theta$, use the sine rule to show that

$$\frac{m}{\sin \alpha} = \frac{DC}{\sin (\theta - \alpha)} \quad \text{and} \quad \frac{n}{\sin \beta} = \frac{DC}{\sin (\theta + \beta)}$$

Hence prove that

$$(m + n)\cot \theta = m\cot \alpha - n \cot \beta$$

11. The square $ABCD$ lies in a horizontal plane. The sides AB, BC, CD and DA each have length a. The points P and Q are at a height h vertically above the points B and D respectively. Show that the triangle APQ has area $\frac{1}{2}(2h^2a^2 + a^4)^{1/2}$.

OX 1983

4.2 The trigonometric functions

Those with a modern, 'scientific' calculator will find that a value can be obtained for, say, $\sin 240°$ or $\cos(-130°)$ and may wish to know how these values are obtained. Whilst we restrict our attention to triangles we only need consider angles ranging between $0°$ and $180°$. However we can interpret angles of any magnitude as follows.

When a line segment OP is rotated about O, an angle θ [pronounced 'theta'] is swept out as measured *anti-clockwise* from a fixed line OX (Fig. 4.14). As OP continues to rotate anti-clockwise the angle increases. At OP', $\theta = 90°$ and at OP'', $\theta = 180°$. When P reaches OP''' the angle swept out is $270°$. By the time P reaches X, $\theta = 360°$, which is not unexpected since there are $360°$ round a point!

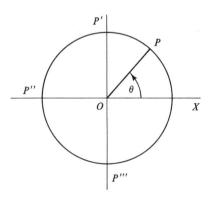

Fig. 4.14

If we allow OP to continue to rotate, the angle swept out is larger than $360°$. For example, if OP rotates once round and reaches the position shown in Fig. 4.15 then the angle swept out is $360° + 50° = 410°$. Clearly angles of any magnitude can be described in this way. On the other hand if OP rotates *clockwise* then the angle is *defined* to be *negative*. Thus in Fig. 4.16, $\theta_1 = -30°$ and $\theta_2 = -240°$.

Notice that OP'' in Fig. 4.16 is in the same position to represent $-240°$ and $120°$ (and in fact many more).

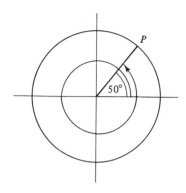

Fig. 4.15

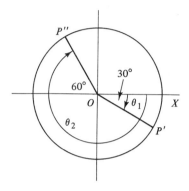

Fig. 4.16

Consider now a circle of *unit* radius. If we superimpose an axis system OX and OY then four quadrants are defined, as indicated in Fig. 4.17.

When P is in the first quadrant the co-ordinates of P (x and y) are both positive.

In the second quadrant the x co-ordinate is negative and the y co-ordinate remains positive. In the third quadrant both co-ordinates are negative and in the fourth quadrant the x co-ordinate is now positive whilst the y co-ordinate is negative. We make the following definitions.

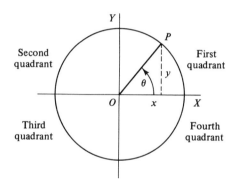

Fig. 4.17

Definition

For *any* angle θ swept out by OP in the *unit* circle measured from OX

$\cos \theta$ = the x co-ordinate of P

$\sin \theta$ = the y co-ordinate of P

$\tan \theta = \dfrac{\sin \theta}{\cos \theta}$ [$\tan \theta$ is not defined when $\cos \theta = 0$.]

NB $\cos \theta$ is short for the cosine of θ. Similarly we write sin for sine and tan for tangent.

The first point to notice is that these definitions agree with the familiar ones used for acute angles in right-angled triangles.

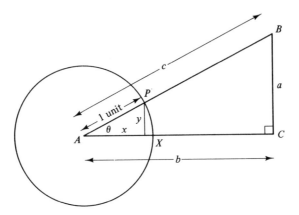

Fig. 4.18

Now in the right-angled triangle, $\triangle ABC$, in Fig. 4.18

$$\sin \theta = \frac{a}{c}$$

and since the triangle with unit hypotenuse is similar to $\triangle ABC$

$$\frac{a}{c} = \frac{y}{1} = y$$

and so

$$\sin \theta = y$$

which is precisely our new definition.

  Now

$$\cos \theta = \frac{b}{c}$$

but

$$\frac{b}{c} = \frac{x}{1} = x$$

so

$$\cos \theta = x$$

again corresponding to our new definition. In the triangle, $\triangle ABC$,

$$\tan \theta = \frac{a}{b} = \frac{a}{b} \times \frac{c}{c}$$

$$= \frac{a}{c} \Big/ \frac{b}{c}$$

$$= \frac{\sin \theta}{\cos \theta}$$

Thus we have *extended* our usual definitions to include angles of any magnitude.

4.3 The graphs of the trigonometric functions

It is useful to draw the graphs of the trigonometric functions, which 'bring to life'
many of their algebraic properties. The discussion of these graphs will be continued
on page 147.

It follows from the definition of
sin θ, and can be seen immediately
from the graph in Fig. 4.19 that

$$-1 \leqslant \sin \theta \leqslant 1 \quad \text{for all } \theta$$

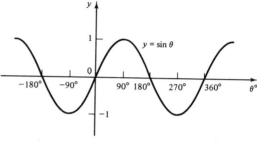

Fig. 4.19

It can similarly be seen that

$$\ldots = \sin(-270°) = 1 = \sin 90° = \sin(360° + 90°) = \sin(720° + 90°) = \ldots$$
and
$$\ldots = \sin(-360°) = \sin(-180°) = 0 = \sin 0° = \sin 180° = \sin 360° = \ldots$$
and
$$\ldots = \sin(-90°) = -1 = \sin 270° = \sin(360° + 270°) = \ldots$$

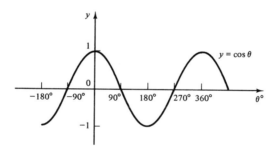

Fig. 4.20

The shape of the graph of $y = \cos \theta$ in Fig. 4.20 is identical to that of $y = \sin \theta$, but
is 'translated' 90° along the horizontal axis.

Of course

$$-1 \leqslant \cos \theta \leqslant 1 \quad \text{for all } \theta$$

but now

$$\ldots = \cos(-360°) = 1 = \cos 0° = \cos 360° = \ldots$$

$$\ldots \cos(-270°) = \cos(-90°) = 0 = \cos 90° = \cos 270° = \ldots$$

$$\ldots = \cos(-180°) = -1 = \cos 180° = \cos 540° = \ldots$$

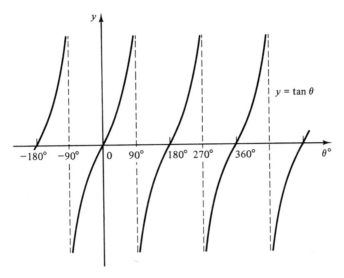

Fig. 4.21

The graph in Fig. 4.21 reflects the fact that the tangent is not defined when $\cos \theta = 0$, that is, when $\theta = 90°$ or $270°$ or .... It is also clear that the tangent can assume any value.

We return now to our formal definitions of the trigonometric functions, or *circular functions* as they are often called. We can readily check that the facts noted in the following table are correct.

	$\sin \theta$	$\cos \theta$	$\tan \theta$
First quadrant	+	+	+
Second quadrant	+	−	−
Third quadrant	−	−	+
Fourth quadrant	−	+	−

In the first quadrant *all* three functions are positive but in each of the following quadrants just one of the circular functions is positive. These observations can be summarised as in Fig. 4.22, where each letter indicates which function is positive — the A indicates that all three functions are positive.

These facts are sometimes recalled using a mnemonic such as *A*ll *S*atisfy *T*his *C*ase.

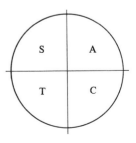

Fig. 4.22

4.4 Evaluation of the trigonometric functions

It is important to realise that the value of the trigonometric functions depends only on the position of P as OP rotates about O.

Thus

$$\sin \theta = \sin (360° + \theta) = \sin (720° + \theta) \qquad \text{and so on.}$$

Similarly

$$\cos \theta = \cos (360° + \theta) = \cos (720° + \theta) \qquad \text{and so on.}$$

We can, however, go further than this. For example, consider $\sin 240°$, as shown in Fig. 4.23. By definition the sine of $240°$ equals minus the length PN. [Notice that $240°$ is in the third quadrant.]

But the length PN is equal to the sine of $60°$. [Recall that we are working in the unit circle.]

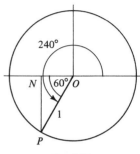

Fig. 4.23

In other words,

$$\sin 240° = -\sin 60°$$

More generally, $\sin \theta$ depends only on

(i) the quadrant in which θ lies — this determines the sign.
(ii) the acute angle OP makes with the x axis.

Exactly the same observation holds for $\cos \theta$ and $\tan \theta$. One immediate advantage of these observations is that one can compute the value of the circular functions for any angle using tables or a calculator which only offer values for angles lying between $0°$ and $90°$.

Example 4.9

Express the following in terms of acute angles.

(a) $\sin 95°$

(b) $\cos 300°$

(c) $\tan 200°$

(d) $\sin (-40°)$

(e) $\tan (-130°)$

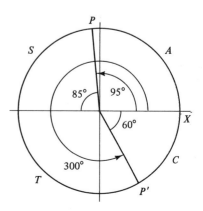

Fig. 4.24

(a) $95°$ is in the second quadrant and so $\sin 95°$ is positive. Now the acute angle OP makes with the x axis is $85°$ and so

$$\sin 95° = \sin 85°$$

(b) $300°$ is in the fourth quadrant and so $\cos 300°$ is positive. Now the acute angle OP' makes with the x axis is $60°$ so

$$\cos 300° = \cos 60°$$

(c) With practice it will be found that a diagram becomes unnecessary. One's thoughts run something like

$$200° \rightarrow \text{third quadrant} \rightarrow \tan \text{positive} \rightarrow \text{acute angle } 20°$$

So

$$\tan 200° = \tan 20$$

(d) $-40° \rightarrow$ fourth quadrant $\rightarrow$ sine negative $\rightarrow$ acute angle $40°$

$$\sin(-40°) = -\sin 40°$$

(e) $-130° \rightarrow$ third quadrant $\rightarrow$ tangent positive $\rightarrow$ acute angle $50°$

$$\tan(-130°) = \tan 50°.$$

All these results can be checked using a modern, scientific calculator!

Example 4.10

Simplify the following, by expressing in terms of acute angles

(a) $\sin 217°$ (b) $\cos 638°$ (c) $\tan(-196.5°)$

(a) $217° \rightarrow$ third quadrant $\rightarrow$ sine negative $\rightarrow$ acute angle $37°$

$$\sin 217° = -\sin 37°$$

(b) $638° \rightarrow 638° - 360° = 278° \rightarrow$ third quadrant $\rightarrow$ cosine positive $\rightarrow$ acute angle $82°$

$$\cos 638° = \cos 82°$$

(c) $-196.1° \rightarrow$ second quadrant $\rightarrow$ tangent negative $\rightarrow$ acute angle $16.1°$

$$\tan(-196.1°) = -\tan 16.1°$$

Some values of the trigonometric functions can be found without the use of tables or a calculator.

These are indicated in the following table.

Function \ Angle	0°	30°	45°	60°	90°
Sine	0	$\dfrac{1}{2}$	$\dfrac{1}{\sqrt{2}}$	$\dfrac{\sqrt{3}}{2}$	1
Cosine	1	$\dfrac{\sqrt{3}}{2}$	$\dfrac{1}{\sqrt{2}}$	$\dfrac{1}{2}$	0
Tangent	0	$\dfrac{1}{\sqrt{3}}$	1	$\sqrt{3}$	undefined

As an aid to obtaining the above results notice each of the following diagrams (Fig. 4.25 (i), (ii)).

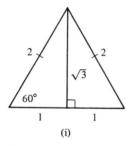
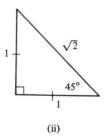

Fig. 4.25

In the Fig. 4.25 (i), the perpendicular bisects the opposite side of an equilateral triangle, with sides of length 2 units. The fact that the perpendicular is $\sqrt{3}$ can be established using Pythagoras' theorem. In the second diagram (Fig. 4.25 (ii)) an isosceles right-angled triangle has base 1 unit and so again by Pythagoras' theorem the hypotenuse is $\sqrt{2}$.

These facts enable one to evaluate the circular functions for many other angles. For example

$$\cos 210° = -\cos 30°$$
$$= -\frac{\sqrt{3}}{2}.$$

Exercise 4.2

1. Sketch the graph of $y = \sin x$ and hence write down the values of

   (a) $\sin 0°$ (b) $\sin 90°$ (c) $\sin(-180°)$ (d) $\sin 270°$

2. Sketch the graph of $y = \cos x$ and hence write down the values of

   (a) $\cos 0°$ (b) $\cos 90°$ (c) $\cos(-90°)$ (d) $\cos 360°$

3. (i) Write down three values of θ for which $\tan \theta$ is undefined.

   (ii) In which quadrants may θ lie if it is known that

   (a) $\tan \theta$ is negative   (b) $\sin \theta$ is positive
   (c) $\cos \theta$ is positive   (d) $\tan \theta < 0$ and $\sin \theta > 0$
   (e) $\cos \theta < 0$ and $\tan \theta > 0$ (f) $\cos \theta > 0$ and $\sin \theta > 0$.

4. Express each of the following in terms of acute angles.

   (a) $\sin 325°$ (b) $-\cos(-50°)$ (c) $\sin 154°$
   (d) $-\tan(-159°)$ (e) $-\cos 165°$ (f) $\sin(-342°)$
   (g) $\tan 468°$ (h) $\cos 725°$ (i) $\tan 138.5°$

5. Evaluate the following without using tables or a calculator.

   (a) $\cos 330°$ (b) $\cos(-30°)$ (c) $\sin 180°$
   (d) $\sin(-180°)$ (e) $\tan(-60°)$ (f) $\tan 300°$
   (g) $\cos 120°$ (h) $\cos(-240°)$ (i) $\sin 135°$
   (j) $\sin 225°$ (k) $\cos 315°$ (l) $\tan(-180°)$

6. Evaluate each of the following correct to 3 d.p.

   (a) $\cos 37°$ (b) $\sin 243°$ (c) $\sin 156°$
   (d) $\tan 125°$ (e) $\tan 263°$ (f) $\cos(-156)°$

4.5   The solution of trigonometric equations

Consider the equation

$$\cos \theta = -\sqrt{3}/2$$

Since the cosine is negative, θ lies in the second or third quadrant. The acute angle whose cosine is $\sqrt{3}/2$ is $30°$. So θ may be $150°$, $210°$ or $-150°$, $-210°$. Then, of course, θ may be $360° + 150° = 510°$ or $360° + 210° = 570°$. In fact there are an infinite number of solutions.

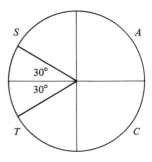

Fig. 4.26

When asked to solve such equations one is often given a range of possible values for θ.
If $0 \leqslant \theta < 360°$ then the solutions are $150°$ or $210°$, but if $-180° < \theta \leqslant 180°$ then the solutions are $-150°$ or $150°$.

Example 4.11

Solve the equation $\tan \theta = 0.53$, in the range $0° \leqslant \theta < 360°$.

The tangent is positive and so θ lies in the first or third quadrant. The acute angle, whose tangent is 0.53, is $27.9°$.

So

$$\theta = 27.9° \quad \text{or} \quad \theta = 207.9°$$

in the required range.

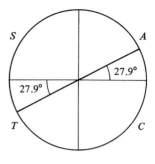

Fig. 4.27

Example 4.12

Solve the equation $3\cos^2\theta = 2\cos\theta$, in the range $-180° < \theta \leqslant 180°$.

This is a quadratic equation in $\cos\theta$. Notice that we write $\cos^2\theta$ rather than $(\cos\theta)^2$ to avoid the use of brackets. Similar conventions are used for higher powers and for the other trigonometric functions.

$$3\cos^2\theta - 2\cos\theta = 0$$

$$\Rightarrow \cos\theta\,(3\cos\theta - 2) = 0$$

$$\Rightarrow \cos\theta = 0 \quad \text{or} \quad \cos\theta = \tfrac{2}{3}$$

When $\cos\theta = 0$, $\theta = -90°$ or $\theta = 90°$ in the required range.

When $\cos\theta = \tfrac{2}{3}$ then since the cosine is positive, θ is in the first or fourth quadrant. The acute angle whose cosine is $\tfrac{2}{3}$ is $48.2°$ Ⓒ.
So

$$\theta = -48.2° \quad \text{or} \quad \theta = 48.2°.$$

Thus the required solutions are

$$\theta = -90° \quad \text{or} \quad \theta = -48.2° \quad \text{or} \quad \theta = 48.2° \quad \text{or} \quad \theta = 90°.$$

4.6 Further trigonometric functions

There are three more functions defined in terms of the three functions already introduced.

Definition

$\sec\theta \equiv \dfrac{1}{\cos\theta}$	$\sec\theta$ is undefined when $\cos\theta = 0$
$\operatorname{cosec}\theta \equiv \dfrac{1}{\sin\theta}$	$\operatorname{cosec}\theta$ is not defined when $\sin\theta = 0$

$$\cot \theta \equiv \frac{\cos \theta}{\sin \theta}$$ $\cot \theta$ is not defined
when $\sin \theta = 0$.

Notice that $\cot \theta \equiv \dfrac{1}{\tan \theta}$

when both are defined

NB $\sec \theta$ is short for secant θ. Similarly we write cosec for cosecant and cot for cotangent.

4.7 Trigonometric identities

No matter in which quadrant θ lies, using Pythagoras' theorem we notice that

$$x^2 + y^2 = 1$$

or, since $\sin \theta = y$ and $\cos \theta = x$

$$\cos^2 \theta + \sin^2 \theta \equiv 1$$

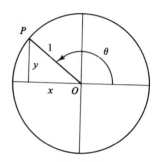

Fig. 4.28

Notice that this is an *identity* as the relationship holds for all values of θ.
    Dividing by $\cos^2 \theta$ we have

$$1 + \frac{\sin^2 \theta}{\cos^2 \theta} = \frac{1}{\cos^2 \theta} \qquad \cos \theta \neq 0$$

or in terms of other functions

$$1 + \tan^2 \theta \equiv \sec^2 \theta \qquad \cos \theta \neq 0$$

These results can be used to solve a further set of trigonometric equations.

Example 4.13

Solve the equations

(a) $2 \sin^2 \theta + 3 \cos \theta = 0$ (b) $4 \cos^2 \theta + 5 \sin \theta = 5$

within the range $0 \leqslant \theta < 360°$.

(a) We can convert this to a quadratic in $\cos \theta$ by noting that

$$\sin^2 \theta \equiv 1 - \cos^2 \theta \qquad \text{[a restatement of our previous result]}$$

$$\Rightarrow 2(1 - \cos^2 \theta) + 3 \cos \theta = 0$$

$$\Rightarrow 2\cos^2\theta - 3\cos\theta - 2 = 0$$

$$\Rightarrow (2\cos\theta + 1)(\cos\theta - 2) = 0$$

$$\Rightarrow \cos\theta = -\tfrac{1}{2} \quad \text{or} \quad \cos\theta = 2$$

If $\cos\theta = -\tfrac{1}{2}$ then $\theta = 120°$ or $240°$ in the required range. There are no solutions for $\cos\theta = 2$ since the maximum value of $\cos\theta$ is 1.

(b) Converting this time to a quadratic in $\sin\theta$, we have

$$4(1 - \sin^2\theta) + 5\sin\theta = 5$$

$$\Rightarrow 4 - 4\sin^2\theta + 5\sin\theta = 5$$

$$\Rightarrow 4\sin^2\theta - 5\sin\theta + 1 = 0$$

$$\Rightarrow (4\sin\theta - 1)(\sin\theta - 1) = 0$$

$$\Rightarrow \sin\theta = \tfrac{1}{4} \quad \text{or} \quad \sin\theta = 1$$

$$\Rightarrow \theta = 14.5° \quad \text{or} \quad 165.5° \quad \text{or} \quad 90° \qquad \boxed{\text{C}}$$

within the required range.

Statement

> For any angle θ
>
> $$\cos^2\theta + \sin^2\theta \equiv 1$$
>
> Or equivalently
>
> $$1 + \tan^2\theta \equiv \sec^2\theta$$
>
> or
>
> $$\cot^2\theta + 1 \equiv \operatorname{cosec}^2\theta$$

Example 4.14

Solve simultaneously the equations

$$x\cos\theta = 3 \qquad \text{①}$$

$$x\sin\theta = -4 \qquad \text{②}$$

where $x > 0$ and $-180 < \theta \leqslant 180°$.

The method of solution of this type of equation is referred to as 'squaring and adding'.

As the phrase suggests we proceed as follows

$$①^2: \qquad x^2\cos^2\theta = 9 \qquad \text{③}$$

$$②^2: \qquad x^2\sin^2\theta = 16 \qquad \text{④}$$

③ + ④: $x^2 \cos^2 \theta + x^2 \sin^2 \theta = 9 + 16$

$\Rightarrow x^2 (\cos^2 \theta + \sin^2 \theta) \; = 25$

$\Rightarrow \qquad\qquad\qquad\qquad x^2 = 25$

$\Rightarrow \qquad\qquad\qquad\qquad x = 5 \qquad$ since $x > 0$

Referring to equation ① and substituting

$$5 \cos \theta = 3$$

$$\Rightarrow \cos \theta = \tfrac{3}{5}$$

$$\Rightarrow \theta = 53.1° \quad \text{or} \quad \theta = -53.1° \qquad \boxed{C}$$

Referring to the second equation it is clear that the negative solution is required, and so we find

$$x = 5 \quad \text{and} \quad \theta = -53.1°.$$

Exercise 4.3

1. Solve each of the following in the range $0 \leqslant \theta < 180°$.

    (a) $\sin \theta = 0.6384$ (b) $\cos \theta = -0.7547$ (c) $\tan \theta = 2.605$

2. Solve each of the following equations in the range $0 \leqslant \theta < 360$.

    (a) $\tan \theta = -\tfrac{1}{2}$ (b) $\tan \theta = \sqrt{3}/2$ (c) $\sin \theta = \sqrt{2}/2$
    (d) $\sin \theta = -\sqrt{2}/2$ (e) $\cos \theta = \tfrac{1}{2}$ (f) $\tan \theta = 1$

3. Solve each of the following equations in the range $-360° < \theta \leqslant 360°$.

    (a) $\sin \theta = 0.814$ (b) $\cos \theta = 0.814$ (c) $\sin \theta = -0.814$
    (d) $\tan \theta = \sin 54°$ (e) $\tan \theta = -\cos(-35)°$ (f) $\cos^2 \theta = \tfrac{1}{4}$
    (g) $\sin^2 \theta = \tfrac{3}{4}$ (h) $\tan^2 \theta = 2$

4. Solve the following equations in the range $-180° < \theta \leqslant 180°$.

    (a) $4 \sin^2 \theta - 9 \sin \theta = 0$ (b) $\cos \theta + 5 \sin \theta \cos \theta = 0$
    (c) $\sqrt{3} \tan \theta = 2 \sin \theta$ (d) $3 \tan^2 \theta + 5 \tan \theta - 2 = 0$
    (e) $6 \cos^2 \theta + \cos \theta - 2 = 0$ (f) $\sin^2 \theta + 3 \sin \theta + 1 = 0$

5. Solve the following equations in the range $0 < \theta \leqslant 360°$.

    (a) $3 \cos^2 \theta + 2 \sin \theta = 0$ (b) $6 \sin^2 \theta + \cos \theta - 2 = 0$
    (c) $4 \sec^2 \theta - 7 \tan \theta - 6 = 0$

6. Given that $\sin \theta = 7/25$ and θ is obtuse one can use Pythagoras' theorem to establish that $\cos \theta = -24/25$ and $\tan \theta = 7/24$

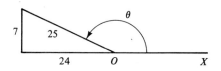

Fig. 4.29

(a) If θ is acute and $\tan \theta = 4/5$ write down the value of $\cos \theta$.

(b) If θ is obtuse and $\tan \theta = -5/12$ write down the value of

   (i) $\sin \theta$ (ii) $\sec \theta$

(c) If θ is reflex and $\cos \theta = 12/13$ write down the value of

   (i) $\cos \theta$ (ii) $\cot \theta$

7. Use the fact that $\sin^2 \theta + \cos^2 \theta \equiv 1$ to show that

$$1 + \cot^2 \theta \equiv \operatorname{cosec}^2 \theta \qquad \text{when } \sin \theta \neq 0.$$

hence solve the equation

$$3 \operatorname{cosec}^2 \theta + 7 \cot \theta - 1 = 0$$

in the range $-180° < \theta \leqslant 180°$.

8. Develop an argument to show that the following results hold for all θ.

(a) $\sin (180° - \theta) \equiv \sin \theta$ (b) $\cos (180° - \theta) \equiv -\cos \theta$

(c) $\sin (-\theta) \equiv -\sin \theta$ (d) $\cos (-\theta) \equiv \cos \theta$

9. Find x and θ satisfying the following simultaneous equations, where in each case $x > 0$ and $-180° < \theta \leqslant 180°$.

(a) $x \cos \theta = 7$ (b) $x \cos \theta = 5$

    $x \sin \theta = 12$ $x \sin \theta = -3$

(c) $x \cos \theta = -2$ (d) $x \cos \theta = -3$

    $x \sin \theta = -4$ $x \sin \theta = 5$

10. Solve the equation

$$3 \sin \theta + 2 \cos \theta = 4 \sin \theta - 3 \cos \theta \qquad \text{where} \quad 0° \leqslant \theta < 180°.$$

11. Find all values of θ between $0°$ and $360°$ for which

$$2 \sin \theta + 8 \cos^2 \theta = 5,$$

giving your answers correct to the nearest $0.1°$ where necessary.

CAMB 1982

4.8 Radians

So far, angles have been measured in degrees. The fact that there are 360° around a point is due to historical reasons. It turns out that there is a far more natural unit of measurement called a radian. The reader will find that this unit of measure is more important for advanced mathematics.

Definition

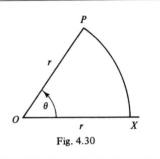

An arc of a circle equal in length to the radius of the circle subtends an angle of 1 radian at the centre

Fig. 4.30

The circumference of a circle is $2\pi r$ and so there are 2π radian about a point. That is

$$2\pi \text{ radians} = 360°$$

$$\Rightarrow \quad \pi \text{ radians} = 180°$$

$$\Rightarrow \quad 1 \text{ radian} = \frac{180°}{\pi}$$

$$\simeq 57.3° \qquad 1 \text{d.p.} \qquad \boxed{C}$$

Statement

It is normally assumed that angles are measured in radians when no units are indicated.

Example 4.15

(a) Express the following in radians.

   (i) 90°   (ii) 120°   (iii) 145°

(b) Express the following in degrees.

   (i) $\dfrac{\pi}{3}$ (ii) $\dfrac{5\pi}{6}$ (iii) $\dfrac{\pi}{8}$

Recall that $\pi = 180°$.

(a) (i) $90° = \dfrac{\pi}{2}$ (ii) $120° = \dfrac{2\pi}{3}$ (iii) $145° = \dfrac{145\pi}{180} = \dfrac{29\pi}{36}$

(b) (i) $\dfrac{\pi}{3} = \dfrac{180°}{3} = 60°$ (ii) $\dfrac{5\pi}{6} = \dfrac{5.180}{6} = 150°$ (iii) $\dfrac{\pi}{8} = \dfrac{180°}{8} = 22.5°$.

Many results are much simplified when angles are measured in radians, though *all trigonometric identities will remain the same*, since the values of the functions do not depend on the units of the angle. Thus

$$\cos\frac{\pi}{3} = \cos 60° = \tfrac{1}{2}$$

Statement

> The length of an arc PQ which subtends an angle of θ—*measured in radians*, in a circle of radius r—is $r\theta$.
>
> The area of a sector POQ containing the angle θ—*measured in radians*, in a circle of radius r—is $\frac{1}{2}r^2\theta$.

Proof

Suppose the length of the arc is l and the area of the sector is A then the ratio of l to the circumference is the same as the ratio of the angle subtended to 2π (since here are 2π radians about a point).

That is

$$\frac{l}{2\pi r} = \frac{\theta}{2\pi} \qquad \Rightarrow l = r\theta$$

Similarly

$$\frac{A}{\pi r^2} = \frac{\theta}{2\pi} \qquad \Rightarrow A = \tfrac{1}{2}r^2\theta$$

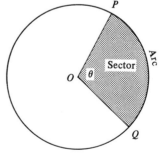

Fig. 4.31

One application of these results is to show that:

Statement

> When θ is measured in radians then
>
> $$\sin\theta \simeq \theta \quad \text{when } \theta \text{ is small}$$
>
> with the approximation improving as smaller values of θ are selected.

Proof

Suppose the chord PQ subtends on angle θ (assumed positive and acute) at the centre of a circle of radius r. Suppose further that the tangent to the circle at Q meets OP (produced) at R. So $\angle OQR = 90°$.

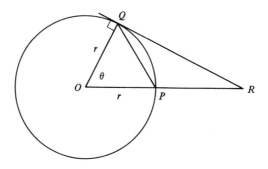

Fig. 4.32

The area of

triangle OPQ is $\frac{1}{2}r^2 \sin \theta$ $[\frac{1}{2}bc \sin A]$

sector OPQ is $\frac{1}{2}r^2 \theta$ since θ is measured in radians.

triangle ORQ is $\frac{1}{2}r^2 \tan \theta$ since the triangle is right-angled and QR is given by

$$\tan \theta = \frac{QR}{r}$$

Comparing areas, it is clear from the diagram that

Area of $\triangle OPQ \leqslant$ Area of sector $OPQ \leqslant$ Area of $\triangle ORQ$

That is

$$\tfrac{1}{2}r^2 \sin \theta \leqslant \tfrac{1}{2}r^2 \theta \leqslant \tfrac{1}{2}r^2 \tan \theta$$

$$\Rightarrow \quad \sin \theta \leqslant \theta \leqslant \tan \theta$$

Dividing throughout by the positive quantity $\sin \theta$ ($\sin \theta \neq 0$) and recalling that

$$\tan \theta \equiv \frac{\sin \theta}{\cos \theta}$$

$$\Rightarrow 1 \leqslant \frac{\theta}{\sin \theta} \leqslant \frac{1}{\cos \theta}$$

Now for small θ

$$\cos \theta \simeq 1$$

and so

$$\frac{\theta}{\sin \theta} \simeq 1$$

$$\Rightarrow \sin \theta \simeq \theta$$

In question 11, Exercise 4.2 we noted that $\sin(-\theta) = -\sin \theta$ so for small θ

$$-\theta \simeq -\sin \theta$$

$$= \sin(-\theta)$$

$$\Rightarrow -\theta \simeq \sin(-\theta)$$

That is, the approximation holds for small θ, either positive or negative.

Modern calculators can be 'programmed' to operate using either degrees or radians. This is indicated on the display. It would be a useful exercise to check the above result on a calculator.

It is, of course, important to realise that one can solve trigonometric equations just as easily if we are working in radians as compared with degrees.

Example 4.16

Solve the equations

(a) $6 \sin^2 \theta + \cos \theta - 4 = 0$ (b) $\cos 2\theta = \dfrac{-\sqrt{3}}{2}$

in the range $-\pi < \theta \leqslant \pi$

(a) Since $\sin^2 \theta \equiv 1 - \cos^2 \theta$ regardless of the unit used to measure θ,

$$6(1 - \cos^2 \theta) + \cos \theta - 4 = 0$$

$$\Rightarrow \qquad 6\cos^2 \theta - \cos \theta - 2 = 0$$

$$\Rightarrow (2\cos \theta + 1)(3\cos \theta - 2) = 0$$

$$\Rightarrow \qquad \cos \theta = -\tfrac{1}{2} \quad \text{or} \quad \cos \theta = \tfrac{2}{3}$$

So in the range $-\pi \leqslant \theta \leqslant \pi$,

when $\cos \theta = -\tfrac{1}{2}$, $\theta = \dfrac{2\pi}{3}$ or $\theta = \dfrac{-2\pi}{3}$

When $\cos \theta = \tfrac{2}{3}$ one solution is $\theta = 0.84$ $\boxed{\text{C}}$
    Alternatively, if working in degrees

$$\theta = 48.19° = 48.19 \times \frac{\pi}{180} \text{ radians}$$

$$= 0.84$$

So the other solution is

$$\theta = -0.84$$

Giving the following solutions for θ:

$$\pm \frac{2\pi}{3}, \qquad \pm 0.84$$

(b) To solve $\cos 2\theta = \frac{-\sqrt{3}}{2}$ for $-\pi < \theta \leqslant \pi$ we must allow $-2\pi < 2\theta \leqslant 2\pi$.

One solution of 2θ is $5\pi/6$, and so in the range $-2\pi < 2\theta \leqslant 2\pi$ solutions for 2θ are

$$-\frac{7\pi}{6}, \quad -\frac{5\pi}{6}, \frac{5\pi}{6}, \frac{7\pi}{6}$$

Giving the following solutions for θ:

$$-\frac{7\pi}{12}, \quad -\frac{5\pi}{12}, \frac{5\pi}{12}, \frac{7\pi}{12}$$

[NB Errors resulting from the use of calculators in trigonometric equations are often the result of failing to work in the correct units!]

4.9 Identities

Suppose that OP sweeps out the angle θ, where P has co-ordinates (x, y), [so $\sin \theta = y$ and $\cos \theta = x$]. If we reflect OP in the x axis onto OQ, we see that the co-ordinates of Q are $(x, -y)$. That is, the x co-ordinate remains unchanged and the y co-ordinate has its sign reversed. But OQ marks out the angle $-\theta$ [since we reflected] and so

$$\sin(-\theta) = -y \quad \text{and} \quad \cos(-\theta) = x$$

Or in other words

$$\sin(-\theta) \equiv -\sin \theta \quad \text{and} \quad \cos(-\theta) \equiv \cos \theta$$

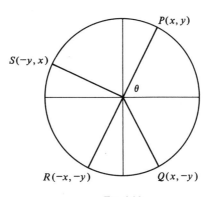

Fig. 4.33

Notice that these are *identities* because the results hold for all values of θ.

Again the angle $\pi + \theta$ is represented by the line segment OR, with co-ordinates $(-x, -y)$.

That is

$$\sin(\pi + \theta) = -y \quad \text{and} \quad \cos(\pi + \theta) = -x$$

$$\sin(\pi + \theta) \equiv -\sin\theta \quad \text{and} \quad \cos(\pi + \theta) \equiv -\cos\theta$$

OR also represents the angle $\theta - \pi$ and so it follows immediately that

$$\sin(\theta - \pi) \equiv -\sin\theta \quad \text{and} \quad \cos(\theta - \pi) \equiv -\cos\theta$$

If OS represents the angle $\theta + \pi/2$ then the co-ordinate of S is $(-y, x)$! [A diagram helps.]

That is

$$\sin\left(\frac{\pi}{2} + \theta\right) \equiv \cos\theta \quad \text{and} \quad \cos\left(\frac{\pi}{2} + \theta\right) \equiv -\sin\theta$$

It is worth drawing a diagram to convince oneself that by a similar argument

$$\sin\left(\frac{\pi}{2} - \theta\right) \equiv \cos\theta \quad \text{and} \quad \cos\left(\frac{\pi}{2} - \theta\right) \equiv \sin\theta$$

These latter results are obvious for acute angles, but it must be stressed that they hold for all angles.

There are of course many more similar results which could be established this way. However, we now consider even more general results.

4.10 Compound angles

Statement

> If A and B represent angles of any magnitude then
>
> $$\sin(A + B) \equiv \sin A \cos B + \cos A \sin B$$
>
> and
>
> $$\cos(A + B) \equiv \cos A \cos B - \sin A \sin B$$

It must be stressed that these results hold irrespective of whether the angles are measured in degrees or radians. That the results are reasonable can be seen by checking some of the identities we have just established.

For example

$$\sin(\pi + \theta) \equiv \sin\pi\cos\theta + \cos\pi\sin\theta$$

Now since

$$\sin\pi = 0 \quad \text{and} \quad \cos\pi = -1$$

we have

$$\sin(\pi + \theta) \equiv -\sin\theta$$

Again

$$\cos\left(\frac{\pi}{2} - \theta\right) = \cos\left(\frac{\pi}{2} + (-\theta)\right) = \cos\frac{\pi}{2}\cos(-\theta) - \sin\frac{\pi}{2}\sin(-\theta)$$

$$= 0.\cos(-\theta) - 1.\sin(-\theta)$$

$$= -\sin(-\theta)$$

and since

$$\sin(-\theta) = -\sin\theta$$

we have

$$\cos\left(\frac{\pi}{2} - \theta\right) = \sin\theta.$$

Proof of the compound angle identities (for acute angles)

The results will be proved in the case that A and B are both acute. The more general results will be proved in Chapter 15. Consider the triangle $\triangle PQR$ where QS is the perpendicular from Q to PR.

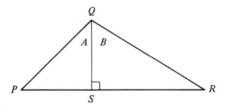

Fig. 4.34

Suppose that $\angle PQS = A$ and $\angle SQR = B$. Now the area of $\triangle PQR$ is

$$\tfrac{1}{2}PQ.QR.\sin(A+B) \qquad [\tfrac{1}{2}ab\sin\theta]$$

But this area is equal to the sum of the areas of $\triangle PQS$ and $\triangle QSR$. That is

$$\tfrac{1}{2}PQ.QR\sin(A+B) = \tfrac{1}{2}PQ.QS\sin A + \tfrac{1}{2}QS.QR\sin B$$

Dividing throughout by $\tfrac{1}{2}PQ.QR$

$$\sin(A+B) = \frac{QS}{QR}\sin A + \frac{QS}{QP}\sin B.$$

But

$$\frac{QS}{QR} = \cos B \quad \text{and} \quad \frac{QS}{QP} = \cos A$$

$$\Rightarrow \sin(A+B) \equiv \sin A\cos B + \cos A\sin B$$

Assuming, now, this result holds for all angles, then in particular it holds for

$$\left(\frac{\pi}{2} + A\right) \quad \text{and} \quad B.$$

That is

$$\sin\left(\left(\frac{\pi}{2} + A\right) + B\right) \equiv \sin\left(\frac{\pi}{2} + A\right)\cos B + \cos\left(\frac{\pi}{2} + A\right)\sin B$$

But we have already shown that

$$\sin\left(\frac{\pi}{2} + A\right) \equiv \cos A$$

and *therefore*

$$\sin\left(\frac{\pi}{2} + (A + B)\right) \equiv \cos(A + B) \quad !!$$

and

$$\cos\left(\frac{\pi}{2} + A\right) \equiv -\sin A.$$

It follows that

$$\cos(A + B) \equiv \cos A \cos B + -\sin A \sin B$$
$$\Rightarrow \cos(A + B) \equiv \cos A \cos B - \sin A \sin B$$

These results are extremely useful and will be further developed in Chapter 7. One immediate application is that they can be used to evaluate trigonometric values for a further class of angles without the use of a calculator.

Example 4.17

Evaluate   (a) $\sin 165°$ (b) $\cos \dfrac{5\pi}{12}$.

(a) $\sin 165° = \sin(120° + 45°)$

$$= \sin 120 \cos 45 + \cos 120 \sin 45$$

$$= \frac{\sqrt{3}}{2} \cdot \frac{1}{\sqrt{2}} + -\frac{1}{2} \cdot \frac{1}{\sqrt{2}}$$

$$= \frac{\sqrt{3}}{2\sqrt{2}} - \frac{1}{2\sqrt{2}} = \frac{\sqrt{3} - 1}{2\sqrt{2}} = \frac{\sqrt{2}}{4}(\sqrt{3} - 1)$$

(b) $\cos\dfrac{5\pi}{12} = \cos\left(\dfrac{3\pi}{12} + \dfrac{2\pi}{12}\right)$

$\qquad = \cos\left(\dfrac{\pi}{4} + \dfrac{\pi}{6}\right)$

$\qquad = \cos\dfrac{\pi}{4} \cdot \cos\dfrac{\pi}{6} - \sin\dfrac{\pi}{4} \cdot \sin\dfrac{\pi}{6}$

$\qquad = \dfrac{1}{\sqrt{2}} \cdot \dfrac{\sqrt{3}}{2} - \dfrac{1}{\sqrt{2}} \cdot \dfrac{1}{2}$

$\qquad = \dfrac{\sqrt{2}}{4}(\sqrt{3} - 1)$!!

Example 4.18

Solve $\cos\theta = \sin(\theta + 30°)$ in the range $0° \leqslant \theta < 360°$.

Expanding

$$\cos\theta = \sin\theta\cos 30° + \cos\theta\sin 30°$$

$$\Rightarrow \cos\theta = \dfrac{\sqrt{3}}{2}\sin\theta + \tfrac{1}{2}\cos\theta$$

$$\Rightarrow \cos\theta = \sqrt{3}\sin\theta$$

$$\Rightarrow \tan\theta = \dfrac{1}{\sqrt{3}}$$

$$\Rightarrow \qquad \theta = 30° \quad \text{or} \quad \theta = 210°$$

Exercise 4.4

[Angles in radians may be left in terms of π.]

1. Express the following in degrees:

   (a) $\dfrac{\pi}{4}$ (b) $\dfrac{\pi}{8}$ (c) $\dfrac{5\pi}{8}$ (d) $\dfrac{2\pi}{3}$ (e) $\dfrac{5\pi}{6}$

   (f) $\dfrac{5\pi}{3}$ (g) 3π (h) 1.5   (i) 1.26   (j) 3.142

2. Express the following in radians:

   (a) 60°   (b) 135°   (c) 270°   (d) 210°   (e) 300°
   (f) 25°   (g) 230°   (h) 46°   (i) 57.6°   (j) 126.3°

3. Evaluate the following:

   (a) $\cos \dfrac{2\pi}{3}$ (b) $\sin \dfrac{5\pi}{3}$ (c) $\tan \pi$ (d) $\cos \dfrac{4\pi}{9}$

   (e) $\sin 1$ (f) $\tan 1.3$

4. Solve each of the following in the range $0 \leqslant \theta < 2\pi$.

   (a) $\cos \theta = -\dfrac{\sqrt{3}}{2}$ (b) $\sin \theta = -\dfrac{1}{\sqrt{2}}$ (c) $\tan \theta = 1.6$

   (d) $\cos \theta = -\sin \dfrac{2\pi}{3}$ (e) $\sin^2 \theta = 1$

5. Solve each of the following equations in the range $-\pi < \theta \leqslant \pi$.

   (a) $15\cos^2 \theta - \sin \theta - 13 = 0$ (b) $3\sec^2 \theta + 5\tan \theta - 2 = 0$

6. A chord divides a circle into a major and a minor segment. Find the area and the perimeter of the minor segment

   (a) if the radius of the circle is $10\,\text{cm}$ and the chord is of length $12\,\text{cm}$,
   (b) if the radius of the circle is $r\,\text{cm}$ and the chord subtends an angle 2θ at the centre, θ measured in radians.

7. Use the fact that $\theta \simeq \sin \theta$ when θ is small (θ measured in radians) to find an approximation to $\sin \theta$ when

   (a) $\theta = 0.6$ radians   (b) $\theta = 5°$

8. Find an approximate value of θ in degrees if

   (a) $\sin \theta = 0.001$ (b) $\sin 2\theta = 0.0052$

9. Use the identity $\cos \theta \equiv 1 - 2\sin^2 \dfrac{\theta}{2}$ to show that when θ, measured in radians, is small, then

   $$\cos \theta \simeq 1 - \frac{\theta^2}{2}.$$

   Use the above to find approximate values for

   (a) $\dfrac{1 - \cos \theta}{\theta^2}$ (b) $\dfrac{1 - \cos 2\theta}{\theta \sin \theta}$ when θ is small.

10. A chord PQ of a circle of radius r subtends an angle θ at O, the centre of the circle. Find in terms of r and θ, the area of the major segment cut off by the chord.

11. In the triangle ABC, $BC = a$, $CA = b$, $AB = c$ and $\angle BAC = \theta$ radians. Given that θ is small, use the cosine rule to show that

$$\theta^2 \simeq \frac{a^2 - (b - c)^2}{bc}.$$

12. AS is a diameter of the circle with centre O and radius r. Two points C and D are selected on the circumference so that $ABCD$ is a trapezium with DC parallel to AB. If angle BOC is θ radians, show that the area of the trapezium may be written as $r^2(\sin\theta + \frac{1}{2}\sin 2\theta)$.

13. Solve the following equations in the range $-\pi < \theta \leqslant \pi$.

(a) $12\sin^2 2\theta - \cos 2\theta - 6 = 0$ (b) $2\csc^2\theta + \cot\theta - 3 = 0$

(c) $\csc\theta = 1.5$ (d) $\cot\theta = -\frac{3}{4}$ (e) $\sec\theta = -1.6$

(f) $\sin\dfrac{\theta}{3} = \frac{1}{4}$ (g) $\cos\dfrac{\theta}{3} = \frac{2}{3}$ (h) $\cos\left(\theta + \dfrac{\pi}{4}\right) = \frac{1}{2}$

(i) $\sin\left(2\theta - \dfrac{\pi}{2}\right) = \dfrac{\sqrt{3}}{2}$ (j) $\cot\left(\theta + \dfrac{\pi}{4}\right) = 1.2$

14. Evaluate the following using the compound angle formulae.

(a) $\sin 75°$ (b) $\cos 165°$ (c) $\sin 225°$ (d) $\sin\dfrac{5\pi}{6}$

(e) $\cos\dfrac{5\pi}{6}$ (f) $\sin\dfrac{7\pi}{6}$

15 Assuming that A and B are acute where $\sin A = \frac{3}{5}$ and $\cos B = \frac{5}{13}$, find

(a) $\tan A$ (b) $\sin B$ (c) $\cos(-A)$ (d) $\sin(-B)$
(e) $\sin(A + B)$ (f) $\cos(A + B)$ (g) $\cos(A - B)$ (h) $\sin(A - B)$

16. By letting $A = \theta$ and $B = -\theta$ in the formula for $\cos(A + B)$ deduce that $\cos^2\theta + \sin^2\theta \equiv 1$.

17. Expand $\sin(A + B + C)$ by setting $B + C$ as B in the formula for the expansion of $\sin(A + B)$.

18. If A, B and C represent the angles of a triangle prove that

(a) $\sin(B + C) = A$ (b) $\sin A = \cos B \sin C + \sin B \cos C$.

19. Solve

(a) $2\cos\theta = \sin(\theta + 30°)$ (b) $\cos(\theta + 45°) = \sqrt{2}\sin\theta$

in the range $0 \leqslant \theta < 360°$.

Chapter 5   Function and graph

5.1   Definition of a function

The value of an expression such as x^2 or $\cos x$ varies according to the choice of x. The relationship between a value of x and the corresponding value of x^2 or $\cos x$ is called a *function*.

If f is a function, then the value of the function at a, is denoted by $f(a)$. Sometimes the function is not defined for certain values. For example, the logarithmic functions have not been defined for negative reals.

Definition

> The set of values for which the function is defined is called the domain of the function.

We may picture a function f as associating with each number x in the domain of f, a *unique* number denoted by $f(x)$.

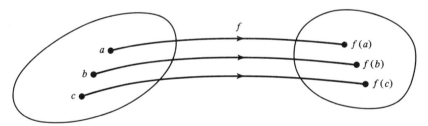

Fig. 5.1

Consider the function g which associates with each real number its square. The function may be defined by

$$g : x \mapsto x^2 \qquad x \in \mathbb{R}$$

or more usually as

$$g(x) \equiv x^2 \qquad x \in \mathbb{R}$$

The domain of g is given explicitly as the set of reals.

Definition

> The set of values of a function is called the range (of the function).

In the above example, the range of g is the set of non negative reals (since the square of any real is non negative).

Notation

> It is common practice to refer to the function f as $f(x)$—to stress that f is expressed in terms of the variable x, for example $\cos x$ or $\lg x$. However, $f(x)$ also means the value of the function at x. The reader should be aware of this potential cause for confusion.

Suppose we define

$$f(x) \equiv \cos x \qquad x \in \mathbb{R}$$

then we may set

$$y = f(x)$$

that is

$$y = \cos x$$

and sketch the graph of this function as in Fig. 5.2.

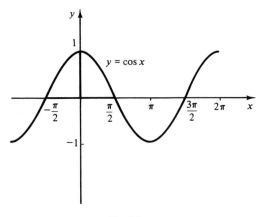

Fig. 5.2

The range of possible values for y represents the range of the function. It can be seen from the graph that the range of f is $[-1, 1]$. However, if the domain of the function is restricted to $[-\pi/2, \pi/2]$ then the range is simply $[0, 1]$.

5.2   Sketch graphs

In order to appreciate the characteristics of a function it is useful to be able to sketch a graph.

By setting the variable y equal to the value of the function f at each x in the domain of f, we obtain the equation.

$$y = f(x)$$

As the value of y depends as the choice of value for x, y is called the *dependent variable* and x is called the *independent variable*.

Each point P on the xy-plane is represented by an ordered pair (x, y), for particular values of x and y. The x co-ordinate is often called the abscissa and the y co-ordinate the ordinate.

In many situations it is not necessary to work very accurately. The salient features of a curve may be observed from a sketch such as Fig. 5.3.

Notice that it is clear from the graph that:

(i) $y \geq 0$ for all x; as noted before, the range is $[0, \infty)$;

(ii) the graph is symmetrical about the y axis;

(iii) large values of y correspond to large values of x (positive or negative);

(iv) the graph passes through the origin $(0, 0)$.

Other examples are shown in Fig. 5.4.

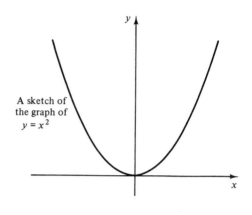

A sketch of the graph of $y = x^2$

Fig. 5.3

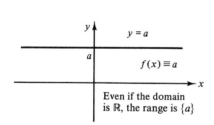

$y = a$

$f(x) \equiv a$

Even if the domain is $\mathbb{R}$, the range is $\{a\}$

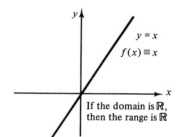

$y = x$

$f(x) \equiv x$

If the domain is $\mathbb{R}$, then the range is $\mathbb{R}$

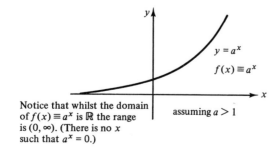

$y = a^x$

$f(x) \equiv a^x$

Notice that whilst the domain of $f(x) \equiv a^x$ is $\mathbb{R}$ the range is $(0, \infty)$. (There is no x such that $a^x = 0$.)

assuming $a > 1$

Fig. 5.4

Exercise 5.1

1. If $f(x) \equiv x^2 - 4$, $x \in \mathbb{R}$, find

   (a) $f(0)$,   (b) $f(3)$,   (c) $f(2)$,   (d) $f(-2)$.

2. Given that

   $$f(x) \equiv \sqrt{x + 2} \qquad x \geqslant -2$$

   (a) Find   (i) $f(7)$ (ii) $f(0)$ (iii) $f(-2)$
   (b) Find x so that   (i) $f(x) = 4$,   (ii) $f(x) = 10$.

3. By drawing suitable graphs find the range of

   (a) $f(x) \equiv 2x$ $x \in [-1, 3]$,
   (b) $f(x) \equiv 1 - x$ $x \in [0, 2]$,
   (c) $f(x) \equiv 2x + 1$ $x \in (1, 2]$.

4. Sketch the graph of $y = \sin x$ and hence find the range of $f(x) \equiv \sin x$ given that the domain is

   (a) $\mathbb{R}$,   (b) $[0, \pi/2]$,   (c) $(-\pi/2, \pi/2)$.

5. Find the range of $f(x) \equiv \tan x$ given that the domain is

   (a) $[0, \pi/2)$,   (b) $(-\pi/2, \pi/2)$.

6. The function f is defined by

   $$f(x) \equiv x^2 + 1 \qquad x \in \mathbb{R}$$

   Find

   (a) $f(-1)$ (b) $f(0)$ (c) $f(1)$ and find the range of f.

7. The function f is defined by

   $$f(x) \equiv (x - 3)^2 + 6 \qquad x \in \mathbb{R}.$$

   Find

   (a) $f(0)$,   (b) $f(3)$,   (c) $f(6)$,   and find the range of f.

8. If $m(x) \equiv 3^x$, $x \in [0, 2]$, find the range of $m(x)$.

9. Consider each of the graphs of functions given in Fig. 5.5 (see facing page) and in each case state the domain and range. (Assume the graphs continue indefinitely.)

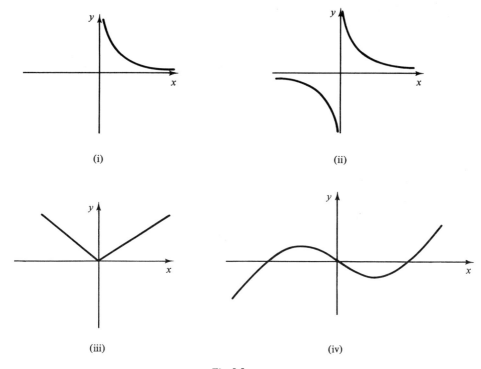

(i)

(ii)

(iii)

(iv)

Fig. 5.5

5.3 New graphs from old

If the graph of $y = f(x)$ is known, it is possible to sketch a number of related graphs with relative ease.

$y = f(-x)$

Suppose the point (a, b) lies on the graph of $y = f(x)$, that is $b = f(a)$. Then the point $(-a, b)$ lies on the graph of $y = f(-x)$. This is because $f(-(-a)) = f(a)$, but $f(a) = b$ so $f(-(-a)) = b$.

It follows that the graph of $y = f(-x)$ is the reflection of the graph of $y = f(x)$ in the y axis. An example is given in Fig. 5.6.

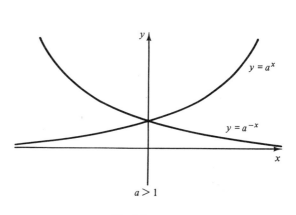

$y = a^x$

$y = a^{-x}$

$a > 1$

Fig. 5.6

$y = -f(x)$

By a similar argument the graph of $y = -f(x)$ can be found by reflecting the graph of $y = f(x)$ in the x axis.

$y = af(x)$

The effect on the graph of $y = f(x)$ when $f(x)$ is multiplied by a constant can be seen from Fig. 5.7. Notice that points on the x axis remain fixed.

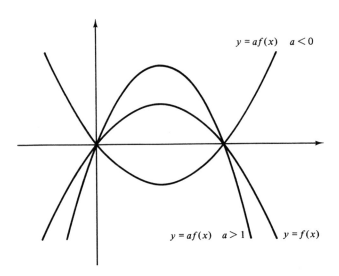

Fig. 5.7

$y + b = f(x + a)$

Suppose the graph of $y = f(x)$ is known, and we wish to sketch the graph of

$$y + b = f(x + a)$$

If we set

$$X = x + a$$

and

$$Y = y + b$$

we have

$$Y = f(X)$$

so we can draw the graph with respect to the X, Y axes.

To locate the new axes we note that

$$X = 0 \quad \text{when} \quad x = -a$$

and

$$Y = 0 \quad \text{when} \quad y = -b$$

so the origin of the X, Y axes is at $(-a, -b)$

Now the graph of

$$y + b = f(x + a)$$

may be obtained simply by drawing the graph of

$$Y = f(X)$$

with respect to the X, Y axes.

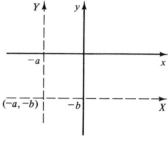

Fig. 5.8

Example 5.1

Sketch each of the following on separate diagrams:

(a) $y = x^2 + 2$,   (b) $y = (x + 1)^2$,   (c) $y = \cos(x - \pi/4) + 1$.

(a) $y = x^2 + 2$

         $\Rightarrow y - 2 = x^2$

    set

         $X = x$

         $Y = y - 2$

    to obtain

         $Y = X^2$

    Now

         $X = 0 \quad \Rightarrow x = 0$

         $Y = 0 \quad \Rightarrow y = 2$

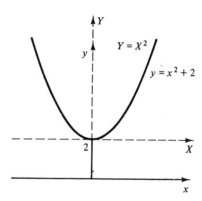

Fig. 5.9

So the origin of the X, Y axes is at $(0, 2)$.

Notice that the effect has been to translate the graph of $y = x^2$ two units 'parallel' to the y axis, to obtain the graph of $y = x^2 + 2$.

(b) For
$$y = (x + 1)^2$$

set

$$X = x + 1$$

$$Y = y$$

to obtain

$$Y = X^2$$

Now

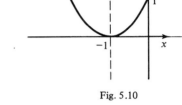

Fig. 5.10

$$X = 0 \qquad \Rightarrow x = -1$$

$$Y = 0 \qquad \Rightarrow y = 0$$

So the origin of the X, Y axes is at $(-1, 0)$.

Notice that the effect has been to translate the graph of $y = x^2$ one unit 'parallel' to the x axis to obtain the graph of $y = (x + 1)^2$

(c)

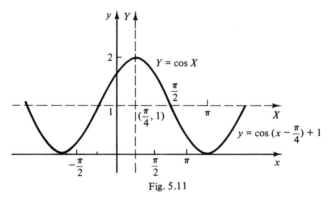

Fig. 5.11

$$y = \cos(x - \pi/4) + 1$$

$$\Rightarrow y - 1 = \cos(x - \pi/4).$$

Set

$$X = x - \pi/4$$

$$Y = y - 1$$

to obtain

$$Y = \cos X.$$

Now

$$X = 0 \qquad \Rightarrow x = \pi/4$$

$$Y = 0 \qquad \Rightarrow y = 1.$$

So the origin of the X, Y axes is located at $(\pi/4, 1)$.

Statement

> The graph of $y = f(x) + b$ may be obtained by translating the graph of $y = f(x)$ b units 'parallel' to the y axis. (Upwards if $b > 0$ and downwards if $b < 0$.)
>
> The graph of $y = f(x + a)$ may be obtained by translating the graph of $y = f(x)$ a units 'parallel' to the x axis (to the left if $a > 0$ and to the right if $a < 0$).

5.4   Odd, even and periodic functions

Definition

> A function is said to be
>
> (i) An *odd* function if
> $f(-x) \equiv -f(x)$ for all x in the domain of f
>
> (ii) an *even* function if
> $f(-x) \equiv f(x)$ for all x in the domain of f
>
> (iii) a *periodic* function with period a, if
> $f(a + x) \equiv f(x)$ for all x in the domain of f.

We have already observed that

$$\sin(-x) \equiv -\sin(x) \qquad x \in \mathbb{R}$$

So the sine function is odd.

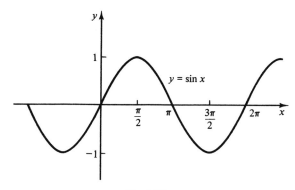

Fig. 5.12

Geometrically, a function is *odd* if its graph is unchanged when reflected in the x axis and then in the y axis (or, equivalently, when 'rotated about the origin' through $180°$).

But

$$\cos(-x) \equiv \cos x \qquad \text{for } x \in \mathbb{R}$$

and so the cosine function is even.

Geometrically, a function is *even* if its graph is unchanged when reflected in the y axis.

Now since

$$\sin(2\pi + x) \equiv \sin x \qquad x \in \mathbb{R}$$

and

$$\cos(2\pi + x) \equiv \cos x \qquad x \in \mathbb{R}$$

both the sine and cosine function are periodic with period 2π. Geometrically, a function is *periodic* with period a if its graph remains unchanged when translated units 'parallel' to the x axis.

Exercise 5.2

1. For each of the graphs in Fig. 5.13 sketch the graphs of

    (a) $y = 2f(x)$ (b) $y = -\tfrac{1}{2}f(x)$ (c) $y = f(-x)$ (d) $y = -f(-x)$

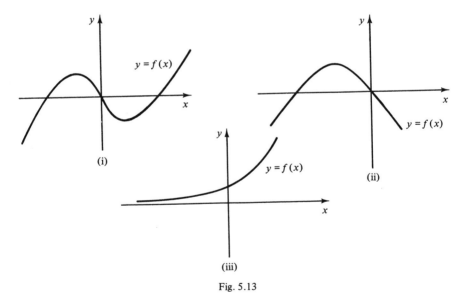

Fig. 5.13

2. Sketch the following pairs of graphs on the same axes.

    (a) $y = 2x, y = -2x$ (b) $y = x^2, y = \tfrac{1}{3}x^2$ (c) $y = x^2, y = -x^2$
    (d) $y = x^2, y = (-x)^2$ (e) $y = x + 1, y = x + 2$
    (f) $y = 2x, y = 2x + 1$ (g) $y = (x - 1)^2, y = (x - 2)^2$
    (h) $y = (x + 1)^2, y = 2(x + 1)^2$ (i) $y = (x + 1)^2, y = -2(x + 1)^2$

3. The graph of $y = x^3$ is given in Fig. 5.14. Sketch the graphs of

(a) $y = -x^3$,
(b) $y = (-x)^3$,
(c) $y = -(-x)^3$,
(d) $y = (x - 2)^3$,
(e) $y = (x + 3)^3$,
(f) $y = 2(x + 3)^3$,
(g) $y = 2(x - 2)^3 + 3$,
(h) $y = (1 - x)^3$,
(i) $y = 2(1 - x)^3 - 1$,
(j) $y = -3(2 - x)^3 + 4$.

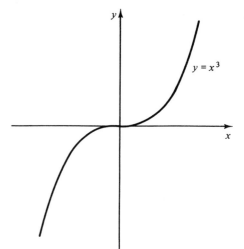

Fig. 5.14

4. Sketch the graphs of

(a) $y = \sin(x + \pi/2)$ (b) $y = \tan(x + \pi/4) + 1$ (c) $y = 2\cos(x + \pi/4)$

5. State which of the functions in Fig. 5.15 are odd, even or periodic.

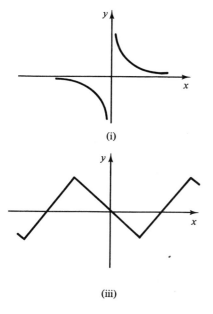

(i)

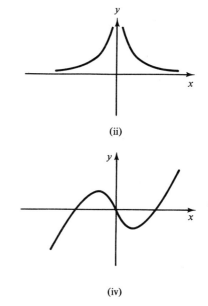

(ii)

(iii)

(iv)

Fig. 5.15

6. Sketch the graphs of

(a) $y = (x - 1)^2 + 3$,   (b) $y = 2(x - 1)^2 + 6$,   (c) $y = 2(x - 1)^2 + 3$.

7. The graph of $y = \dfrac{1}{x}$ is given in Fig. 5.16. Sketch the graphs of

(a) $y = \dfrac{1}{x - 1}$,   (b) $y = \dfrac{2}{x}$,   (c) $y = \dfrac{2}{x + 2}$,   (d) $y = \dfrac{1}{x + 1} + 1$.

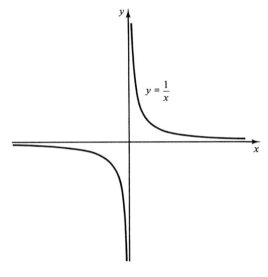

Fig. 5.16

5.5   The general quadratic

We are now in a position to sketch the graph of a quadratic function with relative ease.

Example 5.2
Sketch the graph of $y = 2x^2 - 4x + 5$.

We first complete the square

$$2x^2 - 4x + 5 \equiv 2(x + A)^2 + B$$
$$\equiv 2x + 4Ax + 2A + B$$
$$4A = -4 \qquad \Rightarrow A = -1$$
$$2A^2 + B = 5$$
$$\Rightarrow \quad 2 + B = 5 \qquad \Rightarrow B = 3.$$

So

$$y = 2(x - 1)^2 + 3$$

$$\Rightarrow y - 3 = 2(x - 1)^2.$$

Setting

$X = x - 1 \quad \Rightarrow X = 0$ when $x = 1$

$Y = y - 3 \quad \Rightarrow Y = 0$ when $y = 3$.

We need to sketch the graph of $Y = 2X^2$, with the X, Y origin located at $(1, 3)$. This is shown in Fig. 5.17.

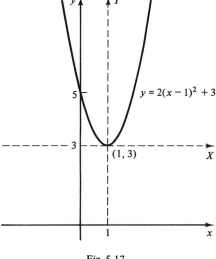

$y = 2(x - 1)^2 + 3$

Fig. 5.17

Statement

It is characteristic of the quadratic function

$$f(x) \equiv ax^2 + bx + c \qquad x \in \mathbb{R}$$

that its graph is of the form shown in Fig. 5.18.

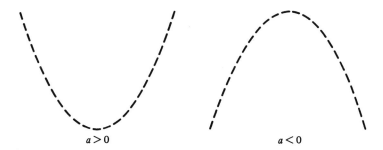

$a > 0$ $a < 0$

Fig. 5.18

These shapes are called parabolas. There is always an axis of symmetry. The function will possess a minimum value when $a > 0$ and a maximum value when $a < 0$. Notice that the line of symmetry passes through the point where this maximum or minimum value occurs.

Example 5.3

Find the maximum value of the function f, and the equation of the line of symmetry of its graph, where

$$f(x) \equiv -x^2 + 3x + 5 \qquad x \in \mathbb{R}$$

and hence find the range of f.

Completing the square we find

$$-x^2 + 3x + 5 \equiv -(x - \tfrac{3}{2})^2 + \tfrac{29}{4}$$

Now

$$(x - \tfrac{3}{2})^2 \geqslant 0 \qquad \text{(since the l.h.s. is a square).}$$

Therefore

$$\Rightarrow -(x - \tfrac{3}{2})^2 \leqslant 0$$

$$\Rightarrow -(x - \tfrac{3}{2})^2 + \tfrac{29}{4} \leqslant \tfrac{29}{4}$$

that is

$$f(x) \leqslant \tfrac{29}{4} \qquad \text{for all } x.$$

It follows that the *maximum* value of the function is $\tfrac{29}{4}$ and this occurs when $x = \tfrac{3}{2}$. Thus the equation of the line of symmetry is $x = \tfrac{3}{2}$.

It follows immediately that the range of f is the set of reals which are less than or equal to $\tfrac{29}{4}$, written as

$$\{x : x \leqslant \tfrac{29}{4}\} \quad \text{or} \quad (-\infty, \tfrac{29}{4}].$$

If a sketch were required we could build up the graph as before. Alternatively, since we now know that the graph is a parabola it is sufficient to plot two or three points.

We have established that the graph of

$$y = -x^2 + 3x + 5$$

passes through the point $A(\tfrac{3}{2}, \tfrac{29}{4})$. The point where the graph crosses the y axis may be found by setting $x = 0$ and noting that $y = 5$. So $B(0, 5)$ lies on the graph. The points where the graph cuts the x axis may be found by setting $y = 0$ and solving

$$0 = -x^2 + 3x + 5$$

as shown in Fig. 5.19.

A particularly useful method of sketching a parabola is available when the quadratic factorises easily.

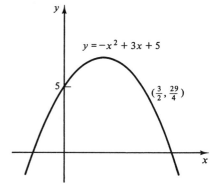

Fig. 5.19

Example 5.4

Sketch the graph of $y = x^2 - 5x + 6$.

Up to three points on the curve may be established by setting $x = 0$ and then $y = 0$.

$$x = 0 \quad \Rightarrow y = 6$$
$$y = 0 \quad \Rightarrow 0 = x^2 - 5x + 6$$
$$\Rightarrow 0 = (x - 2)(x - 3)$$
$$\Rightarrow x = 2 \quad \text{or} \quad x = 3$$

Thus the points $A(0, 6)$, $B(2, 0)$ and $C(3, 0)$ are the curve.

The parabola possesses a minimum which will occur when x is midway between 2 and 3, that is when $x = 2\frac{1}{2}$, by symmetry. Substituting, we have

$$y = (\tfrac{5}{2} - 2)(\tfrac{5}{2} - 3) = -\tfrac{1}{4}.$$

Using all this information we obtain a sketch as indicated in Fig. 5.20.

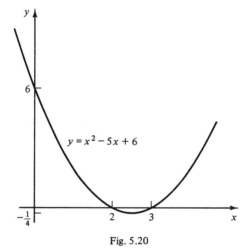

Fig. 5.20

Exercise 5.3

1. By completing the square, sketch the following graphs on separate axes.

   (a) $y = x^2 - 2x - 3$ (b) $y = x^2 + 4x - 1$ (c) $y = x^2 + 3x - 1$
   (d) $y = 2x^2 + 4x + 1$ (e) $y = 2x^2 + 3x - 1$ (f) $y = -x^2 - 6x + 4$
   (g) $y = -2x^2 + 3x - 1$

2. Find the maximum or minimum values of the following quadratics.

   (a) $y = x^2 + 4x + 3$ (b) $f(x) \equiv x(x + 1) + 4$, $x \in \mathbb{R}$
   (c) $y = 6 - 4x - 2x^2$ (d) $f(x) \equiv x(2 - x) + 3$, $x \in \mathbb{R}$

3. Sketch the graphs of the equations and functions in question 2.

4. Sketch the graphs of the following quadratics indicating the points where the graph crosses the axes.

   (a) $y = (x - 1)(x + 2)$ (b) $y = (x + 3)(x + 2)$ (c) $y = 2(x + 1)(x + 3)$
   (d) $y = x^2 + 3x + 2$ (e) $y = x^2 + 6x - 7$

5. Assuming that it is recognised that the graph of $y = ax + b$ represents a straight line, a quick way to draw the graph is to establish where it crosses the axes. Use this observation to draw the graphs of

   (a) $y = 2x + 1$ (b) $y = 3 - x$ (c) $x + y = 1$

   (d) $x + y = 4$ (e) $2x + y = 4$ (f) $\dfrac{x}{a} + \dfrac{y}{b} = 1$

5.6 Inequalities

Graphical techniques are useful when solving inequalities which involve a quadratic.

Example 5.5

Solve the inequalities

(a) $(x - 1)(x + 2) \leqslant 0$ (b) $x^2 > 9$

(a) We first sketch the graph of $y = (x - 1)(x + 2)$. There is no need to find the minimum value. Three points which may be found immediately are $A(0, 2)$, $B(1, 0)$ and $C(-2, 0)$: by setting $x = 0$ and then $y = 0$ in the equation $y = (x - 1)(x + 2)$. The parabola may now be sketched to pass through these points as shown in Fig. 5.21.

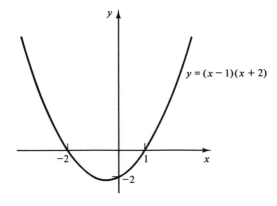

Fig. 5.21

In solving $(x - 1)(x + 2) \leqslant 0$ we are essentially trying to find those values of x for which y is less than or equal to zero; where y is the value of the function defined by

$$f(x) \equiv (x - 1)(x + 2) \qquad x \in \mathbb{R}.$$

Inspection of the graph shows that this occurs when

$$-2 \leqslant x \leqslant 1.$$

We could use the same graph to show that

$$(x - 1)(x + 2) \geqslant 0$$
$$\Rightarrow x \leqslant -2 \quad \text{or} \quad x \geqslant 1.$$

On the other hand

$$(x - 1)(x + 2) > 0$$
$$\Rightarrow x < -2 \quad \text{or} \quad x > 1.$$

(b) To solve $x^2 > 9$ we could sketch the graph of $y = x^2 - 9$ and proceed as above. However, it is fairly obvious that

$$x^2 > 9$$
$$\Rightarrow x > 3 \quad \text{or} \quad x < -3$$

It was noted in Chapter 3 that the quadratic equation

$$ax^2 + bx + c = 0$$

may have two, one double or no roots. Geometrically this corresponds to the graph of the equation $y = ax^2 + bx + c$ crossing the x axis twice, touching the x axis or failing to cross it at all (illustrated in Fig. 5.22).

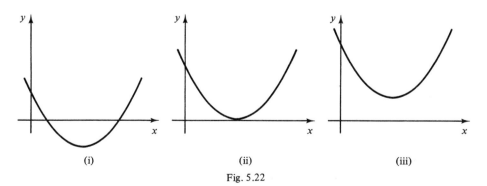

(i)   (ii)   (iii)

Fig. 5.22

Example 5.6

Find the range of values of a for which the quadratic equation

$$x^2 + (a + 3)x + a + 6 = 0$$

has two distinct roots.

For the quadratic to possess two distinct roots the discriminant must be positive. That is

$$(a + 3)^2 - 4(a + 6) > 0$$
$$\Rightarrow a^2 + 6x + 9 - 4a - 24 > 0$$
$$\Rightarrow a^2 + 2a - 15 > 0$$
$$\Rightarrow (a + 5)(a - 3) > 0.$$

Now treating a as though it were a variable we sketch the graph of

$$y = (a + 5)(a - 3).$$

By inspection

$$(a + 5)(a - 3) > 0 \qquad \text{when } a < 5 \text{ or } a > 3.$$

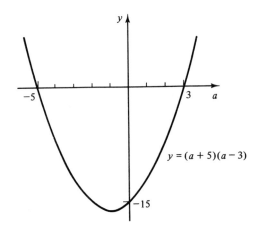

$$y = (a + 5)(a - 3)$$

Fig. 5.23

Notice that when $a = -5$ or $a = 3$ the quadratic equation $x^2 + (a + 3)x + a + 6 = 0$ has repeated roots. When $-5 < a < 3$ there are no roots.

Example 5.7

Solve the inequalities

(a) $|x + 1| \leqslant 4$ (b) $|x + 2| > |x - 3|$

Two points are worth noting.

(i) If $a < b$ and both a and b are positive then $a^2 < b^2$
(ii) For any function f we have

$$|f(x)|^2 = f(x)^2 \qquad x \text{ is the domain of } f.$$

Using these properties

(a) $|x + 1| \leqslant 4$

$\Rightarrow (x + 1)^2 \leqslant 16$

$\Rightarrow x^2 + 2x - 15 \leqslant 0$

$\Rightarrow (x + 5)(x - 3) \leqslant 0$

A sketch of $y = (x + 5)(x - 3)$
is given in Fig. 5.24.

$\Rightarrow -5 \leqslant x \leqslant 3$

(b) $|x + 2| > |x - 3|$

$\Rightarrow (x + 2)^2 > (x - 3)^2$

$\Rightarrow x^2 + 4x + 4 > x^2 - 6x + 9$

$\Rightarrow 10x > 5$

$\Rightarrow x > \frac{1}{2}$

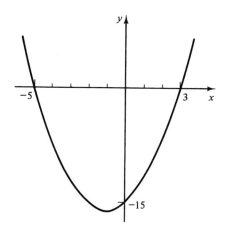

Fig. 5.24

5.7 The graph of $y = |f(x)|$

The graph of $y = |f(x)|$ may be obtained from that of $y = f(x)$ in two stages:

(i) for those values of x for which $f(x) \geqslant 0$ the two graphs coincide,
(ii) for those values of x for which $f(x) < 0$ we reflect the graph of $y = f(x)$ in the
     x axis to obtain the graph of $y = |f(x)|$

The effect of these stages as applied to $y = x^2 - 1$ may be seen in Fig. 5.25.

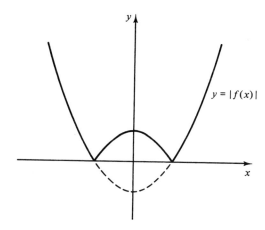

Fig. 5.25

Consider once again part (b) in Example 5.13. In Fig. 5.26 the graph of $y = |x + 2|$ and $y = |x - 3|$ have been drawn. By inspection of the graph

$$|x + 2| > |x - 3|$$
$$\Rightarrow x > \tfrac{1}{2}.$$

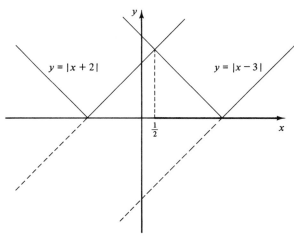

Fig. 5.26

Exercise 5.4

1. Solve the following inequalities using graphical techniques:

    (a) $(x + 1)(x + 2) \leqslant 0$ (b) $(x - 3)(x + 4) > 0$
    (c) $(x + 1)(x - 2) < 0$ (d) $(2x + 1)(3x - 2) \geqslant 0$

2. Solve the following inequalities:

    (a) $x^2 \geqslant 4$ (b) $x^2 + 2x + 1 < 0$
    (c) $x(x + 1) \leqslant -2(2x + 3)$ (d) $5x^2 \leqslant 3x + 2$
    (e) $(x - 1)^2 > 9x^2$

3. Find the range of values of p for which the equation

    $$x^2 + 2px + (p + 2) = 0$$

    has real roots.

4. Find the range of values of p for which the quadratic equation

    $$x^2 + (p + 1)x + p + 1 = 0$$

    has no real roots.

5. Find the range of values of k so that $x^2 > k(2x + 1)$, for all $x \in \mathbb{R}$.

6. By first sketching each of the graphs of $y = f(x)$ given in Fig 5.27, complete the graphs of $y = |f(x)|$

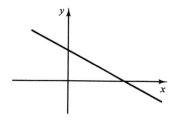

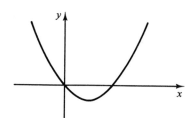

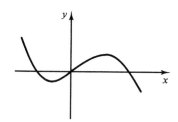

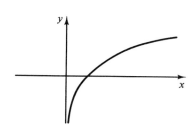

Fig. 5.27

7. Find the range of values of k for which the equation

$$x^2 + x + 1 = k(x + 2)$$

has real roots.

8. Sketch the graphs of

(a) $y = |x - 4|$
(c) $y = |(x - 1)(x + 2)|$

(b) $y = |2x + 1|$
(d) $y = |x^2 + 3x - 10|$

9. Solve the inequalities

(a) $|x - 2| > 1$,
(c) $|x - 2| > |x + 1|$,
(e) $2|x + 2| < |4 - x|$,
(g) $|3x + 1| - 4|x + 1| \geqslant 0$.

(b) $|2x + 1| \leqslant 4$,
(d) $|x + 1| > |2x + 1|$,
(f) $|2 - x| - 2|x + 1| < 0$,

5.8   Function of a function

Consider the function

$$f(x) \equiv (1 + x)^3 \qquad x \in \mathbb{R}$$

If we define

$$g(x) \equiv 1 + x \qquad x \in \mathbb{R}$$

and

$$h(x) \equiv x^3 \qquad x \in \mathbb{R}$$

then

$$f(x) \equiv (g(x))^3 \qquad x \in \mathbb{R}$$

and

$$f(x) \equiv h(1 + x) \qquad x \in \mathbb{R}$$

and combining

$$f(x) \equiv h(g(x)) \qquad x \in \mathbb{R}$$

In other words, to evaluate f at a we first find $g(a)$, that is $1 + a$ and then apply h to this value to obtain $(1 + a)^3$.

When a function is built up in this way from functions g and h then f is called a function of a function or compound function. In such a case f is often denoted by $h \circ g$ (h applied to g, so the order is important). Quite generally:

Definition

> If g and h are functions then
> $$h \circ g(x) \equiv h(g(x))$$
> and
> $$g \circ h(x) \equiv g(h(x))$$
> for x in the domains of g and h.

Sometimes the notation hg is used instead of $h \circ g$.

Usually $g \circ h$ is a distinct function. Using the above example, where

$$g(x) \equiv 1 + x \qquad x \in \mathbb{R}$$

and

$$h(x) \equiv x^3 \qquad x \in \mathbb{R}$$

we have

$$g \circ h(x) \equiv g(h(x))$$
$$\equiv g(x^3)$$
$$\equiv 1 + x^3$$

Example 5.8

If

$$f(x) \equiv \sin x \qquad x \in \mathbb{R}$$

and

$$g(x) \equiv 1 + x^2 \qquad x \in \mathbb{R}$$

simplify the following functions:

(a) $f \circ g$,   (b) $g \circ f$,   (c) $g \circ g$.

(a) $f \circ g(x) \equiv f(g(x))$

$$\equiv f(1 + x^2)$$

$$\equiv \sin(1 + x^2) \qquad x \in \mathbb{R}$$

(b) $g \circ f(x) \equiv g(f(x))$

$$\equiv g(\sin x)$$

$$\equiv 1 + \sin^2 x \qquad x \in \mathbb{R}$$

(c) $g \circ g(x) \equiv g(g(x))$

$$\equiv g(1 + x^2)$$

$$\equiv 1 + (1 + x^2)^2 \qquad x \in \mathbb{R}$$

Example 5.9

Find two functions f and g so that

$$f \circ g = h$$

where

(a) $h(x) \equiv a^{\sin x}$ (b) $h(x) \equiv \sqrt{1 + \sin^2 x}$

(a) Let $f(x) \equiv a^x$ and $g(x) \equiv \sin x$, $x \in \mathbb{R}$, then

$$h(x) \equiv f \circ g(x)$$

$$\equiv f(g(x))$$

$$\equiv f(\sin x)$$

$$\equiv a^{\sin x}$$

as required.

Of course there is always the trivial solution where

$$f(x) \equiv x \qquad \text{and} \qquad g(x) \equiv a^{\sin x}$$

and there may be other solutions.

(b) Let $f(x) \equiv \sqrt{x}$ and $g(x) \equiv 1 + \sin^2 x$, $x \in \mathbb{R}$, then

$$h(x) \equiv f \circ g(x)$$

$$\equiv f(g(x))$$

$$\Rightarrow h(x) \equiv f(1 + \sin^2 x)$$

$$\equiv \sqrt{1 + \sin^2 x}$$

An alternative would be to define

$$f(x) \equiv \sqrt{1 + x^2} \text{ and } g(x) \equiv \sin x \qquad x \in \mathbb{R}$$

5.9 Inverse functions

Suppose

$$f(x) \equiv x^3 + 1 \qquad x \in \mathbb{R}$$

and

$$g(x) \equiv (x - 1)^{1/3} \qquad x \in \mathbb{R}$$

then the compound function $g \circ f$ is defined by

$$g \circ f(x) \equiv g(f(x))$$

$$\equiv g(x^3 + 1)$$

$$\equiv ((x^3 + 1) - 1)^{1/3}$$

$$\equiv x$$

and we are back where we started.
Similarly

$$f \circ g(x) \equiv x$$

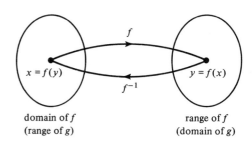

The function g is called the inverse of f, written f^{-1}. This can
be depicted as in Fig. 5.28. In
effect the inverse of f is obtained
by reversing the arrows in the
diagram representing f.

Fig. 5.28

Definition

f $^{-1}$ is the inverse of f, if the domain of f^{-1} is the range of f and

$$f(x) = y \Leftrightarrow f^{-1}(y) = x$$

so that

$$f \circ f^{-1}(y) \equiv y \qquad y \text{ in range of } f$$

and

$$f^{-1} \circ f(x) \equiv x \qquad x \text{ in domain of } f$$

Example 5.10

Find the inverse of the function defined by

(a) $f(x) \equiv \dfrac{1}{1-x}$ $x \neq 1$

(b) $g(x) \equiv a^x$ $x \in \mathbb{R}$

(a) By definition

$$f^{-1}(x) = y \Leftrightarrow f(y) = x$$

$$\Leftrightarrow \frac{1}{1-y} = x \qquad \text{substituting for } f(y)$$

$$\Leftrightarrow y = 1 - \frac{1}{x} \qquad \text{rewriting}$$

It follows that

$$f^{-1}(x) \equiv 1 - \frac{1}{x} \qquad \text{for } x \neq 0$$

$$\text{(0 is not in the range of } f\text{)}$$

(b) $g^{-1}(x) = y \Leftrightarrow g(y) = x$

$$\Leftrightarrow a^y = x$$

$$\Leftrightarrow y = \log_a x \qquad \text{taking logs of both sides}$$

It follows that

$$g^{-1}(x) \equiv \log_a x \qquad x > 0$$

$$\text{(negative numbers are not in the range of } g\text{)}$$

Statement

> If $f(a) = b$ and the inverse f^{-1} exists then
>
> $$a = f^{-1}(b).$$
>
> This is often referred to as 'taking f^{-1} of both sides.' For example, taking 'logs of both sides' or 'square rooting both sides'.

Definition

> A function f is called a one-to-one (1–1) function if no two members of the domain of f are associated with the same member of the range. Otherwise *a* function is said to be many–one.

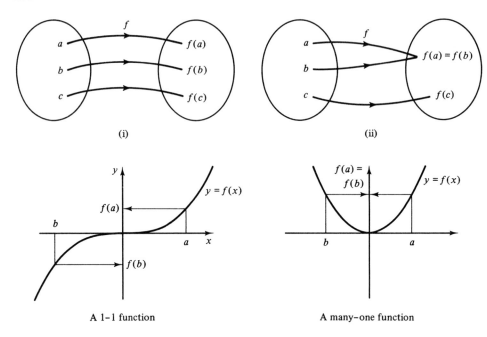

Fig. 5.29

The diagrams in Fig. 5.29 will help fix the distinction. In a 1–1 function (Fig. 5.29 (i)), different members of the domain are associated with different members of the range. In a many–one function there exists at least one pair of numbers a and b such that $f(a) = f(b)$.

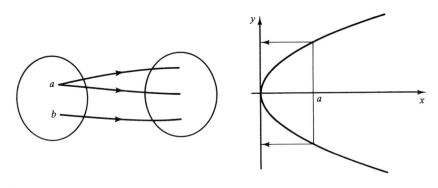

Fig. 5.30

A relationship such as shown in Fig. 5.30 is not a function because a number a has been associated with two (or more) numbers.

Consider the functions f and g defined by

$$f(x) \equiv x^2 \qquad x \in \mathbb{R}$$

and

$$g(x) \equiv x^2 \qquad x \geqslant 0$$

as shown in Fig. 5.31. These are different functions because they have different domains.

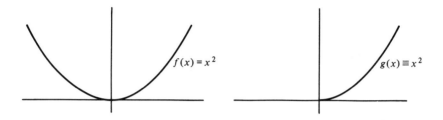

Fig. 5.31

The distinction is important since g is a 1–1 function and f is not. It follows that g will possess an inverse and f will not.

Statement

> It is easy to see that a 1–1 function will possess an inverse which is also 1–1 (simply reverse the arrows on the diagram). A many–one function does not possess an inverse. (Reversing the arrows does not even produce a function!) It is often possible, however, to restrict the domain and so produce a 1–1 function with an inverse.

Example 5.11

Given that

$$f(x) \equiv \frac{1+x}{1-x} \qquad x \neq 1$$

find

(a) $f \circ f(x)$ (written $f^2(x)$)   (b) $f^{-1}(x)$ and obtain the range of f.

(a) $f \circ f(x) \equiv f(f(x))$

$$\equiv f\left(\frac{1+x}{1-x}\right) \qquad x \neq 1$$

$$\equiv \frac{1 + \dfrac{1+x}{1-x}}{1 - \dfrac{1+x}{1-x}} \qquad x \neq 0$$

$$\Rightarrow f \circ f(x) \equiv \frac{(1-x)+(1+x)}{(1-x)-(1+x)}$$

$$\equiv \frac{2}{-2x}$$

$$\equiv -\frac{1}{x} \qquad\qquad x \neq 0, 1$$

(b) $f^{-1}(x) = y \Leftrightarrow f(y) = x$

$$\Leftrightarrow \frac{1+y}{1-y} = x \qquad\qquad y \neq 1$$

$$\Leftrightarrow 1+y = x(1-y)$$

$$\Leftrightarrow xy + y = x - 1$$

$$\Leftrightarrow y = \frac{x-1}{x+1} \qquad\qquad x \neq -1$$

so

$$f^{-1}(x) = \frac{x-1}{x+1} \qquad x \neq -1$$

Since the domain of f^{-1} is the same as the range of f, it follows that the range of f is

$$\{x : x \in \mathbb{R} \text{ and } x \neq -1\}$$

Example 5.12

Show that the inverse of f where

$$f(x) \equiv \frac{1}{x} \qquad x \neq 0$$

is in fact f!

We have

$$f^{-1}(x) = y \Leftrightarrow f(y) = x$$

$$\Leftrightarrow \quad \frac{1}{y} = x$$

$$\Leftrightarrow \quad y = \frac{1}{x}$$

and so

$$f^{-1}(x) \equiv \frac{1}{x} \qquad \text{for } x \neq 0$$

Example 5.13

If f and g are 1–1 functions find the inverse of $f \circ g$.

Since the functions are 1–1, the inverse functions f^{-1} and g^{-1} will exist. Now

$$(f \circ g)^{-1}(x) = y \Leftrightarrow f \circ g(y) = x$$

$$\Leftrightarrow f(g(y)) = x$$

$$\Leftrightarrow \quad g(y) = f^{-1}(x) \qquad \text{'taking } f^{-1} \text{ of both sides'}$$

$$\Leftrightarrow \quad y = g^{-1}(f^{-1}(x)) \qquad \text{'taking } g^{-1} \text{ of both sides'}$$

$$\Leftrightarrow \quad y = g^{-1} \circ f^{-1}(x).$$

It follows that

$$(f \circ g)^{-1}(x) \equiv g^{-1} \circ f^{-1}(x)$$

so the inverse of $f \circ g$ is $g^{-1} \circ f^{-1}$.

Exercise 5.5

1. If

$$f(x) \equiv \cos x \qquad x \in \mathbb{R}$$

and

$$g(x) \equiv 1 + x \qquad x \in \mathbb{R}$$

find in terms of x

(a) $f \circ g$ (b) $f \circ f$ (c) $g \circ f$

2. If

$$f(x) \equiv \sqrt{x} \qquad x \geqslant 0$$

and

$$g(x) \equiv 1 - x^2 \qquad x \in \mathbb{R}$$

find in terms of x

(a) $f \circ g$ (b) $g \circ f$ (c) $g \circ g$

3. (a) If

$$h(x) \equiv \sqrt{(1 + x^2)} \qquad x \in \mathbb{R}$$

find two functions f and g such that

$$f \circ g(x) \equiv h(x).$$

(b) Similarly, find two functions which combine to give

(i) $h(x) \equiv \cos(1 + x)$,   (ii) $h(x) \equiv (4 + x)^3$,
(iii) $h(x) \equiv 5 + x^3$,   (iv) $h(x) \equiv 1 + \tan(x^2)$.

4. Write down the inverse function of each of the following, stating clearly the domain of each function.

(a) $d(x) \equiv x + 1$ $x \in \mathbb{R}$

(b) $e(x) \equiv 2x$ $x \in \mathbb{R}$

(c) $f(x) \equiv x^2 + 1$ $x \geqslant 0$

(d) $g(x) \equiv (x - 1)^2$ $x \geqslant 1$

(e) $h(x) \equiv 2x^2$ $x \geqslant 0$

(f) $i(x) \equiv \lg(x)$ $x > 0$

5. State whether the functions of x represented by the graphs in Fig. 5.32 are 1–1 or not.

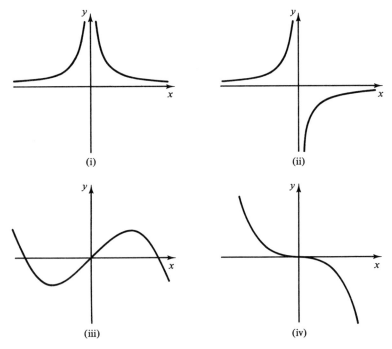

(i)   (ii)

(iii)   (iv)

Fig. 5.32

6. If

$$f(x) \equiv x^2 + 1 \qquad x \in \mathbb{R}$$

find (a) $f \circ f(x)$, (b) f^{-1}, and state its domain

7. Given

$$f(x) = \frac{x^2 + 1}{2x^2 + 1} \qquad x \geqslant 2$$

find an expression for $f^{-1}(x)$ and veryify that $(f^{-1} \circ f)(x) = x$.

W 1982

8. The functions f and g are defined over the real numbers, with $-1, 0, 1$ excluded, by

$$f: x \to \frac{1 + x}{1 - x}, \qquad g: x \mapsto \frac{1}{x}.$$

(a) Given that f^2 denotes $f \circ f$, f^3 denotes $f \circ f^2$ or $f^2 \circ f$, etc., express f^4 and f^7 in the same manner as f.

(b) Show that the function $f \circ g$ is its own inverse.

AEB 1983

9. (a) The function h is defined by $h: x \to x^2 - x \, (x \in \mathbb{R})$. Give a reason to show that h is not one–one. If the domain of h is restricted to the subset of $\mathbb{R}$ for which $x \geqslant A$, find the least value of A for which h is one–one.

(b) Functions f and g are defined as follows:

$$f: x \to e^{-x} \qquad x \in \mathbb{R}_+,$$

$$g: x \to \frac{1}{1 - x} \qquad x \in \mathbb{R}, \quad x < 1.$$

Give the ranges of f, g and $g \circ f$.

Give definitions of the inverse functions f^{-1}, g^{-1} and $(g \circ f)^{-1}$ in a form similar to the above definitions.

CAMB 1983

10. Functions f and g, each with domain $\mathbb{R}$, are defined as follows:

$$f: x \mapsto 3x + 2, \qquad g: x \mapsto x^2 + 1.$$

For each of f and g, state the range of the function and give a reason to show whether or not the function is one–one.

Give explicit definitions, in the above form, of each of the composite functions $f \circ g$ and $g \circ f$, and find the values of x for which $(f \circ g)(x) = (g \circ f)(x)$.

State the domain of the inverse relation $(f \circ g)^{-1}$ and give an explicit definition of this relation. Explain briefly why $(f \circ g)^{-1}$ is not a function.

CAMB 1983

5.10   Graphs of inverse functions

If f is a 1–1 function then there is an inverse function f^{-1} which has the property, represented in Fig. 5.33, that

$$f^{-1}(a) = b \Leftrightarrow f(b) = a$$

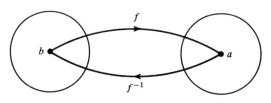

Fig. 5.33

So the point $P(a, b)$ is on the graph of f^{-1} if and only if the point $Q(b, a)$ is on the graph of f.

Now the point $P(a, b)$ may be obtained from $Q(b, a)$ by reflecting $Q(b, a)$ in the line $y = x$.

A few examples may help (see Fig. 5.34):

$$Q(2, 4) \mapsto P(4, 2)$$

$$Q(2, 2) \mapsto P(2, 2)$$

$$Q(-4, 3) \mapsto P(3, -4)$$

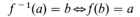

Fig. 5.34

It follows that

Statement

> The graph of f^{-1} may be obtained by reflecting the graph of f in the line $y = x$.

Example 5.14

Sketch the graph of $y = \log_a x$

The inverse of the function defined by

$$f(x) \equiv \log_a x \qquad x \in \mathbb{R}^+$$

is

$$f^{-1}(x) \equiv a^x \qquad x \in \mathbb{R}$$

since $y = \log_a x \Leftrightarrow x = a^y$ by definition.

Now the graph of $y = a^x$ (given on page 14) is shown in Fig. 5.35, where again we suppose $a > 1$.

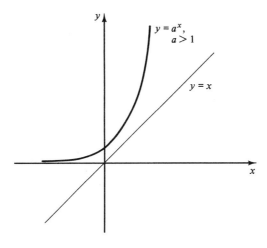

Fig. 5.35

Notice that the graph of $y = x$ and $y = a^x$ fail to intersect since $a^x > x$ for $x \in \mathbb{R}$. So the graph of $y = \log_a x$ may be obtained by reflecting in the line $y = x$ to obtain the graph shown in Fig. 5.36.

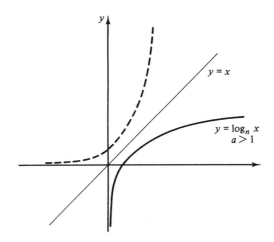

Fig. 5.36

5.11   The graph of $y = 1/f(x)$

Statement

> The graph of $y = 1/f(x)$ can be obtained from that of $y = f(x)$ by noting the following points.
>
> (i) $1/f(x)$ will be positive when $f(x)$ is positive and negative when $f(x)$ is negative.
>
> (ii) If $f(a) = \pm 1$ then $1/f(a) = \pm 1$ (the notation $f(a) = \pm 1$ is short for $f(a) = 1$ or $f(a) = -1$)
>
> (iii) If $f(a) = 0$ then $1/f(a)$ is undefined.
>
> (iv) If $f(a)$ is very large then $1/f(a)$ is very small and vice versa.
>
> (v)   If (a, b) lies in the graph of $f(x)$ then $(a, 1/b)$ lies on the graph of $1/f(x)$.

Example 5.15

Sketch the graph of $y = 1/x$.

Using the points noted above, the graph of $y = 1/x$ is as shown in Fig. 5.37.

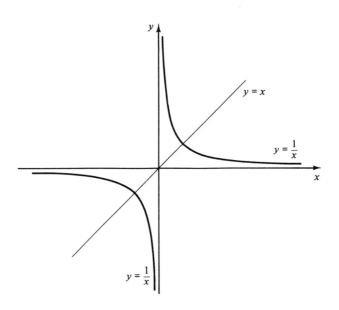

Fig. 5.37

Recall that we have already shown that the graph of $y = 1/x$ is symmetrical about the line $y = x$. For, if f is defined by

$$f(x) = 1/x \qquad x \neq 0,$$

then $f^{-1} = f$ (cf page 138).

Statement

> If the graph of the equation approaches a straight line as x or y becomes very large (positive or negative) then the straight line is called an asymptote to the curve (graph).

Both of the lines representing $x = 0$ and $y = 0$ are asymptotes to the graph of $y = 1/x$

Example 5.16

Sketch the graph of

(a) $y = \dfrac{2}{x - 4}$ (b) $y = \dfrac{1}{(x + 1)(x - 2)}$

(a) The graph of $y = 2/(x - 4)$ can be obtained directly from that of

$$y = \tfrac{1}{2}(x - 4)$$
$$\Leftrightarrow y = \tfrac{1}{2}x - 2$$

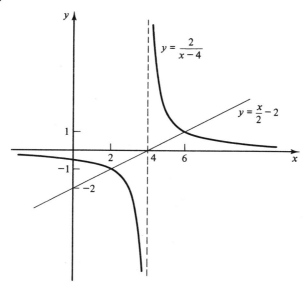

Fig.5.38

The curve crosses the y axis when $x = 0$. Substituting, we have $y = -\tfrac{1}{2}$. This time $x = 4$ and $y = 0$ are asymptotes to the graph of $y = 2/(x - 4)$.

(b) The graph of $y = 1/(x + 1)(x - 2)$ is readily obtained (see Fig. 5.39).

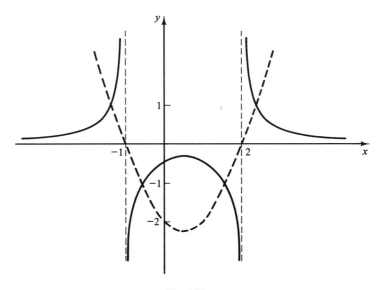

Fig. 5.39

In this case $x = -1$, $x = 2$ and $y = 0$ are asymptotes to the graph of

$$y = \frac{1}{(x + 1)(x - 2)}$$

Exercise 5.6

1. Find the inverse of the functions determined by:

    (a) $y = x + 1$ (b) $y = x^3 - 3$ (c) $y = 2^x$ (d) $y = \lg x$

    In each case sketch the graph of the function and its inverse.

2. Sketch the graphs of each of the following functions together with their inverses.

    (a) $f(x) \equiv x + 3$ $x \in \mathbb{R}$

    (b) $f(x) \equiv 2x - 1$ $x \in \mathbb{R}$

    (c) $f(x) \equiv (x + 1)(x - 1)$ $x > 0$

    (d) $f(x) \equiv x^2 + 3x - 4$ $x > \frac{3}{2}$

3. Sketch the graphs of

    (a) $y = \dfrac{1}{x + 1}$ (b) $y = \dfrac{1}{(x + 1)(x + 2)}$ (c) $y = \dfrac{8}{(x^2 - 4)}$

(d) $y = \dfrac{4}{x^2 + 2x + 1}$ (e) $y = \dfrac{-1}{x^2 + 3x + 2}$

4.   If

$$f(x) \equiv \frac{x - 1}{x + 2} \qquad x \neq -2$$

find $f^{-1}(x)$ and state the domain of the function f^{-1}.

5.12   The graphs of the trigonometric functions

Suppose that a function f has the following properties

  (i) The domain of f is $\mathbb{R}$

 (ii) $f(\pi - x) \equiv f(x) \qquad x \in \mathbb{R}$

(iii) f is odd, that is $f(-x) \equiv -f(x)$

(iv) f is periodic with period 2π, that is

$$f(2\pi + x) \equiv f(x) \qquad x \in \mathbb{R}$$

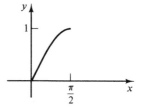

Fig. 5.40

Suppose, further, that part of the graph of $y = f(x)$ is as indicated in Fig. 5.40. We can deduce the shape of the graph of f using the properties given above.

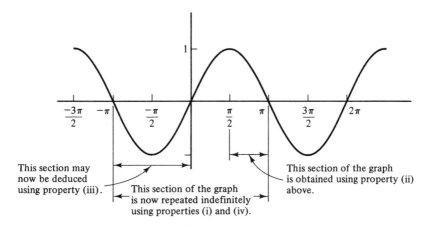

This section may now be deduced using property (iii).

This section of the graph is now repeated indefinitely using properties (i) and (iv).

This section of the graph is obtained using property (ii) above.

Fig. 5.41

In fact, assuming the graph in Fig. 5.40 is that of $y = \sin x$ for $0 \leqslant x \leqslant \pi/2$, then we can show that the graph in Fig. 5.41 is the extension of this—that is the graph of $y = \sin x$. We have already shown that the function $f(x) \equiv \sin x$ for $x \in \mathbb{R}$ (x measured in radians) possesses the properties listed above.

Using similar properties we can deduce the shapes of the graphs of $y = \cos x$ and $y = \tan x$, where x is measured in radians. These graphs are given in Fig. 5.42.

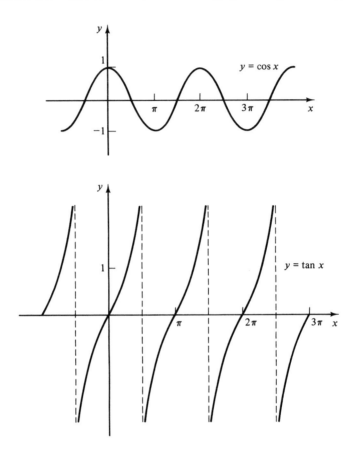

Fig. 5.42

Notice that the function $f(x) \equiv \tan(x)$ has, as its domain, all real numbers except those for which $\cos x = 0$, that is, for $x \neq n\pi + \pi/2$ (where n is an integer).

The graphs of

$$y = \operatorname{cosec} x$$

$$y = \sec x$$

$$y = \cot x$$

may now be obtained immediately by noting that

$$\operatorname{cosec} x \equiv \frac{1}{\sin x} \qquad \text{for all } x \text{ such that } \sin x \neq 0$$

$$\sec x \equiv \frac{1}{\cos x} \qquad \text{for all } x \text{ such that } \cos x \neq 0$$

$$\cot x \equiv \frac{1}{\tan x} \qquad \text{for all } x \text{ such that } \tan x \neq 0$$

For example, the graph of $y = \operatorname{cosec} x$ is given in Fig. 5.43.

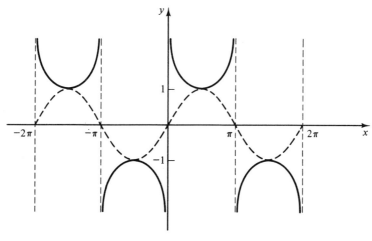

Fig. 5.43

The graph of $y = a \sin(bx + c)$

It is a simple matter to draw the graph of $y = a \sin x$. For example, the graph of $y = \frac{1}{2} \sin x$ is given in Fig. 5.44. Notice that if $y = a \sin x$ then $-a \leqslant y \leqslant a$.

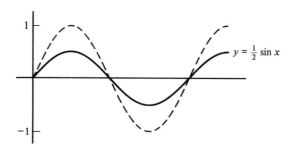

Fig. 5.44

Such graphs are called sine waves and the maximum value is called the *amplitude*.

The graph of $y = \sin(x - c)$ is obtained by translating the graph of $y = \sin x$, c units to the right if c is positive — to the left if c is negative.

Since $\cos x \equiv \sin(x + \pi/2)$ the graph of $y = \cos x$ is obtained by translating the graph of $y = \sin x$, $\pi/2$ units to the left.

The function $f(x) \equiv \sin x$, $x \in \mathbb{R}$, x measured in radians, is periodic with period 2π. However, $g(x) \equiv \sin 2x$, $x \in \mathbb{R}$, is periodic with period π. To see that this is so consider the following table.

x	0	$\pi/4$	$\pi/2$	$3\pi/4$	π	$5\pi/4$	$3\pi/2$	$7\pi/4$	2π
$\sin 2x$	0	1	0	-1	0	1	0	-1	0

The graph of $y = \sin x$ completes one cycle in the range 0 to 2π whilst the graph of $y = \sin 2x$ (Fig. 5.45) completes two such cycles. We say that its frequency is twice that of $y = \sin x$. The graph of $y = \sin 4x$ can be shown to complete 4 cycles in the range 0 to 2π, its frequency is four times that of $y = \sin x$, and its period is $(2\pi/4) = \pi/2$.

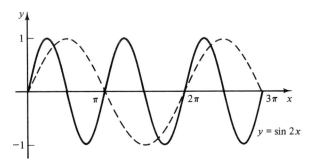

Fig. 5.45

Quite generally, it can be shown that the graph of $y = \sin(ax)$ has frequency a times that of $y = \sin x$ and period $2\pi/a$. Similar observations apply to $y = \cos(ax)$ and $y = \tan(ax)$.

Example 5.17

Sketch on the same axes the graph of

$$y = \cos x$$
$$y = \cos\frac{x}{2} \quad \text{in the range } 0 \leqslant x \leqslant 2\pi.$$

The frequency of the graph of $y = \cos(x/2)$ is one half that of $y = \cos x$. That is sufficient information to sketch the graph (Fig. 5.46).

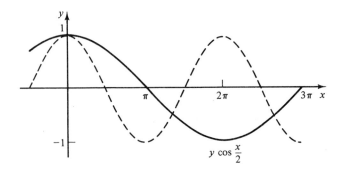

Fig. 5.46

We can now put all these observations together to sketch the graph of

$$y = a \sin (bx + c).$$

Example 5.18

Sketch the graph of $y = 2 \sin \left(\frac{3}{2}x + \frac{1}{4}\pi\right)$ (Fig. 5.47).

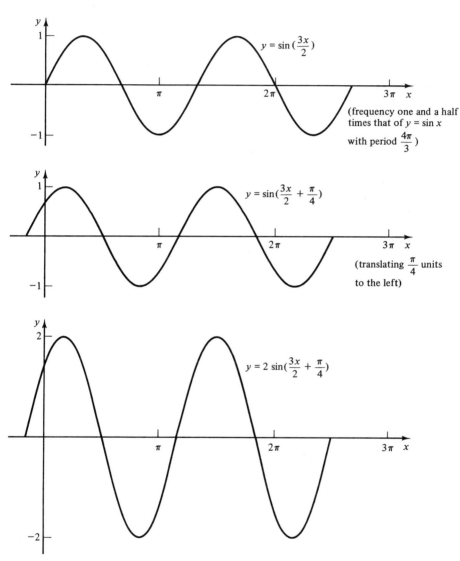

Fig. 5.47

Exercise 5.7

Sketch the graphs of each of the following in the range $0 \leqslant x < 2\pi$

(a) $y = 2\cos x$ (b) $y = \cos\left(x + \dfrac{\pi}{4}\right)$ (c) $y = \cos 2x$

(d) $y = 2\cos\left(2x + \dfrac{\pi}{4}\right)$ (e) $y = 3\sin x$ (f) $y = \sin 3x$

(g) $y = 3\sin 3x$ (h) $y = 3\sin\left(3x + \dfrac{\pi}{3}\right)$ (i) $y = \cos\dfrac{x}{2}$

(j) $y = 2\sin 2x$ (k) $y = 2\cos\dfrac{x}{3}$ (l) $y = \tan 2x$

5.13 Co-ordinate geometry

During the course of this chapter we have increasingly concentrated on the graphs of functions and relations. We continue now by considering some of the geometrical properties of the graphs themselves. It is assumed that a common scale of measurement is used on both axes.

The distance between two points

Example 5.19

Find the distance between the points

(a) $A(7, 4)$, $B(3, 1)$ (b) $B(3, 1)$, $C(-3, 6)$ (c) $D(x_1, y_1)$, $E(x_2, y_2)$

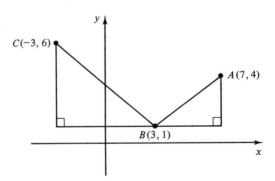

Fig. 5.48

(a) It is evident from plotting the points A and B on Cartesian co-ordinates (Fig. 5.48) that we seek the length of the hypotenuse of a right angled triangle with sides 4 (the difference of the *abscissae* $(7 - 3)$) and 3 (the difference of the *ordinates* $(4 - 1)$).

So, by Pythagoras' theorem, the required

$$\text{distance} = \sqrt{4^2 + 3^2} = 5$$

(b) Again we find the length of a hypotenuse of a right angled triangle with sides 6 (notice that this is still the difference of the abscissae $3 - -3 = 6$) and 5. (This time the difference in the ordinates is $1 - 6 = -5$, but no matter since the negative sign will disappear once it is squared.)

So, by Pythagoras' Theorem, the required

$$\text{distance} = \sqrt{6^2 + 5^2}$$
$$= \sqrt{61}$$
$$\simeq 7.81 \qquad \boxed{C}$$

(c) Generalising from the above we note that the distance between the points $D(x_1, y_1)$ and $E(x_2, y_2)$ is

$$\sqrt{(x_1 - x_2)^2 + (y_1 - y_2)^2}$$

This can be expressed in words.

Statement

> The *square* of the distance between two points with given co-ordinates is the sum of the square of the difference in the abscissae and the square of the difference in the ordinates.

This result enables us to obtain the distance between two points without drawing a diagram.

Example 5.20

Find the distance

(a) from $(-3, -5)$ to the origin   (b) between (a, b) and (x, y)

(a) The origin is the point $(0, 0)$ so the required distance is

$$\sqrt{(-3 - 0)^2 + (-5 - 0)^2}$$
$$= \sqrt{9 + 25}$$
$$= \sqrt{34}$$
$$\simeq 5.83 \qquad \boxed{C}$$

(b) The required distance is

$$\sqrt{(a - x)^2 + (b - y)^2}$$
$$[= \sqrt{(x - a)^2 + (y - b)^2}]$$

The mid-point of the straight line joining two points

The point M midway between the two points $A(x_1, y_1)$ and $B(x_2, y_2)$ can readily be obtained (see Fig. 5.49).

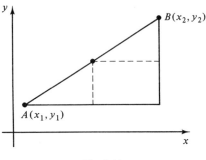

Fig. 5.49

Since the abscissa of M is clearly midway between the abscissae of two points, so too the ordinate is midway between the ordinates of the two points. [If this is not clear then the use of the properties of similar triangles will help.]

Now the number midway between 5 and 7 is 6, that is $\frac{1}{2}(5 + 7)$. Between -3 and 5 it is 1, that is $\frac{1}{2}(-3 + 5)$.

More generally, the number midway between a and c is $\frac{1}{2}(a + c)$ regardless of the signs of a and c.

Statement

The co-ordinates of the point midway between the points $A(x_1, y_1)$ and $B(x_2, y_2)$ are

$$\left(\frac{x_1 + x_2}{2}, \frac{y_1 + y_2}{2} \right)$$

The gradient of a straight line

The gradient of a straight line is a numerical measure of its slope.

Definition

The gradient of a straight line is defined to be the tangent of the positive angle between the line and the x axis (see Fig. 5.50). Notice that the line l_1, which makes an acute angle with the x axis has a positive gradient, and the line l_2 which makes an obtuse angle with the x axis has a negative gradient.

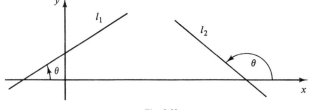

Fig. 5.50

It follows immediately that a horizontal line has zero gradient (tan 0 = 0) and the gradient of a vertical line is undefined (since the tangent of a right-angle is undefined). In such a case we say the gradient is *infinite*.

Example 5.21

Find the gradient of the line passing through the points $A(5, 4)$ and $B(-1, 1)$.

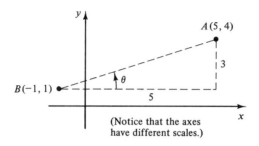

(Notice that the axes have different scales.)

Fig. 5.51

Suppose the line passing through A and B makes an angle θ with the x axis. From the diagram (Fig. 5.51) it is evident that the

$$\text{gradient} = \tan \theta = \frac{AC}{BC} = \frac{4-1}{4--1} = \frac{3}{5}.$$

It is usual to leave the gradient in fractional form.

Statement

> The gradient m of the straight line passing through the points $A(x_1, y_1)$ and $B(x_2, y_2)$ is given by
>
> $$m = \frac{y_2 - y_1}{x_2 - x_1} \quad \left[\text{or } \frac{y_1 - y_2}{x_1 - x_2}\right] \quad x_1 \neq x_2$$
>
> (If $x_1 = x_2$ then the gradient is undefined.)

[The reader may wish to consider points in various quadrants to check the above statement.]

Statement

> (i) Two lines are *parallel* if and only if they have the same gradient.
> (ii) Two lines are *perpendicular* if and only if the product of their gradients is -1, or one has zero gradient and the other an infinite gradient.

Statement (ii) requires proof. Apart from the special case where one has zero gradient it is obvious from geometrical considerations that one of the lines will cut the

x axis in an acute angle, the other an obtuse angle (Fig. 5.52). Suppose l_1 cuts the x axis in an acute angle θ_1 so to $\tan \theta_1 = m_1$. Suppose l_2 cuts the x axis in the obtuse angle θ_2 so that $\tan \theta_2 = m_2$. But then $\tan \theta_3 = -m_2$ using the definition of the tangent function. It follows immediately that

$$\tan \theta_1 = \cot \theta_3 = \frac{1}{\tan \theta_3}$$

$$\Rightarrow \quad m_1 = \frac{-1}{m_2}$$

$$\Rightarrow m_1 m_2 = -1$$

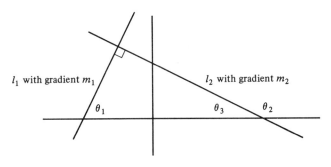

Fig. 5.52

5.14   The general equation of the straight line

A straight line is uniquely determined if it is known to pass through the point (a, b) with gradient m. [If m were infinite then clearly the required line is $x = a$.] We seek a condition on the variable point (x, y) to ensure that it lies on this straight line. Now (x, y) will correspond to a point on the line if and only if the gradient from (x, y) to (a, b) is m. That is, providing

$$m = \frac{y - b}{x - a}$$

$$\Rightarrow y - b = m(x - a)$$

The ordered pairs which satisfy this equation are just those which lie on our given line and so:

Statement

The equation of the straight line which passes through a point (a, b) with gradient m is

$$y - b = m(x - a) \qquad \text{or} \qquad x = a, \text{ if } m \text{ is infinite.}$$

It follows immediately that the line with gradient m which passes through the point $(0, c)$ has equation

$$y - c = m(x - 0)$$

$$\Rightarrow \qquad y = mx + c$$

Example 5.22

Find the equation of the straight line

(a) passing through the point $(3, 2)$ with gradient -1.
(b) passing through the points $A(2, 3)$ and $B(5, 6)$ and show that they are perpendicular to one another.

(a) The equation of the line passing through the point $(3, 2)$ with gradient -1 is

$$y - 2 = -(x - 3)$$

$$\Rightarrow y = -x + 5$$

(b) The gradient of the line passing through $A(2, 3)$ and $B(5, 6)$ is

$$\frac{6 - 3}{5 - 2} = 1$$

So its equation is

$$\begin{array}{ccc} y - 3 = (x - 2) & & \left[\begin{array}{c} y - 6 = (x - 5) \\ \Rightarrow y = x + 1 \quad !! \end{array} \right] \\ \Rightarrow y = x + 1 & or & \end{array}$$

By inspecting the first equation, the gradient is -1 and the second has gradient 1, thus the product is -1. It follows that the two lines are perpendicular.

Statement

> The graph of any equation of the form
>
> $$y = mx + c$$
>
> is a straight line with gradient m, which crosses the y axis at the point $(0, c)$. [We say that the intercept is c.]

Statement

> The general equation of the straight line is
>
> $$ay + bx + c = 0$$
>
> where a, b and c are constants.

If $a = 0$ then the equation reduces to

$$x = \frac{-c}{b}$$

which has infinite gradient. If $a \neq 0$ then the equation can be put into the form

$$y = \frac{-bx}{a} - \frac{c}{a}$$

which has gradient $\frac{-b}{a}$ and intercept $\frac{-c}{a}$.

Example 5.23

Find the equation of the straight line which is perpendicular to the line determined by $3y - 4x + 2 = 0$ which passes through the point $(3, 2)$.

Since

$$3y - 4x + 2 = 0$$

$$\Rightarrow y = \frac{4x}{3} + \frac{2}{3}$$

its gradient is $\frac{4}{3}$. Thus the gradient of the required line is $\frac{-3}{4}$.

So the line has equation

$$y - 2 = \frac{-3}{4}(x - 3)$$

$$\Rightarrow 4y - 8 = -3x + 9$$

$$\Rightarrow 4y + 3x - 17 = 0.$$

Exercise 5.8

1. Find the distance between the points

    (a) $(2, 1)$, $(6, -2)$,   (b) $(-6, -1)$, $(-2, -3)$,   (c) $(3, -1)$, $(4, 6)$,
    (d) $(2, 2)$, $(-3, -5)$,   (e) (a, b), $(0, 0)$,   (f) $(x, -y)$, $(y, -x)$.

2. Find the co-ordinates of the midpoint of the line joining the points A and B, with coordinates:

    (a) $A(-2, 4)$, $B(-3, -2)$ (b) $A(4, 9)$, $B(-3, 6)$
    (c) $A(2, 5)$, $B(6, -4)$ (d) $A(3, 2)$, $B(3, 6)$

3. Find the gradient of the line passing through the points

    (a) $(1, 1)$, $(2, 4)$,   (b) $(0, 0)$, $(-2, 6)$,   (c) $(-1, 0)$, $(4, 1)$,
    (d) $(-1, -1)$, $(2, -1)$,   (e) $(1, 0)$, $(-2, -1)$,   (f) $(-1, -2)$, $(3, -1)$.

4. Write down the gradient of a line perpendicular to the lines with gradient

   (a) 2, (b) -3, (c) a, (d) $\dfrac{1}{m}$.

5. Write down the gradient of the lines determined by

   (a) $y = 2x + 1$, (b) $2y = x + 1$, (c) $2y + 3x + 1 = 0$,
   (d) $y - x = 1$, (e) $x = 2y + 1$.

6. Three points are collinear if they lie on the same straight line. To test whether three points are collinear it is sufficient to check that the gradient between two points is the same as the gradient between two other points.
   Show that the three points given are collinear:

   (a) $(0, 1), (2, 5), (-2, -3)$, (b) $(1, 2), (0, -1), (-1, -4)$.

7. Write down the equation of the straight line in the form $y = mx + c$ which pass through the given point with given gradient.

   (a) $(0, 2); 4$ (b) $(3, 4); 1$ (c) $(a, b); m$

8. Find the equation of the line passing through the points

   (a) $(2, 1), (3, 4)$ (b) $(-2, 6), (4, -2)$ (c) $(x_1, y_1), (x_2, y_2)$

9. Find the equation of the straight line passing through $(2, 1)$ perpendicular to $2y + 3x - 2 = 0$.

10. Find the equation of the perpendicular bisector of the line joining the points $(-2, 4)$ and $(2, -6)$.

11. The points A, B and C have co-ordinates $(1, 2), (-3, -3), (6, -4)$, respectively. Find the equation of the perpendicular to AC passing through B.

12. Show that the points $(a - b, b), (2a, a + b), (2a, b)$ lie at the three corners of a square. Find the co-ordinates of the fourth corner.

13. The co-ordinates of the verticles of the triangle ABC are $A(0, -6)$, $B(5, 6)$, $C(13, -6)$. Show that the triangle is isosceles and find its area.

14. Find the centre of a circle which passes through the points

    (a) $A(0, 0), B(0, 2), C(2, 2)$, (b) $A(-2, -2), B(1, 7), C(4, 2)$.

15. Find the centre and radius of the circle which passes through the points $A(0, -1), B(2, -1), C(2, 1)$.

16. Find the equation of the median AM of the triangle $\triangle ABC$ where the co-ordinates of A, B and C are $(2, -4), (0, 0)$ and $(4, 1)$ respectively.

Chapter 6   Calculus

6.1   Introduction

The study of calculus was first undertaken by the famous 17th-century mathe-
maticians, Newton and Leibnitz. It was found to be an essential tool in solving a
whole range of mathematical problems. Not least, amongst these, is the study of
motion.

At its simplest, suppose a ball is tossed in the air and the height, h, is expressed as a
function of time, t, by $h = f(t)$. If the graph of $h = f(t)$ is drawn (measuring t
horizontally and h vertically) then the *velocity* at any time can be shown to be equal to
the gradient of the tangent to the curve, at the point on the curve corresponding to the
time chosen.

For this reason, amongst others, we seek an efficient method of finding the gradient
of a tangent to a curve at a specified point. At any point on a curve, its gradient is
defined to be the gradient of the tangent at this point.

Consider the problem of finding the gradient of (the tangent to) the curve
determined by

$$y = f(x)$$

at the point A on the curve with co-ordinate $(a, f(a))$.

Definition

> A chord is a line segment which joins two points on a curve.

In Fig. 6.1 it can be seen that
the gradient of the chord AP_1
is

$$\frac{H_1 C}{AC}$$

If we were to select a point P_2
on the curve closer to A, then
the gradient of AP_2 is
$H_2 C / AC$. This is a closer approximation to the gradient of
the tangent at A which is
BC / AC.

Indeed the approximation
improves as points P_3, P_4, ...
are selected closer and closer to
A. (Of course, the length of the
chord diminishes, but this does
not alter the argument.)

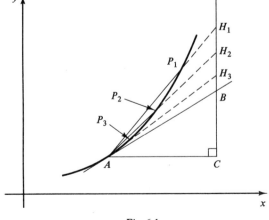

Fig. 6.1

Example 6.1

Find the gradient of the curve defined by $y = x^2$ at the point where $x = a$.

The point A on the curve whose x co-ordinate is a, has co-ordinates (a, a^2). Consider a distinct point B on the curve with co-ordinate (b, b^2).

The gradient of the chord AB is

$$\frac{b^2 - a^2}{b - a} \qquad b \neq a$$

$$= \frac{(b - a)(b + a)}{b - a}$$

$$= b + a.$$

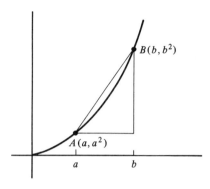

Fig. 6.2

Now as b approaches a, B moves along the curve closer to A. The gradient of the chord AB moves closer to the gradient of the curve at $A(a, a^2)$. But as b approaches a, the gradient of the chord, $b + a$, approaches $2a$. This means that the gradient of the curve is $2a$!

Using an arrow $\rightarrow$ to mean 'tends to' or 'approaches' we can summarise this argument by writing:

$$b \rightarrow a \qquad \Rightarrow b + a \rightarrow 2a$$

and since the gradient of the chord, $b + a$, $\rightarrow$ the gradient of the curve at $A(a, a^2)$, it follows that the gradient of the curve is $2a$.

An alternative to writing

$$b \rightarrow a \qquad \Rightarrow b + a \rightarrow 2a$$

is to write

$$\lim_{b \to a} (b + a) = 2a$$

which reads:

as b tends to a, the limit of the expression $b + a$ is $2a$.

For example,

$$x \rightarrow 2 \qquad \Rightarrow 3x \rightarrow 6$$

is written

$$\lim_{x \to 2} 3x = 6.$$

Notice that we do not let $x = 2$, and so $3x$ is never actually equal to 6. Rather we have in mind a process as indicated in the following table.

x	1	1.5	1.9	1.99	1.9999...
$3x$	3	4.5	5.7	5.97	5.9997...

As values of x are selected closer to 2, so $3x$ approaches 6. There is no other number with this property.

Example 6.2

Obtain

(a) $\lim\limits_{a \to \infty} \left(\dfrac{2a + 1}{a} \right)$ (b) $\lim\limits_{x \to 0} \left(\dfrac{3x + x^2}{x} \right)$ (c) $\lim\limits_{\theta \to 0} \left(\dfrac{\sin \theta}{\theta} \right)$

(a) The symbols $a \to \infty$ reads 'a tends to infinity' which means that a gets indefinitely large.

   Now

$$\lim_{a \to \infty} \left(\frac{2a + 1}{a} \right) = \lim_{a \to \infty} \left(2 + \frac{1}{a} \right) \qquad \text{rearranging the expression}$$

and

$$a \to \infty \qquad \Rightarrow \frac{1}{a} \to 0$$

so

$$\lim_{a \to \infty} \left(2 + \frac{1}{a} \right) = 2.$$

(b) $\dfrac{3x + x^2}{x}$ is not even defined for $x = 0$ but for $x \neq 0$

$$\frac{3x + x^2}{x} \equiv 3 + x.$$

So

$$\lim_{x \to 0} \left(\frac{3x + x^2}{x} \right) = \lim_{x \to 0} (3 + x) = 3.$$

Notice in this case we found

$$\lim_{x \to a} f(x)$$

even when $f(a)$ was undefined!

It often happens, however, that when $f(a)$ is defined

$$\lim_{x \to a} f(x) = f(a).$$

In fact, this forms the basis for a formal definition of a continuous function. See page 167.

(c) On page 103 we showed that for θ measured in radians $\sin \theta \simeq \theta$ for small θ, with the approximation improving as θ approaches zero. In other words

$$\theta \to 0 \qquad \Rightarrow \sin \theta \to \theta$$

which may be written as

$$\theta \to 0 \qquad \Rightarrow \frac{\sin \theta}{\theta} \to 1$$

that is

$$\lim_{\theta \to 0} \left(\frac{\sin \theta}{\theta} \right) = 1.$$

The following properties of limits are important for our subsequent work, though we shall not prove them.

Statement

> If
>
> $$\lim_{x \to a} f(x) = L \quad \text{and} \quad \lim_{x \to a} g(x) = M$$
>
> then, as $x \to a$,
>
> (i) $f(x) + g(x) \to L + M$
>
> (ii) $f(x) \cdot g(x) \to L \cdot M$
>
> (iii) $\dfrac{f(x)}{g(x)} \to \dfrac{L}{M}$ \qquad provided $M \neq 0$

Often there is no limit, for example

$$\lim_{n \to \infty} (n^2)$$

is undefined, since $n \to \infty \Rightarrow n^2 \to \infty$.
Similarly,

$$\lim_{n \to \infty} ((-1)^n)$$

has no limit, as $n \to \infty$.

$$(-1)^n = 1 \quad \text{or} \quad -1$$

depending on whether n is even or odd.

Exercise 6.1

1. Write down the limit of each of the following expressions as $x \to 0$.

   (a) $4x$

   (b) $5 + x^2$

   (c) $\dfrac{(x + 1)(x - 2)}{x + 3}$

   (d) $\dfrac{3}{1 - 2x}$

   (e) $\dfrac{ax - b}{bx - a}$ (a, b non zero constants)

2. Find the limit as $n \to \infty$ of each of the following expressions.

   (a) $3 + \dfrac{4}{n} - \dfrac{5}{n^2}$

   (b) $\dfrac{2n - 1}{n}$

   (c) $\dfrac{n^2 + 6n - 2}{3n^2 - 2n + 1}$

   (d) $\dfrac{3n - 1}{n + 4}$

   (e) $\dfrac{n^3}{5n^2(n - 2)}$

3. Use a calculator to obtain the following limits correct to 2 d.p.

   (a) $\lim\limits_{n \to \infty} \left(1 + \dfrac{1}{n}\right)^n$

   (b) $\lim\limits_{a \to 0} \left(\dfrac{2^a - 1}{a}\right)$

   (c) $\lim\limits_{a \to 0} \left(\dfrac{3^a - 1}{a}\right)$

   (d) $\lim\limits_{\theta \to \pi/2} \left(\dfrac{\cos \theta}{\theta}\right)$ [θ measured in radians]

4. Obtain the following limits

   (a) $\lim\limits_{x \to 0} \left(\dfrac{2x + x^2}{x}\right)$

   (b) $\lim\limits_{x \to 1} \left(\dfrac{x^2 - 1}{x + 1}\right)$

   (c) $\lim\limits_{x \to 1} \left(\dfrac{x^2 - 1}{x - 1}\right)$

   (d) $\lim\limits_{x \to 2} \left(\dfrac{x^2 + x - 6}{x - 2}\right)$

   (e) $\lim\limits_{a \to 0} (a + 2b)$

   (f) $\lim\limits_{a \to 0} (b)$

   (g) $\lim\limits_{n \to \infty} \left(1 + \dfrac{1}{n}\right)\left(2 + \dfrac{1}{n}\right)$

5. By noting that

$$\frac{3x + 1}{2x - 3} \equiv \frac{x\left(3 + \dfrac{1}{x}\right)}{x\left(2 - \dfrac{3}{x}\right)} \equiv \frac{\left(3 + \dfrac{1}{x}\right)}{\left(2 - \dfrac{3}{x}\right)} \qquad x \neq 0$$

it follows that

$$\lim\limits_{x \to \infty} \left(\frac{3x + 1}{2x - 3}\right) = \frac{3}{2}.$$

Find by similar methods

(a) $\lim_{x \to \infty} \left(\dfrac{2x - 1}{5x - 2} \right)$ (b) $\lim_{x \to \infty} \left(\dfrac{1 - 2x}{2 - x} \right)$

(c) $\lim_{x \to \infty} \left(\dfrac{2 + x^2}{(x + 1)(2x - 3)} \right)$ (d) $\lim_{x \to \infty} \left(\dfrac{1 + x^2}{x(2 + x)} \right)$

6.2 The derived function

Definition

> The derived function f', of a function f is defined by
>
> $f'(x)$ = the gradient of the graph of f at the point $P(x, f(x))$, assuming such a gradient exists.
>
> The domain of f' is a subset of the domain of f.

We showed earlier that, for the function f defined by

$$f(x) \equiv x^2 \qquad x \in \mathbb{R}$$

the gradient at the point $P(a, a^2)$ is $2a$. This can be restated as

$$f'(x) \equiv 2x \qquad x \in \mathbb{R}$$

where the gradient at the point $P(x, x^2)$ is $2x$.

More generally, consider the graph of a function f. We seek the gradient to the curve at the general point $P(x, f(x))$ for x in the domain of f.

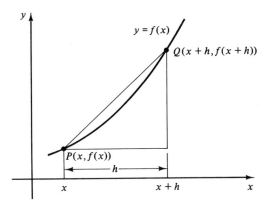

Fig. 6.3

Selecting an abscissa $x + h$, the associated ordinate will be $f(x + h)$ and the gradient of the chord joining the points $P(x, f(x))$ and $Q(x + h, f(x + h))$ is

$$\frac{f(x + h) - f(x)}{h}.$$

So we have a more precise way of defining the derived function.

Definition

$$f'(x) \equiv \lim_{h \to 0}\left[\frac{f(x + h) - f(x)}{h}\right] \qquad \text{for } x \text{ is the domain of } f.$$

It may be that the limit does not exist or depends on the particular values of h selected — in which case $f'(x)$ is not defined.

The process of finding the derived function is called *differentiating*. Finding this limit directly is called *differentiating from first principles*.

Example 6.3

Differentiate from first principles the function f defined by

$$f(x) \equiv \frac{1}{x} \qquad \text{for } x \neq 0.$$

Consider the two points on the graph of f:

$$P\left(x, \frac{1}{x}\right) \quad \text{and} \quad Q\left(x + h, \frac{1}{x + h}\right).$$

The gradient of the chord joining P and Q is

$$\frac{\dfrac{1}{x + h} - \dfrac{1}{x}}{h} = -\frac{h}{h \times (x + h)} = -\frac{1}{x(x + h)}.$$

Now by definition

$$f'(x) \equiv \lim_{h \to 0}\left(\frac{-1}{x(x + h)}\right) \qquad x \neq 0$$

$$= -\frac{1}{x^2} \qquad x \neq 0.$$

Notice that we can now obtain the gradient of the tangent to the curve at the point $(3, \frac{1}{3})$, for this is $f'(3) = -\frac{1}{9}$. For this reason the derived function is often called the gradient function.

Continuity and differentiability

A function is continuous if its graph may be drawn without any breaks in it. The function given by $y = 1/x$ is therefore continuous at every point except when $x = 0$ (see Fig. 6.4).

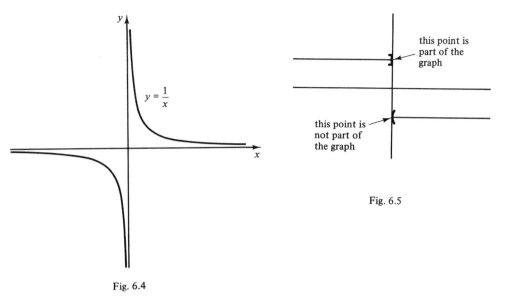

Fig. 6.5

Fig. 6.4

The function whose graph is given in Fig. 6.5 is continuous at every point except when $x = 0$, even though the function is defined at $x = 0$. Formally, a function is continuous at a point $x = a$ if

$$\lim_{x \to a} f(x) = f(a).$$

In the first example $f(0)$ was not defined, and so the equality failed. In the second the limit was not uniquely defined. [A different limit is obtained depending on whether zero is approached from the 'left' or the 'right'.]

The functions met at this level are either continuous everywhere, such as polynomials, or possess points of discontinuity because the function is undefined. For example, the function given by

$$f(x) \equiv \frac{1}{(x-1)(x+2)} \qquad x \neq 1, x \neq -2$$

is continuous everywhere except when $x = 1$ or $x = -2$.

A function is differentiable if it is *smooth* so that a unique tangent may be drawn to the graph of the function at every point. A function f such that $f(x) \equiv |x|$, $x \in \mathbb{R}$, whose graph is given in Fig. 6.6 is not differentiable at the point where $x = 0$. No unique tangent may be drawn.

Notice that f is continuous everywhere. So a function may be continuous at a point but not differentiable. However, if a function is differentiable at the particular point it must be continuous there.

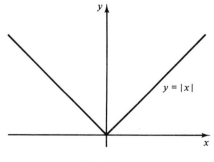

Fig. 6.6

The notation $\dfrac{d}{dx}$

The notation $\dfrac{d}{dx}$, read as 'dee by dee x of', is used to denote the derived function or derivative. So, for example

$$\frac{d}{dx}(x^2) = 2x.$$

If $y = f(x)$ then

$$\frac{df(x)}{dx} = f'(x)$$

which is also written as

$$\frac{dy}{dx} = f'(x)$$

where exceptionally $\dfrac{dy}{dx}$ is read as 'dee y by dee x'.

So, if

$$y = \frac{1}{x} \qquad \textcircled{1}$$

$$\Rightarrow \frac{dy}{dx} = -\frac{1}{x^2} \qquad \textcircled{2}$$

and the process of moving from step ① to step ② is called differentiating.

Similarly, if $s = f(t)$ then $\dfrac{ds}{dt} = f'(t)$.

Differentiating polynomials

Statement

If

$$f(x) \equiv ax^n \qquad\qquad x \in \mathbb{R} \qquad\text{where } n \text{ and } a \text{ are constants}$$

then

$$f'(x) \equiv nax^{n-1} \qquad\qquad x \in \mathbb{R}$$

Equivalently, if $y = ax^n$ then $\dfrac{dy}{dx} = nax^{n-1}$.

This result cannot be proved at this stage, though we shall show that it holds when $n \in \mathbb{N}$.

Suppose then that $y = ax^n$ for some $n \in \mathbb{N}$ and constant a. Two points on the graph of $y = ax^n$ are

$$A(x, ax^n) \quad \text{and} \quad B(x + h, a(x + h)^n)$$

So the gradient of the chord AB is

$$\frac{a(x + h)^n - ax^n}{h}$$

Now $(x + h)^n$ can be expanded using the Binomial Theorem (see page 36) and without going into too much detail we have

$$(x + h)^n = x^n + nhx^{n-1} + \text{terms with a factor of } h^2.$$

So

$$\frac{a(x + h)^n - ax^n}{h} = nax^{n-1} + \text{terms with a factor of } h.$$

So

$$\frac{dy}{dx} = \lim_{h \to 0} \left[\frac{a(x + h)^n - ax^n}{h} \right]$$

$$= nax^{n-1} \qquad\qquad \text{since terms with a factor of } h$$
$$\text{tend to zero as } h \to 0.$$

Example 6.4

Differentiate

(a) $y = 3x^5$ (b) $y = \dfrac{1}{2x^2}$ (c) $y = ax$ (d) $y = b$ (e) $y = 1/\sqrt{x}$

(a) $y = 3x^5 \Rightarrow \dfrac{dy}{dx} = 5.3x^4 = 15x^4$

(b) $y = \dfrac{1}{2x^2} \Rightarrow y = \frac{1}{2}x^{-2} \Rightarrow \dfrac{dy}{dx} = \dfrac{-2x^{-3}}{2} \Rightarrow \dfrac{dy}{dx} = \dfrac{-1}{x^3}$

(c) $y = ax \qquad \Rightarrow y = ax^1 \qquad \Rightarrow \dfrac{dy}{dx} = 1.a.x^0 \qquad \Rightarrow \dfrac{dy}{dx} = a$

(d) $y = b \qquad \Rightarrow y = bx^0 \qquad \Rightarrow \dfrac{dy}{dx} = 0.b.x^{-1} \qquad \Rightarrow \dfrac{dy}{dx} = 0$

(e) $y = \dfrac{1}{\sqrt{x}} \qquad \Rightarrow y = x^{-1/2} \qquad \Rightarrow \dfrac{dy}{dx} = -\frac{1}{2}x^{-3/2} \qquad \Rightarrow \dfrac{dy}{dx} = \dfrac{-1}{2x^{3/2}}$

Statement

> If f and g are functions and
> $$y = af(x) + bg(x)$$
> where a and b are constants, then
> $$\frac{dy}{dx} = af'(x) + bg'(x).$$

This result may seem obvious; however, its proof depends on the earlier results concerned with limits. It does enable us to differentiate many functions with ease including all polynomials.

Example 6.5

Differentiate the following

(a) $y = x^4 + 2x^2 - 6x + 1$ (b) $y = (x-1)(x+2)$ (c) $y = x + \dfrac{1}{x}$

(a) $y = x^4 + 2x^2 - 6x + 1$

$\Rightarrow \dfrac{dy}{dx} = 4x^3 + 4x - 6$

(b) $y = (x-1)(x+2)$

$\Rightarrow y = x^2 + x - 2$

$\Rightarrow \dfrac{dy}{dx} = 2x + 1$

(c) $y = x + \dfrac{1}{x}$

$\Rightarrow y = x + x^{-1}$

$\Rightarrow \dfrac{dy}{dx} = 1 - x^{-2} = 1 - \dfrac{1}{x^2}$

Example 6.6

Find the points on the curve $y = 2x^3 + 3x^2 + 5$ where the gradient is 12 and hence find the tangents to the curve at these points.

If $y = 2x^3 + 3x^2 + 5$ then

$$\frac{dy}{dx} = 6x^2 + 6x.$$

Since the gradient is 12, $\dfrac{dy}{dx} = 12$.

So

$$12 = 6x^2 + 6x$$

$$\Rightarrow 0 = x^2 + x - 2$$

$$\Rightarrow 0 = (x + 2)(x - 1)$$

$$\Rightarrow x = -2 \quad \text{or} \quad x = 1.$$

When

$$x = -2, \quad y = -16 + 12 + 5 = 1$$

$$x = 1, \quad y = 2 + 3 + 5 = 10$$

So the required points are

$$A(-2, 1) \quad \text{and} \quad B(1, 10)$$

The tangent to the curve at $A(-2, 1)$ has gradient 12 so its equation is

$$y - 1 = 12(x - -2) \quad [\text{see page 156}]$$

$$\Rightarrow \quad y = 12x + 25$$

The tangent to the curve at $B(1, 10)$ has gradient 12 also, and its equation is

$$y - 10 = 12(x - 1)$$

$$\Rightarrow \quad y = 12x - 2$$

Example 6.7

If

$$z = \frac{1}{u} - \frac{1}{u^2}$$

find $\dfrac{dz}{du}$ when $u = 2$.

$$z = \frac{1}{u} - \frac{1}{u^2}$$

$$\Rightarrow \quad z = u^{-1} - u^{-2}$$

$$\Rightarrow \frac{dz}{du} = -u^{-2} + 2u^{-3}$$

$$\Rightarrow \frac{dz}{du} = -\frac{1}{u^2} + \frac{2}{u^3}$$

When $u = 2$,

$$\frac{dz}{du} = -\frac{1}{4} + \frac{2}{8} = 0$$

Exercise 6.2

1. Obtain the derived function of each of the following

    (a) x^6 (b) x^{-3} (c) $x^{3/2}$ (d) $\dfrac{1}{x^3}$ (e) $\dfrac{-1}{x}$

    (f) $2x^3$ (g) $3x^{-2}$ (h) $\dfrac{2}{x^2}$ (i) $\dfrac{1}{3\sqrt{x}}$ (j) $\dfrac{5}{3x^3}$

    (k) 14 (l) $9x$ (m) $3x$ (n) $\dfrac{4}{x}$ (o) $5x^5$

    (p) $\dfrac{1}{3x}$ (q) $\dfrac{1}{2x^2}$ (r) $\dfrac{3}{2\sqrt{x}}$ (s) $\dfrac{1}{x^5}$ (t) $\dfrac{3}{2x^4}$

2. Find $\dfrac{dy}{dx}$ when

    (a) $y = x^3 - 2x^2 + 4$ (b) $y = (x - 1)^2$

    (c) $y = x(x^2 + 2)$ (d) $y = (x + 1)(x - 3)(x + 4)$

    (e) $y = (x + 2)^3$ (f) $y = \dfrac{x + 1}{\sqrt{x}}$

    (g) $y = \dfrac{1}{x} + \dfrac{1}{2x^2} + \dfrac{1}{3x^3}$ (h) $y = (2x - 3)(3x + 4)$

    (i) $y = \dfrac{3x^3 + x}{2x^2}$ (j) $y = \left(x + \dfrac{1}{x}\right)^2$

3. Find $\dfrac{dx}{du}$ when

    (a) $x = \dfrac{u - 1}{u^2}$ (b) $x = (3u^2 + 1)^2$

    (c) $u^2x = (u + 1)^2$ (d) $\dfrac{x}{u} = 4.$

4. Find the gradient of the curve at the point indicated.

    (a) $y = x^2 + 1$ when $x = 1$

    (b) $y = \sqrt{x}$ when $x = 2$

    (c) $y = (x + 1)(x + 2)$ when $x = -1$

    (d) $y = (2x + 1)(x - 2)$ at the point where the graph crosses the y axis.

5. Find the points on the curve, where the gradient is as indicated

    (a) $y = \dfrac{4}{x^2}$, -1 (b) $y = 2x^3$, 12 (c) $y = (x + 1)^2$, 1

(d) $y = 2x^3 - 3x^2 - 12x + 4,\quad 0$

(e) $x = \dfrac{u+1}{2\sqrt{u}},\quad 0$

6. At what point in the curve $y = x^2 - 4x + 3$ is the gradient

   (a) zero,   (b) equal to the gradient of the line $y = 2x + 1$?

7. For what values of x is the gradient of $y = x^3 - 6x$

   (a) zero,   (b) positive,   (c) negative?

8. Find the co-ordinates of the point on the curve $y = x + 1/x^2$ where the gradient is zero.

9. At which points on the curve $y = x^3 + 3x^2 - 4$ is the tangent parallel to the line $y - 9x + 2 = 0$?

10. Find the equation of the tangent to the curve $y = 2x^2$ at the point where $x = a$. If the tangent passes through the point $(4, 0)$ obtain two possible values of a.

11. Find the point of intersection of the tangents to the curve $y = x^2$ at the points $(1, 1)$ and $(2, 4)$.

12. If

$$g(x) \equiv \frac{2}{3 - x} \qquad \text{for } x \neq 3$$

find $\dfrac{d}{dx} g^{-1}(x)$ at the point where $x = 2$.

6.3 Tangents and normals

Definition

If a tangent meets a curve at the point $P(a, b)$ then the normal is a straight line perpendicular to the tangent passing through the point $P(ab)$ (Fig. 6.7).

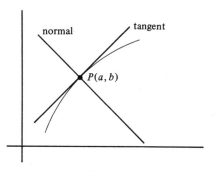

Fig. 6.7

Example 6.8

Find the equation of the normal to the graph of $y = 1/x$ at the point where $x = 2$ (Fig. 6.8). Hence find the points where the normal cuts the curve.

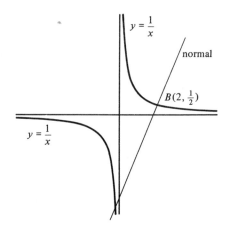

Fig. 6.8

If $y = 1/x$ then

$$\frac{dy}{dx} = -\frac{1}{x^2} \quad \text{(since } y = x^{-1})$$

So the gradient of the curve when $x = 2$ is $-\frac{1}{4}$.

It follows that the gradient of the normal is 4, since the product of the gradient is -1. The normal passes through the point $(2, \frac{1}{2})$ and so the equation of the normal is:

$$y - \tfrac{1}{2} = 4(x - 2)$$
$$\Rightarrow \quad 2y = 8x - 15$$

Solving simultaneously with $y = 1/x$ we have

$$\frac{2}{x} = 8x - 15$$

$$\Rightarrow 8x^2 - 15x - 2 = 0$$

$$\Rightarrow (8x + 1)(x - 2) = 0$$

$$\Rightarrow x = -\tfrac{1}{8} \quad \text{or} \quad x = 2$$

It follows that the points of intersection are $A(-\frac{1}{8}, -8)$ and $B(2, \frac{1}{2})$.

Exercise 6.3

1. Find the equation of the tangent to the following curves at the given points.

    (a) $y = x^2 + 3x + 1$ where $x = 1$

    (b) $y = \dfrac{1}{x}$ where $x = 2$

(c) $y = (x-1)(x+2)$ where $x = \frac{1}{2}$.

2. Find the equations of the normals where the graph of $y = (x-2)(x-3)$ crosses the x axis.

3. Find the equation of the tangent and normal to the graph of $y = x^3$ at the point $(2, 8)$.

4. Find the equation of the normal to the graph of $y = 3x^2$ at the point $(a, 3a^2)$.

5. Find the equation of the tangent to the curve $y = 1/x$ at the point $(t, 1/t)$.

6. Obtain the equations of the normal to the curve $y = x^2 + 1/x^2$ at the points where $y = 2$.

7. Find the equation of the tangent to the curve $y = x(2 - x)$ at the origin. Find a point on the curve where the tangent is perpendicular to the tangent at the origin.

8. Find the equation of the normal at the point where $x = \frac{1}{2}$ on the curve $y = (x-3)(x+2)$.

9. The tangent at the point $Q(4, 16)$ on the curve $y = x^2$ cuts the x axis at P. Find the distance PQ.

10. Obtain the gradient of the curves

$$y = 3x^2 - 1, \quad y = 12x - 13$$

at the point where they meet. Can you draw any geometrical significance from your result?

11. Find the value of a so that $y = 2x + a$ is a tangent to the curve $y = x^2 + 4x - 1$.

12. Find the derived function f' where

$$f(x) \equiv ax^2 + bx + c \qquad x \in \mathbb{R}, \text{ where } a, b \text{ and } c \text{ are constants.}$$

Find in terms of the constants a and b the x co-ordinates of the point on the curve where the gradient is

(a) 0   (b) b (c) $-\dfrac{b}{2a}$

13. Find the points on the curve $y = x^3 - 4x^2 - 6x + 1$, where the gradient is -10.

14. For which values of a are the tangents to the curve $y = x(x - a)$, at the points where the curve crosses the x axis, perpendicular to one another?

6.4 Stationary values

Definition

> A *stationary point*, is a point on the graph of a function where the gradient
> is zero.
>   If $P(a, f(a))$ is a stationary point then $f(a)$ is called a *stationary value*.

Example 6.9

Find the stationary values of the function defined by

$$f(x) \equiv x + \frac{1}{x} \qquad x \neq 0$$

Now

$$f'(x) = 1 - \frac{1}{x^2}.$$

A stationary value occurs when $f'(x) = 0$.
So

$$0 = 1 - \frac{1}{x^2}$$

$$\Rightarrow 0 = x^2 - 1$$

$$\Rightarrow x = 1 \quad \text{or} \quad x = -1$$

when

$$x = 1, \quad f(x) = 2$$

$$x = -1, \quad f(x) = -2.$$

Thus the stationary values of f are 2 and -2.

The stationary points corresponding to those stationary values are $A(1, 2)$ and $B(-1, -2)$. At these points the tangent will be parallel to the x axis.

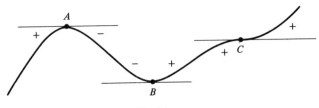

Fig. 6.9

Consider the curve in Fig. 6.9. There are stationary points at A, B and C. The stationary value associated with A is called a *maximum value*. The stationary value associated with B is called a *minimum value*. C is called a *point of inflexion*. A and B are often referred to as turning points.

It should be noted that a maximum value is not necessarily the same as the *greatest value* of a function. It is only a 'local' maximum. Similarly the *least value* of a function may not be identical with a minimum value.

Consider the function, whose graph is as indicated in Fig. 6.10. Clearly there is a minimum at the point $P(2,4)$ but the least value is zero. [It has been assumed that the domain of the function is $[0,6]$. Had the domain been $[0,6)$ then there would not be a least value!]

Fortunately, the maxima, minima and points of inflexion can be distinguished by the behaviour of the derived function close to the stationary points.

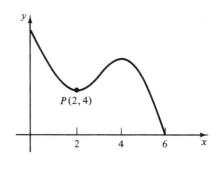

Fig. 6.10

Example 6.10

Find the stationary values of the function defined by

$$f(x) \equiv x^3 + x^2 - x - 1 \qquad x \in \mathbb{R}$$

and distinguish between them. Hence sketch the graph of the function f.

Setting

$$y = x^3 + x^2 - x - 1$$

$$\Rightarrow \frac{dy}{dx} = 3x^2 + 2x - 1$$

For a stationary value $dy/dx = 0$

$$\Rightarrow 0 = 3x^2 + 2x - 1$$
$$\Rightarrow 0 = (3x - 1)(x + 1)$$
$$\Rightarrow x = \tfrac{1}{3} \quad \text{or} \quad x = -1$$

When $x = \tfrac{1}{3}$,

$$y = (\tfrac{1}{3})^3 + (\tfrac{1}{3})^2 - \tfrac{1}{3} - 1$$
$$= \tfrac{1}{27} + \tfrac{1}{9} - \tfrac{1}{3} - 1$$
$$= -1\tfrac{5}{27}$$

and when $x = -1$,

$$y = -1 + 1 + 1 - 1$$
$$= 0.$$

So the stationary values are $-1\tfrac{5}{27}$ and 0.

To distinguish between the stationary values we are concerned with the behaviour of the derived function dy/dx when x is close to $\frac{1}{3}$ and when x is close to -1. Now

$$\frac{dy}{dx} = 3x^2 + 2x - 1$$

from which we can obtain Table 6.1.

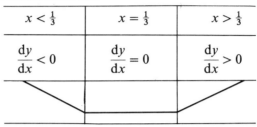

	$x < \frac{1}{3}$	$x = \frac{1}{3}$	$x > \frac{1}{3}$
	$\dfrac{dy}{dx} < 0$	$\dfrac{dy}{dx} = 0$	$\dfrac{dy}{dx} > 0$
graph			

Table 6.1

Table 6.1 has been drawn up by considering the sign of dy/dx for $x < \frac{1}{3}$, $x = \frac{1}{3}$ and $x > \frac{1}{3}$. It suffices to substitute a single value for x 'close' to $\frac{1}{3}$, for example:

$$\text{for } x < \tfrac{1}{3} \quad \text{select} \quad x = 0 \quad \text{and establish that } \frac{dy}{dx} < 0$$

and

$$\text{for } x > \tfrac{1}{3} \quad \text{select} \quad x = 1 \quad \text{and establish that } \frac{dy}{dx} > 0$$

This knowledge about the gradient of the curve close to $x = \frac{1}{3}$ enables one to establish the shape of the graph close to the point $P(\frac{1}{3}, -1\frac{5}{27})$. It is clear that the stationary value $-1\frac{5}{27}$ is a minimum.

Repeating the procedure for $x = -1$, we get Table 6.2. In this case the stationary value 0 is a maximum.

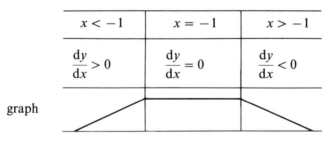

	$x < -1$	$x = -1$	$x > -1$
	$\dfrac{dy}{dx} > 0$	$\dfrac{dy}{dx} = 0$	$\dfrac{dy}{dx} < 0$
graph			

Table 6.2

Fig. 6.11 (see facing page) summarizes the information collected so far, about the graph of the function.

We now establish where the graph of

$$y = x^3 + x^2 - 1$$

cuts the axes.

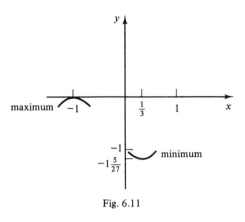

Fig. 6.11

Now

$$x = 0 \quad \Rightarrow y = -1$$
$$y = 0 \quad \Rightarrow 0 = x^3 + x^2 - x - 1.$$

We know that $y = 0$ when $x = -1$ so $x + 1$ is a factor. Therefore

$$\Rightarrow 0 = (x + 1)(x^2 + ax - 1) \quad \text{for some } a$$

and so by division, or by comparing coefficients, we find

$$0 = (x + 1)(x^2 - 1)$$
$$\Rightarrow 0 = (x + 1)^2 (x - 1)$$
$$\Rightarrow x = -1 \quad \text{or} \quad x = 1.$$

We now have sufficient information to sketch the graph (Fig. 6.12).

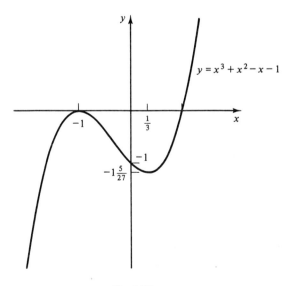

Fig. 6.12

The second derivative

The process of differentiating $y = f(x)$ to obtain $dy/dx = f'(x)$ may be continued as often as we please to obtain $f''(x), f'''(x) \ldots$ sometimes written $f^2(x), f^3(x), \ldots$.
    The notation

$$\frac{d}{dx}\left(\frac{dy}{dx}\right)$$

for the *second derivative* is shortened to

$$\frac{d^2y}{dx^2}$$

which reads 'dee two y by dee x squared', with similar notation for higher derivatives.

Example 6.11

Find the second derivative of $y = x^3 + x^2 + x$

If $y = x^3 + x^2 + x$, then

$$\frac{dy}{dx} = 3x^2 + 2x + 1$$

so

$$\frac{d^2y}{dx^2} = 6x + 2.$$

One reason for our interest in the second derivative is that it gives us a convenient way to distinguish between maximum and minimum in many cases.

Statement

If $f'(a) = 0$ and d^2y/dx^2 is evaluated at $x = a$ then if

$$\frac{d^2y}{dx^2} > 0$$

then $f(a)$ is a minimum value and if

$$\frac{d^2y}{dx^2} < 0$$

then $f(a)$ is a maximum value. But if

$$\frac{d^2y}{dx^2} = 0$$

then no comment can be made and other methods must be used to establish the behaviour of the function at $x = a$.

We suppose that for $x = a$, $f'(a) = 0$, so there is a stationary point at $x = a$.
There are four possibilities, as shown in Fig. 6.13

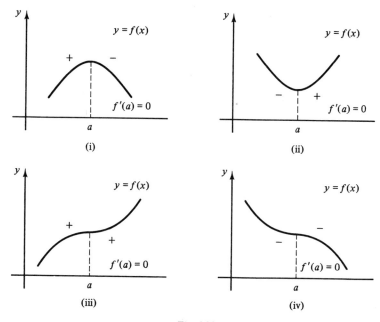

Fig. 6.13

Case 1

Suppose that $f''(a) < 0$. That is the gradient of the derived function is negative at the point where $x = a$.

This means that the graph of the derived function must be of the form given in Fig. 6.14 in which we have drawn a part of the graph of f' close enough to a to be able to approximate by a straight line.

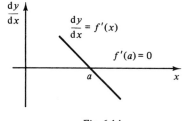

Fig. 6.14

Now from the graph of f' we can deduce that the gradient of f just before a, is positive, and it is negative just above a. This corresponds to possibility 1 which is a maximum.

Thus when $f'(a) = 0$ then

$$f''(a) < 0 \quad \Rightarrow f(a) \text{ is a maximum}$$

Case 2

Suppose that $f''(a) > 0$, then the graph of the derived function will be of the form given in Fig. 6.15. This time, the gradient of f is negative just before a and positive just above a, thus corresponding to possibility 2.

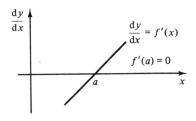

Fig. 6.15

Thus where $f'(a) = 0$ then

$$f''(a) > 0 \qquad \Rightarrow f(a) \text{ is a minimum}$$

Case 3

If $f''(a) = 0$ then the graph of the derived function could have a maximum, minimum or point of inflexion when $x = a$. Each of these cases corresponds in turn to one of the four possibilities above. That is, we cannot tell whether there is a maximum or a minimum or point of inflexion on the graph of f when $x = a$. It follows that other methods must be used to determine the behaviour of the function f.

Example 6.12

Find the stationary values of the function

$$f(x) \equiv x^3 - 3x^2 \qquad x \in \mathbb{R},$$

and hence sketch its graph.

Set

$$y = x^3 - 3x^2$$

$$\Rightarrow \frac{dy}{dx} = 3x^2 - 6x.$$

For a stationary value $dy/dx = 0$

$$\Rightarrow \quad 0 = 3x^2 - 6x$$

$$\Rightarrow \quad 0 = 3x(x - 2)$$

$$\Rightarrow \quad x = 0 \quad \text{or} \quad x = 2.$$

Now since

$$\frac{dy}{dx} = 3x^2 - 6x$$

$$\Rightarrow \frac{d^2y}{dx^2} = 6x - 6$$

So when $x = 0$,

$$\frac{d^2y}{dx^2} < 0 \qquad \Rightarrow \text{maximum}$$

and $y = 0$, upon substitution.
When $x = 2$,

$$\frac{d^2y}{dx^2} > 0 \qquad \Rightarrow \text{minimum}$$

and

$$y = 2^3 - 3.2^2$$
$$= -4.$$

Since

$$y = x^3 - 3x^2$$

setting $y = 0$,

$$\Rightarrow 0 = x^2(x - 3)$$

and so there is a double root
when $x = 0$, and the third root
is $x = 3$.

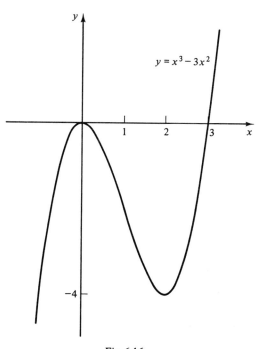

Fig. 6.16

Since we already know where the graph cuts the y axis the sketch can now be given as indicated in Fig. 6.16.

Example 6.13

An open box with a square base has a total surface area of $300 \, \text{cm}^3$. Find the greatest possible volume.

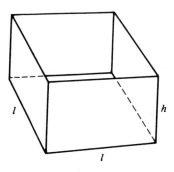

Fig. 6.17

Suppose the dimensions of the box are as indicated in Fig. 6.17 so that the volume V is given by

$$V = l^2 h.$$

The total surface area S is given by

$$S = l^2 + 4hl$$

$$\Rightarrow 300 = l^2 + 4hl.$$

Eliminating h we have

$$300 = l^2 + \frac{4V}{l}$$

$$\Rightarrow \quad V = \frac{l}{4}(300 - l^2).$$

We have expressed V in terms of a single variable. For a stationary value we require $dV/dl = 0$. Now

$$\frac{dV}{dl} = 75 - \frac{3l^2}{4}$$

$$\Rightarrow \quad 0 = 75 - \frac{3l^2}{4}$$

$$\Rightarrow \quad l = 10 \qquad \text{(we do not admit } l = -10 \text{ as a possible value)}$$

Since

$$\frac{d^2V}{dl^2} = -\frac{3l}{2}$$

when $l = 10$

$$\frac{d^2v}{dl^2} < 0 \qquad \Rightarrow \text{a maximum.}$$

When $l = 10$, $h = 5$ and so the maximum possible volume is $500\,\text{cm}^3$.

Example 6.14

Find the minimum value of $14x + y^2$ given that $x + y = 3$

Let $s = 16x + y^2$. Then since

$$x + y = 3$$

we have

$$y = 3 - x$$

so

$$s = 16x + (3 - x)^2$$
$$\Rightarrow s = 9 + 10x + x^2$$

We have expressed s in terms of the single variable x. It would have been just as acceptable to express s in terms of y.

For a turning value $ds/dx = 0$ and since

$$\frac{ds}{dx} = 10 + 2x$$

we require

$$10 + 2x = 0$$
$$\Rightarrow \qquad x = -5$$

Now

$$\frac{d^2s}{dx^2} = 10 > 0 \qquad \Rightarrow \text{minimum}$$

When $x = -5$, $y = 8$, since $x + y = 3$. These values give rise to a minimum value for

$$s = 16x + y^2$$
$$= -80 + 64$$
$$= -16.$$

Exercise 6.4

1. Find the stationary values of the functions given by

    (a) $f(x) \equiv 2x^3 + 3x^2 - 72x + 4 \qquad x \in \mathbb{R}$,
    (b) $f(x) \equiv x^3 - x^2 - x + 1 \qquad x \in \mathbb{R}$,

    (c) $f(x) \equiv 4x^2 + \dfrac{27}{x} \qquad x \neq 0$,

    (d) $f(x) \equiv x + \dfrac{16}{x} \qquad x \neq 0$.

2. Find the turning values of the following curves and distinguish between them

    (a) $y = x^2 + 4x + 1$ \qquad\qquad (b) $y = 5 + 6x - 2x^2$
    (c) $y = x^3 + 5x^2 + 3x - 10$ \qquad (d) $y = 2x^3 + 4x^2 + 2x - 3$

3. Find the position of the point of inflexion on the following curves:

    (a) $y = x^3 - 6x^2 + 12x + 1$, \qquad (b) $y = 4x^3 - 6x^2 + 3x - 1$,
    (c) $y = 8 - 27x + 9x^2 - x^3$.

4. Determine whether the origin is a point of inflexion on the curve

   (a) $y = x^3$,   (b) $y = x^4$,   (c) $y = x^5$,   (d) $y = x^6$.

5. Obtain the stationary points on the graph of the function
$$f(x) \equiv x^4 - 12x^2 + 27 \qquad x \in \mathbb{R}$$
   and distinguish between them.

6. Sketch the graph of f and f' on the same axes where
$$f(x) \equiv x^3 \qquad x \in \mathbb{R}.$$

7. Find the stationary values of the functions given by

   (a) $f(x) \equiv (x + 1)^3$ $x \in \mathbb{R}$,
   (b) $g(x) \equiv x^2(x - 2)^3$ $x \in \mathbb{R}$.

   and distinguish between them. Hence sketch the graphs of the functions.

8. Find the turning value of
$$y = \frac{x + 2}{\sqrt{x}} \qquad \text{where } x \neq 0$$
   and determine whether it corresponds to a maximum value or a minimum.

9. Prove that the function f, given by $f(x) \equiv x^2 - 3x + 3$, is positive for all values of x.

10. If $x^2y = x + 1$, find the value of d^2y/dx^2 when $x = 2$.

11. If $uv = c$ where c is a constant, find dv/du when $u = 2c$. Show that there are no stationary values when v is expressed as a function of u.

12. Sketch the curve $y = (x - 2)^3$.

13. The function $y = ax^3 + bx^2 + c$ has turning points at $(0, 4)$ and $(-1, 5)$. Sketch the graph of the function.

14. Find the maximum possible value of x^2y if $x + y = 6$.

15. A closed box with a square base has a total surface area of $36\,\text{m}^2$, what is the maximum possible volume?

16. A cylinder with an open top, contains $512\,\pi\,\text{cm}^3$. Obtain the surface area of the cylinder in terms of the radius, and hence find the minimum surface area. What will the dimensions of the cylinder be in this case?

17. Find the minimum value of $x^2 + y^2$ if $x + y = 10$.

18. If

$$\frac{d^2y}{dx^2} = -\frac{8}{x^3}$$

find y in terms of x, given that $dy/dx = 3$ and $y = 3$ when $x = 2$.

19. The curve $y = ax^2 + bx + c$ has a minimum when $x = +\frac{1}{2}$ and passes through the point $(2, 0)$ and $(1, -3)$. Find the value of a, b and c.

20. Show that the derivative of

$$y = 12x^5 - 20x^3 + 15x + 1$$

is non negative for all values of x and hence show that $y \geqslant 1$ for $x \geqslant 0$.

21. The curve $y = ax^3 + bx^2 + cx + d$ passes through the points $(1, 3)$ and $(0, 1)$. If the gradient is zero at $(2, 1)$ find the stationary points and sketch the curve.

22. Find the maximum value and also the greatest value of the expression

$$(x^2 + x + 1)(x - 2)$$

in the range $-1 \leqslant x \leqslant 2$.

23. The sum of two numbers a and b is 16. Find the maximum value of the product $a \cdot b$.

24. A wire of length $4a$ is bent to enclose a sector of a circle of radius r. Find this radius if the area of the sector is a maximum.

25. A sector of angle θ radians is cut from a circular disc of radius 4π cm and used to make the complete curved surface of a right circular cone, with no overlap.

    (a) Prove that the volume of this cone is $\dfrac{8\pi}{3}(4\pi^2\theta^4 - \theta^6)^{1/2}$.

    (b) Find the value of θ for which the volume of this cone is a maximum. (You need not test for the maximum.)

                                                                                                    AEB 1982

6.5 Rates of change

Suppose the distance s of a particle from a fixed point is given as a function of time t by

$$s = f(t)$$

A change in t usually results in a change in the value of s. Using the notation δx (delta x) to indicate a small change in x then as time changes from t to $t + \delta t$ the distance changes from $s\,[=f(t)]$ to $s + \delta s\,[=f(t + \delta t)]$.

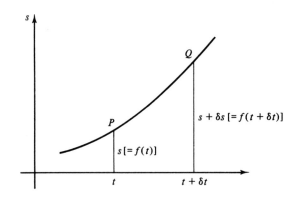

Fig. 6.18

Now $\delta s/\delta t$ represents the gradient of the chord joining $P(t, s)$ to $Q(t + \delta t, s + \delta s)$. By the, now familiar, techniques of differentiating

$$\lim_{\delta t \to 0} \frac{\delta s}{\delta t} = \frac{ds}{dt}$$

In other words ds/dt may be viewed as a measure of *the rate of change of s with respect to t.*

In particular, the rate of change of distance w.r.t. time is called velocity, v. [w.r.t. is short for 'with respect to'].

The rate of change of velocity w.r.t. time, dv/dt, is called the acceleration, a, of the particle. Notice that

$$a = \frac{dv}{dt} = \frac{d^2 s}{dt^2} \qquad \text{if } s \text{ represents distance.}$$

For

$$v = \frac{ds}{dt}$$

$$\Rightarrow \frac{dv}{dt} = \frac{d}{dt}\left(\frac{ds}{dt}\right) = \frac{d^2 s}{dt^2}.$$

The phrase 'the rate of change of v w.r.t. r' means dv/dr. If dv/dr is positive then v is increasing as r increases. However when dv/dr is negative then v decreases as r increases.

If we simply refer to the rate of change of v then we mean dv/dt where t represents time.

Example 6.15

A point moves in a straight line so that its displacement s m from a fixed point at time t s is given by

$$s = 2t^2 - 3t + 1$$

Obtain the velocity of the particle when (a) $t = 0$ s and (b) $t = 1$ s. Find the time when the velocity is zero and obtain the acceleration at this time.

$$s = 2t^2 - 3t + 1$$

$$\Rightarrow v = \frac{ds}{dt} = 4t - 3$$

So when

(a) $\qquad t = 0, v = -3\,\mathrm{m\,s}^{-1}$

and when

(b) $\qquad t = 1, v = 1\,\mathrm{m\,s}^{-1}$

Therefore

$$v = 0 \text{ when } 4t - 3 = 0$$

$$\Rightarrow t = \tfrac{3}{4}\,\mathrm{s}$$

$$a = \frac{dv}{dt} = 4$$

The acceleration is $4\,\mathrm{m\,s}^{-2}$ at all times, in particular when $t = \tfrac{3}{4}$ s, i.e. when $v = 0\,\mathrm{m\,s}^{-1}$.

Example 6.16

Find the rate of change of the volume $v\,\mathrm{cm}^3$ of a sphere w.r.t. the radius r when $r = 4.0$ cm.

Since

$$v = \tfrac{4}{3}\pi r^3$$

$$\frac{dv}{dr} = 4\pi r^2.$$

When $r = 4$ cm, $dv/dr = 64\pi\,\mathrm{cm}^2$.

Exercise 6.5

1. A particle moves so that its displacement from a fixed point is given by

$$s = t^3 - 2t$$

Obtain the velocity and acceleration when (a) $t = 0\,\text{s}$ (b) $t = 1\,\text{s}$. Find the time when the velocity is zero and hence obtain the acceleration at this time.

2. A ball is thrown vertically into the air so that its initial velocity is $40\,\text{m s}^{-1}$. Its height h above the ground at time t is given by

$$h = 5t(8 - t)$$

(a) How long is it before the ball returns to its starting point?
(b) Find the velocity of the ball when (i) $t = 3\,\text{s}$ (ii) $t = 4\,\text{s}$.
(c) When is the velocity zero?
(d) Find the height the ball reaches.

3. A particle moves so that its velocity is given by

$$v = Rt - \frac{k}{t^2}$$

where R and k are constants. Find:

(a) the acceleration of the particle when $t = \dfrac{k}{2R}\,\text{s}$

(b) the time when the acceleration is zero.

4. A rectangular enclosure is to be made in a field using $100\,\text{m}$ of fencing for three sides, the fourth being part of a boundary wall. Express the area A of the enclosure in terms of the length x of the side opposite the wall. Obtain the rate of change of A w.r.t. x and hence find the dimensions of the field to maximise A.

6.6 The chain rule

Suppose $y = f(u)$ and $u = g(x)$, then the rate of change of y w.r.t. u is dy/du $[=f'(u)]$ and the rate of change of u w.r.t. x is $du/dx\,[=g'(x)]$.

Clearly, a small change δx in x results in a small change δu in u, which in turn results in a small change δy in y. What then will the rate of change of y be, with respect to x? Now

$$\frac{\delta y}{\delta x} = \frac{\delta y}{\delta u} \cdot \frac{\delta u}{\delta x} \qquad \text{since we are dealing with fractions.}$$

So

$$\lim_{\delta x \to 0} \frac{\delta y}{\delta x} = \lim_{\delta x \to 0} \frac{\delta y}{\delta u} \cdot \frac{\delta u}{\delta x}.$$

But as $\delta x \to 0$ so $\delta u \to 0$ and using the properties of limits which were stated on page 163 we have

$$\lim_{\delta x \to 0} \frac{\delta y}{\delta x} = \lim_{\delta u \to 0} \frac{\delta y}{\delta u} \cdot \lim_{\delta x \to 0} \frac{\delta u}{\delta x}$$

That is

$$\frac{dy}{dx} = \frac{dy}{du} \cdot \frac{du}{dx}.$$

It must not be thought that dy/dx is a fraction, it is not; it is the limit of a fraction. However, as a matter of interest, in many respects it behaves as though it were a fraction and so with this in mind the above result is not too surprising.

Statement — The chain rule

If $y = f(u)$ and $u = g(x)$ then

$$\frac{dy}{dx} = \frac{dy}{du} \cdot \frac{du}{dx}$$

assuming the derived functions exist.

The chain rule may be extended so that

$$\frac{dy}{dx} = \frac{dy}{du} \cdot \frac{du}{dt} \cdot \frac{dt}{dx}$$

assuming the variables are suitably functionally related.

The chain rule has many applications. For example, if a, v, s represent the acceleration, velocity and distance of a particle from a fixed point, then by the chain rule,

$$a = \frac{dv}{dt} = \frac{dv}{ds} \cdot \frac{ds}{dt}$$

and since $ds/dt = v$ we note that

$$a = v\frac{dv}{ds}.$$

The chain rule can also be used to broaden the range of functions we can differentiate with ease.

Example 6.17
Differentiate w.r.t. x

(a) $y = (x^2 + 1)^5$ (b) $y = \sqrt{1 - x}$ (c) $y = \dfrac{1}{1 + 3x}$

(a) By expanding the bracket we could differentiate directly. However, suppose $u = x^2 + 1$ then $y = u^5$ and so

$$\frac{dy}{du} = 5u^4 \quad \text{and} \quad \frac{du}{dx} = 2x$$

Now since

$$\frac{dy}{dx} = \frac{dy}{du} \cdot \frac{du}{dx}$$

we have

$$\frac{dy}{dx} = 5u^4 \cdot 2x$$

$$= 10(x^2 + 1)^4 x \quad \text{since } u = x^2 + 1$$

(b) For $y = \sqrt{1 - x}$, let $u = 1 - x$. Then $y = \sqrt{u}$.

$$\Rightarrow \frac{du}{dx} = -1 \quad \text{and} \quad \frac{dy}{du} = \tfrac{1}{2}u^{-1/2}$$

$$= \frac{1}{2(1 - x)^{1/2}}$$

and since

$$\frac{dy}{dx} = \frac{dy}{du} \cdot \frac{du}{dx}$$

$$\frac{dy}{dx} = -1 \cdot \frac{1}{2(1 - x)^{1/2}} = \frac{-1}{2\sqrt{1 - x}}$$

(c) With practise much of the working is done mentally, so

$$y = \frac{1}{1 + 3x}$$

$$\Rightarrow y = (1 + 3x)^{-1}$$

$$\frac{dy}{dx} = -1(1 + 3x)^{-2} \cdot 3 \qquad \text{letting } u = 1 + 3x$$

$$[-1 \, (\text{bracket})^{-2} \, (\text{derivative of bracket})]$$

$$= \frac{-3}{(1 + 3x)^2}$$

Statement

If

$$y = [f(x)]^n \qquad n \in \mathbb{Q}$$

then

$$\frac{dy}{dx} = n[f(x)]^{n-1} \cdot f'(x)$$

For suppose

$$y = [f(x)]^n$$

and we set $u = f(x)$ then

$$y = u^n \qquad \text{and} \qquad u = f(x)$$

$$\Rightarrow \frac{dy}{du} = nu^{n-1} \quad \text{and} \quad \frac{du}{dx} = f'(x)$$

By the chain rule

$$\frac{dy}{dx} = \frac{dy}{du} \cdot \frac{du}{dx}$$

$$= nu^{n-1} f'(x)$$

$$= n(f(x))^{n-1} f'(x) \qquad \text{since } u = f(x)$$

Example 6.18

Differentiate   (a) $y = (2x + x^3)^5$ (b) $y = \dfrac{1}{\sqrt{(1 + 2x)}}$

(a) $y = (2x + x^3)^5$

$$\Rightarrow \frac{dy}{dx} = 5(2x + x^3)^4 \cdot (2 + 3x^2)$$

(b) $y = \dfrac{1}{\sqrt{(1 + 2x)}}$

$$\Rightarrow y = (1 + 2x)^{-1/2}$$

$$\Rightarrow \frac{dy}{dx} = -\tfrac{1}{2}(1 + 2x)^{-3/2} \cdot 2$$

$$\Rightarrow \frac{dy}{dx} = -\frac{1}{\sqrt{(1 + 2x)^3}}$$

Statement

$$\boxed{\begin{array}{c} \dfrac{dy}{dx} = \dfrac{1}{dx/dy} \\[2mm] \text{assuming } dx/dy \text{ exists} \end{array}}$$

This follows immediately from the chain rule. For

$$\frac{dy}{dy} = \frac{dy}{dx} \cdot \frac{dx}{dy}$$

but $dy/dy = 1$, this is the case for all variables, and the result then follows.

Example 6.19

Find dy/dx if $x = (3y^2 + 1)^3$.

Now

$$\frac{dx}{dy} = 3(3y^2 + 1)^2 \cdot 6y \qquad \text{by the chain rule,}$$

$$= 18y(3y^2 + 1).$$

It follows that

$$\frac{dy}{dx} = \frac{1}{dx/dy} = \frac{1}{18y(3y^2 + 1)}.$$

Exercise 6.6

1. Use the chain rule to differentiate each of the following

   (a) $y = (x + 2)^3$ (b) $y = (3x + 1)^5$
   (c) $y = 2(1 - 2x)^4$ (d) $y = (2x^2 - 3x + 1)^3$
   (e) $y = \sqrt{(1 + x)}$ (f) $y = \sqrt{(1 + x^2)}$

   (g) $y = \sqrt{(2 - x)}$ (h) $y = \dfrac{1}{x + 1}$

   (i) $y = \dfrac{3}{x^2 + 2}$ (j) $y = \dfrac{4}{2x^2 - 3}$

   (k) $y = \dfrac{3}{(1 + x)^2}$ (l) $y = \dfrac{5}{(6 + x)^3}$

   (m) $y = \dfrac{2}{\sqrt{(1 + 2x)}}$ (n) $y = \dfrac{6}{\sqrt{(6 - x)}}$

   (o) $f(x) \equiv \dfrac{1}{\sqrt{(x^2 + 1)}}$ (p) $f(x) \equiv \dfrac{1}{(3x - 1)^4}$

   (q) $y = 3(x + 1)^{4/5}$ (r) $y = 6(2 + x^2)^3$

   (s) $y = (3x + 1)^4 - 2(x + 1)^3$ (t) $y = \dfrac{1}{(2x - 1)^3} + \sqrt{2x - 1}$

   (u) $y = \dfrac{1}{(x + 1)^2} - \dfrac{1}{2x + 1}$

2. Find dx/dy if

   (a) $y = 3x + 1$ (b) $y = (x - 1)^3$ (c) $y^2 = 3x$ (d) $y^3 = (x - 2)^2$

3. Find dy/dx if   (a) $x = (2y + 1)^2$ (b) $x = \sqrt{(1 + y^2)}$

Further applications of the chain rule

Suppose u is a function of x and we are asked to differentiate u^3 with respect to x.
Now

$$y = u^3$$

$$\Rightarrow \frac{dy}{du} = 3u^2$$

Also

$$\frac{dy}{dx} = \frac{dy}{du} \cdot \frac{du}{dx}$$

$$= 3u^2 \frac{du}{dx}$$

In words: to differentiate u^3 w.r.t. x, first differentiate w.r.t. u and multiply by du/dx.
More generally:

> If u is a function of x, then
> $$\frac{d}{dx}[f(u)] = f'(u)\frac{du}{dx}$$

Example 6.20
Differentiate the following expressions w.r.t. x.

(a) θ^3 (b) y^2 (c) $\dfrac{1}{t}$ (d) $(u^2 + 1)^3$

(a) $\dfrac{d}{dx}(\theta^3) = \dfrac{d}{d\theta}(\theta^3) \cdot \dfrac{d\theta}{dx}$

$$= 3\theta^2 \frac{d\theta}{dx}$$

(b) $\dfrac{dy^2}{dx} = \dfrac{dy^2}{dy} \cdot \dfrac{dy}{dx}$

$$= 2y \cdot \frac{dy}{dx}$$

Notice that $\dfrac{dy^2}{dx}$ is *not* the same as $\dfrac{d^2y}{dx^2}$.

(c) $\dfrac{d}{dx}\left(\dfrac{1}{t}\right) = \dfrac{d}{dt}\left(\dfrac{1}{t}\right) \cdot \dfrac{dt}{dx}$

$$= -\frac{1}{t^2}\frac{dt}{dx}$$

(d) $\dfrac{d}{dx}((u^2 + 1)^3) = \dfrac{d}{du}((u^2 + 1)^3) \cdot \dfrac{du}{dx}$

$$= 3(u^2 + 1)^2 \cdot 2u \cdot \dfrac{du}{dx}$$

The derivative of $(u^2 + 1)^3$ was itself differentiated by applying the chain rule mentally $[3 \,(\text{bracket})^2 \cdot \text{derivative of bracket}]$.

Example 6.21

A cylinder of base radius a cm is being filled with water at the rate of $kh\,\mathrm{cm^3\,s^{-1}}$ where h is the height of the liquid in the cylinder at time t, and k is a constant. Find the rate at which the height is increasing.

At time t the volume of water in the cylinder is given by

$$v = \pi a^2 h$$

$$\Rightarrow \dfrac{dv}{dh} = \pi a^2 \qquad \text{since } \pi \text{ and } a \text{ are constants.}$$

We require to find dh/dt, but

$$\dfrac{dh}{dt} = \dfrac{dh}{dv} \cdot \dfrac{dv}{dt}$$

Now

$$\dfrac{dh}{dv} = \dfrac{1}{dv/dh} = \dfrac{1}{\pi a^2}$$

and

$$\dfrac{dv}{dt} = kh$$

It follows that

$$\dfrac{dh}{dt} = \dfrac{1}{\pi a^2} \cdot kh$$

$$= \dfrac{kh}{\pi a^2}$$

6.7   Parametric equations

If two variables x and y, say, are each equal to functions of a third variable t, say, then the relationship between x and y is said to be defined parametrically or in terms of a parameter.

For example, consider the equations

$$y = t + \frac{1}{t} \quad \text{and} \quad x = t^3$$

We could eliminate the parameter t to obtain the relationship between x and y directly. However if we simply require dy/dx we have

$$y = t + \frac{1}{t} \quad \text{and} \quad x = t^3$$

$$\Rightarrow \frac{dy}{dt} = 1 - \frac{1}{t^2} \qquad \Rightarrow \frac{dx}{dt} = 3t^2$$

Now using the chain rule

$$\frac{dy}{dx} = \frac{dy}{dt} \cdot \frac{dt}{dx}$$

$$. = \left(1 - \frac{1}{t^2}\right) \cdot \frac{1}{3t^2} \qquad \text{recalling that } \frac{dt}{dx} = \frac{1}{dx/dt}$$

$$= \frac{t^2 - 1}{3t^4}$$

Of course dy/dx has been expressed in terms of t, but if x and y are known then t can be obtained from the relevant equations and substituted to find dy/dx.

If required, we can find $\dfrac{d^2y}{dx^2}$ as follows

$$\frac{d^2y}{dx^2} = \frac{d}{dx}\left(\frac{dy}{dx}\right) = \frac{d}{dx}\left(\frac{t^2 - 1}{3t^4}\right)$$

$$= \frac{d}{dt}\left(\frac{t^2 - 1}{3t^4}\right) \cdot \frac{dt}{dx}$$

$$= \frac{d}{dt}\left(\frac{1}{3t^2} - \frac{1}{3t^4}\right) \cdot \frac{dt}{dx}$$

$$= \left(-\frac{2}{3t^3} + \frac{4}{3t^5}\right)\frac{1}{3t^2}$$

$$= \frac{2(2 - t^2)}{9t^7}$$

Exercise 6.7

1. Differentiate each of the following functions w.r.t. x.

    (a) $(u + 1)^3$ (b) $2u^2$ (c) $\sqrt{1 - u}$ (d) $(u - 1)^3$

(e) $(2u + 1)^3$ (f) y^3 (g) $2y^2$ (h) $(y^2 + 1)^2$

(i) $\dfrac{1}{y^2}$ (j) $\dfrac{1}{y - 1}$ (k) $\dfrac{1}{\sqrt{1 - \theta^2}}$ (l) $3(2\theta^2 + 1)^{-1}$

2. A spherical balloon is blown up so that the radius is increasing at the rate $2\,\text{cm s}^{-1}$. Find the rate at which the volume is increasing w.r.t. r and also find dv/dt when $r = 3\,\text{cm}$.

3. The surface area of a cube is increasing at the rate $3\,\text{cm s}^{-1}$. Find the rate at which the volume is increasing when the surface area is $6\,\text{cm}^2$.

4. Find dy/dx if

(a) $x = t^2$ and $y = t^3$ (b) $x = t^3 + 1$ and $y = t^2$

(c) $x = t^2$ and $y = 2t + 1$ (d) $x = t$ and $y = \dfrac{1}{t}$ $t \neq 0$

(e) $x = t - \dfrac{1}{t}$ and $y = t + \dfrac{1}{t}$ $t \neq 0$

(f) $x = (t^2 + 1)^3$ and $y = \dfrac{1}{t + 1}$ $t \neq -1$

5. The slant edge of a circular cone is $6\,\text{cm}$. Find the height of the cone when the volume is a maximum.
    This cone is inverted and water is poured in at the constant rate of $2\,\text{cm}^3\,\text{s}^{-1}$. Find the time taken to fill the cone to the nearest second.

6. If $x = ct$ and $y = c/t$ where c is a constant find dy/dx in terms of t. Find the equation of the tangent when (a) $t = 2$, (b) $t = t_1$.

7. A curve is given by $x = at^2$ and $y = 2at$. Obtain the equation of the tangent and normal at the point where $t = t_1$.

8. If $x = t^2$ and $y = t^3$, find both $\dfrac{dy}{dx}$ and $\dfrac{d^2y}{dx^2}$ and so find

(a) $\dfrac{d^2y}{dx^2}$ when $t = 2$ (b) $\dfrac{dy}{dx}$ when $x = 4$ (there are two values).

9. The radius, r, of a circle is increasing at the rate of $2/r^2\,\text{m s}^{-1}$. Obtain the rate at which the area is increasing when $r = 4$.

10. If

$$f(x) \equiv x^2 \qquad \text{for } x \in \mathbb{R}$$

and

$$g(x) \equiv (2 - x) \qquad \text{for } x \in \mathbb{R}$$

obtain $f'(g(x)) \cdot g'(x)$ in terms of x, and show that this is

$$\frac{d}{dx}(f \circ g(x))$$

by first obtaining $f \circ g$ in terms of x.

11. If f is an even function show that f' is odd.

12. If $x = 2t(1 - t^2)$ and $y = 1 - t^2$, find dy/dx and find the co-ordinates of the point where

    (a) the gradient is zero   (b) the gradient is undefined

    Find also the equations of the tangents at the origin.

13. If $y = \sqrt{ax^2 + 3}$, show that

$$y^3 \frac{d^2y}{dx^2} + a^2 x^2 = 0.$$

14. A spherical balloon is inflated by gas being pumped in at the constant rate of $200 \, \text{cm}^3$ per second.
    What is the rate of increase of the surface area of the balloon when its radius is $100 \, \text{cm}$?

    (Surface area of sphere $= 4\pi r^2$, volume of sphere $= \frac{4}{3}\pi r^3$.)

    SU 1982

15. The radius r and height h of a cylinder vary in such a way that the surface area s remains a constant.
    Find dh/dr when $h = 2r$.

16. The volume of a right circular cone increases at a rate of $120 \, \text{cm}^3$ s in such a way that the height of the cone is always half the radius of the base.
    Find the rate at which the radius is increasing when the volume of the cone is $36\,000\pi \, \text{cm}^3$.
    Find also, at the same instant, the rate at which the area of the curved surface of the cone is increasing.
    (The area of the curved surface of a cone is $\pi r l$, where r is the radius of the base and l is the slant height.)

    AEB 1982

17. The line AB is perpendicular to a horizontal
plane through the point B and $AB = 5\,\text{m}$. A
variable point P moves on the plane along a
straight line through B, and when $BP = x$
metres the angle PAB is θ radians, as shown in
the diagram. Show that when $BP = (x + \delta x)$
metres and δx is small, the angle PAB is
$\theta + \delta\theta$, where

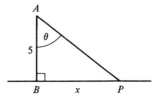

Fig. 6.19

$$\delta\theta \approx \frac{\delta x}{5}\cos^2\theta.$$

Given that P is moving towards B with a constant speed of $3\,\text{m s}^{-1}$, find the rate
of change of θ with respect to time when $x = 5$.

JMB 1983

6.8 Integration

If it is known that a derived function is given by

$$f'(x) \equiv 2x$$

what can be said of the function f?
   Clearly $f(x) \equiv x^2$ is a possibility, but then so is

$$f(x) \equiv x^2 + 1$$

or $$f(x) \equiv x^2 + 2$$
$$\vdots$$

Indeed there are a whole family of possible functions of the form

$$f(x) \equiv x^2 + c$$

with the property that

$$f'(x) \equiv 2x.$$

Again, suppose that

$$\frac{dy}{dx} = x^3$$

then

$$y = \frac{x^4}{4} + c \qquad \text{where } c \text{ is a constant.}$$

Statement

> The reverse process of differentiation is called integration. The constant
> which appears is called the *constant of integration*.

The notation

$$\int f(x)\,dx$$

reads 'the integral of $f(x)$ w.r.t. x' and is equal to that function whose derivative is $f(x)$.

Thus

$$\int 2x\,dx = x^2 + c$$

and

$$\int x^3\,dx = \frac{x^4}{4} + c$$

Statement

$$\int f(x)\,dx = F(x) + c \Leftrightarrow F'(x) \equiv f(x)$$

In particular

$$\int ax^n\,dx = \frac{ax^{n+1}}{n+1} + c \qquad \text{providing } n \neq -1$$

In other words, in order to check that the integral of a given function is found, it suffices to differentiate the integral to obtain the original function. In particular, if

$$y = \frac{ax^{n+1}}{n+1} + c$$

then

$$\frac{dy}{dx} = \frac{a}{n+1}(n+1)x^n$$

$$= ax^n$$

giving our stated result.

Clearly the rule fails to work when $n = -1$ and other methods are needed to obtain

$$\int \frac{1}{x}\,dx \qquad \text{(see page 343)}.$$

Example 6.22

Integrate the following functions w.r.t. x.

(a) $3x^4$ (b) $2x^3 + 6x + 2$ (c) $\dfrac{1}{x^{1/2}}$

(a) $\displaystyle\int 3x^4\,dx = \frac{3x^5}{5} + c$ by the formula.

(b) $\displaystyle\int 2x^3 + 6x + 2\,dx = \frac{2x^4}{4} + \frac{6x^2}{2} + 2x + c$

$$= \frac{x^4}{2} + 3x^2 + 2x + c$$

which may be checked by differentiation.
Notice that:

$$\int a\,dx = \int a x^0\,dx$$

$$= \frac{ax^1}{1} + c$$

$$= ax + c.$$

(c) $\displaystyle\int \frac{1}{x^{1/2}}\,dx = \int x^{-1/2}\,dx$

$$= \frac{x^{1/2}}{\frac{1}{2}} + c$$

$$= 2\sqrt{x} + c.$$

Statement

> The integral of the sum of two expressions may be found by integrating each expression in turn and adding the result. That is
>
> $$\int f(x) + g(x)\,dx = \int f(x)\,dx + \int g(x)\,dx$$

Example 6.23

Evaluate $\displaystyle\int \frac{(x+1)^2}{\sqrt{x}}\,dx.$

Now

$$\int \frac{(x+1)^2}{\sqrt{x}}\,dx = \int \frac{x^2 + 2x + 1}{\sqrt{x}}\,dx$$

$$= \int x^{3/2} + 2x^{1/2} + x^{-1/2}\,dx$$

$$\Rightarrow \int \frac{(x+1)^2}{\sqrt{x}}\,dx = \frac{x^{5/2}}{\frac{5}{2}} + \frac{2x^{3/2}}{\frac{3}{2}} + \frac{x^{1/2}}{\frac{1}{2}} + c$$

$$= \frac{2}{5}x^{5/2} + \frac{4}{3}x^{3/2} + 2x^{1/2} + c.$$

Exercise 6.8

1. Find each of the following integrals.

(a) $\displaystyle\int 2x^5\,dx$ (b) $\displaystyle\int \frac{1}{x^2}\,dx$

(c) $\displaystyle\int 3x^{-4}\,dx$ (d) $\displaystyle\int 2x^2 - x\,dx$

(e) $\displaystyle\int (1-x)^2\,dx$ (f) $\displaystyle\int \tfrac{1}{4}(x+2)^3\,dx$

(g) $\displaystyle\int (\sqrt{x}-1)^2\,dx$ (h) $\displaystyle\int \frac{x-1}{\sqrt{x}}\,dx$

(i) $\displaystyle\int \frac{1}{x^2} + \frac{2}{x^3}\,dx$ (j) $\displaystyle\int \frac{(1+2\sqrt{x})^2}{\sqrt{x}}\,dx$

(k) $\displaystyle\int (1+x)^2\,dx$ (l) $\displaystyle\int (2+x)(x-1)\,dx$

(m) $\displaystyle\int \frac{3}{2x^3}\,dx$ (n) $\displaystyle\int \frac{(1-x)^2}{x^5}\,dx$

(o) $\displaystyle\int \frac{(x^3+1)^2}{\sqrt{x}}$ (p) $\displaystyle\int (1+x)x^2\,dx$

6.9   Integration by guesswork

An important skill in integration is to be able to guess the answer and then check by differentiating.

Example 6.24

Find $\displaystyle\int (2x+1)^5\,dx$.

A guess might be

$$\frac{(2x+1)^6}{6}$$

but its derivative is

$$\frac{6(2x + 1)^5}{6} \cdot 2 = 2(2x + 1)^5.$$

We are a factor of 2 too large so

$$\int (2x + 1)^5 \, dx = \frac{(2x + 1)^6}{12} + c,$$

a result which can be checked by differentiating.

Example 6.25

Find $\int (ax + b)^{1/2} \, dx$.

We might try

$$(ax + b)^{3/2}.$$

Now its derivative is

$$\tfrac{3}{2}(ax + b)^{1/2} \cdot a,$$

so we need to divide by $\tfrac{3}{2}a$, giving

$$\int (ax + b)^{1/2} \, dx = \frac{2}{3a}(ax + b)^{3/2} + c.$$

Remember the constant of integration.

This method will not always be successful. For example, if we are asked to find

$$\int (x^2 + 1)^{1/2} \, dx$$

we might be tempted to try

$$(x^2 + 1)^{3/2}$$

which gives on differentiating

$$\frac{3}{2}(x^2 + 1)^{1/2} \cdot 2x.$$

It appears that we need to divide by $3x$ to give

$$\frac{(x^2 + 1)^{3/2}}{3x}$$

but this time the derivative is not $(x^2 + 1)^{1/2}$.

Only when we need to divide by a constant can we hope to guess the answer in this way. However, consider the following example.

Example 6.26

Find

$$\int x(x^2 + 1)^{1/2}\,dx.$$

This time the derivative of (the guessed) $(x^2 + 1)^{3/2}$ is

$$\frac{3}{2}(x^2 + 1)^{1/2} \cdot 2x$$

$$= 3x(x^2 + 1)^{1/2} \quad \text{and then we only need divide by 3.}$$

Thus

$$\int x(x^2 + 1)^{1/2}\,dx = \frac{(x^2 + 1)^{3/2}}{3} + c.$$

Statement

> Any multiple of an expression of the form $f'(x)[f(x)]^n$ may be integrated by this method.

Exercise 6.9

1. Find the following — remembering the constant of integration.

(a) $\displaystyle\int x^3\,dx$

(b) $\displaystyle\int 2x\,dx$

(c) $\displaystyle\int (x - 1)^2\,dx$

(d) $\displaystyle\int dx$

(e) $\displaystyle\int \frac{1}{x^2}\,dx$

(f) $\displaystyle\int \frac{3}{x^2}\,dx$

(g) $\displaystyle\int \frac{1}{x^2} + \frac{1}{x^3} + \frac{1}{x^4}\,dx$

(h) $\displaystyle\int x^{3/2}\,dx$

(i) $\displaystyle\int \frac{4}{x^{1/3}}\,dx$

(j) $\displaystyle\int (2x + 1)^3\,dx$

(k) $\displaystyle\int x\sqrt{x}\,dx$

(l) $\displaystyle\int (\sqrt{x} - 1)^2\,dx$

2. Integrate each of the following

(a) $\displaystyle\int 4t^3\,dt$

(b) $\displaystyle\int (u + 3)^2\,du$

(c) $\displaystyle\int \frac{v - 1}{\sqrt{v}}\,dv$

(d) $\displaystyle\int \frac{(s - 1)^2}{s^5}\,ds$

(e) $\displaystyle\int (2w - 1)\,dw$

3. Find the following integrals by first guessing the answer and checking by differentiation.

(a) $\int (2x + 1)^5 \, dx$

(b) $\int (1 - x)^3 \, dx$

(c) $\int 2x(x^2 + 1)^5 \, dx$

(d) $\int (2x - 1)(x^2 - x)^3 \, dx$

(e) $\int \dfrac{x}{(x^2 + 1)^2} \, dx$

(f) $\int x^2 \sqrt{x^2 + 4} \, dx$

(g) $\int \dfrac{3x}{\sqrt{1 - x^2}} \, dx$

(h) $\int \dfrac{1}{(1 + x)^2} - \dfrac{x}{(1 + x^2)^2} \, dx$

(i) $\int (2x + 3)(x^2 + 3x - 2)^2 \, dx$

(j) $\int x(x^2 + 3)^{3/2} \, dx$

(k) $\int (3x - 2)^3 \, dx$

(l) $\int \dfrac{x}{2}(-x^2)^{1/2} \, dx$

4. The acceleration of a particle may be found by differentiating the velocity equation w.r.t. time. It follows that the velocity may be found by integrating the acceleration equation. If $a = t - 1$ then $v = \frac{1}{2}t^2 - t + c$. The value of the constant may be established if we know the velocity at some particular time. If $v = 5 \, \text{m s}^{-1}$ when $t = 0$ it follows by substitution that $c = 5$. We may then find the velocity at any subsequent time. Again, if the velocity equation is integrated w.r.t. t then we can find the displacement equation w.r.t. time.

(a) If $a = 3 \, \text{m s}^{-1}$ find v when $t = 3$ if $v = 4 \, \text{m s}^{-1}$ when $t = 0$.
(b) If $a = 2t - 5$ find v when $t = 0$ if $v = 0$ when $t = 5$.
(c) A ball is dropped from a high tower so that its acceleration is $g \, \text{m s}^{-2}$ where g is a constant. Find the velocity and the distance travelled after 2 seconds, in terms of g.

5. A curve $y = f(x)$ which passes through the point $(1, 0)$, has gradient given by

$$\frac{dy}{dx} = 3x^2 - 2x + 3.$$

Find the equation of the curve.

Chapter 7 Trigonometry II

7.1 Multiple angles

It was stated in Chapter 4 that

$$\sin (A + B) \equiv \sin A \cos B + \cos A \sin B \qquad \text{①}$$

and

$$\cos (A + B) \equiv \cos A \cos B - \sin A \sin B \qquad \text{②}$$

for all A and B.
  We then deduced that

$$\sin (A - B) \equiv \sin A \cos B - \cos A \sin B \qquad \text{③}$$

and

$$\cos (A - B) \equiv \cos A \cos B + \sin A \sin B \qquad \text{④}$$

  By direct division of the first two identities, we obtain

$$\tan (A + B) \equiv \frac{\sin A \cos B + \cos A \sin B}{\cos A \cos B - \sin A \sin B} \qquad \text{provided } \cos (A + B) \neq 0.$$

Then dividing each term on the r.h.s. by $\cos A \cdot \cos B$ we obtain

$$\tan (A + B) \equiv \frac{\dfrac{\sin A \cos B}{\cos A \cos B} + \dfrac{\cos A \sin B}{\cos A \cos B}}{\dfrac{\cos A \cos B}{\cos A \cos B} - \dfrac{\sin A \sin B}{\cos A \cos B}} \qquad \text{provided } \cos A \cdot \cos B \neq 0.$$

So

$$\tan (A + B) \equiv \frac{\tan A + \tan B}{1 - \tan A \tan B} \qquad \text{⑤}$$

It can be similarly shown that

$$\tan (A - B) \equiv \frac{\tan A - \tan B}{1 + \tan A \tan B} \qquad \text{⑥}$$

Example 7.1
Find $\tan 105°$ without using tables or a calculator.

$$\tan 105° = \tan (60° + 45°) = \frac{\tan 60° + \tan 45°}{1 - \tan 60° \cdot \tan 45°}$$

$$= \frac{\sqrt{3} + 1}{1 - \sqrt{3} \cdot 1} \qquad \tan 60° = \sqrt{3}$$

$$\Rightarrow \tan 105^\circ = \frac{\sqrt{3}+1}{1-\sqrt{3}} \qquad \tan 45^\circ = 1$$

Setting $A = B$ in equations ①, ② and ⑤ above, we obtain:

The double angle formulae

$$
\begin{aligned}
&\sin 2A \equiv 2 \sin A \cos A && ⑦ \\
&\cos 2A \equiv \cos^2 A - \sin^2 A && ⑧
\end{aligned}
\Big\} \quad \text{for all } A
$$

$$
\tan 2A \equiv \frac{2 \tan A}{1 - \tan^2 A} \qquad ⑨ \qquad \tan A \neq \pm 1
$$

Equation ⑧ can be expressed in the form

$$
\begin{aligned}
&\cos 2A \equiv 1 - 2 \sin^2 A \\
\text{or} \quad &\cos 2A \equiv 2 \cos^2 A - 1 \qquad \text{using the fact that } \cos^2 A + \sin^2 A \equiv 1
\end{aligned}
$$

These may be shown as follows

$$
\begin{aligned}
\cos 2A &= \cos^2 A - \sin^2 A \\
&= (1 - \sin^2 A) - \sin^2 A \quad \text{or} \quad \cos^2 A - (1 - \cos^2 A) \\
&= 1 - 2 \sin^2 A \quad \text{or} \quad 2 \cos^2 A - 1
\end{aligned}
$$

Example 7.2
Express $\cos 70^\circ$ in terms of $\cos 35^\circ$.

$$\cos 2A = 2 \cos^2 A - 1$$

setting $A = 35$, we have

$$\cos 70^\circ = 2 \cos^2 35^\circ - 1$$

Example 7.3
Show that $\sin 3\theta \equiv 3 \sin \theta - 4 \sin^3 \theta$

$$
\begin{aligned}
\text{Now} \qquad \sin 3\theta &\equiv \sin(\theta + 2\theta) \\
&\equiv \sin \theta \cos 2\theta + \cos \theta \sin 2\theta \\
&\equiv \sin \theta (1 - 2 \sin^2 \theta) + \cos \theta . 2 \sin \theta \cos \theta \\
&\equiv \sin \theta - 2 \sin^3 \theta + 2 \sin \theta \cos^2 \theta \\
&\equiv \sin \theta - 2 \sin^3 \theta + 2 \sin \theta (1 - \sin^2 \theta) \\
&\equiv 3 \sin \theta - 4 \sin^3 \theta.
\end{aligned}
$$

It can be similarly shown that $\cos 3\theta \equiv 4\cos^3\theta - 3\cos\theta$.

Example 7.4

Solve the equations

(a) $\cos 2\theta = \sin\theta$ (b) $\tan 2\theta + \tan\theta = 0$ in the range $0 \leqslant \theta < 360°$

(a) $\cos 2\theta = 1 - 2\sin^2\theta$, so

$$\cos 2\theta = \sin\theta$$
$$\Rightarrow 1 - 2\sin^2\theta = \sin\theta$$
$$\Rightarrow 2\sin^2\theta + \sin\theta - 1 = 0 \qquad \text{a quadratic in } \sin\theta$$
$$\Rightarrow (2\sin\theta - 1)(\sin\theta + 1) = 0$$
$$\Rightarrow \sin\theta = \tfrac{1}{2} \quad \text{or} \quad \sin\theta = -1$$
$$\Rightarrow \theta = 30° \text{ or } 150° \text{ or } 270° \qquad \text{in the required range.}$$

(b) $\tan 2\theta + \tan\theta = 0$

$$\Rightarrow \frac{2\tan\theta}{1 - \tan^2\theta} + \tan\theta = 0$$
$$\Rightarrow 2\tan\theta + \tan\theta(1 - \tan^2\theta) = 0$$
$$\Rightarrow 3\tan\theta - \tan^3\theta = 0$$
$$\Rightarrow \tan\theta(3 - \tan^2\theta) = 0$$
$$\Rightarrow \tan\theta = 0 \quad \text{or} \quad \tan\theta = +\sqrt{3} \quad \text{or} \quad \tan\theta = -\sqrt{3}$$
$$\Rightarrow \theta = 0° \text{ or } \theta = 180° \text{ or } \theta = 60° \text{ or } \theta = 120° \text{ or } \theta = 240° \text{ or } \theta = 300°$$

7.2 The half angle identities

We have already shown that

$$\tan 2A \equiv \frac{2\tan A}{1 - \tan^2 A} \qquad \text{provided } \tan A \neq \pm 1$$

It is possible to express both $\sin 2A$ and $\cos 2A$ in terms of $\tan A$. For

$$\sin 2A \equiv 2\sin A \cos A$$
$$\equiv 2\tan A \cos^2 A$$
$$\equiv \frac{2\tan A}{\sec^2 A} \qquad \text{since} \quad \frac{1}{\sec^2 A} \equiv \cos^2 A$$
$$\equiv \frac{2\tan A}{1 + \tan^2 A} \qquad \text{since} \quad 1 + \tan^2 A \equiv \sec^2 A$$

Again

$$\cos 2A \equiv \cos^2 A - \sin^2 A$$

$$\equiv (1 - \tan^2 A)\cos^2 A$$

$$\equiv \frac{1 - \tan^2 A}{\sec^2 A}$$

$$\equiv \frac{1 - \tan^2 A}{1 + \tan^2 A}$$

Now setting $2A = \theta$ and $\tan A = \tan \theta/2 = t$ we have the half angle identities shown in Fig. 7.1.

The half angle identities

$$\tan \theta \equiv \frac{2t}{1 - t^2}$$

$$\sin \theta \equiv \frac{2t}{1 + t^2}$$

$$\cos \theta \equiv \frac{1 - t^2}{1 + t^2}$$

where $\tan \dfrac{\theta}{2} = t$

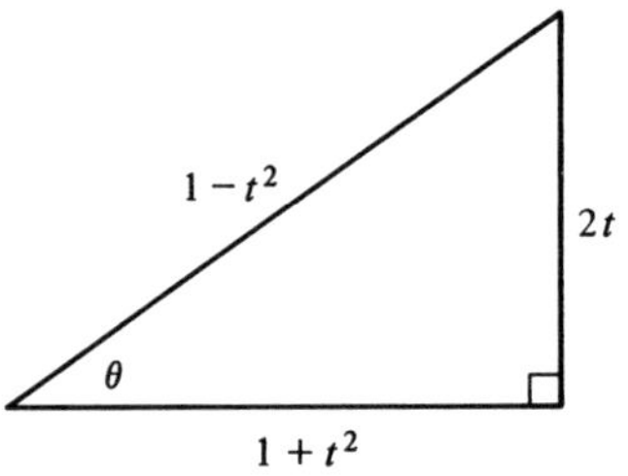

Fig. 7.1

These identities may be used to convert an expression involving any of the trigonometric functions of an angle into one which involves just t, the tangent of the 'half' angle.

Example 7.5

Express $\dfrac{\cos \theta}{1 + \sin \theta}$ in terms of tangents.

Direct substitution of the half-angle identities gives

$$\frac{\cos \theta}{1 + \sin \theta} \equiv \frac{1 - t^2}{1 + t^2} \bigg/ 1 + \frac{2t}{1 + t^2}$$

$$\equiv \frac{1 - t^2}{1 + t^2} \bigg/ \frac{1 + t^2 + 2t}{1 + t^2}$$

$$\equiv \frac{1 - t^2}{1 + 2t + t^2}$$

$$\equiv \frac{(1 - t)(1 + t)}{(1 + t)(1 + t)}$$

$$\Rightarrow \frac{\cos\theta}{1+\sin\theta} \equiv \frac{1-t}{1+t}$$

$$\equiv \frac{1-\tan\theta/2}{1+\tan\theta/2} \qquad \text{since } t = \tan\theta/2$$

Example 7.6

Solve the equation

$$\sin 2\theta + 2\cos 2\theta = 1 \qquad \text{for} \quad 0 < \theta \leqslant 2\pi.$$

The half-angle identities may be used for $\sin 2\theta$ and $\cos 2\theta$ by setting $t = \tan\theta$. Thus

$$\frac{2t}{1+t^2} + \frac{2(1-t^2)}{1+t^2} = 1$$

$$\Rightarrow 2t + 2 - 2t^2 = 1 + t^2$$

$$\Rightarrow 3t^2 - 2t - 1 = 0$$

$$\Rightarrow (3t+1)(t-1) = 0$$

$$\Rightarrow t = -\tfrac{1}{3} \quad \text{or} \quad t = 1$$

$$\Rightarrow \tan\theta = -\tfrac{1}{3} \quad \text{or} \quad \tan\theta = 1$$

$$\Rightarrow \theta = -0.322 \quad \text{or} \quad \theta = 2.820 \quad \boxed{\text{C}}$$

or

$$\theta = \frac{\pi}{4} \quad \text{or} \quad \theta = \frac{5\pi}{4}$$

Exercise 7.1

1. Find without using a calculator

   (a) $\cos 75°$ (b) $\sin 105°$ (c) $\tan 225°$ (d) $\tan 15°$

2. If $\sin\theta = \tfrac{4}{5}$ and θ is acute, find without using a calculator

   (a) $\cos\theta$ (b) $\tan\theta$ (c) $\sin 2\theta$ (d) $\cos 2\theta$ (e) $\tan 2\theta$

3. If $\sin\theta = \tfrac{12}{13}$ and θ is obtuse, find without using a calculator

   (a) $\cos 2\theta$ (b) $\tan 2\theta$

4. If $\cos 2\theta = \tfrac{7}{25}$ find

   (a) $\sin\theta$ (b) $\cos\theta$ (c) $\sin 2\theta$

5. If $\tan(x + 45°) = a$, find $\tan x$ in terms of a.

6. Show that $\tan(\theta + 45°) \equiv \dfrac{1 + \tan\theta}{1 - \tan\theta}$.

7. If $\tan A = \frac{1}{2}$ and $\tan B = \frac{1}{3}$, evaluate

    (a) $\tan(A + B)$ (b) $\tan 2A$

8. Find $\tan 75°$ without using tables or a calculator.

9. By expressing $\tan(45°) = \tan(22\frac{1}{2} + 22\frac{1}{2})$ and expanding, show that
$$\tan^2(22\frac{1}{2}°) + 2\tan(22\frac{1}{2}°) - 1 = 0.$$
    Hence show that $\tan 22\frac{1}{2}° = \sqrt{2} - 1$.

10. Express each of the following in terms of $\tan\theta$.

    (a) $\sqrt{\dfrac{1 + \cos 2\theta}{1 - \sin 2\theta}}$ (b) $\dfrac{1}{1 + \cos 2\theta}$ (c) $3\sin 2\theta + \cos 2\theta$

11. Prove that $\cos 3\theta \equiv 4\cos^3\theta - 3\cos\theta$.

12. Show that $\dfrac{\sin\theta + \sin 2\theta}{1 + \cos\theta + \cos 2\theta} \equiv \tan\theta$.

13. Express $\sin(A + B + C)$ in terms of trigonometric functions of A, B and C, by first expanding $\sin((A + B) + C)$.

14. Show that $\cos\theta \equiv 1 - 2\sin^2(\theta/2)$. Hence deduce that if θ (measured in radians) is small then $\cos\theta \simeq 1 - \theta^2/2$.

15. Show that $\tan 15° = 2 - \sqrt{3}$ without using tables.
    (*Hint*: Express $\tan 30° = \tan(15° + 15°)$, then expand.)

16. Solve the following equations in the range $0 \leqslant \theta < 360°$.

    (a) $3\sin 2\theta = \cos\theta$ (b) $\cos 2\theta = \cos\theta$.

17. If $\sin\theta = \dfrac{1 - x}{1 + x}$, find in terms of x

    (a) $\tan\theta$ (b) $\tan\theta/2$.

18. Given that A is acute and that $\tan 2A = \frac{4}{3}$, calculate, without using tables or a calculator, the value of

    (a) $\cos 2A$,   (b) $\sin A$,   (c) $\tan A$,   (d) $\tan 3A$,   (e) $\cos 4A$

AEB 1983

19. The equation $1 + \sin^2 \phi° = a \cos 2\phi°$ has a root 30. Find the value of a and all the roots in the range 0 to 360.

SU 1983

20. Find all solutions of the equation

$$4 \tan 2x = \cot x,$$

for which $0° < x < 360°$, giving your answer to the nearest $0.1°$.

CAMB 1983

21. The acute angles A, B and C are such that $\tan A = 3$, $\tan B = 4$ and $\tan C = 4\frac{1}{2}$. Without using a calculator or tables, show that $\tan (A + B + C) = 1$ and deduce the value of $A + B + C$ in radians, expressing your answer as a rational multiple of π.

22. Given that $A + B + C = 2\pi$, prove that

   (i) $\sin \frac{1}{2}A = \sin \frac{1}{2}(B + C)$,
   (ii) $\cos \frac{1}{2}A = -\cos \frac{1}{2}(B + C)$,
   (iii) $\sin A + \sin B + \sin C = 4 \sin \frac{1}{2}A \sin \frac{1}{2}B \sin \frac{1}{2}C$.

AEB 1982

23. Show that

$$\operatorname{cosec} 2x - \cot 2x = \tan x.$$

Deduce, or find otherwise, the exact value of $\tan 75°$, expressing your answer in the form $a + b\sqrt{3}$, where a and b are integers.

CAMB 1983

24. Assuming that

$$\frac{\cot A}{1 + \cot A} \times \frac{1}{1 + \tan(45° - A)}$$

is defined and is not equal to zero, show that its value is $\frac{1}{2}$.

W 1983

25. Show that

$$\tan 4\theta = \frac{4t(1 - t^2)}{1 - 6t^2 + t^4}, \quad \text{where} \quad t = \tan \theta.$$

[You may quote, without proof, any standard results.]

Deduce that, when $\tan \theta = \frac{1}{5}$,

$$\tan (4\theta - \tfrac{1}{4}\pi) = \tfrac{1}{239}.$$

CAMB 1982

26. Given that $\sin 3\theta = 3 \sin \theta - 4 \sin^3\theta$, show that

$$\frac{\sin 3\theta}{\sin \theta} - \frac{\sin 6\theta}{\sin 2\theta} \equiv 2(\cos 2\theta - \cos 4\theta),$$

assuming that $\sin \theta \neq 0$ and $\sin 2\theta \neq 0$.

                                                                                        W 1981

27. If $\sin x = \frac{3}{5}$ find the possible values of $\cos 2x$ without using tables.

28. Find the possible values of $\tan x$ given that
$$16 \sin x - 3 \cos x + 11 = 0.$$

Use the 'half-tan' substitution.

7.3   The expression $a \cos \theta + b \sin \theta$

An expression of the form $a \cos \theta + b \sin \theta$ can be reduced to the form $r \cos (\theta \pm \alpha)$ or $r \sin (\theta \pm \alpha)$ where r is positive and both r and α depend only on the values of a and b.

Example 7.7
Express $5 \cos \theta + 12 \sin \theta$ in the form $r \cos (\theta - \alpha)$ and hence solve the equation
$$5 \cos \theta + 12 \sin \theta = 5$$
in the range $0° \leqslant \theta < 360°$.

Recall that
$$r \cos (\theta - \alpha) \equiv r \cos \theta \cos \alpha + r \sin \theta \sin \alpha$$

Thus $$r \cos (\theta - \alpha) \equiv 5 \cos \theta + 12 \sin \theta$$

if $r \cos \alpha = 5$ ①

and $r \sin \alpha = 12$ ②   (by comparing coefficients)

Squaring and adding (①^2 + ②^2) we obtain
$$r^2 \cos^2 \alpha + r^2 \sin^2 \alpha = 5^2 + 12^2$$
$$\Rightarrow r^2 = 169 \qquad \text{since } \cos^2 \alpha + \sin^2 \alpha = 1$$
$$\Rightarrow r = 13 \qquad \text{since } r > 0$$

Thus by substitution into ① and ②
$$\cos \alpha = \tfrac{5}{13} \quad \text{and} \quad \sin \alpha = \tfrac{12}{13}$$
$$\Rightarrow \alpha = 67.38 \quad \boxed{C} \quad \text{(taking the acute solution).}$$

Hence
$$5 \cos \theta + 12 \sin \theta \equiv 13 \cos (\theta - 67.38°).$$

The equation we are asked to solve reduces to

$$13 \cos (\theta - 67.38°) = 5$$

$$\Rightarrow \cos (\theta - 67.38°) = \frac{5}{13}$$

$$\Rightarrow \theta - 67.38° = 67.38° \quad \text{or} \quad \theta - 67.38 = -67.38°$$

$$\Rightarrow \theta = 134.76 \qquad\qquad \text{or} \quad \theta = 0.$$

This equation could also have been solved using the half-angle identities.

Example 7.8

Find the maximum and minimum values of the function f defined by

$$f(\theta) \equiv 3 \sin \theta - 4 \cos \theta \qquad (\theta \text{ measured in radians})$$

Suppose

$$3 \sin \theta - 4 \cos \theta \equiv r \sin (\theta - \alpha)$$

(It would have been as convenient to choose $r \cos (\theta + \alpha)$—other choices would lead to an obtuse angle for α.)
  Now since

$$r \sin (\theta - \alpha) \equiv r \sin \theta \cos \alpha - r \cos \theta \sin \alpha$$

we require

$$r \cos \alpha = 3 \quad \text{①}$$

$$r \sin \alpha = 4 \quad \text{②}$$

squaring and adding

$$r^2 = 3^3 + 4^2$$

$$\Rightarrow \quad r = 5 \qquad \text{since } r > 0$$

so

$$\cos \alpha = \tfrac{3}{5} \quad \text{and} \quad \sin \alpha = \tfrac{4}{5}$$

$$\Rightarrow \quad \alpha = 0.927 \text{ (radians)} \qquad [\text{assuming } \alpha \text{ is acute}] \qquad \boxed{C}$$

So $f(\theta) \equiv 5 \sin (\theta - 0.927°)$.
  The sine or cosine of an angle has a maximum value of 1 and a minimum value of -1. It follows immediately that the maximum value of f is 5, which occurs whenever $\sin (\theta - 0.927) = 1$. For example, when

$$\theta - 0.927 = \frac{\pi}{2}$$

$$\Rightarrow \theta = 2.498 \qquad 3 \text{ d.p.} \qquad \boxed{C}$$

Similarly, the minimum value is -5, which occurs, for example, when

$$\theta = -2.498 \qquad 3\,\text{d.p.} \qquad \boxed{\text{C}}$$

The graph of f is given in Fig. 7.2.

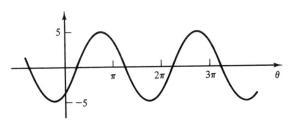

Fig. 7.2

Exercise 7.2

1. Express each of the following in the form $r\cos(\theta - \alpha)$.

   (a) $3\cos\theta + 5\sin\theta$ (b) $2\cos\theta + 9\sin\theta$

2. Express each of the following in the form $r\sin(\theta - \alpha)$.

   (a) $5\sin\theta - 2\cos\theta$ (b) $3\sin\theta - 5\cos\theta$

3. Use your answers to questions 1 and 2 to solve the following equations in the range $0° \leqslant \theta < 360°$.

   (a) $3\cos\theta + 5\sin\theta = 2$ (b) $2\cos\theta + 9\sin\theta = 6$
   (c) $5\sin\theta - 2\cos\theta = 4$ (d) $3\sin\theta - 5\cos\theta = 3$

4. Find the maximum and minimum values of the functions defined below.

   (a) $f(x) \equiv 3\cos x + 5\sin x$ (b) $g(x) \equiv 5\sin x - 2\cos x + 1$
   (c) $h(x) \equiv 2\sin x + 3\sin x$ (d) $h(x) \equiv \sin 2x + 4\cos 2x$

5. Solve each of the following equations in the range $-180° < \theta \leqslant 180°$. (Try $r\cos(\theta + \alpha)$.)

   (a) $4\cos\theta - 3\sin\theta = 2$ (b) $5\sin\theta - 2\cos\theta = 3$
   (c) $3\cos\theta + 5\sin\theta = 2$ (d) $6\cos 2\theta - 3 = 5\sin 2\theta$

6. Solve the equation $\sin 2x + \sin^2 x = 0$ in the range $-\pi < x \leqslant \pi$.

7. Express $\sin 2\theta - \cos 2\theta$ in the form $r\cos(2\theta + \alpha)$ and hence find the minimum value of $\sin 2\theta - \cos 2\theta + 1$.
   Solve the equation $\sin 2\theta - \cos 2\theta + 1 = 0$ for $0 \leqslant \theta < 180°$.

8. Express $4 \cos x - 3 \sin x$ in the form $R \cos (x + \alpha)$, where $R > 0$ and $0 < \alpha < \pi/2$ and hence find the range of values of the function defined by

$$f(x) \equiv 4 \cos x - 3 \sin x \qquad x \in \mathbb{R}$$

9. Sketch each of the graphs of the functions given in question 4; for values of x from $0°$ to $360°$.

10. Express $\sin \theta - \sqrt{3} \cos \theta$ in the form $R \sin (\theta - \alpha)$, where R is positive and $0 < \alpha < \frac{1}{2}\pi$. Hence find all values of θ in the range $0 \leqslant \theta < 2\pi$ for which $\sin \theta - \sqrt{3} \cos \theta = 1$.

OX 1983

11. Given that $y = 3 \sin \theta + 3 \cos \theta$, express y in the form $R \sin (\theta + \alpha)$ where $R > 0$ and $0° < \alpha < 90°$.

   Hence find

   (a) the greatest and least values of y^2,

   (b) the values of θ in the interval $0°$ to $90°$ for which $y = \dfrac{3\sqrt{6}}{2}$.

AEB 1983

12. If $4 \sin \theta + 3 \cos \theta \equiv R \sin (\theta + \alpha)$ where R is positive and α acute, find the values of R and x.

   Hence, or otherwise, solve the equation $2 \sec \theta - 4 \tan \theta = 3$, giving roots in the range $0° < \theta < 360°$.

SU 1982

13. Solve the equation

$$3 \cos \theta + 2 \sin \theta = 1$$

   for $0° < \theta < 360°$.

14. Given that

$$2 \sin 2x + \cos 2x = 2,$$

   find all possible values of x which lie between $0°$ and $360°$ inclusive, giving your answers to the nearest $0.1°$.

CAMB 1982

15. Given that $0 \leqslant \theta \leqslant 360$, find all the solutions of the equation

$$\sin \theta° + 2 \cos \theta° = k,$$

   when (a) $k = 0$, (b) $k = 1$.
   Find the range of values of k for which the equation has real solutions for θ.

AEB 1983

16. Find the solutions of the equation $\cos 2x + 1 = \sin 2x$ in the range $0 \leqslant x < \pi$.

17. For those values of x which satisfy the equation $8 \sin x + \cos x = 4$, calculate

(a) the possible values of $\tan \frac{1}{2}x$,
(b) the possible values of $\cos x$.

If, in addition, $-\pi/2 < x < \pi/2$, calculate, in radians, the value of x which satisfies the given equation, giving your answer to two decimal places.

AEB 1982

18. Solve for x in the range $-\pi < x \leqslant \pi$

(a) $2 \cos x + \sin x = 1$ (b) $5 \cos 2x - \sin 2x = 2$.

7.4   The factor formulae

Consider again the compound identities

$$\sin (A + B) \equiv \sin A \cos B + \cos A \sin B$$

and $$\cos (A + B) \equiv \cos A \cos B + \sin A \sin B \qquad \text{for all } A \text{ and } B.$$

We have

$$\sin(A + B) + \sin (A - B) \equiv 2 \sin A \cos B. \qquad \text{①}$$

Now setting $A + B = P$ and $A - B = Q$ it is easy to show that

$$A = \tfrac{1}{2}(P + Q) \quad \text{and} \quad B = \tfrac{1}{2}(P - Q).$$

Then substituting into ① we obtain

$$\sin P + \sin Q \equiv 2 \sin \frac{(P + Q)}{2} \cos \frac{(P + Q)}{2} \qquad \text{⑤}$$

In this way the sum of two sines can be factorised, i.e. expressed as a product. Equation ① is useful for expressing a product as a sum.

Using similar techniques we can establish the following factor formulae.

$$2 \sin A \cos B \equiv \sin (A + B) + \sin (A - B) \qquad \text{①}$$
$$2 \cos A \sin B \equiv \sin (A + B) - \sin (A - B) \qquad \text{②}$$
$$2 \cos A \cos B \equiv \cos (A + B) + \cos (A - B) \qquad \text{③}$$
$$-2 \sin A \sin B \equiv \cos (A + B) - \cos (A - B) \qquad \text{④}$$

Then by setting $A + B = P$ and $A - B = Q$

$$\sin P + \sin Q \equiv 2 \sin \frac{(P + Q)}{2} \cos \frac{(P - Q)}{2} \qquad \text{⑤}$$

$$\sin P - \sin Q \equiv 2 \cos \frac{(P + Q)}{2} \sin \frac{(P - Q)}{2} \qquad \text{⑥}$$

$$\cos P + \cos Q \equiv 2\cos\frac{(P+Q)}{2}\cos\frac{(P-Q)}{2} \qquad ⑦$$

$$\cos P - \cos Q \equiv -2\sin\frac{(P+Q)}{2}\sin\frac{(P-Q)}{2} \qquad ⑧$$

These identities are most easily learned in words, so for example equation ⑤ may be recalled as

sine + sine = twice sin (half-sum) cos (half-difference).

Example 7.9

Solve the equations

(a) $\sin 3\theta - \sin \theta = 0$ (b) $\cos\left(\theta + \dfrac{2\pi}{3}\right) - 1 = \cos\theta$

in the range $0 \leqslant \theta < 2\pi$

(a) Using equation ⑥

$$\sin 3\theta - \sin\theta \equiv 2\cos 2\theta . \sin\theta$$

so

$$\sin 3\theta - \sin\theta = 0$$

$$\Rightarrow 2\cos 2\theta . \sin\theta = 0$$

$$\Rightarrow \cos 2\theta = 0 \quad \text{or} \quad \sin\theta = 0$$

$$\Rightarrow 2\theta = \frac{\pi}{2}, \frac{3\pi}{2}, \frac{5\pi}{2}, \frac{7\pi}{2} \quad \text{or} \quad \theta = 0 \text{ or } \pi$$

$$\Rightarrow \theta = 0, \frac{\pi}{4}, \frac{3\pi}{4}, \pi, \frac{5\pi}{4} \text{ or } \frac{7\pi}{4}$$

(b) $\cos\left(\theta + \dfrac{2\pi}{3}\right) - 1 = \cos\theta$

$$\Rightarrow \cos\left(\theta + \frac{2\pi}{3}\right) - \cos\theta = 1$$

$$\Rightarrow -2\sin\left(\theta + \frac{\pi}{3}\right)\sin\frac{\pi}{3} = 1$$

$$\Rightarrow \sin\left(\theta + \frac{\pi}{3}\right) = \frac{-1}{\sqrt{3}} \qquad \text{since} \quad \sin\frac{\pi}{3} = \frac{\sqrt{3}}{2}$$

$$\Rightarrow \theta + \frac{\pi}{3} = \pi + 0.615 \quad \text{or} \quad \theta + \frac{\pi}{3} = 2\pi - 0.615$$

$$\Rightarrow \theta = 2.71 \quad \text{or} \quad \theta = 4.62.$$

Example 7.9a

Solve $\cos\theta + \cos 2\theta + \cos 3\theta + \cos 4\theta = 0$ in the range $0 < \theta < \pi$.

In this example we may use equation ⑦ for both $\cos\theta + \cos 2\theta$ and $\cos 3\theta + \cos 4\theta$ to obtain

$$2\left(\cos\frac{3\theta}{2}\cos\frac{-\theta}{2}\right) + 2\left(\cos\frac{7\theta}{2}\cos\frac{-\theta}{2}\right)1 = 0$$

$$\Rightarrow \qquad 2\cos\left(-\frac{\theta}{2}\right)\left(\cos\frac{3\theta}{2} + \cos\frac{7\theta}{2}\right) = 0$$

Applying equation ⑦ again

$$2\cos\left(-\frac{\theta}{2}\right)2\left(\cos\frac{5\theta}{2}\cos(-\theta)\right) = 0$$

$$\Rightarrow \cos\frac{\theta}{2}\cos\frac{5\theta}{2}\cos\theta = 0 \qquad \text{(since } \cos(-\theta/2) = \cos\theta/2 \text{ and}$$
$$\cos(-\theta) = \cos\theta)$$

$$\Rightarrow \cos\frac{\theta}{2} = 0 \quad \text{or} \quad \cos\frac{5\theta}{2} = 0 \quad \text{or} \quad \cos\theta = 0.$$

Since $0 < \theta < \pi$, then $0 < \theta/2 < \pi/2$ and there are no solutions of $\cos\theta/2 = 0$ in the required range.
  However if

$$\cos\frac{5\theta}{2} = 0 \qquad \text{where } 0 \leqslant \frac{5\theta}{2} < \frac{5\pi}{2}$$

$$\Rightarrow \qquad \frac{5\theta}{2} = \frac{\pi}{2} \quad \text{or} \quad \frac{3\pi}{2}$$

$$\Rightarrow \qquad \theta = \frac{\pi}{5} \quad \text{or} \quad \frac{3\pi}{5}$$

Finally, if $\cos\theta = 0$ then $\theta = \pi/2$. Thus the solutions in the required range are

$$\theta = \frac{\pi}{5}, \frac{\pi}{2} \text{ and } \frac{3\pi}{5}.$$

Exercise 7.3

1. Factorise each of the following expressions.

    (a) $\sin 3\theta + \sin 5\theta$ (b) $\sin 5\theta - \sin 3\theta$ (c) $\cos 7\theta + \cos 3\theta$
    (d) $\cos 5x - \cos 3x$

2. Solve each of the following equations in the range $0° \leqslant \theta < 180°$.

    (a) $\sin 5\theta - \sin 3\theta = 0$ (b) $\sin\theta + \sin 3\theta = 0$ (c) $\cos 3\theta + \cos\theta = 0$

3. Factorise $\cos 5x + \cos x$ and hence solve the equation $\cos 5x + \cos x = \cos 3x$ in the range $0° \leqslant x < 360°$.

4. Solve the equations $\cos 2\theta + \cos 4\theta + \cos 6\theta = 0$ in the range $-180° < \theta \leqslant 180°$.

5. By noting that $\cos A = \sin(90° - A)$ solve the equation $\sin 3x + \cos x = 0$ in the range $0 \leqslant x < 360$.

6. Solve the equation

$$\cos \theta - 2 \cos 2\theta + 2 \cos 4\theta - \cos 5\theta = 0$$

for $0 \leqslant \theta \leqslant \pi$.

7. Show that, for all θ and x,

$$\cos \theta + \cos(\theta + \alpha) + \cos(\theta + 2\alpha) = (2 \cos \alpha + 1) \cos(\theta + \alpha).$$

Given that $0 < \alpha < \frac{1}{2}\pi$, find, in terms of α, all values of θ in the interval $0 < \theta < 2\pi$, for which

$$\cos \theta + \cos(\theta + \alpha) + \cos(\theta + 2\alpha) = 0.$$

W 1983

8. Find all solutions of

$$\sin \theta - \sin 2\theta + \sin 3\theta - \sin 4\theta = 0$$

lying between $0°$ and $180°$ inclusive.

W 1983

9. (a) Find all values of x in the range $0 \leqslant x \leqslant \pi$ for which $\tan 4x + \tan x = 0$.
   (b) Determine the set of values of x in the range $0 \leqslant x \leqslant 2\pi$ for which $\sin x > \cos x$.

O&C 1983

10. Solve for x in the range $0 \leqslant x \leqslant \pi$

    (a) $\cos x + \cos 5x = \cos 2x$
    (b) $\sin x + \sin 3x = \sin 2x + \sin 4x$.

7.5 Further identities

Example 7.10
Show that

$$\frac{1 - \cos 2\theta}{\sin 2\theta} \equiv \tan \theta$$

When establishing an identity, it is usual to start with one side and 'break it down' to the other side. Now

$$\frac{1 - \cos 2\theta}{\sin 2\theta} \equiv \frac{1 - (\cos^2 \theta - \sin^2 \theta)}{2\cos \theta \sin \theta}$$

$$\equiv \frac{1 - (1 - 2\sin^2 \theta)}{2\cos \theta \sin \theta}.$$

This variant of the expansion of $\cos 2\theta$ is used to ensure that the constant term in the numerator is eliminated.

$$\equiv \frac{2\sin^2 \theta}{2\cos \theta \sin \theta}$$

$$\equiv \tan \theta$$

Example 7.11
Show that

$$\cot (A - B) \equiv \frac{1 + \cot A \cot B}{\cot B - \cot A}.$$

Now

$$\cot (A - B) \equiv 1/\tan (A - B)$$

$$\equiv 1 \bigg/ \frac{\tan A - \tan B}{1 + \tan A \tan B}$$

$$\equiv \frac{1 + \tan A \tan B}{\tan A - \tan B}$$

In order to reintroduce cotangents, multiply both the numerator and the denominator by $\cot A \cdot \cot B$.

$$\frac{\cot A \cot B + 1}{\cot B - \cot A}$$

which is our required result.

Example 7.12
Express $\tan (A + B + C)$ in terms of $\tan A$, $\tan B$ and $\tan C$. Hence show that when A, B and C are the angles of a triangle then

$$\tan A + \tan B + \tan C \equiv \tan A \cdot \tan B \cdot \tan C.$$

Now

$$\tan (A + B + C) \equiv \frac{\tan A + \tan (B + C)}{1 - \tan A \cdot \tan (B + C)}$$

$$\Rightarrow \tan(A + B + C) \equiv \frac{\tan A + \left(\dfrac{\tan B + \tan C}{1 - \tan B \tan C}\right)}{1 - \tan A \left(\dfrac{\tan B + \tan C}{1 - \tan B \tan C}\right)}$$

$$= \frac{\tan A(1 - \tan B \tan C) + (\tan B + \tan C)}{(1 - \tan B \tan C) - \tan A(\tan B + \tan C)}$$

$$= \frac{\tan A + \tan B + \tan C - \tan A \tan B \tan C}{1 - \tan B \tan C - \tan A \tan B - \tan A \tan C}$$

If A, B and C represent the angles of a triangle then $A + B + C = \pi$, and so $\tan(A + B + C) = 0$. Thus the numerator of r.h.s. is zero, that is

$$\tan A + \tan B + \tan C - \tan A \tan B \tan C = 0$$

$$\Rightarrow \tan A + \tan B + \tan C = \tan A \tan B \tan C \quad \text{!!}$$

Exercise 7.4

1. Prove that $\cot \theta - 2 \cot 2\theta \equiv \tan \theta$.

2. If A, B and C are the angles of a triangle show that

    (a) $\sin(B + C) \equiv \sin A$ (b) $\cos(B + C) \equiv -\cos A$.

3. Show that

    $$\frac{\sin 7\theta - \sin \theta}{\cos 7\theta - \cos 5\theta} \equiv \tan \theta.$$

4. Show that

    $$\frac{\tan \theta + \cot \theta}{\sec \theta + \operatorname{cosec} \theta} \equiv \frac{1}{\sin \theta + \cos \theta}.$$

5. Show that

    $$\cos 2\theta + \sin 2\theta \equiv \frac{(1 + \tan \theta)^2 - 2\tan^2 \theta}{1 + \tan^2 \theta}.$$

6. Show that

    $$(\sec \theta - \tan \theta)^2 \equiv \frac{1 - \sin \theta}{1 + \sin \theta}.$$

7. Prove the identity

    $$\cos 3\theta = 4\cos^3 \theta - 3\cos \theta.$$

Find all the solutions in the interval $-\pi < \theta \leqslant \pi$ of the equation

$$\cos 3\theta = 4\cos^2 \theta.$$

JMB 1982

8. Prove that

$$\left(\frac{\sin 3A}{\sin A}\right)^2 - \left(\frac{\cos 3A}{\cos A}\right)^2 \equiv 8\cos 2A.$$

9. Prove that

$$\cot 2\theta + \tan \theta \equiv \operatorname{cosec} 2\theta$$

for those values of θ for which each of the above expressions are defined.

Exercise 7.5 (miscellaneous)

1. (a) Express $\dfrac{\sin 4A}{\sin A}$ as a polynomial in $\cos A$.

    (b) Express $\sin^2 (A + B) - \sin^2 (A - B)$ in terms of $\sin 2A$ and $\sin 2B$.
    (c) If $\tan A = \frac{1}{3}$ and $\sin (A + B) = 2\cos (A - B)$, find $\tan B$ without using a calculator (or other calculating aid).

    O&C 1982

2. Show that $\cos 3\theta = 4\cos^3 \theta - 3\cos \theta$ and hence, or otherwise, find all values of θ between $0°$ and $360°$ satisfying $4\cos 3\theta = \cos^2 \theta$.

    W 1982

3. Show, without using tables, that $\tan 22\frac{1}{2}° = \sqrt{2} - 1$.

4. (a) Show, without using tables or calculator, that if $\alpha = \frac{1}{5}\pi$ then $\sin 3\alpha = \sin 2\alpha$.
    (b) Given that $\sin 3\theta = \sin 2\theta$, show that either $\sin \theta = 0$, or $4\cos^2 \theta - 2\cos \theta - 1 = 0$.
    (c) Using the results of (a) and (b), or otherwise, obtain the exact value of $\cos \frac{1}{5}\pi$.
    (d) Show that $\cos \frac{2}{5}\pi = \frac{1}{4}(\sqrt{5} - 1)$.

    CAMB 1983

5. Solve the following equations, giving your answers in the range $0°$ to $360°$.

    (a) $\sec^2 \theta - 3\tan \theta - 5 = 0$,
    (b) $\cos 4x + \cos 2x - \sin 4x + \sin 2x = 0$.

    AEB 1982

6. (a) Solve for θ the equation $3\cos 2\theta - \sin 2\theta = 2$, giving all solutions between $0°$ and $360°$, correct to the nearest degree.

(b) Prove that $\sin 3x = 3 \sin x - 4 \sin^3 x$. Hence solve the equation $\sin 3x = \sin^2 x$, giving all solutions for x (radians) such that $0 \leqslant x \leqslant 2\pi$.

7. Show that $(2t - 1)$ is a factor of the expression $30t^3 - 11t^2 - 4t + 1$ and find the other factors. Hence find all values of θ in the interval $-180° < \theta < 180°$ for which $30 \tan^3 \theta - 11 \tan^2 \theta - 4 \tan \theta + 1 = 1$, giving each answer to the nearest degree.

AEB 1983

8. Using the result

$$\tan 3x = \frac{3 \tan x - \tan^3 x}{1 - 3 \tan^2 x},$$

show that the equation

$$11 \tan 3x = 13 \tan x - 24$$

becomes

$$7t^3 - 18t^2 + 5t + 6 = 0,$$

where $t = \tan x$.

Find a positive integer satisfying this cubic equation in t and hence find the three values of t.

Find all angles x, such that $0° < x < 180°$, satisfying the equation

$$11 \tan 3x = 13 \tan x - 24$$

giving your answers to the nearest $0.1°$.

CAMB 1983

9. Prove that $\operatorname{cosec} \theta - \cot \theta \equiv \tan \dfrac{\theta}{2}$.

10. Solve for x in the range $0 \leqslant x \leqslant 2\pi$

$$5 \cos^2 x - 5 \sin x \cos x + 2 \sin^2 x = 1.$$

(*Hint*: divide by $\cos^2 x$.)

11. Show that $\cos^4 \theta + \sin^4 \theta \equiv 1 - 2 \sin^2 \theta \cos^2 \theta$.

12. Solve for x, where $-180° < x \leqslant 180°$, the equations

(a) $\sin (x + 60°) = \cos x$,   (b) $5 \cos x = 2 + 3 \sin x$,

giving your answers to the nearest $0.1°$.

CAMB 1983

13. Given that $y = \dfrac{\sin \theta - 2 \sin 2\theta + \sin 3\theta}{\sin \theta + 2 \sin 2\theta + \sin 3\theta}$, prove that $y = -\tan^2 \dfrac{\theta}{2}$. Find

(a) the exact value of $\tan^2 15°$ in the form $p + q\sqrt{r}$, where p, q and r are integers,

(b) the values of θ between $0°$ and $360°$ for which $2y + \sec^2 \dfrac{\theta}{2} = 0$.

AEB 1983

14. By quoting any necessary formulae, show that, if $\cos \theta$ is denoted by c, then
$\cos 3\theta = 4c^3 - 3c$.

Given that $\theta = 2\pi/5$ satisfies the equation $\cos 3\theta = \cos 2\theta$, deduce that $\cos 2\pi/5$ is a root of the equation $4c^3 - 2c^2 - 3c + 1 = 0$.

Solve this cubic equation and hence show that the value of $\cos 2\pi/5$ can be expressed as $\frac{1}{4}(\sqrt{5} - 1)$.

SU 1982

15. A cube has its base $ABCD$ in contact with a horizontal plane, the edges AA', BB', CC', DD' being vertical. Each edge of the cube is of length a. The cube is rotated about the edge AD through an angle β, where $0° < \beta < 45°$, and is maintained in this position.

(a) Find, in terms of a and β, the heights of C and C' above the horizontal plane.

(b) Show that the diagonal AC of the base of the cube is inclined to the horizontal at an angle θ, where

$$\sin \theta = \frac{\sin \beta}{\sqrt{2}}.$$

(c) Show that the diagonal AC' of the cube is inclined to the horizontal at an angle ϕ, where

$$\sin \phi = \frac{\sin \beta + \cos \beta}{\sqrt{3}}.$$

CAMB 1983

16. (a) Find the four smallest positive angles, in radians, satisfying $\sin^2 2x = \frac{1}{2}$. (You may leave the angles in terms of π.)

(b) Sketch on the same axes the graphs of $y = \sin x$ and $y = \tan(\frac{1}{2}x + \frac{1}{2}\pi)$ in the range $-2\pi \leqslant x \leqslant 2\pi$ and hence write down the solutions of $\sin x = \tan(\frac{1}{2}x + \frac{1}{2}\pi)$ in that range.

SU 1980

17. (a) A circle has centre O and radius $2\,\text{cm}$ and AB is a chord of the circle. Given that the angle ABO is 0.4 radians, calculate the area of the major segment of the circle cut off by AB.

(b) Find all the values of x between 0 and 360 which satisfy the equation

$$\sin 2x° = \cos 36°.$$

18. Express $8\cos^4\theta$ in the form

$$a\cos 4\theta + b\cos 2\theta + c,$$

giving the numerical values of the constants a, b and c.

CAMB 1982

19. Two equal vertical poles AB, DC have their lowest points A and D on horizontal ground. The point O on the ground is such that the angles of elevation of B and C from O are each equal to α. It is given that $A\hat{O}D = 2\beta$. The mid-point of BC is N.

    (a) Denoting the angle BOC by 2θ, prove that $\sin\theta = \cos\alpha\sin\beta$.
    (b) Denoting the angle of elevation of N from O by ϕ, prove that $\tan\phi = \tan\alpha\sec\beta$.

CAMB 1982

20. A pyramid $VABCD$ stands on a square vase $ABCD$ and has four identical sloping faces meeting at the vertex V. Given that VA is 60 cm and $A\hat{V}B = 80°$, calculate

    (a) the length of AB,
    (b) the length of VM, where M is the mid-point of AB,
    (c) the angle that the face VAB makes with the horizontal,
    (d) the perpendicular distance from A to VB,
    (e) the angle between the faces VAB and VBC.

AEB 1983

21. An aircraft flying in a straight line is climbing at an angle ϕ to the horizontal. It is observed from a point O on horizontal ground. The aircraft passes through a point A' vertically above a point A on the ground due east of O and, at a later time, through a point B' vertically above a point B on the ground due north of O, where $OA = OB$. When the aircraft is at A', its angle is elevation from O is α and when at B', its angle of elevation from O is β. Prove that

$$\tan\phi = \frac{1}{\sqrt{2}}(\tan\beta - \tan\alpha).$$

Show further that, when the aircraft is vertically above the mid-point of AB, its angle of elevation, γ, from O is given by

$$\tan\gamma = \frac{1}{\sqrt{2}}(\tan\alpha + \tan\beta).$$

CAMB 1983

22. The top H of a hill is 200 m above the horizontal plane through a point P and is due north of P. The angle of elevation of H from P is α, where $\tan\alpha = \frac{2}{5}$.

    The top T of a tower is 40 m above the horizontal plane through P and such that H is on a bearing 015° from T. The angle of elevation of H from T is β, where $\tan\beta = \frac{1}{2}$.

Calculate the horizontal distance of

(a) *P* from *H*   (b) *T* from *H*   (c) *P* from *T*.

AEB 1982

23. Find all the solutions, in terms of π, in the interval $0 \leqslant \theta \leqslant 2\pi$ of **each** of the following equations:

(a) $\cos 2\theta + \cos \theta = 0$;
(b) $\tan 2\theta - 3 \tan \theta = 0$;
(c) $\sqrt{3} \cos \theta + \sin \theta + 1 = 0$.

JMB 1983

7.6   Differentiation of the trigonometric functions

Suppose $f(x) \equiv \sin x$ for all x, where x is measured in *radians* and we wish to find the derived function $f'(x)$.

Recall that

$$f'(x) \equiv \lim_{\delta x \to 0} \left[\frac{f(x + \delta x) - f(x)}{\delta x} \right]$$

where δx is a small change in x. So, in this case

$$f'(x) \equiv \lim_{\delta x \to 0} \left[\frac{\sin(x + \delta x) - \sin x}{\delta x} \right]$$

Now, by factorising we find

$$\sin(x + \delta x) - \sin x \equiv 2\cos\text{ (half sum) sin (half difference)}$$

$$\equiv 2\cos\left(x + \frac{\delta x}{2}\right)\sin\left(\frac{\delta x}{2}\right).$$

That is

$$f'(x) \equiv \lim_{\delta x \to 0} \left[\frac{2\cos\left(x + \dfrac{\delta x}{2}\right)}{\delta x} \sin\left(\frac{\delta x}{2}\right) \right]$$

$$\equiv \lim_{\delta x \to 0} \left[\frac{\cos\left(x + \dfrac{\delta x}{2}\right)\sin\dfrac{\delta x}{2}}{\dfrac{\delta x}{2}} \right]$$

$$\equiv \lim_{\delta x \to 0} \left[\cos\left(x + \frac{\delta x}{2}\right) \right] \cdot \lim_{\delta x \to 0} \left[\sin\frac{\delta x}{2} \bigg/ \frac{\delta x}{2} \right]. \qquad ①$$

using the property of limits stated on page 163.

However, since x is measured in radians, then

$$\sin\frac{\delta x}{2} \simeq \frac{\delta x}{2} \text{ with the approximation improving as } \delta x \to 0$$

or in other words

$$\lim_{\delta x \to 0} \frac{\sin(\delta x/2)}{(\delta x/2)} = 1 \qquad ②$$

and clearly

$$\lim_{\delta x \to 0}\left[\cos\left(x + \frac{\delta x}{2}\right)\right] = \cos x. \qquad ③$$

So from ① ② and ③

$$f'(x) \equiv \cos x.$$

Statement

> If x is measured in radians
>
> $$\frac{d}{dx}(\sin x) = \cos x$$
>
> $$\frac{d}{dx}(\cos x) = -\sin x$$
>
> $$\frac{d}{dx}(\tan x) = \sec^2 x$$

The proof that $\dfrac{d}{dx}(\cos x) \equiv -\sin x$ is left as an exercise.

To show that $\dfrac{d}{dx}(\tan x) \equiv \sec x$, suppose that

$$f(x) \equiv \tan x \qquad \text{for } \cos x \neq 0$$

then

$$f'(x) \equiv \lim_{\delta x \to 0}\left[\frac{\dfrac{\sin(x + \delta x)}{\cos(x + \delta x)} - \dfrac{\sin x}{\cos x}}{\delta x}\right]$$

$$\equiv \lim_{\delta x \to 0}\left[\frac{\sin(x + \delta x)\cos x - \sin x \cos(x + \delta x)}{\delta x \cos(x + \delta x)\ \cos x}\right]$$

$$\equiv \lim_{\delta x \to 0}\left[\frac{\sin \delta x}{\delta x \cos(x + \delta x)\cos x}\right] \qquad \text{since } \sin \delta x \equiv \sin((x + \delta x) - x)$$

$$\Rightarrow f'(x) \equiv \lim_{\delta x \to 0}\left[\frac{\sin \delta x}{\delta x}\right] . \lim_{\delta x \to 0}\left[\frac{1}{\cos (x + \delta x)\cos x}\right]$$

Now

$$\lim_{\delta x \to 0}\left[\frac{\sin \delta x}{\delta x}\right] = 1$$

and

$$\lim_{\delta x \to 0}\left[\frac{1}{\cos (x + \delta x)\cos x}\right] = \frac{1}{\cos^2 x}$$

$$= \sec^2 x$$

It follows that

$$f'(x) \equiv \sec^2 x$$

that is

$$\frac{d}{dx}(\tan x) \equiv \sec^2 x$$

Example 7.13

Differentiate the following functions w.r.t. x

(a) $y = \sin 2x$ (b) $y = \sec x$ (c) $y = \tan^2 x$

(a) Let $u = 2x$ so $y = \sin u$.

$$\Rightarrow \frac{du}{dx} = 2 \quad \text{and} \quad \frac{dy}{du} = \cos u$$

$$\Rightarrow \frac{dy}{dx} = \frac{dy}{du} . \frac{du}{dx}$$

$$= (\cos u) . 2$$

$$= 2\cos 2x \qquad \text{since } u = 2x$$

More generally, if $y = \sin (ax)$ then

$$\frac{dy}{dx} = a\cos (ax)$$

(b) If $y = \sec x$ we could differentiate this from first principles, as we did for $y = \sin x$. However

$$y = \frac{1}{\cos x}$$

$$y = (\cos x)^{-1}.$$

Now setting $u = \cos x$ we have

$$y = u^{-1} \quad \text{and} \quad u = \cos x$$

$$\Rightarrow \frac{dy}{du} = \frac{-1}{u^2} \quad \text{and} \quad \frac{du}{dx} = -\sin x$$

so

$$\frac{dy}{dx} = \frac{dy}{du} \cdot \frac{du}{dx}$$

$$= \frac{-1}{u^2}(-\sin x)$$

$$= \frac{\sin x}{u^2}$$

$$= \frac{\sin x}{\cos^2 x} \qquad \text{since } u = \cos x$$

$$= \tan x \sec x$$

(c) $y = \tan^2 x$. Let $u = \tan x$ so $y = u^2$

$$\Rightarrow \frac{du}{dx} = \sec^2 x \quad \text{and} \quad \frac{dy}{du} = 2u$$

$$\Rightarrow \frac{dy}{dx} = \frac{dy}{du} \cdot \frac{du}{dx}$$

$$= 2u \cdot \sec^2 x$$

$$= 2 \tan x \sec^2 x.$$

Statement

If x is measured in radians then

$$\frac{d}{dx}(\operatorname{cosec} x) \equiv -\operatorname{cosec} x \cot x$$

$$\frac{d}{dx}(\sec x) \equiv \sec x \tan x$$

$$\frac{d}{dx}(\cot x) \equiv -\operatorname{cosec}^2 x$$

It is worth noting in general that

Statement

> If
>
> $$y = \sin(f(x))$$
>
> then
>
> $$\frac{dy}{dx} = \cos(f(x)) f'(x)$$
>
> Further, if
>
> $$y = f(\sin x)$$
>
> then
>
> $$\frac{dy}{dx} = f'(\sin x) \cos x$$

Similar results hold for the other trigonometric functions.

Example 7.14

Differentiate

(a) $y = \cos(x^3 + 1)$,   (b) $y = \sqrt{1 + \sin^2 x}$

(a) $\dfrac{dy}{dx} = -\sin(x^3 + 1) \cdot 3x^2$

$\qquad = -3x^2 \sin(x^3 + 1)$.

(b) $\qquad y = \sqrt{1 + \sin^2 x}$

$\Rightarrow \dfrac{dy}{dx} = \tfrac{1}{2}(1 + \sin^2 x)^{-1/2} \cdot 2 \sin x \cos x$

$\qquad = \sin x \cos x (1 + \sin^2 x)^{-1/2}$

$\qquad = \dfrac{\sin 2x}{2\sqrt{1 + \sin^2 x}}$

Exercise 7.6

1. Differentiate each of the following w.r.t. x, where a and b are constants

(a) $\sin 3x$	(b) $\sin ax$	(c) $\sin(x + \pi)$
(d) $\sin(ax + b)$	(e) $\cos 5x$	(f) $\cos(-x)$
(g) $\cos(x - \pi/2)$	(h) $\cos(ax + b)$	(i) $\tan(ax + b)$

(j) $\sin^2 x$ (k) $\sin^3 x$ (l) $\sin^9 x$

(m) $\cos^3 x$ (n) $\tan^3 x$ (o) $\sin(x^2)$

(p) $\sin(x^2 + 1)$ (q) $\sin\sqrt{1 + x}$ (r) $\cos\sqrt{x^2 + 1}$

(s) $\tan\left(\dfrac{1}{x^2}\right)$ (t) $\cos[(1 + x)^2]$

2. Find dy/dx using stated results, where

    (a) $y = \operatorname{cosec} x$ (b) $y = \cot(x^2)$ (c) $y = \cot(\sqrt{x})$

    (d) $y = \sin(\cos x)$ (e) $y = \sqrt{1 + \sin x}$ (f) $y = (1 - \cos x)^5$

    (g) $y = 1/(1 - \tan x)$ (h) $y = (1 - \sin^2 x)^{1/2}$

3. Differentiate the following from first principles.

    (a) $\cos x$ (b) $\operatorname{cosec} x$

4. By changing x to radians find dy/dx where

    (a) $y = \sin x°$ (b) $y = \cos x°$ (c) $y = \tan x°$

5. By writing the derivation of $f(x)$ as $f'(x)$ find the derivative of the function s given by

    (a) $s(x) \equiv \sin(f(x))$ (b) $s(x) \equiv \cos(f(x))$

6. Find $\dfrac{dy}{dx}$ and $\dfrac{d^2y}{dx^2}$ if $y = \cos x - 2\sin x$.

7. Use the methods of calculus to find the stationary values of the functions given by

$$f(\theta) \equiv 5\sin\theta + 12\cos\theta$$
$$g(\theta) \equiv 3\sin\theta - 4\cos\theta$$
$$h(\theta) \equiv 24\cos\theta - 7\sin\theta$$

In each case determine whether the stationary value is a maximum or a minimum.

[It is worth comparing this approach with the one developed on page 214]

8. Differentiate

    (a) $y = \sin^5 x$,   (b) $y = \sqrt{\cos x}$,   (c) $y = (1 + \sin^2 x)^3$,

    (d) $y = \sqrt{\cos x + \sin x}$.

7.7   The inverse trigonometric functions

The function

$$f(x) \equiv \sin x \qquad x \in \mathbb{R} \ (x \text{ measured in radians})$$

maps each real number x onto the number $\sin x$, where

$$-1 \leqslant \sin x \leqslant 1.$$

It is a many–one function and so does not possess an inverse.
   However, we may define a new function by

$$s(x) \equiv \sin x \qquad -\pi/2 \leqslant x \leqslant \pi/2$$

where by restricting the domain we obtain a 1–1 function, which will, therefore, possess an inverse, see Fig. 7.3.

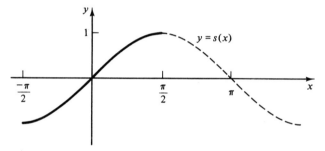

Fig. 7.3

   The inverse of s will map a number β, where $-1 \leqslant \beta \leqslant 1$ onto another number α where $\sin \alpha = \beta$, α is called the $\sin^{-1}$ of β or more generally

$$s^{-1}(x) \equiv \sin^{-1}(x) \qquad -1 \leqslant x \leqslant 1.$$

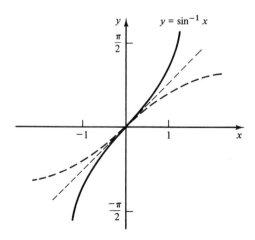

Fig.7.4

For example,

$$\sin^{-1} 1 = \pi/2 \qquad \text{since} \quad \sin \pi/2 = 1$$

$$\sin^{-1} \tfrac{1}{2} = \pi/6 \qquad \text{since} \quad \sin \pi/6 = \tfrac{1}{2}.$$

Thus $\beta = \sin \alpha \Leftrightarrow \sin^{-1} \beta = \alpha$.

Now the graph of $y = \sin^{-1} x$ (Fig. 7.4) may be obtained from that of

$$y = \sin x \qquad \text{for} \quad -\pi/2 \leqslant x \leqslant \pi/2$$

by reflecting in the line $y = x$. Notice that

$$\sin(\sin^{-1} x) \equiv x$$

[in words: the sine of (the angle whose sin is x) $= x$!]. Similarly

$$\sin^{-1}(\sin x) = x.$$

Thus if

$$y = \sin^{-1} x \qquad -\pi/2 \leqslant x \leqslant \pi/2$$

then $\sin y = x.$

Again the cosine function is many–one, but by restricting its domain we have

$$c(x) \equiv \cos x \qquad 0 \leqslant x \leqslant \pi$$

which is a 1–1 function.

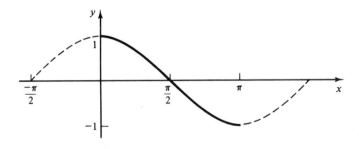

Fig.7.5

The inverse function is given by

$$c^{-1}(x) \equiv \cos^{-1} x \qquad -1 \leqslant x \leqslant 1$$

where $\cos^{-1} x$ is the angle whose cosine is x and so the range of c^{-1} is from 0 to π. The graph of the inverse function is obtained by reflecting in the line $y = x$. Thus

$$y = \cos x \Leftrightarrow \cos^{-1} y = x \qquad \text{for} \quad 0 \leqslant x \leqslant \pi$$

and

$$y = \cos^{-1} x \Leftrightarrow \cos y = x \qquad \text{for} \quad -1 \leqslant x \leqslant 1$$

See Fig. 7.6 on the next page.

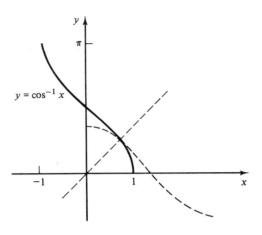

Fig. 7.6

Similarly we consider the 1–1 function

$$t(x) \equiv \tan x \qquad -\pi/2 \leqslant x \leqslant \pi/2$$

where the inverse

$$t^{-1}(x) \equiv \tan^{-1} x$$

where $\tan^{-1} x$ is the angle whose tangent is x.

Thus

$$y = \tan x \Leftrightarrow \tan^{-1} y = x \qquad \text{for} \quad -\pi/2 \leqslant x \leqslant \pi/2$$

and

$$y = \tan^{-1} x \Leftrightarrow \tan y = x \qquad \text{for } x \in \mathbb{R}$$

The graph of $y = \tan^{-1} x$ may readily be found (see Fig. 7.7).

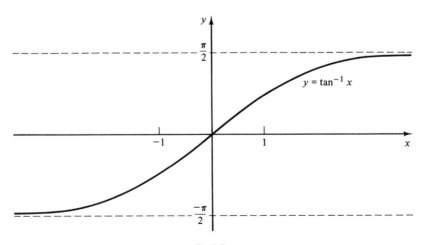

Fig.7.7

Differentiation of the inverse trigonometric functions

Suppose

$$y = \sin^{-1} x \qquad -1 \leqslant x \leqslant 1$$

then $\sin y = x.$

So differentiating with respect to y

$$\cos y = \frac{dx}{dy}$$

$$\Rightarrow \quad \frac{dy}{dx} = \frac{1}{\cos y} \qquad \text{since } \frac{dy}{dx} = \frac{1}{dx/dy}.$$

But

$$\cos y \equiv \sqrt{1 - \sin^2 y}$$

$$\equiv \sqrt{1 - x^2} \qquad \text{since } \sin y = x.$$

[*Note*: the positive root was taken since the gradient of $y = \sin^{-1} x$ is positive for all relevant x.]

Thus

$$\frac{dy}{dx} = \frac{1}{\sqrt{1 - x^2}}.$$

It can be shown that

Statement

	notice that the gradient is:
$y = \sin^{-1} x \Rightarrow \dfrac{dy}{dx} = \dfrac{1}{\sqrt{1 - x^2}}$	positive over the domain
$y = \cos^{-1} x \Rightarrow \dfrac{dy}{dx} = \dfrac{-1}{\sqrt{1 - x^2}}$	negative over the domain
$y = \tan^{-1} x \Rightarrow \dfrac{dy}{dx} = \dfrac{1}{1 + x^2}$	positive over the domain

Example 7.15

Differentiate

(a) $y = \sin^{-1} 3x,$ (b) $y = \tan^{-1} x^2,$ (c) $y = \cos^{-1}(2x + 4).$

(a) $y = \sin^{-1} 3x$. Let $u = 3x$, so $y = \sin^{-1} u$.

$$\frac{du}{dx} = 3 \quad \text{and} \quad \frac{dy}{dx} = \frac{1}{\sqrt{1 - u^2}}.$$

Now

$$\frac{dy}{dx} = \frac{dy}{du} \cdot \frac{du}{dx}$$

$$= 3 \cdot \frac{1}{\sqrt{1 - u^2}}$$

$$= \frac{3}{\sqrt{1 - 9x^2}} \quad \text{since } u = 3x.$$

Notice that if $y = \sin^{-1} 3x$ then $-1 \leqslant 3x \leqslant 1$.
    It then follows that $1 - 9x^2$ will be positive, and so the square root is defined.

(b) $y = \tan^{-1} x^2$. Let $u = x^2$, so $y = \tan^{-1} u$.

$$\frac{du}{dx} = 2x \quad \text{and} \quad \frac{dy}{du} = \frac{1}{1 + u^2}$$

Now

$$\frac{dy}{dx} = \frac{dy}{du} \cdot \frac{du}{dx}$$

$$= \frac{2x}{1 + u^2}$$

$$= \frac{2x}{1 + x^4} \quad \text{since } u = x^2.$$

(c) $y = \cos^{-1}(2x + 4)$. Let $u = 2x + 4$, so $y = \cos^{-1} u$.

$$\frac{du}{dx} = 2 \quad \text{and} \quad \frac{dy}{dx} = \frac{-1}{\sqrt{1 - u^2}}$$

Now

$$\frac{dy}{dx} = \frac{dy}{du} \cdot \frac{du}{dx}$$

$$= \frac{-2}{\sqrt{1 - u^2}}$$

$$= \frac{-2}{\sqrt{1 - (2x + 4)^2}}$$

Remark

> The function $\sin^{-1} x$ is sometimes referred to as $\arcsin x$. Similarly, $\arccos x$ is used instead of $\cos^{-1} x$. The notation is unfortunate since it could be mistaken for $1/\sin x$. This would be written as $(\sin x)^{-1}$.
> The confusion arises since $(\sin x)^n$ may be written as $\sin^n x$ for $n \neq -1$.

Exercise 7.7

1. Differentiate each of the following w.r.t. x.

   (a) $\sin^{-1} 6x$ (b) $\sin^{-1}(x + 4)$

   (c) $\sin^{-1}(4x - 1)$ (d) $\cos^{-1} x^2$

   (e) $\cos^{-1}(\sqrt{x})$ (f) $\cos^{-1}(1/\sqrt{x})$

   (g) $\tan^{-1}(1/x)$ (h) $\tan^{-1}(\sqrt{1 - x})$

   (i) $\tan^{-1}[(x^2 - 2)^2]$ (j) $\sqrt{\sin^{-1} x}$

2. Find the gradient at the origin of the curve

   (a) $y = \sin^{-1} x$,   (b) $y = \tan^{-1} x$,

3. Evaluate

   (a) $\sin^{-1}\left(\dfrac{1}{\sqrt{2}}\right)$ (b) $\cos^{-1}\left(\dfrac{-1}{\sqrt{2}}\right)$ (c) $\tan^{-1}\left(\dfrac{1}{\sqrt{3}}\right)$

   (d) $\tan^{-1}(\sqrt{3})$ (e) $\sin^{-1}\left(\dfrac{-\sqrt{3}}{2}\right)$ (f) $\sin^{-1} 1$

   (g) $\sin^{-1}\frac{1}{2}$ (h) $\tan^{-1} 2$.

4. Simplify each of the following.

   (a) $\sin^{-1}(\sin x)$
   (b) $\cos(\sin^{-1} x)$ [recall $\cos\theta = \sqrt{1 - \sin^2\theta}$]
   (c) $\tan(\tan^{-1} x + \tan^{-1} y)$

5. Find $\dfrac{d^2 y}{dx^2}$ if $y = \tan^{-1} x$.

6. Prove that

   $$\tan^{-1} A + \tan^{-1} B = \tan^{-1}\left(\frac{A + B}{1 - AB}\right)$$

   and hence show that $\tan^{-1}\frac{1}{2} + \tan^{-1}\frac{1}{3} = \dfrac{\pi}{4}$.

7. Given that $\sin^{-1} x$, $\cos^{-1} x$ and $\sin^{-1} (1 - x)$ are acute angles,

    (a) prove that $\sin \left[\sin^{-1} x - \cos^{-1} x \right] = 2x^2 - 1$,

    (b) solve the equation $\sin^{-1} x - \cos^{-1} x = \sin^{-1} (1 - x)$.

AEB 82

8. Starting from the formulae for $\sin (A + B)$ and $\cos (A + B)$, prove that

$$\tan (A + B) = \frac{\tan A + \tan B}{1 - \tan A \tan B}.$$

Find, to 2 decimal places, the positive value of x which satisfies the equation

$$\tan^{-1} x + \tan^{-1} 2x = \frac{\pi}{4}.$$

AEB 1983

7.8 Integration of trigonometric and related functions

It was noted on page 201, that

$$F'(x) = f(x) \Leftrightarrow \int f(x)\,dx = F(x) + c$$

Now, since

$$\frac{d}{dx}(\sin x) = \cos x$$

it follows that

$$\int \cos x\,dx = \sin x + c$$

assuming x is measured in radians.

    The following results may be similarly obtained.

Statement

$$\int \sin x\,dx = -\cos x + c$$

$$\int \cos x\,dx = \sin x + c$$

$$\int \sec^2 x\,dx = \tan x + c$$

assuming x is measured in radians.

Example 7.16
Integrate each of the following w.r.t. x.

(a) $\sin 2x$ (b) $\cos (4x + \pi/3)$ (c) $3 \sec^2(3x - \pi/4)$

(a) Since $\dfrac{d}{dx}(\cos 2x) = -2 \sin 2x$ it follows that

$$\int -2 \sin 2x \, dx = \cos 2x + c.$$

$$\Rightarrow \quad \int \sin 2x \, dx = -\frac{\cos 2x}{2} + c,$$

a fact which may be checked by differentiating the r.h.s.

(b) $\displaystyle\int \cos (4x + \pi/3) \, dx = \dfrac{\sin (4x + \pi/3)}{4} + c$

[This problem may have been tackled by first guessing the answer to be $\sin (4x + \pi/3)$. Now since

$$\frac{d}{dx}(\sin (4x + \pi/3)) = 4 \cos (4x + \pi/3)$$

the result follows.]

(c) $\displaystyle\int 3 \sec^2 (3x - \pi/4) \, dx = \tan (3x - \pi/4) + c$

$$\left[\frac{d}{dx}(\tan (3x - \pi/4)) = 3 \sec^2 (3x - \pi/4) \right]$$

Generally

$$\int \sin ax \, dx = -\frac{\cos ax}{a} + c$$

$$\int \cos (ax + b) \, dx = \frac{\sin (ax + b)}{a} + c$$

Example 7.17
Find

(a) $\displaystyle\int \sin^2 x \, dx,$ (b) $\displaystyle\int \sin^2 x \cos x \, dx,$ (c) $\displaystyle\int \sin^5 x \, dx.$

(a) This important integral, together with the related integral of $\cos^2 x$, may be found by using the double angle formula.

Recall that

$$\cos 2x \equiv 1 - 2\sin^2 x \quad \text{or} \quad 2\cos^2 x - 1$$

$$\Rightarrow \quad \sin^2 x \equiv \tfrac{1}{2}(1 - \cos 2x) \quad \text{and} \quad \cos^2 x \equiv \tfrac{1}{2}(1 + \cos 2x).$$

In particular

$$\int \sin^2 x \, dx = \int \left(\frac{1}{2} - \frac{\cos 2x}{2}\right) dx,$$

$$= \frac{x}{2} - \frac{\sin 2x}{4} + c.$$

(b) Since

$$\frac{d}{dx} \sin^3 x = 3\cos x \sin^2 x,$$

it follows that

$$\int \sin^2 x \cos x \, dx = \frac{\sin^3 x}{3} + c.$$

(c) The above method is useful for integrating an odd power of $\sin x$ or $\cos x$ as the following method illustrates.

$$\sin^5 x = \sin^4 x \sin x$$

$$= \sin^2 x \sin^2 x \sin x$$

$$= (1 - \cos^2 x)(1 - \cos^2 x)\sin x$$

$$= (1 - 2\cos^2 x + \cos^4 x)\sin x$$

$$= \sin x - 2\cos^2 x \sin x + \cos^4 x \sin x.$$

Thus

$$\int \sin^5 x \, dx = \int \sin x - 2\cos^2 x \sin x + \cos^4 x \sin x \, dx$$

$$= -\cos x + 2\frac{\cos^3 x}{3} - \frac{\cos^5 x}{5} + c.$$

The above methods may be adapted to deal with many other integrals.

Example 7.18

Find $\displaystyle\int \sin^4 x \, dx$.

$$\int \sin^4 x \, dx = \int \sin^2 x \sin^2 x \, dx$$

$$= \int \tfrac{1}{2}(1 - \cos 2x) \cdot \tfrac{1}{2}(1 - \cos 2x) \, dx$$

$$= \tfrac{1}{4} \int 1 - 2\cos 2x + \cos^2 2x \, dx$$

$$= \tfrac{1}{4} \int 1 - 2\cos 2x + \tfrac{1}{2}(1 + \cos 4x) \, dx$$

$$= \frac{x}{4} - \frac{1}{4}\sin 2x + \frac{x}{8} + \frac{1}{32}\sin 4x + c$$

Example 7.19

Find $\displaystyle\int \sin 2x \cos 4x \, dx$.

Recall from page 218 that

$$\sin mx \cos nx = \tfrac{1}{2}(\sin(m+n)x + \sin(m-n)x).$$

So

$$\int \sin 2x \cos 4x \, dx = \tfrac{1}{2} \int \sin(6x) + \sin(-2x) \, dx$$

$$= \tfrac{1}{2}\left[-\frac{\cos 6x}{6} + \frac{\cos(-2x)}{2} \right] + c.$$

Exercise 7.8

1. Integrate the following where a and b are constants

   (a) $2\sin x$

   (b) $\pi \cos x$

   (c) $\dfrac{\sec^2 x \, dx}{2}$

   (d) $a\cos x + b\sin x$

   (e) $3\cos(3x)$

   (f) $a\sec^2(bx)$

   (g) $\sec^2(2x - 1)$

   (h) $\sin(1 - x)$

2. Find

   (a) $\displaystyle\int \sin^3 x \cos x \, dx$

(b) $\displaystyle\int \cos^4 x \sin x \, dx$

(c) $\displaystyle\int \sin^3 x \, dx$ [*note*: $\sin^2 \theta \equiv 1 - \cos^2 \theta.$]

(d) $\displaystyle\int \cos^3 x \, dx$

(e) $\displaystyle\int \cos^2 x \, dx$ [*note*: $\cos^2 x \equiv \frac{1}{2}(1 - \cos 2x).$]

(f) $\displaystyle\int \sin^2 x \, dx$ [*note*: $\sin^2 x \equiv \frac{1}{2}(1 + \cos 2x).$]

(g) $\displaystyle\int \cos^2 3x \, dx$

(h) $\displaystyle\int a \sin^2 (bx) \, dx$

(i) $\displaystyle\int \sec^2 x \tan^3 x \, dx$

(j) $\displaystyle\int \sec^2 3x \tan^3 3x \, dx$

3. By converting to radians find

(a) $\displaystyle\int \sin x° \, dx$ (b) $\displaystyle\int \cos x° \, dx$ (c) $\displaystyle\int \sec^2 x° \, dx$

4. Find

(a) $\displaystyle\int \sin 3x \cos x \, dx$ (b) $\displaystyle\int \cos 2x \cos 5x \, dx$

(c) $\displaystyle\int 2 \sin 5x \sin 2x \, dx$ (d) $\displaystyle\int \cos x \cos 2x \, dx$

5. Find

(a) $\displaystyle\int \sin^3 x \cos^2 x \, dx$ (b) $\displaystyle\int \cos^5 x \, dx$

Revision Exercise B

1. Find an expression for $\log_{25} 200$ in terms of $\log_{10} 2$. Hence evaluate $\log_{25} 200$.

W 1983

2. The expression $2x^2 + 3x + C$, where C is a constant, has a minimum value of $2\frac{7}{8}$. Find the value of C.

CAMB 1982

3. If

$$f(x) \equiv \sin x \qquad -\pi/2 \leqslant x \leqslant \pi/2$$

find

(a) $f(1)$ (b) $f^{-1}(1)$.

4. The product of two numbers is 29 more than their sum, and the difference between the numbers is 1. Find the possible pairs of numbers.

SU 1983

5. Show that $x = 2$ is a root of the equation $x^3 + 3x - 14 = 0$.
   Given that the other roots are α and β, show that $\alpha + \beta = -2$ and find the value of $\alpha\beta$.
   Find the equation with numerical coefficients whose roots are

(a) $\alpha + 3$ and $\beta + 3$,
(b) 5, $\alpha + 3$ and $\beta + 3$.

AEB 1982

6. Given

$$4x^3 + 3x^2y + y^3 = 8,$$
$$2x^3 - 2x^2y + xy^2 = 1$$

and $y = mx$, show that m satisfies the equation

$$m^3 - 8m^2 + 19m - 12 = 0.$$

Hence or otherwise, find the real values of x and y, leaving your answers in surd form.

W 1983

7. The cubic polynomial

$$2x^3 - 3x^2 + cx - 60$$

has a factor $(x + 2)$. Find the value of the constant c.
   With this value of c,

(a) express the polynomial as the product of three linear factors,
(b) obtain the set of values of x for which the polynomial is negative.

CAMB 1982

8. Obtain the stationary points of the function

$$f(x) \equiv 1 + 3\cos x + \cos 3x \qquad 0 \leqslant x < 2\pi$$

and determine whether each is a maximum, minimum or a point of inflexion.

9. The functions f and g are defined by

$$f : x \mapsto 2 + x - x^2, \quad x \in \mathbb{R},$$

$$g : x \mapsto \frac{1}{1 + \tan x}, \qquad 0 \leqslant x < \pi/2.$$

Determine the range of each function and state, in each case, whether or not an inverse function exists.

LOND 1983

10. Find the real values of k for which the equation $x^2 + (k + 1)x + k^2 = 0$ has (a) real roots, (b) one root double the other.

AEB 1983

11. Express the polynomial

$$f(x) = 2x^4 + x^3 - x^2 + 8x - 4$$

as the product of two linear factors and a quadratic factor $q(x)$.
    Prove that there are no real values of x for which $q(x) = 0$.

AEB 1983

12. In a triangle ABC, $AB = 9\,\text{cm}$, $BC = 10\,\text{cm}$, and $AC = 7\,\text{cm}$. A point O inside the triangle is such that $\angle OBC = 10°$ and $\angle OCB = 40°$. Calculate the length of OB and the size of $\angle AOB$.

W 1983

13. A function f is defined by

$$f(x) = \sin x \qquad \left(-\frac{\pi}{2} \leqslant x \leqslant \frac{\pi}{2} \right).$$

State briefly why f has an inverse function f^{-1}, giving the domain and range of f^{-1}. Find $f^{-1}(\tfrac{1}{2})$. Given further that $g(x) \equiv \cos x$ for all x, find

$$(g \circ f^{-1}) \left(\frac{\sqrt{3}}{2} \right) \qquad \text{and} \qquad (f^{-1} \circ g) \left(\frac{\pi}{3} \right).$$

W 1983

14. The diagram (Fig. 7.8) shows part of the graph of the function f. Draw as much of the graph as possible given that (a) f is odd, (b) f is even, (c) f is odd with period $2a$.

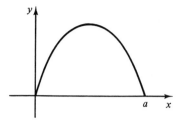

Fig. 7.8

15. The curve C has equation $xy = 2x^2 + 3$ and the line L has equation $y = kx + 3k + 2$, where k is a constant. Find the set of values of k for which

   (a) L is a tangent to C,
   (b) L does not meet C.

   AEB 1983

16. Solve the simultaneous equations,

$$x^2 + 2xy = 3,$$

$$2x - y = 1.$$

   OX 1983

17. (a) Given that $2^x/4^y = 1$ and $3^x(\sqrt{3})^y = 9$, obtain two linear equations in x and y. Hence determine x and y.
   (b) Given that $\log_{10} t = p$, express, in terms of p,

   (i) $\log_{10} \dfrac{10}{t^2}$,   (ii) $\log_t \dfrac{10}{t^2}$,   (iii) $\dfrac{\log_t 10}{\log_t t^2}$.

   AEB 1983

18. Determine whether the following functions are odd, even or periodic.

$$f(x) \equiv \frac{1}{\sqrt{1 - x^2}} \qquad x \in \mathbb{R}$$

$$g(x) \equiv (\cos x)^2 \qquad x \in \mathbb{R}$$

$$h(x) \equiv \tan x \qquad x \in \mathbb{R}$$

19. The quadratic function $f(x)$ takes the value 20 when $x = 1$ and $x = 5$ and takes the value 14 when $x = 2$. Obtain $f(x)$ in the form $ax^2 + bx + c$. Express $f(x)$ in the form $a(x - p)^2 + q$ and *hence* find the least possible value of $f(x)$. Draw a rough sketch of the graph of $y = f(x)$ and state its relation to the graph of $y = ax^2$. The line through the origin and the point $(2, 14)$ meets the graph of $y = f(x)$ again at the point P. Find the coordinates of P.

   SU 1980

20. Given that α and β are the roots of

$$4x^2 + 9x + 1 = 0,$$

find the values of $\alpha + \beta$ and $\alpha\beta$. Show that

$$\left(\sqrt{\frac{\alpha}{\beta}} - \sqrt{\frac{\beta}{\alpha}}\right)^2 = \frac{65}{4}.$$

W 1983

21. Find the set of values of k for which the equation

$$(2k - 1)x^2 + (2k + 1)x + (k + 1) = 0$$

has real roots.

Given that the sum of the roots of the equation

$$(2k - 1)x^2 + (2k + 1)x + (k + 1) = 0$$

is 7, find k and hence, or otherwise, find the numerical value of the product of these roots.

AEB 1982

22. Factorise $(a + b)^6 - (a - b)^6$.

Chapter 8   Further Differentiation

8.1   Exponential functions

We cannot as yet differentiate an exponential function of the form

$$f(x) \equiv a^x \qquad x \in \mathbb{R}$$

or

$$y = a^x.$$

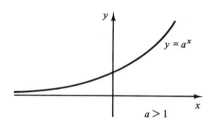

Fig. 8.1

As usual, we must attempt to differentiate from first principles. Now we know

$$\frac{\mathrm{d}y}{\mathrm{d}x} = \lim_{\delta x \to 0} \left[\frac{f(x + \delta x) - f(x)}{\delta x} \right]$$

$$= \lim_{\delta x \to 0} \left[\frac{a^{x+\delta x} - a^x}{\delta x} \right]$$

$$= \lim_{\delta x \to 0} \left[\frac{a^x(a^{\delta x} - 1)}{\delta x} \right]$$

$$= a^x \lim_{\delta x \to 0} \left[\frac{a^{\delta x} - 1}{\delta x} \right] \qquad \text{since } a^x \text{ does not depend on } \delta x.$$

The value of

$$\lim_{\delta x \to 0} \left[\frac{a^{\delta x} - 1}{\delta x} \right]$$

will depend on the choice of a.

Some approximate values are given in Table 8.1. These approximations may be checked on a calculator, selecting ever smaller values for δx.

a	approximate limit
1	0
2	0.69
3	1·09
4	1·39

Table 8.1

It is evident from these data that there will be a value, between 2 and 3, for a, when the limit will equal 1. This number is called e and is approximately 2.718. This number plays an important role in the further development of mathematics.

So we have

$$\lim_{\delta x \to 0} \left[\frac{e^{\delta x} - 1}{\delta x} \right] = 1.$$

It follows that

$$\boxed{\begin{array}{c} y = e^x \\ \Rightarrow \dfrac{dy}{dx} = e^x. \end{array}}$$

Example 8.1

Differentiate

(a) e^{2x} (b) $e^{\sin x}$ (c) $e^{f(x)}$.

(a) Set

$$y = e^{2x} \quad \text{and} \quad u = 2x.$$

So

$$y = e^u$$

$$\Rightarrow \frac{dy}{du} = e^u \quad \text{and} \quad \frac{du}{dx} = 2.$$

Now

$$\frac{dy}{dx} = \frac{dy}{du} \cdot \frac{du}{dx}$$

$$= e^u \cdot 2$$

$$= 2e^{2x}, \qquad \text{since } u = 2x.$$

(b) Set

$$y = e^{\sin x} \quad \text{and} \quad u = \sin x.$$

Thus

$$y = e^u$$

$$\Rightarrow \frac{dy}{du} = e^u \quad \text{and} \quad \frac{du}{dx} = \cos x.$$

Now

$$\frac{dy}{dy} = \frac{dy}{du} \cdot \frac{du}{dx}$$

$$= e^u \cos x$$

$$= \cos x\, e^{\sin x}, \qquad \text{since } u = \sin x.$$

(c) Set

$$y = e^{f(x)} \quad \text{and} \quad u = f(x)$$

Thus

$$y = e^u$$

$$\Rightarrow \frac{dy}{du} = e^u \quad \text{and} \quad \frac{du}{dx} = f'(x) \quad \text{[the derivative of } f(x)\text{]}$$

Now

$$\frac{dy}{dx} = \frac{dy}{du} \cdot \frac{du}{dx}$$

$$= e^u \cdot f'(x)$$

$$= f'(x)\, e^{f(x)}, \qquad \text{since } u = f(x).$$

So, quite generally,

$$\boxed{\begin{array}{c} y = e^{f(x)} \\[2mm] \dfrac{dy}{dx} = f'(x)\, e^{f(x)}. \end{array}}$$

We may now return to our original problem

$$y = a^x$$

$$\Rightarrow \ln y = \ln a^x \qquad \text{taking logs to base e}$$

$$[\textit{Note}: \ln x \equiv \log_e x]$$

$$\Rightarrow \ln y = x \ln a$$

$$\Rightarrow e^{\ln y} = e^{x \ln a}$$

$$\Rightarrow \quad y = e^{x \ln a} \qquad \text{since } e^{\ln y} = y$$

$$\Rightarrow \frac{dy}{dx} = \ln a\, e^{x \ln a}$$

$$= \ln a \cdot a^x \qquad \text{since } e^{x \ln a} = y$$

$$\text{and} \qquad y = a^x \quad !$$

It is worth noting that

$$f(x) \equiv e^x \qquad x \in \mathbb{R}$$

is called *the* exponential function; $f(x) \equiv a^x$ is called *an* exponential function.

Example 8.2.

Differentiate   (a) $\cos(e^{x^2})$,   (b) $\cos^2(e^x)$.

(a) Set

$$y = \cos(e^{x^2})$$

$$\frac{dy}{dx} = -\sin(e^{x^2}) \cdot \frac{d}{dx}(e^{x^2})$$

$$= -\sin(e^{x^2}) \cdot 2x e^{x^2}$$

$$= -2x e^{x^2} \sin(e^{x^2}).$$

(b) Set

$$y = \cos^2(e^x)$$

$$= [\cos(e^x)]^2$$

$$\Rightarrow \frac{dy}{dx} = 2\cos(e^x) \cdot \frac{d}{dx}(\cos(e^x))$$

$$= 2\cos(e^x) \cdot -\sin(e^x) \cdot \frac{d}{dx}(e^x)$$

$$= -2\cos(e^x)\sin(e^x) \cdot e^x$$

$$= -\sin(2e^x) \cdot e^x.$$

Exercise 8.1

1. Differentiate each of the following functions.

(a) $y = e^{2x}$
(b) $y = e^{-3x}$
(c) $y = e^{4x} + e^{-x}$
(d) $y = e^x - e^{-x}$
(e) $y = e^{x/2}$
(f) $y = e^{-x/3}$
(g) $y = e^{x^2}$
(h) $y = e^{-x^2/2}$
(i) $y = e^{3x+4}$
(j) $y = e^{2x-3}$
(k) $y = e^{-(x+3)}$
(l) $y = e^{2x^2-1}$
(m) $y = e^{(x-1)^2}$
(n) $y = e^{-2(x-3)^2}$
(o) $y = e^{-(x-2)^2}$
(p) $y = e^{\cos x}$
(q) $y = e^{\cos(x^2)}$
(r) $y = e^{\sin^2(x)}$
(s) $y = \cos(e^x)$
(t) $y = \tan(e^{2x})$
(u) $y = e^{e^x}$

(v) $y = \dfrac{e^{2x-1}}{e^x}$
(w) $y = (1 - e^x)^2$
(x) $y = \dfrac{(1 + e^{2x})^2}{e^x}$

(y) $y = e^x \sin^2 x + e^x \cos^2 x$!

2. Solve the equations

   (a) $e^{2x} - 4e^x + 3 = 0$ (b) $e^{2x} - 2e^x - 8 = 0$.

3. Find the equations of the tangents to the curve $y = e^{x^2}$ which pass through the origin.

JMB 1982

8.2   Differentiating logarithms

Suppose

$$y = \ln x \qquad [\text{Recall } \ln x \equiv \log_e (x).]$$

then

$$e^y = e^{\ln x}$$

$$\Rightarrow \quad e^y = x$$

$$\Rightarrow \quad e^y = \frac{dx}{dy}$$

$$\Rightarrow \quad \frac{dy}{dx} = \frac{1}{e^y} \qquad \left[\text{Recall that } \frac{dy}{dx} = \frac{1}{dx/dy} \right]$$

$$\qquad = \frac{1}{x} \qquad \text{since } x = e^y.$$

Statement

$$
\boxed{
\begin{array}{ll}
\text{If} & y = \ln (f(x)) \\[2mm]
\text{then} & \dfrac{dy}{dx} = \dfrac{f'(x)}{f(x)}
\end{array}
}
$$

For suppose $u = f(x)$ then $y = \ln u$ and so

$$\frac{dy}{du} = \frac{1}{u} \quad \text{and} \quad \frac{du}{dx} = f'(x).$$

Now

$$\frac{dy}{dx} = \frac{dy}{du} \cdot \frac{du}{dx}$$

$$= \frac{1}{u} \cdot f'(x)$$

$$= \frac{f'(x)}{f(x)} \qquad \text{since } u = f(x).$$

Example 8.3

Differentiate

(a) $y = \ln(1 + x)$,   (b) $y = \ln(\sin x)$,   (c) $y = \ln(x \sin x)$.

(a) $y = \ln(1 + x)$

$$\Rightarrow \frac{dy}{dx} = \frac{1}{1 + x}$$

(b) $y = \ln(\sin x)$

$$\Rightarrow \frac{dy}{dx} = \frac{\cos x}{\sin x} \equiv \cot x \qquad \sin x \neq 0.$$

(c) $y = \ln(x \sin x)$

$$\Rightarrow \quad y = \ln x + \ln(\sin x) \qquad \text{using the usual rules for logarithms}$$

$$\Rightarrow \frac{dy}{dx} = \frac{1}{x} + \cot x.$$

Example 8.4

Find the derived function where

(a) $f(x) \equiv \ln \sqrt{\dfrac{x}{1 - x}} \qquad 0 < x < 1$

(b) $g(x) \equiv \log_{10} x \qquad x > 0$

(a) By the usual rules for logarithms

$$f(x) \equiv \tfrac{1}{2}(\ln x - \ln(1 - x)) \qquad 0 < x < 1$$

$$\Rightarrow f'(x) \equiv \tfrac{1}{2}\left(\frac{1}{x} - \frac{-1}{1 - x}\right)$$

$$\equiv \tfrac{1}{2}\left(\frac{1}{x} + \frac{1}{1 - x}\right)$$

$$\equiv \frac{1}{2x(1 - x)} \qquad 0 < x < 1.$$

(b) If $y = \log_{10} x$ then

$$10^y = x$$

$$\Rightarrow y \ln(10) = \ln x$$

$$\Rightarrow \qquad y = \frac{\ln x}{\ln(10)}$$

$$\Rightarrow \qquad \frac{dy}{dx} = \frac{1}{x \ln(10)}$$

Alternatively, changing base

$$y = \log_{10} x$$

$$\Rightarrow \qquad y = \frac{\ln x}{\ln(10)}$$

$$\Rightarrow \qquad \frac{dy}{dx} = \frac{1}{x \ln(10)}$$

Exercise 8.2

1. Differentiate each of the following

   (a) $y = \ln(2x)$ (b) $y = \ln(ax)$ (c) $y = \ln(2x - 1)$
   (d) $y = \ln(1 - x)$ (e) $y = \ln(x^3)$ (f) $y = \ln(1 + 2x)^2$
   (g) $y = \ln(\cos x)$ (h) $y = \ln(\tan x)$ (i) $y = \ln(x^2 + 4)$
   (j) $y = \ln(e^x + e^{-x})$ (k) $y = \ln(x + 3x^2)$
   (l) $y = \ln\sqrt{x^2 - 1}$ (m) $y = \ln(x + \sqrt{x^2 + 1})$

   (n) $y = \ln\left(\dfrac{x}{1 - x}\right)$ (o) $y = \ln\left(\dfrac{2x}{2x + 1}\right)$

   (p) $y = \ln\sqrt{\cos x}$ (q) $y = \ln(x e^x)$

   (r) $y = \ln(x^2 \sin x)$ (s) $y = \ln\left(\dfrac{x + 1}{x - 1}\right)$

   (t) $y = \ln\left(\dfrac{x^2 + 1}{1 - 3x}\right)$ (u) $y = \ln\left(\dfrac{1}{(1 - x)(2 - x)}\right)$

   (v) $y = \log_2 x$ (w) $y = \log_a x$

   (x) $y = \ln(e^x \cos x)$ (y) $y = \ln(\sin^{-1} x)$

   (z) $y = \ln(\tan^{-1} x)$

2. $y = \ln\left(\dfrac{2 + \cos x}{2 - \sin x}\right)$

3. Show that if $y = \ln f(x)$, then $\dfrac{dy}{dx} = \dfrac{f'(x)}{f(x)}$.

4. Differentiate the following w.r.t. x

   (a) $y = 3^x$ (b) $y = a^x$ (c) $y = a^{-x}$ (d) $y = 10^x$

5. If $y = a \ln(x + b)$, show that

$$a\frac{d^2 y}{dx^2} + \left(\frac{dy}{dx}\right)^2 = 0.$$

6. If $y = \ln(\sec x + \tan x)$, find dy/dx when $x = 0$.

7. Show that if $y = \ln(\sin \sqrt{x})$ then

$$\frac{dy}{dx} = \frac{1}{2\sqrt{x}} \cot \sqrt{x}.$$

8.3 The product and quotient rules

Suppose $u = f(x)$ and $v = g(x)$ over a specified domain, then $y = u \cdot v$ is composed of the product of two functions. We seek dy/dx. Now a change δx in x results in a change δu, δv and δy in u, v and y respectively, so that

$$y + \delta y = (u + \delta u)(v + \delta v)$$

$$\Rightarrow \qquad \delta y = uv + u\delta v + v\delta u + \delta u \delta v - y$$

$$= u\delta v + v\delta u + \delta u \delta v \qquad \text{since } y = uv.$$

So

$$\frac{\delta y}{\delta x} = u\frac{\delta v}{\delta x} + v\frac{\delta u}{\delta x} + \delta u\frac{\delta v}{\delta x}.$$

Now as $\delta x \to 0$, δu, δv and $\delta y \to 0$ but

$$\frac{\delta y}{\delta x} \to \frac{dy}{dx}, \quad \frac{\delta v}{\delta x} \to \frac{dv}{dx} \quad \text{and} \quad \frac{\delta u}{\delta x} \to \frac{du}{dx}$$

and so in the limit we have (using the property of limits stated on page 163).

$$\frac{dy}{dx} = u\frac{dv}{dx} + v\frac{du}{dx} \qquad \left[\text{(the term } \delta u\frac{\delta v}{\delta x} \to 0)\right]$$

Statement: the product rule

> If $y = u \cdot v$ where $u = f(x)$ and $v = g(x)$ then
>
> $$\frac{dy}{dx} = u\frac{dv}{dx} + v\frac{du}{dx}.$$

Example 8.5
Differentiate each of the following.

(a) $y = x^3 \sin x$ (b) $y = (1 + 2x)^2 \cdot \tan^3 x$

(a) $y = x^3 \sin x$. Let

$$u = x^3 \qquad \Rightarrow \frac{du}{dx} = 3x^2$$

$$v = \sin x \qquad \Rightarrow \frac{dv}{dx} = \cos x$$

Now

$$\frac{dy}{dx} = u\frac{dv}{dx} + v\frac{du}{dx}$$

$$= x^3 \cos x + 3x^2 \sin x.$$

(b) $y = (1 + 2x)^2 \tan^3 x$. Let

$$u = (1 + 2x)^2 \qquad \Rightarrow \frac{dy}{dx} = 2(1 + 2x).2$$

$$= 4(1 + 2x)$$

and

$$v = \tan^3 x \qquad \Rightarrow \frac{dv}{dx} = 3\tan^2 x \sec^2 x.$$

So

$$\frac{dy}{dx} = (1 + 2x)^2 . 3\tan^2 x \sec^2 x + 4(1 + 2x)\tan^3 x$$

$$= (1 + 2x)\tan^2 x[3(1 + 2x)\sec^2 x + 4\tan x]$$

The quotient rule

> If $y = u/v$ where $u = f(x)$ and $v = g(x)$ then
>
> $$\frac{dy}{dx} = \left(v\frac{du}{dx} - u\frac{dv}{dx}\right)\bigg/ v^2 \qquad \text{for } v \neq 0$$

Proof

As stated before, an increase δx in x results in increases δu, δv and δy in u, v and y respectively. So if $y = u/v$, then

$$y + \delta y = \frac{u + \delta u}{v + \delta v}.$$

So

$$\delta y = \frac{u + \delta u}{v + \delta v} - \frac{u}{v} \qquad\qquad \text{since } y = \frac{u}{v}$$

$$\Rightarrow \delta y = \frac{v(u + \delta u) - u(v + \delta v)}{(v + \delta v)v}$$

$$= \frac{v\delta u - u\delta v}{(v + \delta v)v}.$$

So

$$\frac{\delta y}{\delta x} = \left(v\frac{\delta u}{\delta x} - u\frac{\delta v}{\delta x}\right)\Big/ (v + \delta v)v.$$

In the limit as $\delta x \to 0$, we have

$$\frac{dy}{dx} = \left(v\frac{du}{dx} - u\frac{dv}{dx}\right)\Big/ v^2$$

since $v + \delta v \to v$ as $\delta v \to 0$.

Example 8.6
Differentiate

(a) $y = \dfrac{x}{1 + x^2}$ (b) $y = \sec x$

(a) $y = \dfrac{x}{1 + x^2}.$ Let

$$u = x \qquad \Rightarrow \frac{du}{dx} = 1$$

and

$$v = 1 + x^2 \qquad \Rightarrow \frac{dv}{dx} = 2x.$$

Now

$$\frac{dy}{dx} = \left(v\frac{du}{dx} - u\frac{dv}{dx}\right)\Big/ v^2$$

$$= \frac{(1 + x^2).1 - x.2x}{(1 + x^2)^2}$$

$$= \frac{1 - x^2}{(1 + x^2)^2}.$$

(b) $y = \sec x$

$$\Rightarrow y = \frac{1}{\cos x}$$

Let

$$u = 1 \qquad \Rightarrow \frac{du}{dx} = 0,$$

and

$$v = \cos x \qquad \Rightarrow \frac{du}{dx} = -\sin x.$$

$$\Rightarrow \frac{dy}{dx} = \frac{\cos x . 0 - 1 . -\sin x}{\cos^2 x}$$

$$= \frac{\sin x}{\cos^2 x}$$

$$= \tan x \sec x.$$

Exercise 8.3

1. Find dy/dx in each of the following.

(a) $y = x \cos x$
(b) $y = x^2 \sin x$
(c) $y = x(x + 1)^{1/2}$

(d) $y = x e^x$
(e) $y = x \ln x$
(f) $y = (1 + x) \tan x$

(g) $y = \cos x \sin 2x$
(h) $y = e^x \ln x$
(i) $y = \dfrac{x}{1 + x}$

(j) $y = \dfrac{\cos x}{x}$
(k) $y = \dfrac{\sin x}{x^2}$
(l) $y = \dfrac{1 - x}{1 + x}$

(m) $y = \dfrac{x^2}{1 + x}$
(n) $y = \dfrac{x^3 + 1}{x}$ [Divide?]

2.

> Show that
>
> (a) $\dfrac{d}{dx}(\cot x) = -\operatorname{cosec}^2 x$
>
> (b) $\dfrac{d}{dx}(\sec x) = \sec x \tan x$
>
> (c) $\dfrac{d}{dx}(\operatorname{cosec} x) = -\operatorname{cosec} x \cot x.$

3. Find the turning values of the following functions.

(a) $f(x) \equiv \dfrac{x}{x^2 + 1}$

(b) $g(x) \equiv 1 - \dfrac{1}{x} \qquad x \neq 0$

4. Find dy/dx in terms of y where

(a) $\tan y = x$ (b) $\sec y = x$ (c) $\cot y = x$.

Now find dy/dx in terms of x.

5. Find dy/dx if

(a) $y = \sqrt{\dfrac{x}{1 - x^2}}$ (b) $y = \dfrac{x}{\sqrt{1 - x^2}}$ (c) $y = \operatorname{cosec} x \,.\, \cos x$

(d) $y = \sqrt{x \,.\, \ln x}$ (e) $y = \sqrt{x \sin x}$ (f) $y = x \,.\, e^x \,.\, \sin x$

6. Find $\dfrac{d^2y}{dx^2}$ when $x = \dfrac{1}{\sqrt{3}}$ if $y = x \tan^{-1} x$.

7. Differentiate (a) $y = e^{x \sin x}$, (b) $y = x \sin^{-1} x$.

8. If $y = e^{-x} \cos x$, show that

$$\frac{d^2y}{dx^2} + 2\frac{dy}{dx} + 2y = 0.$$

9. If $y = a e^{-bx} \sin bx$, show that

$$\frac{d^2y}{dx^2} + 2b\frac{dy}{dx} + 2b^2y = 0.$$

10. Find the tangent to the curve

$$y = \frac{x^2 + 1}{x - 2}$$

at the point $(3, 10)$.

11. Find $\dfrac{d^2y}{dx^2}$ in terms of $\dfrac{dy}{dx}$ and y if $y = (a + bx)e^x$.

12. Two polynomials P and Q are such that
$$P(x) \equiv (x - a)^2 Q(x) \qquad \text{for some constant } a.$$
Prove that $P(a) = P'(a) = 0$. Deduce that the x axis is a tangent to the curve.

13. Find the stationary values of the function defined by
$$g(x) \equiv (x - 2)^2(x - 3)^3 \qquad x \in \mathbb{R}$$

14. If $y = x \sin x$, show that

$$\frac{d^2y}{dx^2} = 2 \cos x - y.$$

15. Find the stationary points on the curve

   (a) $y = x - \ln x$ (b) $y = (x - 2)^2(x + 3)^2$

16. A polynomial $P(x)$ has a factor $(x - a)^2$. Show that $x - a$ is a factor of $P'(x)$. Find the possible values of the constant k given that

$$x^3 - 4x^2 - 3x + k = 0$$

   has two equal roots.

17. Find dy/dx given that

   (a) $y = \operatorname{cosec} x \tan x$ (b) $y = \sqrt{\tan x}$

18. Prove that

$$\frac{d}{d\theta}\left(\log_e \tan \frac{\theta}{2}\right) = \frac{1}{\sin \theta}.$$

   Given that $dy/dx = 2 \sin^2 x \sin y$ and that $y = \pi/2$ when $x = 0$, prove that $y = 2 \tan^{-1}(e^x)$ when $x = \pi$.

AEB 1982

19. Given that $y = \sec x \tan^2 x$, show that

$$y\frac{dy}{dx} = a \tan^7 x + b \tan^5 x + c \tan^3 x,$$

   where a, b and c are constants to be determined.

JMB 1983

20. Find $f' [= df/dx]$ in terms of g' if

   (a) $f(x) \equiv g(x + a)$ a is a constant, $x \in \mathbb{R}$
   (b) $f(x) \equiv g(x + g(x))$ $x \in \mathbb{R}$
   (c) $f(x) \equiv g(xg(x))$.

8.4 Implicit differentiation

If we assume that y is a function of x then we may differentiate, say, y^2 w.r.t. x using the chain rule:

$$\frac{d}{dx}(y^2) = \frac{d}{dy}(y^2) \cdot \frac{dy}{dx}$$

$$= 2y\frac{dy}{dx}.$$

More generally, if $f(y)$ is a function of y then

$$\frac{d}{dx}f(y) = \frac{d}{dy}(f(y)) \cdot \frac{dy}{dx}$$

$$\Rightarrow \frac{d}{dx} f(y) = f'(y) \cdot \frac{dy}{dx}.$$

So, for example

$$\frac{d}{dx} \sin y = \cos y \frac{dy}{dx}.$$

In each case it is assumed implicitly that y is a function of x.

Consider now the expression $x^2 y^3$.

We may differentiate this w.r.t. x if we use the product rule. For

$$\frac{d}{dx}(x^2 y^3) = \frac{d}{dx}(x^2) \cdot y^3 + x^2 \cdot \frac{d}{dx}(y^3)$$

$$= 2xy^3 + x^2 \cdot 3y^2 \frac{dy}{dx}$$

In the equation

$$xy^2 + x^2 = 2$$

y is said to be expressed implicitly in terms of x. (The equation could be rearranged as

$y = \dfrac{\sqrt{2 - x^2}}{x}$, in which case y is expressed explicitly in terms of x.) We may obtain

dy/dx by differentiating each term w.r.t. x, to obtain

$$\left(1 \cdot y^2 + x \cdot 2y \frac{dy}{dx}\right) + 2x = 0$$

$$\Rightarrow \qquad\qquad 2xy \frac{dy}{dx} = -(2x + y^2)$$

$$\Rightarrow \qquad\qquad \frac{dy}{dx} = -\frac{(2x + y^2)}{2xy}$$

In particular, at the point $(1, 1)$ on the curve given by $xy^2 + x^2 = 2$, we have

$$\frac{dy}{dx} = -\frac{((2 \cdot 1) + 1)}{2 \cdot 1 \cdot 1}$$

$$= -\frac{3}{2}$$

However, *differentiating implicitly*, as this process is called, is extremely useful when it is difficult or impossible to express y explicitly in terms of x.

Example 8.7

Find dy/dx in terms of x and y by differentiating implicitly each of the following.

(a) $y^2 + y = x$ (b) $x \ln y = x + 1$ (c) $xy^2 + x^2 y = 4$

(a) $y^2 + y = x$.

Differentiating both sides w.r.t. x term by term, and recalling that

$$\frac{dy^2}{dx} = 2y\frac{dy}{dx}$$

by the chain rule. We have

$$2y\frac{dy}{dx} + \frac{dy}{dx} = 1$$

$$\Rightarrow \qquad \frac{dy}{dy} = \frac{1}{2y + 1}$$

(b) $x \ln y = x + 1$.

Differentiating w.r.t. x and noting that

$$\frac{d}{dx}\ln y = \frac{1}{y} \cdot \frac{dy}{dx}$$

we have, using the product rule

$$x\frac{1}{y}\frac{dy}{dx} + \ln y = 1$$

$$\Rightarrow \qquad \frac{dy}{dx} = y\frac{(1 - \ln y)}{x}$$

(c) $\quad xy^2 + x^2y = 4$

$$\Rightarrow x2y\frac{dy}{dx} + y^2 + x^2\frac{dy}{dx} + 2xy = 0$$

$$\Rightarrow \quad (2xy + x^2)\frac{dy}{dx} + y^2 + 2xy = 0$$

$$\Rightarrow \qquad \frac{dy}{dx} = -\frac{(y^2 + 2xy)}{2xy + x^2} \qquad \text{for } 2xy + x^2 \neq 0$$

Example 8.9

If $xy + y^2 = 0$, show that

$$(x + 2y)\frac{d^2y}{dx^2} + 2\frac{dy}{dx}\left(\frac{dy}{dx} + 1\right) = 0.$$

If $xy + y^2 = 0$, then differentiating w.r.t. x,

$$x\frac{dy}{dx} + y + 2y\frac{dy}{dx} = 0.$$

Differentiating again w.r.t. x,

$$\left(x\frac{d^2y}{dx^2} + \frac{dy}{dx}\right) + \frac{dy}{dx} + \left(2y\frac{d^2y}{dx^2} + 2\frac{dy}{dx}\cdot\frac{dy}{dx}\right) = 0$$

$$\Rightarrow (x + 2y)\frac{d^2y}{dx^2} + 2\frac{dy}{dx}\left(1 + \frac{dy}{dx}\right) = 0.$$

Exercise 8.4

1. Find dy/dx in each of the following.

(a) $xy = 1$ (b) $x^2 + y^2 = 4$ (c) $9x^2 - 4y^2 = 36$
(d) $xe^y = 2$ (e) $xy + y^2 \sin x = 0$ (f) $x \sin y + y \sin x = 1$

2. Find the equation of the tangent to the curve $a + y = xy$ at the point $(2, 2)$.

3. Find the equation of the tangent to the following curves at the point (x_1, y_1)

(a) $y^2x = 1$ (b) $x^2 + y^2 = 4$ (c) $\frac{x^2}{a^2} + \frac{y^2}{b^2} = 1$

where a and b are constants.

4. Find the maximum and minimum value of y if

$$x^2 + 2xy = 3y - 1.$$

5. Find the gradient of the tangent to the curve

$$x^3 - 3x^2y + y^2 = -1$$

at the point $(1, 1)$.

6. Find dy/dx in each of the following cases.

(a) $\sin(x + y) = xy$ (b) $e^{x+y} = 2x$

7. If $y = x \ln y$ prove that

$$\frac{d^2y}{dx^2} = \frac{y^2(y - 2x)}{x(y - x)^3} \qquad x \neq y \quad \text{and} \quad x \neq 0.$$

8. Show that all the points on the curve

$$3x^2 = 7xy + 9y^2 = 2$$

for which $dy/dx = 0$ lie on the line $6x - 7y = 0$.

9. By first taking logs to base e, differentiate

(a) $y = 2^x$ (b) $y = x^x$ (c) $y = x^{\ln x}$

10. Given that $\ln xy = x^2 + y^2$, find dy/dx.

11. Given that $x^3 - 3xy + y^3 = k$, where k is a constant, find dy/dx in terms of x and y.

8.5 Logarithmic differentiation (revisited)

It is often convenient to 'take logarithms' before differentiating a function; as the following example illustrates.

Example 8.10

Differentiate (a) $y = x e^{\tan x}$, (b) $y = \dfrac{x}{1 - x^2}$.

(a) $y = x e^{\tan x}$

$\Rightarrow \ln y = \ln(x e^{\tan x})$

$= \ln x + \ln e^{\tan x}$

$= \ln x + \tan x$ since $\ln e = 1$

Differentiating w.r.t. x,

$$\frac{1}{y}\frac{dy}{dx} = \frac{1}{x} + \sec^2 x$$

$$\frac{dy}{dx} = y\left(\frac{1}{x} + \sec^2 x\right)$$

$$= x e^{\tan x}\left(\frac{1}{x} + \sec^2 x\right)$$

$$= e^{\tan x}(1 + x\sec^2 x).$$

(b) $y = \dfrac{x}{1 - x^2}$

$\Rightarrow \ln y = \ln x - \ln(1 - x^2)$

$\Rightarrow \dfrac{1}{y}\dfrac{dy}{dx} = \dfrac{1}{x} - \dfrac{-2x}{1 - x^2}$

$\Rightarrow \dfrac{dy}{dx} = \dfrac{x}{1 - x^2}\left(\dfrac{1}{x} + \dfrac{2x}{1 - x^2}\right)$

$= \dfrac{1 + x^2}{(1 - x^2)^2}.$

Example 8.11

Use the method of taking logarithms to derive the product rule.

Suppose

$$y = u \cdot v$$

where u and v are functions of x. Then

$$\ln y = \ln u \cdot v$$

$$\Rightarrow \quad \ln y = \ln u + \ln v$$

$$\Rightarrow \quad \frac{1}{y}\frac{dy}{dx} = \frac{u'}{u} + \frac{v'}{v} \qquad \text{where } u' \text{ represents the derived function of } u.$$

$$\Rightarrow \quad \frac{dy}{dx} = y\left[\frac{vu' + uv'}{uv}\right].$$

But

$$y = u \cdot v$$

$$\Rightarrow \frac{dy}{dx} = vu' + uv'.$$

Exercise 8.5

1. Use the method of taking logarithms to differentiate the following.

    (a) $y = x e^x$

    (b) $y = x^2 e^{-x}$

    (c) $y = x e^{(-x^2/2)}$

    (d) $y = 3x^2(x + 4)^4$

    (e) $y = (1 - x)^2(1 + x)^5$

    (f) $y = \dfrac{x^2}{1 + x}$

    (g) $y = \dfrac{2x}{(1 + x)(2x + 1)}$

    (h) $y = \dfrac{(x - 1)(x + 3)}{(x - 4)(x + 5)}$

    (i) $y = \dfrac{\tan x}{e^x}$

2. Use the technique of taking logarithms to find dy/dx if $y = u/v$, where u and v are functions of x.

3. If $y = x^n e^{ax}$ show that

$$\frac{dy}{dx} - ay = \frac{ny}{x}.$$

4. If $x^y = y^x$ where $x, y > 0$, find dy/dx at the point $(1, 1)$.

5. Find dy/dx if

(a) $y = x^n \ln \dfrac{1}{x}$ (b) $y = \dfrac{x e^{x^2}}{(1 + x) e^x}$.

6. Find the stationary values of the function

$$f(x) \equiv \frac{(x - 4)^2}{x - 3} \qquad x \neq 3$$

distinguishing between maximum and minimum values.

7. If

$$y = \frac{e^{2x}}{(x + 2)^4}$$

find dy/dx when $x = 0$.

8.6 Parametric equations

Suppose $y = t^3$ and $x = t^2$, then x and y are themselves related. A particular value of t will fix values for x and y. For example:

t	-2	-1	0	1	2
x	4	1	0	1	4
y	-8	1	0	1	8

When x and y are related via a third variable t, then t is called a *parameter*. The equations are called *parametric equations*. The more familiar Cartesian equation may often be obtained by eliminating the parameter.

For

$$y = t^3 \quad \text{and} \quad x^2 = t^2$$
$$\Rightarrow y^2 = t^6 \quad \text{and} \quad x^3 = t^6$$

So

$$y^2 = x^3.$$

However, it is not necessary to obtain y explicitly in terms of x to find dy/dx.

Example 8.12

Find dy/dx if $y = at^2$ and $x = 2at$ where a is a constant.

Now

$$y = at^2$$

$$\Rightarrow \frac{dy}{dt} = 2at$$

and

$$x = 2at$$

$$\Rightarrow \frac{dx}{dt} = 2a \qquad \Rightarrow \frac{dt}{dx} = \frac{1}{2a}$$

Since

$$\frac{dy}{dx} = \frac{dy}{dt} \cdot \frac{dt}{dx}$$

by the chain rule, then

$$\frac{dy}{dx} = 2at \cdot \frac{1}{2a}$$

$$\Rightarrow \frac{dy}{dx} = t.$$

Of course, if we wished to evaluate dy/dx for a particular value of x, we could use the equation $x = 2at$ to establish the corresponding value of t and thus find dy/dx.

Notice that since $x = 2at$ then $t = \dfrac{x}{2a}$ and so

$$y = at^2$$

$$\Rightarrow y = \frac{x^2}{4a}$$

$$\Rightarrow \frac{dy}{dx} = \frac{x}{2a}$$

$$\Rightarrow \frac{dy}{dx} = t$$

resulting in the same answer.

Statement

> It is often convenient to express a curve *parametrically*, that is in the form
>
> $$y = f(t) \quad \text{and} \quad x = g(t).$$
>
> In such a case t is called a parameter. To find the usual Cartesian equation we simply eliminate the parameter.

Example 8.13

Find the Cartesian equation of the curve given parametrically as

$$x = a \cos \theta \quad \text{and} \quad y = b \sin \theta.$$

Find also the equation of the tangent to the curve at the point on the curve where $\theta = \pi/3$.

Since $\cos \theta = x/a$ and $\sin \theta = y/b$, squaring and adding gives

$$\cos^2 \theta + \sin^2 \theta = \frac{x^2}{a^2} + \frac{y}{b^2}$$

$$\Rightarrow \frac{x^2}{a^2} + \frac{y^2}{b^2} = 1 \qquad \text{the Cartesian form}$$

To find the equation of the tangent notice that

when $\qquad \theta = \pi/3, \quad \cos \theta = \tfrac{1}{2} \quad \text{and} \quad \sin \theta = \frac{\sqrt{3}}{2}.$

Thus when $\qquad \theta = \pi/3, \quad x = \frac{a}{2} \quad \text{and} \quad y = \frac{b\sqrt{3}}{2}.$

By differentiating the Cartesian form w.r.t. x we could find the gradient at the required point and hence write down the equation of the tangent.

Alternatively, the following method may be used. This method does not require us to find the Cartesian form.

$$x = a \cos \theta \qquad\qquad y = b \sin \theta$$

$$\Rightarrow \frac{dx}{d\theta} = -a \sin \theta \qquad \frac{dy}{d\theta} = b \cos \theta$$

By the chain rule

$$\frac{dy}{dx} = \frac{b \cos \theta}{-a \sin \theta}$$

$$= \frac{-b}{a} \cot \theta.$$

When $\theta = \pi/3$, $\cot \theta = 1/\sqrt{3}$ and so at this point

$$\frac{dy}{dx} = \frac{-b}{a\sqrt{3}}.$$

The required equation is then

$$y - \frac{b\sqrt{3}}{2} = \frac{-b}{a\sqrt{3}}\left(x - \frac{a}{2}\right)$$

$$\Rightarrow a\sqrt{3}\, y + bx - 2ab = 0.$$

Example 8.14

If $x = a(1 - \cos \theta)$ and $y = a \sin \theta$, find d^2y/dx^2 in terms of θ.

Now

$$x = a - a \cos \theta \quad \text{and} \quad y = a \sin \theta$$

$$\Rightarrow \frac{dx}{d\theta} = a \sin \theta \qquad\qquad \frac{dy}{d\theta} = a \cos \theta$$

By the chain rule,

$$\frac{dy}{dx} = \frac{a \cos \theta}{a \sin \theta}$$

$$= \cot \theta$$

and

$$\frac{d^2y}{dx^2} = \frac{d}{dx}\left(\frac{dy}{dx}\right)$$

$$= \frac{d}{d\theta}\left(\frac{dy}{dx}\right) \cdot \frac{d\theta}{dx} \qquad\qquad \text{by the chain rule}$$

$$= -\csc^2 \theta \cdot \frac{1}{a \sin \theta} \qquad\qquad \text{since the derivative of } \cot \theta \text{ is } -\csc^2 \theta$$

$$= \frac{-1}{a \sin^3 \theta}$$

$$= \frac{-\csc^3 \theta}{a}.$$

Exercise 8.6

1. Given that

$$x = \frac{a}{t^2} \quad \text{and} \quad y = \frac{2a}{t}$$

    obtain the equation of the tangent and normal at the point where $t = t_1$.

2. Find dy/dx if $x = a(\theta - \sin \theta)$ and $y = a(1 - \cos \theta)$.

3. Prove that the tangent to the curve

$$x = \frac{\cos t}{1 + \sin t} \qquad y = \frac{\sin t}{1 + \cos t}$$

    at the point where $t = t_1$ is

$$x(1 + \sin t_1) + y(1 + \cos t_1) = \cos t_1 + \sin t_1.$$

4. Find d^2y/dx^2 in terms of t where $y = t^3 + 3t$ and $x = t^2 + 2$.

5. If $x = a\cos\theta$ and $y = a\sin\theta$, find

   (a) $\dfrac{dy}{dx}$ (b) $\dfrac{d^2y}{dx^2}$ (c) the tangent to the curve at the point where $\theta = \pi/3$.

6. Find $\dfrac{d^2y}{dx^2}$ if $x = a\cos^3\theta$ and $y = a\sin^3\theta$.

7. If

$$x = \frac{t^2}{1 + t^3} \quad \text{and} \quad y = \frac{t^3}{1 + t^3}$$

   show that

$$\frac{dy}{dx} = \frac{3t}{2 - t^3}$$

8. If $x = te^t$ and $y = t^2 e^t$ find dy/dx in terms of t.
   Show that $y/x = t$ and hence that

$$y = x\ln\left(\frac{x^2}{y}\right)$$

9. Given that $x = 2t + \sin 2t$ and $y = \cos 2t$, show that

   (a) $\dfrac{dy}{dx} = -\tan t$ (b) $\dfrac{d^2y}{dx^2} = -\tfrac{1}{4}\left(1 + \left(\dfrac{dy}{dx}\right)^2\right)^2$

10. Given that

$$x = \frac{2t}{1 - t^2} \quad \text{and} \quad y = \frac{1 + t^2}{1 - t^2}$$

    express dy/dx and d^2y/dx^2 in terms of t.

11. The parametric equations of a curve are $x = \log_e(1 + t)$, $y = e^{t^2}$ for $t > -1$.
    Find dy/dx and d^2y/dx^2 in terms of t.
        Prove that the curve has only one turning point and that it must be a minimum.

AEB 1982

12. Given that $y = e^{2t}\cos 3^t$ and $\log_e x = \sin 3t$, where t is a parameter, prove that

    (a) $\dfrac{dy}{dx} = \left(\dfrac{2}{3} - \tan 3t\right)e^{(2t - \sin 3t)}$,

(b) if $\dfrac{dy}{dx} = 0$, then $x = e^{2/\sqrt{13}}$ or $x = e^{-2/\sqrt{13}}$.

AEB 1982

13. Given that $\cos\theta = \dfrac{x}{x+1}$ and $\sin\theta = \dfrac{y}{y+1}$, find $\dfrac{dy}{dx}$.

14. Given that $x = \dfrac{1}{1+t^2}$ and $y = \dfrac{t}{1+t^2}$, show that

$$\frac{dy}{dx} = \frac{t^2 - 1}{2t}$$

and find d^2y/dx^2 in terms of t.

15. Show that the equation of the tangent at the point where $t = t_1$ on the curve

$$x = \frac{\cos t}{1 + \sin t} \qquad y = \frac{\sin t}{1 + \cos t}$$

is $x(1 + \sin t_1) + y(1 + \cos t_1) = \cos t_1 + \sin t_1$.

Exercise 8.7 (miscellaneous)

1. Differentiate $\log_e(\tan(3x + 5))$, $0 < 3x + 5 < \pi/2$.

W 1982

2. Given $y = x\tan^{-1}x$, show that $\dfrac{d^2y}{dx^2} = \dfrac{2}{(1 + x^2)^2}$.

W 1982

3. Find $\dfrac{d^2y}{dx^2}$ when $y = \sin^{-1}x - x\sqrt{(1 - x^2)}$, expressing your answer as simply as possible.

W 1981

4. Differentiate with respect to x:

   (a) $\tan^4 2x$ (b) $\dfrac{x-1}{2x-3}$ (c) $x^2\log_e x$.

SU 1982

5. (a) Find dy/dx, given that $y = e^{-2x}\cos 3x$.

   (b) Find the value of dy/dx at the point $(2, 3)$ on the curve $xy^2 = 18$.

CAMB 1982

6. (a) Given that $y = \dfrac{x+2}{x^2+1}$, find dy/dx.

(b) The parametric equations of a curve are

$$x = a\left(t + \frac{1}{t}\right), \quad y = a\left(t - \frac{1}{t}\right),$$

where a is a constant. Find the gradient of the tangent to the curve at the point where $t = 2$.

7. A function $y = f(x)$ satisfies $\dfrac{dy}{dx} = x^2 - y, f(1) = 0$.

(a) Show that $\dfrac{d^2y}{dx^2} = 2x - x^2 + y$ and find a similar expression for $\dfrac{d^3y}{dx^3}$.

(b) Evaluate $\dfrac{dy}{dx}, \dfrac{d^2y}{dx^2}$ and $\dfrac{d^3y}{dx^3}$ when $x = 1$.

8. (a) Find dy/dx when

(i) $y = \log_e(\sec x + \tan x)$,
(ii) $xy - y + 2x = 3$.

(b) Given that

$$x = t + 2\log_e t \quad \text{and} \quad y = 2t^2 e^t,$$

show that $\dfrac{dy}{dx} = y$ and find $\dfrac{d^2y}{dx^2}$.

AEB 1982

9. Given that $y = \sec x \tan x + \log_e(\sec x + \tan x)$, prove that $dy/dx = k\sec^3 x$, stating the value of the constant k.

AEB 1982

10. Find the coordinates of the turning point on the curve

$$y = 2e^{3x} + 8e^{-3x},$$

and determine the nature of this turning point.

CAMB 1983

11. (a) Differentiate with respect to x

(i) $3^{\log_e x}$,
(ii) $\sin^{-1}\sqrt{(1 - x^2)}$.

(b) A point P has coordinates given parametrically by $x = \cos^{1/3}\theta$, $y = \sin^{1/3}\theta$ $(0 \leqslant \theta \leqslant \pi/2)$. Show that the maximum value of OP^2, where O is the origin, occurs when $\tan\theta = 8$. Find this maximum value.

W 1983

12. From a large thin plane sheet of metal it is required to make an open rectangular box with a square base so that the box would contain a given volume V. Express the area of the sheet used in terms of V and x, where x is the length of a side of the square base.

   Hence find the ratio of the height of the box to x in order that the box consists of a minimum area of the metal sheet.

LOND 1984

13. (a) Differentiate the following with respect to x, simplifying your answers where possible.

   (i) $\dfrac{1}{4 - x^3}$,   (ii) $e^{\tan kx}$,   (iii) $(x^2 + 1) \log_e (x^2 + 1)$.

   (b) Find the gradient of the curve $y^2 = 2xy + 8$ at the point $(1, -2)$.

   (c) A body moves in a straight line Ox such that at time t seconds it is at a distance x metres from O and its speed is v metres per second where $v^2 = 100 - 25x^2$. Find the distance from O at which the body is instantaneously at rest and show that it cannot move further away from O than this distance. Find the magnitude of the acceleration of the body when it is at rest.

SU 1983

14. The gradient of a curve at the point (x, y) is given by

$$\frac{dy}{dx} = \frac{x^4 - 16}{x^3}.$$

Find the equation of the curve given that it passes through the point $(2, 4)$. Find the coordinates of the two stationary points on the curve and determine whether they are maximum or minimum points. State, with reason, whether or not the curve has a point of inflection. Give a rough sketch of the curve.

SU 1983

15. Given that $y^2 - 5xy + 8x^2 = 2$, prove that $\dfrac{dy}{dx} = \dfrac{5y - 16x}{2y - 5x}$.

   The distinct points P and Q on the curve $y^2 - 5xy + 8x^2 = 2$ each have x-coordinate 1. The normals to the curve at P and Q meet at the point N. Calculate the coordinates of N.

AEB 1983

16. (a) Given that a is a constant, differentiate with respect to x

   (i) $e^{(a-x)^2}$,   (ii) $x/\sqrt{(x^2 + a^2)}$.

   (b) If $x = 2t + 4t^3$, $y = t^2 + 3t^4$, find $\dfrac{dy}{dx}$ and $\dfrac{d^2y}{dx^2}$ in terms of t.

O&C 1983

17. Show that, if

$$f(x) = \frac{ax + b}{x^2 + c},$$

then

$$(x^2 + c)\frac{d^2f}{dx^2} + 4x\frac{df}{dx} + 2f = 0.$$

Show that when $b = 0$ and $a > 0$ the graph of the function $f(x)$ has three points of inflexion if c is positive, and one point of inflexion if c is negative.

OX 1983

Chapter 9 Sequences and series

9.1. Arithmetic series

Consider a function whose domain is $\mathbb{N}^+$ and whose range is a subset of the reals. The values of the function may be written in the natural order

$$u(1), u(2), u(3), u(4), \ldots u(r), \ldots$$

Such a listing is called a *sequence*, where $u(n)$, the nth term of the sequence, is often written u_n.

The function u given by

$$u(n) \equiv 2n \qquad n \in \mathbb{N}^+.$$

generates the sequence

$$2, 4, 6, 8, 10, \ldots 2n, \ldots.$$

The sequence

$$1, 2, 4, 8, 16, 32, \ldots$$

is generated by u where

$$u(n) \equiv 2^{n-1} \qquad n \in \mathbb{N}^+.$$

The sum of the terms of a sequence is called an (*infinite*) *series* and the sum of the first n terms is called a (*finite*) *series*.

A sequence where each term (other than the first) is obtained by adding a constant to the previous term is called an *arithmetic sequence*. If the first term is a, and the constant, called the common difference, is d, then the sequence has the form

$$a, a + d, a + 2d, a + 3d, \ldots$$

The nth term of this sequence is

$$u_n = a + (n - 1)d.$$

The following are examples of arithmetic sequences.

(i) $2, 5, 8, 11, \ldots, 2 + (n - 1)3, \ldots$
$\qquad$ [the nth term is $a + (n - 1)d$,
$\qquad$ where $a = 2$ and $d = 3$]

(ii) $3, -2, -8, -13, \ldots, 3 + (n - 1)(-5), \ldots$
$\qquad$ [the nth term is $a + (n - 1)d$,
$\qquad$ where $a = 3$ and $d = -5$]

Example 9.1

The third term of an arithmetic sequence is 11 and the ninth term is 35. Obtain the sixth term.

$$u_3 = a + 2d = 11 \qquad ①$$
$$u_9 = a + 8d = 35 \qquad ②$$

② − ① gives

$$6d = 24$$
$$\Rightarrow d = 4$$
$$\Rightarrow a = 3 \qquad \text{from } ①.$$

Now

$$u_6 = a + 5d$$
$$= 3 + 5 \times 4$$
$$= 23.$$

Suppose we wish to find the sum, S_n, of the first n natural numbers

$$S_n = 1 + 2 + 3 + 4 + \ldots + n.$$

Reversing the order gives

$$S_n = n + (n - 1) + \ldots + 2 + 1.$$

Adding (vertically)

$$2S_n = (1 + n) + (2 + (n - 1)) + \ldots + ((n - 1) + 2) + (n + 1)$$
$$= (n + 1) + (n + 1) + \ldots + (n + 1) + (n + 1)$$
$$= n(n + 1) \quad \text{since there are } n \text{ terms.}$$

So

$$S_n = \frac{n}{2}(n + 1).$$

It follows then that

$$1 + 2 + 3 + \ldots + 10 = \frac{10}{2}(10 + 1)$$
$$= 55$$

and

$$1 + 2 + 3 + \ldots + 100 = \frac{100}{2}(101)$$
$$= 5050.$$

This argument may be used to obtain the sum of the first n terms of a general arithmetic series. (*An arithmetic series is often called an arithmetic progression or AP for short*).

If

$$S_n = a + (a + d) + (a + 2d) + \ldots + (l - d) + l$$

where $l = a + (n - 1)d$ is the nth term (the *last* term) then reversing

$$S_n = l + (l - d) + \ldots + (a + d) + a.$$

Adding (vertically)

$$2S_n = (a + l) + (a + l) + \ldots + (a + l) + (a + l)$$

$$= n(a + l) \quad \text{since there are } n \text{ terms,}$$

$$\Rightarrow S_n = \frac{n}{2}(a + l)$$

$$= \frac{n}{2}(2a + (n - 1)d).$$

In summary:

Statement

> If S_n represents the sum of the first n terms of the AP
>
> $$a + (a + d) + (a + 2d) + \ldots + (a + (n - 1)d)$$
>
> then
>
> $$S_n = \frac{n}{2}(2a + (n - 1)d)$$
>
> $$= \frac{n}{2}(a + l) \quad \text{where } l = a + (n - 1)d \text{ is the last term.}$$

Example 9.2

Find the sum of all the natural numbers divisible by 6 between 50 and 150 inclusive.

The first natural number divisible by 6 is 54 and the last is 150, so we require

$$54 + 60 + 66 + \ldots + 150$$

where $l = a + (n - 1)d = 150$. But

$$a = 54 \quad \text{and} \quad d = 6$$

so

$$54 + (n - 1)6 = 150 \qquad \Rightarrow n = 17.$$

Since

$$S_n = \frac{n}{2}(a + l),$$

$$S_{17} = \frac{17}{2}(54 + 150)$$

$$= 1734.$$

Notice that in order to find S_n we require a, n and either d or l.

Example 9.3

Find three numbers in arithmetic progression whose sum is 24 and whose product is 480.

Taking the three numbers to be $a - d$, a, and $a + d$, we have

$$(a - d) + a + (a + d) = 24 \qquad ①$$

$$(a - d) \times a \times (a + d) = 480 \qquad ②$$

Now from ① we obtain

$$3a = 24$$

$$\Rightarrow a = 8.$$

Substituting into ②

$$(8 - d)\, 8\, (8 + d) = 480$$

$$\Rightarrow (64 - d^2) = 60$$

$$\Rightarrow d = \pm 2.$$

Thus the three numbers are

$$6, 8, 10 \qquad \text{(or } 10, 8, 6\text{)}.$$

When three numbers are known to form an arithmetic sequence it is often convenient to take the numbers to be $a - d$, a, and $a + d$. The reader may like to rework this example using a, $a + d$, and $a + 2d$.

Exercise 9.1

1. Write down the fifth, tenth and nth terms of the arithmetic sequence $2, 7, 12, \ldots$

2. Find the number of terms in the APs
    (a) $1 + 59 + \ldots + 101$,
    (b) $2 + 5 + 8 + \ldots + 245$,
    (c) $16 + 15 + \ldots + (-1)$.

3. Find the first and fifth terms of an arithmetic progression whose third term is 6 and seventh is 12.

4. Write down the first four members of the sequence whose nth term is

   (a) $2n + 1$,   (b) $3n - 2$.

5. Find the tenth term and the sum to ten terms of the following arithmetic sequence.
   (a) $3, 5, 7, 9, \ldots$ (b) $\frac{1}{5}, 0, -\frac{1}{5}, \ldots$ (c) $a - d, a, a + d, \ldots$
   (d) $x, 2x, 3x, \ldots$ (e) $(x + 1), 2(x + 1), 3(x + 1), \ldots$

6. How many multiples of 3 are there between 100 and 300 inclusive? Find the sum of these multiples.

7. Find the sum of the integers which are *not* divisible by 6, between 1 and 200.

8. Find n, if the sum to n terms of the arithmetic sequence 4, 6, 8, ... is 180.

9. How many terms of the AP $3 + 5 + 7 + \ldots$ are required

   (a) for the sum to equal 483,   (b) for the last term to equal 483.

10. Find the sum of the integers from n to $2n$ inclusive.

11. How many terms of the arithmetic sequence, 4, 8, 12, ... must be taken for the sum to exceed 5000?

12. Find the sum of the first eight terms of the arithmetic sequences

    (a) $a, b, 2b - a, \ldots$,   (b) $n - 1, 1, \ldots$.

13. The second term of an AP is 6 and the sum of the first three terms is 18, find the nth term and the sum to n terms.

14. The third term of an AP is -20 and the eleventh term is 20. Show that the sum to 13 terms is zero.

15. The ninth term of an AP is 38. Find the sum of the first 17 terms. Find also the fourth term if it is three times the first.

16. An arithmetic progression has first term a and common difference d. Its tenth term is 69 and the sum of its first 30 terms is four times the sum of its first 10 terms. Find the values of a and d.

17. An arithmetical progression is such that the sum of the first n terms is $2n^2$ for all positive integral values of n. Find (by substituting two values of n, or otherwise) the first term and the common difference.

                                                                                    SU 1980

18. The first three terms of an arithmetic series are $-3\frac{1}{8}$, $-1\frac{7}{8}$, $-\frac{5}{8}$. Find the sum of the first 70 terms of this series.

CAMB 1983

19. Show that the sum of the odd numbers from 1 to $(2n - 1)$ inclusive is n^2.
     Show that the sum of those positive odd numbers which are less than 1002 and *not* divisible by 3 is 6×167^2.

CAMB 1982

20. In the triangle ABC, $BC = a$, $CA = b$, $AB = c$ and the angle C is obtuse. Prove that

$$c^2 = a^2 + b^2 - 2ab\cos C.$$

The perimeter of a triangle is $3l$ and its largest angle is $120°$. Given that the lengths of the sides of the triangle are in arithmetic progression, find, in terms of l, the length of each side.

JMB 1982

9.2 Geometric series

A sequence where each term (other than the first) is obtained by multiplying the previous term by a constant is called a *geometric* sequence. If the first term is a and the constant, called the *common ratio*, is r then the sequence has the form

$$a, ar, ar^2, ar^3, \ldots$$

The nth term in this sequence is $u_n = ar^{n-1}$.
  The following are examples of geometric sequences.

(i) $1, 2, 4, 8, 16, \ldots, 2^{n-1}$

[the nth term is ar^{n-1} where $a = 1$ and $r = 2$]

(ii) $3, -\frac{3}{2}, \frac{3}{4}, -\frac{3}{8}, \ldots, \dfrac{3}{(-2)^{n-1}}$

[the nth term is ar^{n-1} where $a = 3$ and $r = -\frac{1}{2}$]

Example 9.4

Find a positive value of x, and the fourth term if $1 + 2x$, $4x - 1$, $6 + 3x$ are the first three terms in a geometric sequence.

Comparing with a, ar, ar^2 the common ratio may be expressed as

$$\frac{(4x-1)}{(1+2x)} \quad \text{and as} \quad \frac{(6+3x)}{(4x-1)} \quad \left[\text{that is} \quad \frac{ar}{a} \quad \text{or} \quad \frac{ar^2}{ar}\right].$$

So

$$\frac{4x - 1}{1 + 2x} = \frac{3(2 + x)}{4x - 1}$$

$$\Rightarrow \quad (4x - 1)^2 = 3(2 + x)(1 + 2x)$$

$$\Rightarrow \quad 16x^2 - 8x + 1 = 6 + 15x + 6x^2$$

$$\Rightarrow \quad 10x^2 - 23x - 5 = 0$$

$$\Rightarrow \quad (2x - 5)(5x + 1) = 0$$

$$\Rightarrow x = \tfrac{5}{2} \quad \text{or} \quad x = -\tfrac{1}{5}$$

$$\Rightarrow x = \tfrac{5}{2} \quad \text{since } x > 0.$$

Substituting, the three terms are

$$6, 9, 13.5.$$

It follows that $a = 6$, $r = \tfrac{3}{2}$ and the fourth term $ar^3 = 81/4$.

The sum of the first n terms of a geometric series may be found as follows.
  Suppose

$$S_n = a + ar + ar^2 + \ldots + ar^{n-1}.$$

Multiplying both sides by r

$$rS_n = ar + ar^2 + ar^3 + \ldots + ar^n.$$

Subtracting

$$(1 - r)S_n = a - ar^n$$

$$= a(1 - r^n)$$

$$\Rightarrow S_n = \frac{a(1 - r^n)}{(1 - r)}.$$

In summary:

Statement

If S_n represents the sum of the first n terms of a geometric series with first
term a and common ratio r then

$$S_n = \frac{a(1 - r^n)}{1 - r}$$

A geometric series is often referred to as a *geometric progression* or GP for
short.

Example 9.5

Find the sum of the first ten terms of each of the following GPs.

(a) $1 + \frac{1}{2} + \frac{1}{4} + \frac{1}{8} + \ldots$ (b) $r + ar + a^2r + a^3r + \ldots$

(a) $a = 1$ and $r = \frac{1}{2}$. So

$$S_{10} = \frac{(1 - (\frac{1}{2})^{10})}{1 - \frac{1}{2}}$$

$$= 2(1 - (\frac{1}{2})^{10})$$

$$= 1.998 \qquad\qquad 3\,\text{d.p.} \qquad \boxed{C}$$

(b) In this example there is the obvious opportunity for confusing the notation. We suppose the GP is of the form $A + AR + AR^2 + \ldots$ Then

$$S_n = \frac{A(1 - R^n)}{1 - R}$$

Now in this example $A = r$ and $R = a$, so

$$S_{10} = \frac{r(1 - a^{10})}{1 - a}$$

Example 9.6

The fifth term of a GP is 80 and the sum of the third and fourth terms is -20. Find the sum of the first eight terms.

Assuming the GP is of the form

$$a + ar + ar^2 + \ldots$$

we have

$$u_5 = ar^4 \qquad\qquad = 80 \qquad ①$$

$$u_3 + u_4 = ar^2 + ar^3 = -20 \qquad ②$$

From ① we obtain:

$$a = \frac{80}{r^4}.$$

Substituting into ②,

$$\frac{80}{r^2} + \frac{80}{r} = -20$$

$$\Rightarrow \qquad 4 + 4r = -r^2$$

$$\Rightarrow r^2 + 4r + 4 = 0$$

$$\Rightarrow \qquad (r + 2)^2 = 0$$

$$\Rightarrow \quad r = -2$$

$$\Rightarrow \quad a = 5 \quad \text{substituting into } ① .$$

Now

$$S_n = \frac{a(1 - r^n)}{1 - r}$$

$$\Rightarrow S_8 = \frac{5(1 - (-2)^8)}{1 - (-2)}$$

$$= \tfrac{5}{3}(1 - 2^8)$$

$$= -425. \qquad \boxed{C}$$

Example 9.7

How many terms of the GP

$$1 + \tfrac{1}{2} + \tfrac{1}{4} + \tfrac{1}{8} + \dots$$

are required for the sum to exceed 1.99? Show that no matter how many terms are taken the sum will never exceed 2.

We have $a = 1$, $r = \tfrac{1}{2}$ and

$$S_n = \frac{a(1 - r^n)}{1 - r}$$

$$= 2(1 - (\tfrac{1}{2})^n)$$

If $S_n > 1.99$, then

$$2(1 - (\tfrac{1}{2})^n) > 1.99$$

$$\Rightarrow 2 - (\tfrac{1}{2})^{n-1} > 1.99$$

$$\Rightarrow \quad -(\tfrac{1}{2})^{n-1} > -0.01$$

$$\Rightarrow \quad (\tfrac{1}{2})^{n-1} < 0.01 \qquad \text{reversing the sign!}$$

$$\Rightarrow \quad 100 < 2^{n-1}$$

$$\Rightarrow \quad \lg 100 < \lg 2^{n-1} \qquad \text{taking logs, notice that if} \\ a < b \text{ then } \lg a < \lg b$$

$$\Rightarrow \quad \lg 100 < (n - 1)\lg 2$$

$$\Rightarrow \quad \frac{\lg 100}{\lg 2} < n - 1$$

$$\Rightarrow \quad n > 1 + \frac{2}{\lg 2} \qquad \text{since } \lg 100 = 2$$

$$\Rightarrow \quad n > 7.64.$$

However, since n is an integer, eight or more terms are required for the sum to exceed 1.99.

Recall that

$$S_n = 2(1 - (\tfrac{1}{2})^n) = 2 - (\tfrac{1}{2})^{n-1}$$

No matter the value of n, $(\tfrac{1}{2})^{n-1} > 0$, so $S_n < 2$.

9.3 The arithmetic and geometric means

Definition

> If three numbers a, b and c are in arithmetic progression, then b is called the *arithmetic mean* of a and c. However, if the three numbers are in geometric progression then b is called the *geometric mean* of a and c.

Example 9.8

Find the arithmetic mean of 3 and 7.

If m is the arithmetic mean then 3, m, 7 are in arithmetic progression. So

$$m - 3 = 7 - m$$

since there is a common difference.

$$\Rightarrow 2m = 10$$

$$\Rightarrow m = 5$$

More generally, if m is the arithmetic mean of a and b, then

$$m - a = b - m$$

$$\Rightarrow \quad 2m = a + b$$

$$\Rightarrow \quad m = \frac{a+b}{2}$$

Now if g is the geometric mean of a and c then a, g and c are in geometric progression. So

$$\frac{g}{a} = \frac{c}{g}$$

since there is a common ratio.

$$\Rightarrow g^2 = ac.$$

Thus a and c will have the same sign.

$$\Rightarrow g = \pm\sqrt{ac}.$$

The positive root is taken if a and c are positive, the negative root if a and c are both negative.

Statement

> The *arithmetic mean*, m, of two numbers a and c is given by
> $$m = \frac{a + c}{2}$$
> The *geometric mean*, g, of two numbers a and c is given by
> $$g = \pm\sqrt{ac}$$

Example 9.9

Show that the geometric mean of two distinct positive numbers is less than the arithmetic mean.

Suppose the two numbers are a and b, we wish to show that

$$\sqrt{ab} < \frac{(a + b)}{2}.$$

$$\Leftrightarrow \quad 2\sqrt{ab} < (a + b)$$
$$\Leftrightarrow \quad 4ab < (a + b)^2 \qquad [a < b \Leftrightarrow a^2 < b^2$$
$$\Leftrightarrow \quad 4ab < a^2 + 2ab + b^2 \qquad \text{only because } a \text{ and}$$
$$\Leftrightarrow \quad 0 < (a - b)^2 \qquad b \text{ are both positive.}]$$

but this last statement must be true, and so it follows that the original statement is also true. [This style of argument is quite difficult. A safer technique — though more difficult to remember, is to begin with

$$0 < (a - b)^2$$

and then deduce that

$$\left.\sqrt{ab} < \frac{a + b}{2}.\right]$$

Exercise 9.2

1. Write down the fourth, seventh and nth terms of the geometric sequence

    $$3, 6, 12, \ldots$$

2. Find the first and second term of the GP whose third term is 18 and whose fourth is 54.

3. Find the eighth term and the sum to eight terms of the geometric sequences:

   (a) $8, 4, 2, \ldots$ (b) $4, -2, 1, -\frac{1}{2}, \ldots$ (c) $1, x, x^2, x^3, \ldots$

   (d) $n, \dfrac{n^2}{a}, \dfrac{n^3}{a^2}, \ldots$

4. Find the value of n, if the sum of n terms of the GP $2 + 4 + 8 + \ldots$ is 510.

5. Find the sum of the following GPs.

   (a) $1 + 4 + 16 + \ldots + 1024$
   (b) $25 + 5 + 1 + \ldots 10$ terms
   (c) $a + ar + ar^2 + \ldots + ar^{n+1}$
   (d) $a + ar + ar^2 + \ldots + ar^{2n-1}$

6. The angles of a triangle are in GP, and the smallest angle is $20°$. Obtain the largest angle to the nearest degree.

7. $x + 2, x, x - 4$ are the first three terms of a GP. Find the fourth term.

8. The third, fifth and eighth terms of a GP are a, b, c, respectively. Show that $c^2 a^3 = b^5$.

9. Find the sum of n terms of the GP

$$g + p + \ldots$$

10. Find the sum of six terms of the GP whose fourth term is 7 and whose seventh term is 4.

11. If the population of a city grows at the rate of 5% per year. How many years are required before the population doubles?

12. The first term of an arithmetic series is 2 and the terms of the series are not all equal. The first, third and eleventh terms of this series are also the first, second and third terms, respectively, of a geometric series. Find the common difference of the arithmetic series.

   JMB 1984

13. The sum of the first n terms of a series is $\frac{1}{2}n(n + 15)$ for all positive integers n. Determine

   (a) the third term,
   (b) the nth term,
   (c) the difference between the $(n + 1)$th term and the nth term.

   JMB 1982

9.4   Sigma notation

The symbol Σ which means 'the sum of' may be used to express a series concisely.
For example,

$$\sum_{s=1}^{5} 2s = 2 + 4 + 6 + 8 + 10$$

Where the terms of the form $2s$ are summed for values of s from 1 to 5 inclusive. Now

$$\sum_{s=1}^{\infty} \frac{1}{s^2} = 1 + \tfrac{1}{4} + \tfrac{1}{9} + \tfrac{1}{16} + \dots$$

where the use of the symbol ∞ indicates that the summation continues indefinitely, to
form an infinite series.

It is worth noting that

$$\sum_{s=1}^{n} (as + b) = (a \cdot 1 + b) + (a \cdot 2 + b) + \dots + (an + b)$$

$$= a(1 + 2 + 3 + \dots + n) + nb$$

$$= a \sum_{s=1}^{n} s + nb$$

assuming a and b are constants.

Example 9.10

Rewrite each of the following series, using sigma notation.

(a) $1 + 4 + 9 + 16 + \dots + 81$
(b) $5 + 9 + 13 + 17 + \dots + 69$
(c) $1 - x + x^2 - x^3 + \dots + (-x)^{10}$

(a) Each term is of the form s^2, with the first term corresponding to $s = 1$ and the last
to $s = 9$. So the series is

$$\sum_{s=1}^{9} s^2.$$

(b) Each term is of the form $5 + 4s$ with s ranging from 0 to 16 (solve $5 + 4s = 69$).
So the series is

$$\sum_{s=0}^{16} 5 + 4s.$$

We could have represented each term in the form $1 + 4s$ where this time s ranges
from 1 to 17 so the same series is

$$\sum_{s=1}^{17} 1 + 4s.$$

Notice also that s is a so-called dummy variable, it could be changed to a new
variable without altering the meaning.

(c) Each term is of the form $(-x)^s$, with the first term corresponding to $s = 0$ and the last to $s = 10$. So the series is

$$\sum_{s=0}^{10} (-x)^s = \sum_{s=0}^{10} (-1)^s x^s = \sum_{s=1}^{11} (-1)^{s+1} x^{s-1}$$

for example.

The following results will be proved on page 298.

Statement

$$1^2 + 2^2 + \ldots + n^2 = \sum_{s=1}^{n} s^2 = \frac{n}{6}(n + 1)(2n + 1)$$

$$1^3 + 2^3 + \ldots + n^3 = \sum_{s=1}^{n} s^3 = \left[\frac{n}{2}(n + 1)\right]^2 = \frac{n^2}{4}(n + 1)^2$$

These results may be used together with the result already proved that

$$\sum_{s=1}^{n} s = \frac{n}{2}(n + 1)$$

to sum further series.

Example 9.11

Sum each of the following series.

(a) $1.2 + 2.3 + 3.4 + \ldots + n(n + 1)$ (b) $\displaystyle\sum_{s=1}^{10} \lg 2^{s^3}$

(a) The general term of this series has the form $s(s + 1)$ when s ranges from 1 to n. So the series is

$$\sum_{s=1}^{n} s(s + 1) = \sum_{s=1}^{n} (s^2 + s)$$

$$= \sum_{s=1}^{n} s^2 + \sum_{s=1}^{n} s$$

using the fact that

$$\sum_{s=1}^{n} (f(s) + g(s)) = \sum_{s=1}^{n} f(s) + \sum_{s=1}^{n} g(s)$$

(a result which the reader may care to prove if it is not obvious). Then

$$\sum_{s=1}^{n} s(s + 1) = \frac{n}{6}(n + 1)(2n + 1) + \frac{n}{2}(n + 1)$$

$$= \frac{n(n + 1)}{6}[2n + 1 + 3]$$

$$\Rightarrow \sum_{s=1}^{n} s(s+1) = \frac{n(n+1)(n+2)}{3}$$

(b) $\displaystyle\sum_{s=1}^{10} \lg 2^{s^3} = \sum_{s=1}^{10} s^3 \lg 2$

$$= \lg 2 \sum_{s=1}^{10} s^3 \qquad \text{using the fact that } \sum_{s=1}^{10} af(s) = a \sum_{s=1}^{10} f(s)$$

$$= \lg 2 \cdot \frac{10^2}{4}(10+1)^2 \qquad \text{where } a \text{ is a constant}$$

$$= 3025 \lg 2$$

$$= 910.6 \qquad 1 \text{ d.p.} \quad \boxed{C}$$

Use of partial fractions

Suppose we are asked to find

$$\sum_{s=1}^{n} \frac{1}{s(s+1)} = \frac{1}{1.2} + \frac{1}{2.3} + \frac{1}{3.4} + \ldots + \frac{1}{n(n+1)}.$$

We may first find A and B so that

$$\frac{1}{s(s+1)} \equiv \frac{A}{s} + \frac{B}{s+1}$$

$$\Rightarrow \qquad 1 \equiv A(s+1) + Bs.$$

Setting $s = 0$,

$$1 = A + 0 \qquad \Rightarrow A = 1.$$

Setting $s = -1$,

$$1 = 0 - B \qquad \Rightarrow B = -1.$$

Thus

$$\frac{1}{s(s+1)} \equiv \frac{1}{s} - \frac{1}{s+1}.$$

It follows that

$$\sum_{s=1}^{n} \frac{1}{s(s+1)} = \sum_{s=1}^{n} \left(\frac{1}{s} - \frac{1}{s+1} \right).$$

$$= \quad \tfrac{1}{1} - \cancel{\tfrac{1}{2}} \qquad\qquad \text{for } s = 1$$

$$+ \cancel{\tfrac{1}{2}} - \cancel{\tfrac{1}{3}} \qquad\qquad \text{for } s = 2$$

$$+ \cancel{\tfrac{1}{3}} - \cancel{\tfrac{1}{4}} \qquad\qquad \text{for } s = 3$$

$$\vdots \quad \vdots \qquad\qquad\qquad \vdots$$

$$\vdots \qquad \vdots \qquad\qquad \vdots$$

$$+ \frac{1}{n-1} - \frac{1}{n} \qquad \text{for } s = n - 1$$

$$+ \frac{1}{n} - \frac{1}{n+1} \qquad \text{for } s = n$$

$$= 1 - \frac{1}{n+1} \qquad \text{since other terms cancel!}$$

Exercise 9.3

1. Write out each of the following in full

   (a) $\sum_{s=1}^{4} s^2$

   (b) $\sum_{s=1}^{3} s(s-1)$

   (c) $\sum_{s=1}^{4} 2^s + 1$

   (d) $\sum_{s=0}^{4} (-1)^{s+1}s$

   (e) $\sum_{s=0}^{3} 2(s+1)^2$

2. Write down in Σ notation each of the following.

   (a) $1 + 2 + 3 + 4 + 5 + 6$
   (b) $1 - 2 + 3 - 4 + 5 - 6 + 7 - 8$
   (c) $-1 + 2 - 3 + 4 - 5$
   (d) $1 + 4 + 9 + \ldots + n^2$
   (e) $1.3 + 2.4 + 3.5 + \ldots + n.(n+2)$
   (f) $x + 2x^2 + 3x^3 + \ldots + nx^n$
   (g) $-1 + x - x^2 + x^3 + \ldots - x^{10}$

3. Evaluate each of the following.

   (a) $\sum_{s=1}^{4} (s+2)$

   (b) $\sum_{s=1}^{3} s^3$

4. Use the appropriate formulae to evaluate each of the following.

   (a) $\sum_{s=1}^{2n} s$

   (b) $\sum_{s=1}^{t} s^3$

   (c) $\sum_{s=1}^{n} s(s+2)$

   (d) $\sum_{s=1}^{n} s(s+1)(s+2)$

5. Find $\sum_{r=1}^{n} (4r+1)$.

SU 1981

6. Evaluate $\sum_{r=1}^{n} (3r+1)^2$.

W 1981

7. The rth term of a series is $2^r + 3r - 2$. Find a formula for the sum of the first n terms.

SU 1983

8. Find the sum of the series

$$1 \times 2\tfrac{1}{2} + 2 \times 3\tfrac{1}{4} + 3 \times 4\tfrac{1}{6} + \ldots + n\left(n + 1 + \frac{1}{2n}\right).$$

W 1983

9   Find, from first principles, a formula for the sum of the integers $1, 2, 3, \ldots, n$.
    Prove that

$$\sum_{r=0}^{n} \log_a (2a^r) = \tfrac{1}{2}(n + 1)(n + \log_a 4).$$

JMB 1983

10. Given that $\{r(r + 1)\}^2 - \{(r - 1)r\}^2 \equiv 4r^3$ find the sum of the series

$$1^3 + 2^3 + 3^3 + \ldots + n^3.$$

Deduce the sum of the series

$$1^3 - 2^3 + 3^3 - 4^3 + \ldots + (2n + 1)^3.$$

11. Find

(a) $\displaystyle\sum_{s=1}^{n} \frac{1}{s} - \frac{1}{s + 1}$ (b) $\displaystyle\sum_{s=1}^{n} \frac{1}{s} - \frac{1}{s + 2}$ (c) $\displaystyle\sum_{s=1}^{n} \frac{1}{(s + 1)(s + 2)}$

(d) $\displaystyle\sum_{s=1}^{n} \frac{1}{(s + 3)(s + 4)}$

9.5   The sum to infinity of a geometric progression

We have stated that $\displaystyle\sum_{s=1}^{\infty} u_s$ represents

$$u_1 + u_2 + u_3 + \ldots + u_n + \ldots$$

However, this infinite sum may have no meaning.
    Consider, for example

$$1 + 2 + 3 + \ldots + n + \ldots$$

For any given n, the sum to n terms is

$$S_n = \frac{n}{2}(n + 1)$$

As $n \to \infty$ so $S_n \to \infty$ and the infinite sum is undefined. Such a series is said to be
divergent.
    In general we have

$$\sum_{s=1}^{\infty} u_s = \lim_{n \to \infty} \sum_{s=1}^{n} u_s \quad \text{when the limit exists.}$$

When the limit exists, the series is said to be *convergent*.

In example 9.7 it was shown that the sum of n terms of the GP

$$1 + \tfrac{1}{2} + \tfrac{1}{4} + \tfrac{1}{8} + \ldots$$

never exceeded 2, no matter how large n was. In fact,

$$\sum_{s=1}^{n} (\tfrac{1}{2})^{s-1} = S_n = 2(1 - (\tfrac{1}{2})^n).$$

We have

$$\lim_{n \to \infty} \sum_{s=1}^{n} (\tfrac{1}{2})^{s-1} = \lim_{n \to \infty} S_n = \lim_{n \to \infty} 2(1 - (\tfrac{1}{2})^n)$$

$$= 2 \qquad \text{since } \lim_{n \to \infty} (\tfrac{1}{2})^{n-1} = 0$$

So

$$\sum_{s=1}^{\infty} (\tfrac{1}{2})^{s-1} = \lim_{n \to \infty} S_n = 2$$

Suppose, more generally, that we wish to find the sum to infinity of the GP $a + ar + ar^2 + \ldots$.

$$\sum_{s=1}^{n} ar^{s-1} = S_n = \frac{a(1 - r^n)}{1 - r} = \frac{a}{1 - r}(1 - r^n)$$

so

$$\sum_{s=1}^{\infty} ar^{s-1} = \lim_{n \to \infty} S_n$$

$$= \lim_{n \to \infty} \frac{a}{1 - r}(1 - r^n)$$

$$= \frac{a}{1 - r} + \frac{a}{1 - r} \lim_{n \to \infty} (r^n) \quad \text{using the property of limits on page 163.}$$

If $|r| > 1$, then $\lim_{n \to \infty} r^n$ is undefined since r^n increases indefinitely in magnitude as $n \to \infty$. However, if $|r| < 1$, *then* $\lim_{n \to \infty} r^n = 0$. [See Ex. 9.4, question 9, for a further discussion of this point.]

So we find:

Statement

> The sum of the GP, $a + ar + ar^2 + \ldots$ is given by
>
> $$\frac{a}{1 - r} \qquad \underline{\text{provided } |r| < 1}$$

Example 9.12

The sum of an infinite geometric progression is equal to three times the first term. Find the common ratio.

Suppose the geometric progression is

$$a + ar + ar^2 + ar^3 + \ldots$$

then

$$S_\infty = \sum_{s=1}^{\infty} ar^{s-1} = \frac{a}{1-r}$$

Now

$$\frac{a}{1-r} = 3a$$

$$\Rightarrow \quad 1 = 3(1-r)$$

$$\Rightarrow \quad r = \tfrac{2}{3}.$$

Example 9.13

Express the recurring decimal $0.\dot{3}\dot{4}$ as a fraction.

$$0.\dot{3}\dot{4} = 0.3434343434\ldots$$

$$= \frac{34}{100} + \frac{34}{10000} + \frac{34}{100000} + \ldots$$

$$= \frac{34}{100} + \frac{34}{100} \cdot \frac{1}{100} + \frac{34}{100} \cdot \left(\frac{1}{100}\right)^2 + \ldots$$

This is a geometric series, with first term $\dfrac{34}{100}$ and common ratio $\dfrac{1}{100}$.

The sum to infinity is

$$\frac{34}{100} \bigg/ \left(1 - \frac{1}{100}\right) = \frac{34}{99},$$

so

$$0.\dot{3}\dot{4} = \frac{34}{99}.$$

It is a generalisation of this argument which forms the basis of the claim made on page 1 that every recurring decimal represents a rational.

Exercise 9.4

1. Write down the sum to infinity of the following GPs and where appropriate write down the conditions for the sum to exist.

(a) $4 + \dfrac{16}{5} + \dfrac{64}{25} + \ldots$ (b) $3 - 1 + \frac{1}{3} + \frac{1}{9} + \ldots$ (c) $r + ar + a^2 r + \ldots$

(d) $na + nr + \dfrac{nr^2}{a}$ (e) $1 + 2x + 4x^2 + \ldots$

2. Express the following recurring decimals as fractions.

(a) $0.\dot{1}$ (b) $0.34\dot{2}$ (c) $0.\dot{5}\dot{4}$ (d) $0.36\dot{1}$

3. The sum to infinity of a GP is 8 and the second term is 2. Find the first term and the common ratio.

4. Find the sum to infinity of the geometric series
$$1, \ -\tfrac{1}{2}, \ +\tfrac{1}{4}, \ -\tfrac{1}{8}, \ +\tfrac{1}{16}, \ \ldots.$$
Find the least value of n for which the sum to n terms differs from the sum to infinity by less than 0.01.

5. A ball is dropped onto the ground and rebounds to five-eighths of the previous height. Find the total distance moved by the ball when it is dropped from a height of 3 m.

6. The sum of the first n terms of a GP is $\dfrac{2}{3^n}(3^n - 1)$. Obtain the first three terms and the sum to infinity.

7. The first four terms of a GP sum to 5 and the sum of the first and fourth term is 3. Find two series satisfying these conditions, and find the sum to infinity of the convergent series.

8. Find the value for r for which the sum to infinity of
$$1 + r + r^2 + \ldots + r^{n-1} + \ldots$$
is double the sum to infinity of $1 - r + r^2 - r^3 + \ldots.$

9. In order to show that
$$\lim_{n \to \infty} r^n = 0 \qquad \text{when } |r| < 1$$
it suffices to show that we can make r^n as small as we please, simply by taking n large enough. For example, suppose $r = 0.9$ and we require
$$(0.9)^n < 0.1$$
$$\Rightarrow n \lg 0{\cdot}9 < \lg 0.1 \qquad \text{taking logs.}$$
Complete this argument to find the least value of n satisfying the inequality. Find the least value of n for which $0.99^n < 0.001$.

10. A tortoise challenges a hare to a race of 100 m. The hare gives the tortoise a 10 m start, since he runs ten times faster than the tortoise — who has been training!
    Curiously, the tortoise is odds on favourite, because he argues as follows:

    'The race is in the bag, for the hare will not overtake me. When the hare has travelled 10 m, I will be 1 m ahead. When he covers the next metre I will be 10 cm ahead. When he travels a further 10 cm I will be 1 cm ahead. I agree that he is catching me up, but this argument continues indefinitely and I will always be ahead of the hare.'

    This argument worried the Greeks, because they could not fault it!

    (a) Where does the hare overtake the tortoise?
    (b) What is wrong with the tortoise's point of view?

11. Find the sum to infinity of the series.

    $$1 + 2\cos^2 \theta + 4\cos^4 \theta + \ldots + \ldots$$

    stating clearly the values of θ for which the sum exists.
    Show that the sum may be expressed as $-\sec 2\theta$.

12. Find

    $$S_a = 1 + \cos \theta + \cos^2 \theta + \ldots$$

    $$S_b = 1 + \cos 2\theta + \cos^2 2\theta + \ldots$$

    for $|\cos \theta| \neq 1, 0$. Show that

    $$S_b = \frac{S_a^2}{2(S_a - 1)}.$$

13. Define a rational number and prove that $\sqrt{2}$ is not rational. Express the recurring decimal $0.0545454\ldots$ as a fraction in its simplest form.

    JMB 1982

14. Find the sum to infinity of the series

    $$1 + 2\cos^2 \theta + 4\cos^4 \theta + \ldots + 2^r \cos^{2r} \theta + \ldots$$

    and state the range of values of θ within the limits $-\pi < \theta \leqslant \pi$ for which the sum exists.

15. An arithmetic progression has a first term of 100 and the common difference is 200. A geometric progression has a first term of 1 and its common ratio is 2. Find

    (a) the sum, A_n, of n terms of the arithmetic progression,
    (b) the sum, G_n, of n terms of the geometric progression.

    By trial and error find the least value of n for which G_n is greater than A_n.

    AEB 1983

16. The first term of an infinite geometrical progression is 20 and the sum to infinity is 40. Find the common ratio and the sum of the first 10 terms.

SU 1980

17. Find the least number of terms of the geometrical progression

$$2 + 3 + 4.5 + \ldots$$

that will have a sum in excess of 10 000.

SU 1983

18. An arithmetic progression has first term a and common difference d, where $d \neq 0$. The 1st, 3rd and 6th terms are in geometric progression. Obtain an expression for d in terms of a and hence find the common ratio of the geometric progression.

AEB 1983

19. The first three terms of a geometric series are $2, -\frac{1}{2}, \frac{1}{8}$. Find the sum to infinity of this series.
    Find also the least value of n for which the sum of the first n terms of this series differs from the sum to infinity by less than 10^{-5}.

CAMB 1983

20. The sum of the first n terms of a series is $1 - (\frac{1}{2})^n$. Obtain the values of the first three terms of this series. What is the sum to infinity of this series?

SU 1983

21. (a) The sum of the first nine terms of an arithmetic progression is 75 and the twenty-fifth term is also 75. Find the common difference and the sum of the first hundred terms.
    (b) A geometric series has common ratio $(a - b)/a$ where $a > 0$. Show that the series has a sum to infinity provided that $0 < b < 2a$.
    Given that the first term is a, find the sum to infinity of the series.

22. An arithmetic series has first term 6, common difference 8 and the sum of the first n terms is S_n. Express S_n in terms of n and show that it is the product of two consecutive integers. Deduce that S_n is not an integral power of 2.
    The sum of the first n terms of a geometric series is $\frac{4}{3}(4^n - 1)$ and the nth term is u_n. Express u_n in terms of n and show that u_n is an integral power of 2.

AEB 1983

9.6 Proof by induction

Suppose it has been conjectured that the expression $n^3 - n$ is divisible by 6 for all positive integers n. We might test this for the first few integers (Table 9.1 — see next page) but no matter how long we make the table we can never hope to prove the conjecture in this way.

n	$n^3 - n$
1	0
2	6
3	24
4	60
5	120

Table 9.1

Suppose now that we assume the conjecture to hold for $n = k$, that is '$k^3 - k$ is divisible by 6'. What can be said for the case $n = k + 1$? Is $(k + 1)^3 - (k + 1)$ divisible by 6?

Now

$$(k + 1)^3 - (k + 1) = k^3 + 3k^2 + 3k + 1 - k - 1$$
$$= k^3 - k + 3k(k + 1)$$

but $3k(k + 1)$ is divisible by 6 since $k(k + 1)$ must be even. So if $k^3 - k$ is divisible by 6 then so is $(k + 1)^3 - (k + 1)$!

But now we can claim that the conjecture has been proved. For, suppose someone doubted, we could argue that it was demonstrably true for $n = 1$. Now if it is true for $k = 1$ we have shown that it is true for $k + 1 = 2$. And if it is true for $k = 2$ it must be true for $k + 1 = 3$. And if it is true for $k = 3 \ldots$ We may continue until all doubts cease!

The basic argument involves showing that the conjecture is true for some initial value of n (say $n = 1$), then, on the *assumption* that it is true for $n = k$, *proving* that it will then be true for $n = k + 1$. If we succeed then we have proved the result holds for all n greater than or equal to the initial value, by induction.

Example 9.14

Prove by induction, that

$$\sum_{r=1}^{n} r^2 = \frac{n}{6}(n + 1)(2n + 1) \qquad \text{for } n \in \mathbb{N}^+.$$

For $n = 1$,

$$\left.\begin{array}{l} \text{l.h.s.} = \sum_{r=1}^{1} r^2 = 1 \\[2mm] \text{r.h.s.} = \frac{1}{6}(2)(3) = 1. \end{array}\right\} \qquad \text{The result is true for } n = 1.$$

Assume the result is true for $n = k$, that is

$$\sum_{r=1}^{k} r^2 = \frac{k}{6}(k + 1)(2k + 1).$$

We wish to prove that

$$\sum_{r=1}^{k+1} r^2 = \frac{(k + 1)}{6}(k + 2)(2k + 3) \qquad [\text{replacing } k \text{ with } k + 1]$$

Now

$$\sum_{r=1}^{k+1} r^2 = \sum_{r=1}^{k} r^2 + (k + 1)^2$$

$$= \frac{k}{6}(k + 1)(2k + 1) + (k + 1)^2 \qquad \text{by the assumption}$$

$$= \frac{(k + 1)}{6}[k(2k + 1) + 6(k + 1)]$$

$$= \frac{(k + 1)}{6}[2k^2 + 7k + 6]$$

$$= \frac{(k + 1)}{6}(2k + 3)(k + 2)$$

So the result follows, by induction!!

Exercise 9.5

1. Prove each of the following assertions, using the method of induction.

(a) $\displaystyle\sum_{r=1}^{n} 2r - 1 = n^2$

(b) $\displaystyle\sum_{r=1}^{n} r^3 = \frac{1}{4}n^2(n + 1)^2$

(c) $\displaystyle\sum_{r=1}^{n} r(r + 1) = \frac{1}{3}n(n + 1)(n + 2)$

(d) $a + (a + d) + (a + 2d) + \ldots + a + (n - 1)d = \dfrac{n}{2}(2a + (n - 1)d)$

(e) $a + ar + ar^2 + \ldots + ar^{n-1} = \dfrac{a(1 - r^n)}{1 - r}$

(f) $\displaystyle\sum_{r=1}^{n} (r^2 + 1)r! = n(n + 1)! \qquad [n! = n(n - 1) \times (n - 2) \times \ldots \times 2 \times 1]$

(g) $\displaystyle\sum_{r=1}^{n} \frac{1}{r(r + 1)} = \frac{n}{n + 1}$

(h) $x^n - 1$ is divisible by $x - 1$ \qquad [Consider $x^{k+1} - x + x - 1$.]

(i) $16^n - 1$ is divisible by 15.

2. Prove by induction that $4^n \geqslant 3n^2 + 1$.

3. Prove by induction that
$$1^3 + 2^3 + 3^3 + \ldots + (2n)^3 = n^2(2n + 1)^2.$$

4. Prove by induction that

$$\frac{(a + b)^n}{2} \leqslant \frac{a^n + b^n}{2} \qquad \text{for } a, b > 0.$$

5. Prove by induction that

$$(1 + x)^n > 1 + nx \qquad x > -1$$

Find a counter-example for $x < -1$.

6. Show by induction, or otherwise, that

$$1 \times 3 + 3 \times 5 + 5 \times 7 + \ldots + (2n - 1) \times (2n + 1) =$$

$$\tfrac{1}{6}(2n - 1)(2n + 1)(2n + 3) + \tfrac{1}{2}.$$

OX 1983

7. Prove by mathematical induction that

$$\frac{3}{4} + \frac{5}{36} + \ldots + \frac{2n + 1}{n^2(n + 1)^2} = 1 - \frac{1}{(n + 1)^2},$$

for all positive integers n.

W 1981

8. Prove by mathematical induction, or otherwise, that $3^{2n} - 1$ is divisible by 8 for all positive integers n.

W 1982

9. Let $f(n) = 9^{2n} - 5^{2n}$, where n is a non-negative integer.

    (a) Evaluate $f(0)$ and $f(1)$.
    (b) Write down the value of $f(n + 1)$.
    (c) Prove that $f(n + 1) - 25f(n) = 56(9^{2n})$.
    (d) Hence, using induction, prove that $f(n)$ is always divisible by 7.

O & C 1982

9.7 The binomial theorem

The binomial theorem was introduced on page 36 in the form

$$(a + b)^n = a^n + na^{n-1}b + \frac{n(n - 1)}{2.1.}a^{n-2}b^2 + \ldots + b^2$$

It is convenient to introduce the following definitions.

Definition

Factorial n, written $n!$, is defined as

$$n! = n \times (n - 1) \times (n - 2) \times \ldots \times 3 \times 2 \times 1,$$

where n is a positive integer and $0! = 1$.

For example,

$$4! = 4 \times 3 \times 2 \times 1$$
$$= 24$$

Notice that $(n + 1)! = (n + 1)n!$

Definition

$$\binom{n}{r} \equiv {}^nC_r \equiv \frac{n!}{(n - r)!\, r!} \qquad \text{for integers, } n \geqslant r$$

The following observations will justify our interest in this definition.

$$\binom{n}{0} = {}^nC_0 = \frac{n!}{(n - 0)!\,.\,0!} = 1$$

Then

$$\binom{n}{1} = {}^nC_1 = \frac{n!}{(n - 1)!\,.\,1!} = n,$$

since $n! = n(n - 1)!$ if $n \geqslant 1$.

$$\binom{n}{2} = {}^nC_2 = \frac{n!}{(n - 2)!\,2!} = \frac{n(n - 1)}{2!},$$

since $n! = n(n - 1)(n - 2)!$ if $n \geqslant 2$.

$$\binom{n}{3} = {}^nC_3 = \frac{n!}{(n - 3)!\,3!} = \frac{n(n - 1)(n - 2)}{3!}$$

since $n! = n(n - 1)(n - 2)(n - 3)!$ if $n \geqslant 3$.

And more generally

$$\binom{n}{r} = {}^nC_r = \frac{n!}{(n - r)!\, r!} = \frac{n(n - 1)(n - 2)(n - 3)\ldots(n - r + 1)}{r!} \qquad \text{for } n \geqslant r.$$

Statement

> So the binomial theorem may be restated in the form
>
> $$(a + b)^n = \binom{n}{0}a^n + \binom{n}{1}a^{n-1}b + \binom{n}{2}a^{n-2}b^2 + \ldots$$
>
> $$\ldots + \binom{n}{r}a^{n-r}b^r + \ldots + b^n$$
>
> or more succinctly as
>
> $$\sum_{r=0}^{n}\binom{n}{r}a^{n-r}b^r = (a + b)^n.$$

Notice the form of the general term

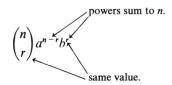

powers sum to n.

same value.

Example 9.15

Show that

$$\binom{n}{r} + \binom{n}{r+1} = \binom{n+1}{r+1}.$$

Now

$$\text{r.h.s.} = \frac{(n+1)!}{(n-r)!\,(r+1)!}$$

and

$$\text{l.h.s.} = \frac{n!}{(n-r)!\,r!} + \frac{n!}{(n-r-1)!\,(r+1)!}$$

$$= \frac{n!}{(n-r-1)!\,r!}\left[\frac{1}{n-r} + \frac{1}{r+1}\right]$$

$$= \frac{n!}{(n-r-1)!\,r!}\left[\frac{n+1}{(n-r)(r+1)}\right]$$

$$= \frac{(n+1)!}{(n-r)!\,(r+1)!} \quad !!$$

So the result follows.

Proof of the binomial theorem

We shall use the method of induction to prove that

$$\sum_{r=0}^{n} \binom{n}{r} a^{n-r} b^r = (a + b)^n.$$

When $n = 1$,

$$\text{r.h.s.} = a + b$$

and

$$\text{l.h.s.} = \sum_{r=0}^{1} \binom{1}{r} a^{1-r} b^r = \binom{1}{0} ab^0 + \binom{1}{1} a^0 b^1$$

$$= a + b$$

Assume that the result holds for $n = k$, that is

$$(a+b)^k = \sum_{r=0}^{k} \binom{k}{r} a^{k-r} b^r$$

Now

$$(a + b)^{k+1} = (a + b)(a + b)^k$$

$$= (a + b) \sum_{r=0}^{k} \binom{k}{r} a^{k-r} b^r$$

$$= (a + b)\left(a^k + ka^{k-1}b + \ldots + \binom{k}{r} a^{k-r} b^r + \binom{k}{r+1} a^{k-r-1} b^{r+1}\right.$$

$$\left. + \ldots + kab^{k-1} + b^k\right)$$

$$= a^{k+1} + ka^k b + \ldots + \binom{k}{r+1} a^{k-r} b^{r+1} + \ldots + ka^2 b^{k-1} + ab^k$$

$$+ a^k b + \ldots + \binom{k}{r} a^{k-r} b^{r+1} + \ldots + kab^k + b^{k+1}$$

$$= a^{k+1} + (k + 1)a^k b + \ldots + \left(\binom{k}{r+1} + \binom{k}{r}\right) a^{k-r} b^{r+1}$$

$$+ \ldots + (k + 1)ab^k + b^{k+1}.$$

But from the previous example $\quad \binom{k}{r+1} + \binom{k}{r} = \binom{k+1}{r+1}$

$$\Rightarrow (a + b)^{k+1} = a^{k+1} + (k + 1)a^k b + \ldots + \binom{k+1}{r+1} a^{(k+1)-(r+1)} b^{r+1}$$

$$+ \ldots + (k + 1)ab^k + b^{k+1}$$

$$= \sum_{r=0}^{k+1} \binom{k+1}{r} a^{k+1-r} b^r$$

and the result follows for $n = k + 1$, thus the binomial theorem has been proved by induction. An alternative proof may be found on page 438.

Example 9.16

Write down the sixth term in the expansion of $\left(2 + \dfrac{x}{3}\right)^8$ in ascending powers of x.

The sixth term is the term involving x^5, that is

$$\binom{8}{5} 2^3 \left(\frac{x}{3}\right)^5$$

Now

$$\binom{8}{5} = \frac{8!}{3! \, 5!} = \frac{8 \times 7 \times 6}{3 \times 2 \times 1} = 56$$

So the sixth term is

$$\frac{56 . 2^3 . x^5}{3^5} = \frac{7 . 2^6 x^5}{3^5}$$

Example 9.17

Find the coefficient of x^r in the expansion of $(2 + x^2)(1 + x)^n$.

Each term in the expansion of $(1 + x)^n$ is to be multiplied by 2 and by x^2. Two terms will then contain x^r.
    Namely,

$$2 \times \binom{n}{r} x^r \quad \text{and} \quad x^2 \times \binom{n}{r-2} x^{r-2}$$

So the coefficient of x is

$$2\binom{n}{r} + \binom{n}{r-2}$$

Example 9.18

Obtain an approximation of $(1.01)^{10}$ correct to 4 d.p. without using a calculator or other aids.

First consider the expansion of

$$(1 + x)^{10} = 1 + 10x + \frac{10.9}{2!} x^2 + \frac{10.9.8}{3!} x^3 + \dots$$

Now set $x = 0.01 = \dfrac{1}{100}$

$$(1.01)^{10} = 1 + 10 \times \frac{1}{100} + 45 \times \left(\frac{1}{100}\right)^2 + 120 \times \left(\frac{1}{100}\right)^3 + \dots$$

$$= 1 + 0.1 + 0.0045 + 0.00012 + \dots$$

$$= 1.10462\dots$$

$$\simeq 1.1046 \qquad \text{4 d.p.}$$

Exercise 9.6

1. Obtain the first four terms in the binomial expansion of

   (a) $(1 - 2x)^5$ (b) $(x + y)^6$ (c) $\left(1 + \frac{1}{x}\right)^7$.

2. Find the coefficient of x^4 in the binomial expansion of

   (a) $(1 - x)^8$ (b) $(2 - 3x)^7$ (c) $\left(x + \frac{1}{x}\right)^6$ (d) $(1 - x)(1 + x)^6$.

3. Expand and simplify $(1 + \sqrt{3})^5 + (1 - \sqrt{3})^5$.

4. Find the middle term in the binomial expansion of $(1 + 2x)^6$.

5. Find the term independent of x in the expansion of $\left(x - \frac{1}{x^2}\right)^9$.

6. Expand and simplify

   (a) $(1 - x)^7 + (1 + x)^7$ (b) $\left(x + \frac{2}{x}\right)^4 - \left(x - \frac{2}{x}\right)^4$.

7. Find the term in x^3 in the expansions of

   (a) $(x + 1)^{10}$ (b) $(3 - 2x)^7$
   (c) $(1 + x)(1 + 2x)^5$ (d) $(1 + x)^2(1 + 2x)^6$

8. Expand each of the following up to and including the term in x^3.

   (a) $(1 + x(x - 1))^5$ (b) $(2 - x - 2x^2)^4$

9. Use the first three terms in the expansion of an appropriate binomial to obtain approximations to

   (a) $(1.01)^8$ (b) $(1.02)^7$ (c) $(2.001)^6$

10. Assuming that x is so small that powers of x of 3 and above may be neglected, find a quadratic approximation of each of the following.

(a) $(1 - x)^{10}$ (b) $\left(2 + \dfrac{x}{3}\right)^8$ (c) $(1 + x)(2 + x)^9$

(d) $(1 - x)^5 . (1 + 2x)^4$ (e) $(2 + x)^6 . \left(1 - \dfrac{x}{2}\right)^9$

11. Find the values of the constants a and b if the coefficients of x and x^2 are both zero in the expansion of

$$(1 + 2x^2)^3 (1 - x)^{10} - (1 + bx^2)^4 (1 + ax)^5.$$

9.8   The binomial theorem for any index

The GP, $1 + x + x^2 + x^3 + \ldots$, has sum $\dfrac{1}{1 - x}$ provided $|x| < 1$. That is

$$\dfrac{1}{1 - x} \equiv (1 - x)^{-1}$$

$$\equiv 1 + x + x^2 + x^3 + \ldots$$

$$\equiv 1 + (-1)(-x) + \dfrac{(-1)(-2)}{2!}(-x)^2 + \dfrac{(-1)(-2)(-3)}{3!}(-x)^3 + \ldots$$

provided $|x| < 1$. In this example, it seems that we can apply the binomial theorem to the case where $n = -1$, provided $|x| < 1$.

Statement

> The binomial expansion
>
> $$(1 + x)^n = 1 + nx + \dfrac{n(n - 1)}{2!}x^2 + \dfrac{n(n - 1)(n - 2)}{3!}x^3 + \ldots$$
>
> is valid for any $n \in \mathbb{Q}$, provided that $|x| < 1$.

The proof of this result will be found on page 438.

Notice that the expansion in the bracket must be of the form $(1 + f(x))$ where $|f(x)| < 1$.

Example 9.19

Find the first four terms in the expansion of

(a) $\left(1 + \dfrac{x^2}{2}\right)^{-1}$ (b) $(1 + 2x)^{1/2}$ (c) $\dfrac{1}{(2 - x)}$

(a) $\left(1 + \dfrac{x^2}{2}\right)^{-1} = 1 + (-1)\left(\dfrac{x^2}{2}\right) + \dfrac{(-1)(-2)}{2!}\left(\dfrac{x^2}{2}\right)^2 + \dfrac{(-1)(-2)(-3)}{3!}\left(\dfrac{x^2}{2}\right)^3 + \ldots$

$$\Rightarrow \left(1 + \frac{x^2}{2}\right)^{-1} = 1 - \frac{x^2}{2} + \frac{x^4}{4} - \frac{x^6}{8} + \dots$$

provided

$$\left|\frac{x^2}{2}\right| < 1 \Leftrightarrow |x^2| < 2 \Leftrightarrow |x| < \sqrt{2}.$$

(b) $(1 + 2x)^{1/2} = 1 + (\frac{1}{2})(2x) + \frac{(\frac{1}{2})(-\frac{1}{2})}{2!}(2x)^2 + \frac{(\frac{1}{2})(-\frac{1}{2})(-\frac{3}{2})}{3!}(2x)^3 + \dots$

$$= 1 + x - \frac{1}{8} \cdot 4x^2 + \frac{1}{16} \cdot 8x^3 + \dots$$

$$= 1 + x - \frac{x^2}{2} + \frac{x^3}{2} + \dots$$

provided $|2x| < 1 \Leftrightarrow |x| < \frac{1}{2}$.

(c) $\dfrac{1}{2 - x} = (2 - x)^{-1}$

$$= 2^{-1}\left(1 - \frac{x}{2}\right)^{-1}$$

$$= \frac{1}{2}\left[1 + (-1)\left(\frac{-x}{2}\right) + \frac{(-1)(-2)}{2!}\left(\frac{-x}{2}\right)^2 + \frac{(-1)(-2)(-3)}{3!}\left(\frac{-x}{2}\right)^3 + \dots\right]$$

$$= \frac{1}{2}\left[1 + \frac{x}{2} + \frac{x^2}{4} + \frac{x^3}{8} + \dots\right]$$

$$= \frac{1}{2} + \frac{x}{4} + \frac{x^2}{8} + \frac{x^3}{16} + \dots$$

Example 9.20

Find the first three terms of the expansion in ascending powers of x of each of the following.

(a) $\dfrac{1}{(1 - x)(x - 2)}$ (b) $\sqrt{\dfrac{1 - x}{1 + x}}$

(a) We could attempt this problem by expanding $(1 - x)^{-1}$ and $(x - 2)^{-1}$ and multiplying the results. We can avoid this difficulty by first expressing in partial fractions. Now if

$$\frac{1}{(1 - x)(x - 2)} \equiv \frac{A}{(1 - x)} + \frac{B}{(x - 2)}$$

then

$$1 \equiv A(x - 2) + B(1 - x)$$

from which we deduce that $B = -1$ and $A = -1$.

So

$$\frac{1}{(1-x)(x-2)} \equiv \frac{-1}{(1-x)} + \frac{-1}{(x-2)}.$$

Now

$$(1-x)^{-1} = 1 + (-1)(-x) + \frac{(-1)(-2)}{2!}(-x)^2 + \dots$$

$$= 1 + x + x^2 + \dots$$

and

$$(x-2)^{-1} = (-2)^{-1}\left(\frac{-x}{2} + 1\right)^{-1}$$

$$= -\tfrac{1}{2}\left(1 - \frac{x}{2}\right)^{-1}$$

$$= -\tfrac{1}{2}\left[1 + (-1)\left(\frac{-x}{2}\right) + \frac{(-1)(-2)}{2!}\left(\frac{-x}{2}\right)^2 + \dots\right]$$

$$= -\tfrac{1}{2}\left[1 + \frac{x}{2} + \frac{x^2}{4} + \dots\right]$$

$$= -\tfrac{1}{2} - \frac{x}{4} - \frac{x^2}{8} + \dots$$

So

$$\frac{1}{(1-x)(x-2)} = -[1 + x + x^2 + \dots] - \left[-\tfrac{1}{2} - \frac{x}{4} - \frac{x^2}{8} + \dots\right]$$

$$= -\tfrac{1}{2} - \frac{3x}{4} - \frac{7x^2}{8} - \dots$$

Notice that the expansion is valid provided

$$|-x| < 1 \quad \text{and} \quad \left|\frac{-x}{2}\right| < 1,$$

that is $|x| < 1$ and $|x| < 2$, which occurs if $|x| < 1$.

Sometimes we have to face the prospect of multiplying two infinite series, as in the next example.

(b) $\sqrt{\dfrac{1-x}{1+x}} \equiv (1-x)^{1/2}(1+x)^{-1/2} \qquad x \neq -1$

$$= \left[1 + (\tfrac{1}{2})(-x) + \frac{(\tfrac{1}{2})(-\tfrac{1}{2})}{2!}(-x)^2 + \dots\right]$$

$$\times \left[1 + (-\tfrac{1}{2})x + \frac{(-\tfrac{1}{2})(-\tfrac{3}{2})}{2!}x^2 + \dots\right]$$

$$\Rightarrow \sqrt{\frac{1-x}{1+x}} = \left[1 - \frac{x}{2} - \frac{x^2}{8} + \ldots\right] \times \left[1 - \frac{x}{2} + \frac{3x^2}{8} + \ldots\right]$$

$$= 1 - \frac{x}{2} + \frac{3x^2}{8}$$

$$- \frac{x}{2} + \frac{x^2}{4} - \frac{3x^3}{16}$$

$$- \frac{x^2}{8} + \frac{x^3}{16} - \frac{3x^4}{64} \cdots$$

$$= 1 - x + \frac{x^2}{2} + \ldots$$

The expansion is valid provided $|-x| < 1$ and $|x| < 1$, that is $|x| < 1$.

Exercise 9.7

1. Expand each of the following up to the term involving x^4. In each case state the values of x for which the expansion is valid.

   (a) $(1 + 2x)^{-1}$ (b) $(1 - x)^{-2}$ (c) $(1 + x)^{1/2}$

   (d) $(2 - x)^{1/3}$ (e) $\dfrac{1}{3 + x}$ (f) $\dfrac{1}{\sqrt{3 - x}}$

2. Use the binomial theorem to show that if $|x| < 1$,

$$\frac{1 + x^2}{(1 - x)^3} = 1 + 3x + 3x^2 + 13x^3 + \ldots$$

$$\left[Hint: \quad \frac{1 + x^2}{(1 - x)^3} = (1 + x^2)(1 - x)^{-3}.\right]$$

3. Find the first four terms in the expansion of

   (a) $\dfrac{1 - x}{1 + x}$ (b) $\dfrac{1 + x}{1 - 2x}$

   where $|x| < \frac{1}{2}$

4. If x is small so that powers of x higher than x^5 may be neglected show that

$$\frac{1}{1 + x^2} = 1 - x^2 + x^4$$

   By taking $x = \dfrac{1}{10}$, show that $\dfrac{100}{101} \simeq 0.99\,01$ to 4 d.p.

5. Find the first four terms in the expansion of $(100 + x)^{1/2}$, stating when the expansion is valid. By choosing a suitable value for x, find an approximation to $(101)^{1/2}$ correct to 3 d.p.

6. Find an approximate value of $\sqrt[3]{1001}$ correct to 3 d.p.

7. Express each of the following in partial fractions and hence find the first three terms in the expansion in ascending powers of x. State in each case when the expansion is valid.

   (a) $\dfrac{5 - 7x}{(1 - x)(1 - 2x)}$ (b) $\dfrac{3x - 1}{(2 - x)(x + 3)}$

8. Show that if x is small and measured in radians that

$$\cos x \simeq 1 - \frac{x^2}{2} + \ldots.$$

   [*Hint*:   expand $(1 - \sin^2 x)^{1/2}$ and use the approximation $\sin x \simeq x$.]

9. Find the value of a and b if

$$(1 + ax)^b \simeq 1 + 2x + 6x^2$$

   for small x.

10. Show that

$$\sqrt{\frac{1 - x}{1 + x}} \equiv \frac{1 - x}{\sqrt{1 - x^2}} \qquad |x| < 1$$

   and hence find the first four terms in the expansion of $\sqrt{\dfrac{1 - x}{1 + x}}$ in ascending powers of x.

11. Show that

$$\left(\frac{1 + x}{1 - x}\right)^3 \simeq 1 + 6x + 18x^2 + \ldots$$

12. Find the possible values of a if the coefficient of x^2 in the expansion of $(1 + ax)^4$ is the same as the coefficient of x^4 in the expansion of $(1 + x)^6$.

   W 1981

13. Obtain the term independent of x in the expansion of $\left(2x - \dfrac{1}{x}\right)^{20}$.

   You may leave the answer in terms of factorials.

   SU 1980

14. State the binomial expansion of $(x - y)^4$, giving the coefficients as integers. Given that $x - y = p$ and $xy = q$, express $x^4 + y^4$ in terms of p and q.

CAMB 1983

15. (a) Find the expansion of

$$(1 + 3x + x^2)^4$$

in ascending powers of x up to and including the term in x^2.

(b) Calculate the term independent of x in the expansion of $\left(x - \dfrac{2}{x^2} \right)^{15}$.

O&C 1982

16. Expand $(1 + x)^{-1/2}$ in ascending powers of x, giving the first four terms and the general term. Hence, without using tables or calculator, obtain the value of $1/\sqrt{101}$, correct to 6 decimal places, showing all working.

SU 1980

17. Show that, for $|6x| < 1$, the first three terms in the expansion in ascending powers of x of $(1 + 6x)^{1/3}$ are identical to the first three terms in the expansion in ascending powers of x of $(1 + 4x)/(1 + 2x)$.

LOND 1983

18. Show that

$$26 \left(1 - \frac{1}{26^2} \right)^{1/2} = n\sqrt{3},$$

where n is an integer, whose value is to be found.

Given that $|x| < 1$, expand $(1 - x)^{1/2}$ as a series of ascending powers of x, up to and including the term in x^2, simplifying the coefficients.

By using the first *two* terms of the expansion of $\left(1 - \dfrac{1}{26^2} \right)^{1/2}$, obtain an approximate value for $\sqrt{3}$ in the form p/q, where p and q are integers.

LOND 1983

19. Write down the first four terms in the expansion in ascending powers of x of $(1 + ax)^p$.

Given that the first three terms in this expansion are $1, 2x$ and $\frac{11}{4}x^2$, show that $2pa(pa - a) = 11$ and find the values of a and p.

State the range of values of x for which this series expansion is valid.

AEB 1982

20. Write down the first three terms of the binomial expansion for $(1 + 2y)^n$. Show that if y is greater than 2 then $(1 + 2y)^n + (1 - 2y)^n$ is greater than $16n^2 - 16n + 2$.

N 1982

21. Use the binomial theorem to expand

$$\sqrt{\frac{1 + x}{1 - x}}$$

as a series of ascending powers of x up to and including the term in x^2, where $|x| < 1$.

By putting $x = \frac{1}{10}$ in your result, show that $\sqrt{11}$ is approximately equal to $\frac{663}{200}$.

CAMB 1983

22. Express

$$\frac{10 - 17x + 14x^2}{(2 + x)(1 - 2x)^2}$$

in partial fractions of the form

$$\frac{A}{2 + x} + \frac{B}{1 - 2x} + \frac{C}{(1 - 2x)^2}.$$

Hence, or otherwise, obtain the expansion of

$$\frac{10 - 17x + 14x^2}{(2 + x)(1 - 2x)^2}$$

in ascending powers of x, up to and including the term in x^3. State the restrictions which must be imposed on x for the expansion in ascending powers of x to be valid.

CAMB 1982

23. Why cannot $f(x)$, where

$$f(x) = \frac{x^2}{(x + 1)(x - 2)},$$

be expressed in the form

$$\frac{a}{x - 1} + \frac{b}{x - 2}$$

where a and b are constants? Express $f(x)$ in partial fractions and hence expand $f(x)$ in ascending powers of x as far as the term in x^3. Give the range of values of x for which the expansion is valid.

SU 1983

24. Given that

$$y = \frac{8 - 11x + 4x^2}{(1 - x)^2(2 - x)},$$

express y as the sum of three partial fractions.

The expression for y is to be expanded in ascending powers of x, where $|x| < 1$. Find the terms in this expansion up to and including the term in x^2.

JMB 1984

25. An attempt to construct an equilateral triangle of side 4 cm results in a triangle ABC for which angle BAC is $60°$, AB is $(4 + h)$ cm and AC is $(4 - h)$ cm. Show that the length of BC is a cm where $a^2 = 16 + 3h^2$.

Show that, provided h is small, the length of BC is approximately $(4 + th^2)$ cm, where t is a constant whose value is to be found.

JMB 1983

26. In the triangle ABC (Fig. 9.1), $AB = 1$ unit, $AC = 3$ units and the angle BAC is θ radians. D is the point on AC such that the angle ABD is $3\pi/4$. Prove that

(a) $BC = (10 - 6\cos\theta)^{1/2}$ (b) $AD = 1/(\cos\theta - \sin\theta)$

Given that θ is small enough for the approximations

$$\sin\theta \approx \theta \quad \text{and} \quad \cos\theta \approx 1 - \tfrac{1}{2}\theta^2$$

to apply, show that

(c) $BC \approx p + q\theta^2$ (d) $AD \approx 1 + r\theta + s\theta^2$

where p, q, r and s are constants to be determined.

JMB 1983

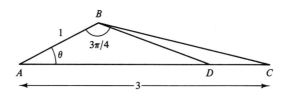

Fig. 9.1

Chapter 10 Integration

10.1 The area under a curve

An ancient problem in mathematics was to find the area or volume of a shape with curved sides or surfaces. It is unlikely, for example, that the reader has yet seen a proof that the area of a circle of radius r is πr^2 (see page 338). Such proofs as are available, without using the methods of calculus, usually assume the formula for the circumference of a circle.

Consider the problem of finding the area shaded in Fig. 10.1, bounded by the line $x = 1$, the x axis and the curve $y = x^2$. Such an area is referred to as the area under the curve (defined by $y = x^2$) between the limits $x = 0$ and $x = 1$.

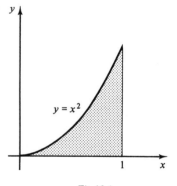

Fig.10.1

One way of finding an approximate answer is to plot the graph accurately and then 'count the squares'. Another way, more useful for our purposes, is to approximate the area by a number of rectangles with equal widths, where the heights depend on the equation of the curve. If we take four strips (rectangles) then we have the situation indicated in Fig. 10.2

In Fig. 10.2 (i) we have taken *inner* rectangles, the total area of which gives an underestimate of the actual area required. In Fig. 10.2 (ii) we have taken *outer* rectangles which gives an over estimate of the actual area. In each case the error (the

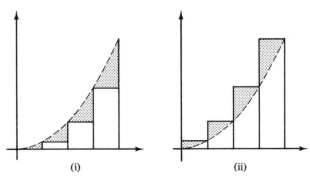

(i)   (ii)

Fig.10.2

difference between the actual area and our estimate) is the shaded area. The error may be made as small as we please by taking more strips. This remains true whether we consider inner or outer rectangles (see Fig. 10.3).

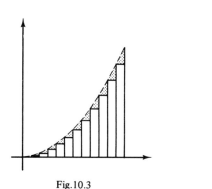

Fig.10.3

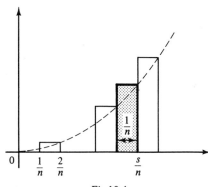

Fig.10.4

Returning to our problem, suppose we decide to take n strips, so that the width of each one is $1/n$. Considering the outer rectangles, the height of the sth strip is $(s/n)^2$ (see Fig. 10.4).

So the area of the sth strip is

$$\frac{1}{n} \times \left(\frac{s}{n}\right)^2 = \frac{s^2}{n^3}$$

To find our estimate of the total area we need to find the area of all n rectangles. That is, we need to find

$$\sum_{s=1}^{n} \frac{s^2}{n^3}$$

$$= \frac{1}{n^3} \sum_{s=1}^{n} s^2$$

$$= \frac{1}{n^3} \frac{n}{6}(n+1)(2n+1)$$

$$= \frac{1}{n^3} \frac{n^3}{6}\left(1 + \frac{1}{n}\right)\left(2 + \frac{1}{n}\right)$$

$$= \tfrac{1}{6}\left(1 + \frac{1}{n}\right)\left(2 + \frac{1}{n}\right).$$

We can improve our approximation by letting $n \to \infty$. So the actual area will be

$$\lim_{n \to \infty} \tfrac{1}{6}\left(1 + \frac{1}{n}\right)\left(2 + \frac{1}{n}\right) = \tfrac{1}{6}.1.2 = \tfrac{1}{3}.$$

Example 10.1

Use the method of summing areas to find the area under the curve $y = 2x$ between the limits 0 and b.

The required area is indicated in Fig. 10.5. This is, of course, a right-angled triangle, with base b and height $2b$, and so we know from elementary considerations that the answer is b^2.

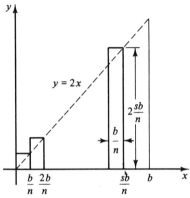

Fig.10.5

Now, considering n rectangles of width b/n, we find that the area of the sth rectangle (again considering outer rectangles) is

$$\frac{b}{n} \times 2\left(\frac{sb}{n}\right) = \frac{2b^2}{n^2} s$$

So we wish to find

$$\sum_{s=1}^{n} \frac{2b^2}{n^2} s = \frac{2b^2}{n^2} \sum_{s=1}^{n} s$$

$$= \frac{2b^2}{n^2} \frac{n}{2}(n + 1)$$

$$= \frac{b^2}{n^2} n^2 \left(1 + \frac{1}{n}\right)$$

$$= b^2 \left(1 + \frac{1}{n}\right).$$

Now the actual area is

$$\lim_{n \to \infty} b^2 \left(1 + \frac{1}{n}\right) = b^2 \quad \text{as required.}$$

This method of finding the areas under a curve, will be further developed in Chapter 17.

An alternative approach

Consider now, quite generally, that we wish to find the area under the curve $y = f(x)$ between the limits 0 and x as indicated in Fig. 10.6.

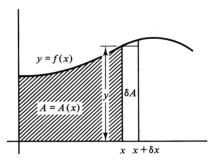

Fig.10.6

The area A, of this shape will depend on x. In fact, A is a function of x, where $A(x)$ represents the area shaded with lines in Fig. 10.6. If x is increased by an amount δx then A increases by an amount δA, which will be approximately equal to the area of the rectangle shaded with dots in Fig. 10.6.

The area of this rectangle is

$$y\delta x = f(x)\,\delta x$$

and so

$$\delta A \simeq f(x)\,\delta x \qquad \text{i.e. } \delta A \simeq y\delta x$$

that is

$$\frac{\delta A}{\delta x} \simeq f(x) \qquad \text{i.e. } \frac{\delta A}{\delta x} \simeq y$$

with the approximation improving as $\delta x \to 0$.

So

$$\lim_{\delta x \to 0} \frac{\delta A}{\delta x} = f(x)$$

and so finally

$$\frac{dA}{dx} = f(x).$$

But this means that

$$A = \int f(x)\,dx \qquad \text{i.e. } A = \int y\,dx$$

(though there is a slight complication due to the presence of the constant of integration).

Example 10.2

Find the area under the curve $y = x^2$ between

(a) the limits $x = 0$ and $x = 1$
(b) the limits $x = a$ and $x = b$, where $b > a$.

(a) We have shown that

$$A = \int x^2 \, dx$$

$$= \frac{x^3}{3} + c.$$

When $x = 0$, $A = 0$ (since we have the area of a mere line) from which it follows that $c = 0$. So

$$A = \frac{x^2}{3}.$$

When $x = 1$,

$$A = \tfrac{1}{3} \quad \text{as found earlier!!!}$$

This is an amazing discovery; integration, which was introduced as the reverse of differentiating, can be used to find the areas of certain shapes. Indeed the elongated

$\displaystyle\int$, which is the integral sign, owes its origin to the fact that integration is a form

of summation.

(b) $A = \displaystyle\int x^2 \, dx$

$\Rightarrow A = \dfrac{x^3}{3} + c$

When $x = a$, $A(a) = 0$ and so

$$c = -\frac{a^3}{3}$$

$$\Rightarrow A = \frac{x^3}{3} - \frac{a^3}{3}.$$

Now when $x = b$,

$$A = \frac{b^3}{3} - \frac{a^3}{3}$$

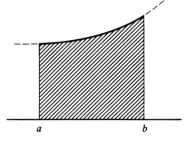

Fig.10.7

The above working is usually written

$$A = \int_a^b x^2 \, dx \qquad \text{step 1}$$

$$= \left[\frac{x^3}{3}\right]_a^b \qquad \text{step 2}$$

$$= \frac{b^3}{3} - \frac{a^3}{3} \qquad \text{step 3}$$

Step 1

$$\int_a^b x^2 \, dx$$

is called the *definite integral* of x^2 from a to b. (An integral without the limits is called the *indefinite integral* — because of the constant of integration.)

Step 2
To evaluate a definite integral, integrate in the normal way (without adding the constant of integration).

Step 3
Evaluate $F(b) - F(a)$, where $F(x)$ is the resultant integral and b and a are the upper and lower limits respectively.

Example 10.3
Evaluate the definite integrals

(a) $\displaystyle\int_0^1 x(x^2 - 1) \, dx$ (b) $\displaystyle\int_{-1}^1 x(x^2 - 1) \, dx$

and hence find the area between the curve and the x axis.

(a) Now

$$\int_0^1 x(x^2 - 1) \, dx$$

$$= \int_0^1 x^3 - x \, dx$$

$$= \left[\frac{x^4}{4} - \frac{x^2}{2}\right]_0^1$$

$$= (\tfrac{1}{4} - \tfrac{1}{2}) - (0)$$

$$= -\tfrac{1}{4}$$

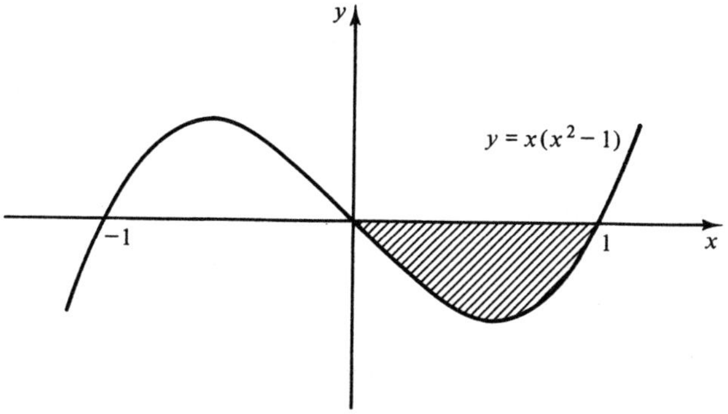

$y = x(x^2 - 1)$

Fig.10.8

The reason for the negative sign is apparent from an inspection of Fig. 10.8. The element of area $y \, dx$ has a negative sign, because over the range of interest y is negative. Now although the definite integral is negative, the area is $\tfrac{1}{4}$.

(b)
$$\int_{-1}^{1} x(x^2 - 1)\,dx = \int_{-1}^{1} x^3 - x\,dx$$

$$= \left[\frac{x^4}{4} - \frac{x^2}{2}\right]_{-1}^{1}$$

$$= (\tfrac{1}{4} - \tfrac{1}{2}) - (\tfrac{1}{4} - \tfrac{1}{2})$$

$$= -\tfrac{1}{4} - (-\tfrac{1}{4})$$

$$= 0$$

This answer to part (b) emphasises the need to sketch a diagram and deal separately with regions above and below the x axis. With what has been found so far it is obvious that

$$\int_{-1}^{0} x(x^2 - 1)\,dx = \tfrac{1}{4}$$

and so the required area is $\tfrac{1}{4} + \tfrac{1}{4} = \tfrac{1}{2}$.

Example 10.4
Find the shaded area given in Fig. 10.9.

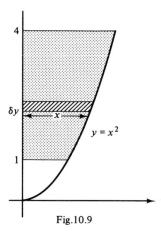

Fig.10.9

In this example, it is convenient to consider the horizontal strips with area $x\,\delta y$. The required area is given by

$$A = \int_{1}^{4} x\,dy$$

since $y = x^2$, $x = \sqrt{y}$, taking the positive root. Therefore

$$A = \int_{1}^{4} \sqrt{y}\,dy$$

$$= [\tfrac{2}{3}y^{3/2}]_{1}^{4}$$

$$= \tfrac{2}{3}.8 - \tfrac{2}{3}$$

$$\Rightarrow A = \tfrac{14}{3}$$
$$= 4\tfrac{2}{3}.$$

Example 10.5

Find the area enclosed by the curves given by $y = x^2$ and $y = x(2 - x)$.

The required area is shaded in Fig. 10.10. Solving simultaneously, we find that the curves intersect when $x = 0$ and when $x = 1$. So we require

$$\int_0^1 x(2 - x)\,dx - \int_0^1 x^2\,dx$$

$$= \left[x^2 - \frac{x^3}{3} \right]_0^1 - \left[\frac{x^3}{3} \right]_0^1$$

$$= \tfrac{2}{3} - \tfrac{1}{3}$$

$$= \tfrac{1}{3}$$

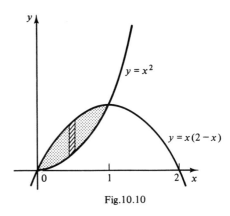

Fig. 10.10

More conveniently we may write

$$\int_0^1 x(2 - x)\,dx - \int_0^1 x^2\,dx$$

$$= \int_0^1 2x - x^2 - x^2\,dx$$

$$= \int_0^1 2(x - x^2)\,dx$$

$$= \left[x^2 - \frac{2x^3}{3} \right]_0^1$$

$$= \tfrac{1}{3} \qquad \text{as before.}$$

It will be found that combining integrals in this way will give the correct area even if part of the required area is below the x axis!

Exercise 10.1

1. Evaluate the following definite integrals.

(a) $\displaystyle\int_1^2 4x^3\,dx$ 　　　　　(b) $\displaystyle\int_0^3 5x^4\,dx$ 　　　　　(c) $\displaystyle\int_1^3 (x + 1)^3\,dx$

(d) $\displaystyle\int_0^\pi \sin x \, dx$ (e) $\displaystyle\int_0^\pi \cos x \, dx$ (f) $\displaystyle\int_{-1}^1 3x^2 - 1 \, dx$

(g) $\displaystyle\int_8^{15} (1 + x)^{1/2} \, dx$ (h) $\displaystyle 2\int_0^{\pi/2} \sin^3 x \cos x \, dx$ (i) $\displaystyle\int_0^1 x(3 + x^2)^{1/2} \, dx$

(j) $\displaystyle\int_0^\pi \cos^2 x \, dx$ (k) $\displaystyle\int_0^\pi \sin^2 x \, dx$ (l) $\displaystyle\int_{-\pi/2}^{\pi/2} \tan^3 x \, dx$

(m) $\displaystyle\int_1^9 \frac{x + 1}{x} \, dx$ (n) $\displaystyle\int_0^{\pi/4} \sec^2 x \, dx$

2. Find the areas bounded by the given lines and curves

   (a) $y = x^2$, $x = 3$, $x = 6$, $y = 0$
   (b) $y = 1 - x^2$, $y = 0$
   (c) $y = (x + 1)(x - 2)$ and the x axis

   (d) $y = x - \dfrac{1}{x^2}$, $x = 1$, $x = 2$, $y = 0$

   (e) $y = 8 - x^3$, the x and y axes.

3. Find the area of the segment of the curve $y = 3x - x^2$ cut off by the line $y = 2$.

4. Find the area enclosed by the curve $y = x^2$, $y = 1$, $y = 4$ and the y axis.

5. Find the area under the curve $y = \sin 2x$ between the limits $x = \pi/4$ and $x = \pi/2$.

6. Solve the equation

$$\int_1^x 2t - 4 \, dt = 0.$$

7. Find the number h for which the line $x = h$ bisects the area enclosed between the curve $y = 2\sqrt{x}$, $x = 4$ and $y = 0$.

8. Find the area between the curve $y = 5x - 2x^2$ and the line $y = x$.

9. Find the area in the first quadrant between the curves $y = 3x^2 + 4x$ and $y = 3x^3 - 2x$.

10.2 Some properties of definite integration

If $F'(x) \equiv f(x)$ then we know that

$$F(x) \equiv \int f(x) \, dx$$

at least to within a constant. It follows by definition that

$$\int_a^b f(x)\,dx = F(b) - F(a)$$

Now

$$\int_b^a f(x)\,dx = F(a) - F(b) = -[F(b) - F(a)]$$

and so

$$\int_b^a f(x)\,dx = -\int_a^b f(x)\,dx$$

The following observations, which are almost self evident, are worth noting.

$$\int_a^b kf(x)\,dx = k\int_a^b f(x)\,dx \qquad \text{provided } k \text{ is a } constant,$$

$$\int_a^b f(x)\,dx = \int_a^b f(y)\,dy.$$

In other words, the variable may be changed to any other variable without altering the definite integral.

Another version of this is

$$\int_0^x f(t)\,dt = F(x) + c$$

where $F'(x) \equiv f(x)$ and c is a constant. Considering definite integration as an area under a curve, it is clear that

$$\int_a^b f(x)\,dx = \int_a^c f(x)\,dx + \int_c^b f(x)\,dx$$

If f is an *even function*, so that its graph (Fig. 10.11) is symmetrical about the y axis then

$$\int_{-a}^a f(x)\,dx = 2\int_0^a f(x)\,dx.$$

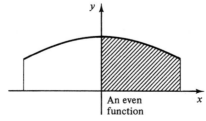

An even function

Fig.10.11

However if g is an *odd function* (Fig. 10.12) then

$$\int_{-a}^{0} g(x)\,\mathrm{d}x = -\int_{0}^{a} g(x)\,\mathrm{d}x$$

or, in other words

$$\int_{-a}^{a} g(x)\,\mathrm{d}x = 0.$$

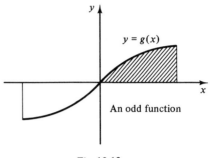

An odd function

Fig. 10.12

Some special cases

Particular attention needs to be paid when attempting to integrate close to a point where a function is undefined.

Suppose we wish to find

$$\int_{a}^{1} \frac{1}{\sqrt{x}}\,\mathrm{d}x \qquad \text{where } 0 < a < 1$$

$$= \int_{a}^{1} x^{-1/2}\,\mathrm{d}x$$

$$= [2x^{1/2}]_{a}^{1}$$

$$= 2 - 2\sqrt{a}$$

so as $a \to 0$,

$$\int_{a}^{1} \frac{1}{\sqrt{x}}\,\mathrm{d}x \to 2$$

We *define*

$$\int_{0}^{1} \frac{1}{\sqrt{x}}\,\mathrm{d}x = \lim_{a \to 0} \int_{a}^{1} \frac{1}{\sqrt{x}}\,\mathrm{d}x$$

$$= 2 \qquad \text{in this case.}$$

However,

$$\int_{a}^{1} \frac{1}{x^2}\,\mathrm{d}x = \int_{a}^{1} x^{-2}\,\mathrm{d}x \qquad a > 0$$

$$= [-x^{-1}]_{a}^{1}$$

$$= \frac{1}{a} - 1.$$

Now as $a \to 0$, $\dfrac{1}{a} - 1 \to \infty$ and so

$$\lim_{a \to 0} \int_a^1 \frac{1}{x^2} \, dx$$

is undefined.

This shows the importance of sketching a graph. No attempt should be made to integrate between limits which includes a point where the function is undefined, without due care.

Suppose we wish to find

$$\int_1^n \frac{1}{x^2} \, dx \qquad \text{for some } n \text{ where } n > 1$$

$$= \int_1^n x^{-2} \, dx$$

$$= [-x^{-1}]_1^n$$

$$= 1 - \frac{1}{n}$$

When $n \to \infty$,

$$\int_1^n \frac{1}{x^2} \, dx \to 1.$$

We define

$$\int_1^\infty \frac{1}{x^2} \, dx = \lim_{n \to \infty} \int_1^n \frac{1}{x^2} \, dx$$

$$= 1 \quad \text{in this case.}$$

However

$$\int_1^\infty \frac{1}{\sqrt{x}} \, dx = \lim_{n \to \infty} \int_1^n \frac{1}{\sqrt{x}} \, dx$$

$$= \lim_{n \to \infty} \int_1^n x^{-1/2} \, dx$$

$$= \lim_{n \to \infty} [2x^{1/2}]_1^n$$

$$= \lim_{n \to \infty} [2\sqrt{n} - 2]$$

and this limit does not exist!

Exercise 10.2

1. Show by direct integration that

(a) $\displaystyle\int_0^1 x^3 \, dx = -\int_1^0 x^3 \, dx$

(b) $\displaystyle\int_0^{\pi/2} k \sin x \, dx = k \int_0^{\pi/2} \sin x \, dx$, where k is a constant.

2. The function f is defined by $f(x) \equiv \sin x \cos x$, $x \in \mathbb{R}$.

(a) Show that f is an odd function.

(b) Show by direct integration that

$$\int_{-\pi/2}^{\pi/2} f(x) \, dx = 0.$$

3. By first checking whether the integral is odd or even obtain

(a) $\displaystyle\int_{-1}^1 \frac{x}{1+x^2} \, dx$ \qquad (b) $\displaystyle\int_{-\pi}^{\pi} \sin^2 x \, dx$ \qquad (c) $\displaystyle\int_{-17}^{17} x^{17} \, dx$

(d) $\displaystyle\int_{-\pi/4}^{\pi/4} \frac{\tan x}{1 + \cos^2 x} \, dx$

4. Solve the equation $\displaystyle\int_0^x t(t-1) \, dt = 0$.

10.3   Integration as a summation

Suppose we wish to find the area under the curve $y = f(x)$ between the limits $x = a$ and $x = b$.

The area of the strip located at x of width δx and height $y \,(= f(x))$ is $y\delta x$. The sum of all these rectangles may be written as

$$\sum_{\substack{\text{all } \delta x \text{ from} \\ a \text{ to } b}} y\,\delta x$$

where the notation is to imply that we sum the areas of all the rectangles determined by values of x from

$$a, a + \delta x, a + 2\delta x, \ldots$$

until b is reached.

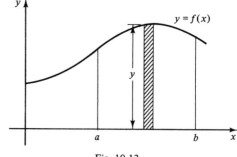

Fig. 10.13

By what has previously been stated the actual area is given by

$$\lim_{\delta x \to 0} \sum_{\substack{\text{all } \delta x \text{ from} \\ a \text{ to } b}} y\,\delta x$$

and also by

$$\int_a^b y\,dx$$

(assuming y is positive throughout).

So we have

Statement

$$\lim_{\delta x \to 0} \sum_{\substack{\text{all } \delta x \text{ from} \\ a \text{ to } b}} y\,\delta x = \int_a^b y\,dx$$

Notice that this equality makes no specific mention of area and does not in fact depend on the idea of area.

10.4 Volumes of revolution and other applications of integration

If the shaded region in Fig. 10.14 is rotated about the x axis through 2π radians it would sweep out a volume of space called a *volume of revolution*. In this section we develop techniques for finding such volumes.

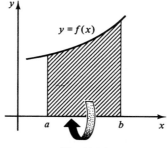

Fig. 10.14

Consider the volume of revolution determined by $y = x^2$ between the limits 0 and 1. This volume may be approximated by a number of discs. The approximation will improve as the number of discs is increased.

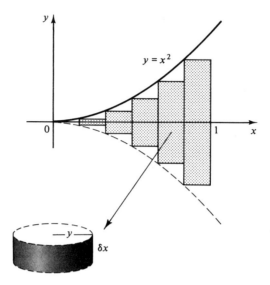

Fig. 10.15

A typical disc located a distance x from the origin will have dimensions as indicated in Fig. 10.15, where in this case $y = x^2$. Such a disc has volume: $\pi y^2 \delta x$. So the required volume is given approximately by

$$\text{Vol} \simeq \sum_{\substack{\text{all } \delta x \text{ from} \\ 0 \text{ to } 1}} \pi y^2 \, \delta x$$

and so

$$\text{Vol} = \lim_{\delta x \to 0} \sum_{\substack{\text{all } \delta x \text{ from} \\ 0 \text{ to } 1}} \pi y^2 \, \delta x$$

$$= \int_0^1 \pi y^2 \, dx \qquad \text{using the result on page 327}$$

$$= \pi \int_0^1 x^4 \, dx \qquad \text{since } y = x^2$$

$$= \pi \left[\frac{x^5}{5} \right]_0^1$$

$$= \frac{\pi}{5}.$$

Example 10.6

Find the volume of a sphere of radius r.

The first problem is to find the equation for a circle of radius r with centre the origin.

The point $P(x, y)$ lies on the circle if and only if $x^2 + y^2 = r^2$ by Pythagoras' theorem (Fig. 10.16).

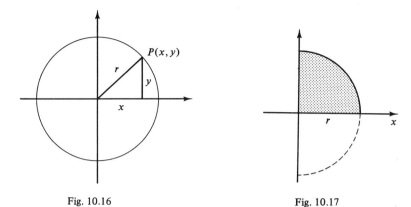

Fig. 10.16   Fig. 10.17

If we rotate the portion of the curve given by $y = \sqrt{r^2 - x^2}$ between the limits 0 and r then we obtain a hemisphere. The volume of this hemisphere is given by

$$\int_0^r \pi y^2 \, dx$$

by an argument similar to that in the previous example.

$$\text{Vol} = \pi \int_0^r (r^2 - x^2) \, dx \qquad \text{since } y = \sqrt{r^2 - x^2}$$

$$= \pi \left[r^2 x - \frac{x^3}{3} \right]_0^r$$

$$= \pi \left[\left(r^3 - \frac{r^3}{3} \right) - (0) \right]$$

$$= \frac{2\pi r^3}{3}.$$

The volume of the sphere of radius r is twice this, that is $\dfrac{4\pi r^3}{3}$.

The mean value

Definition

> The mean value of the function $f(x)$ between the values $x = a$ and $x = b$ is defined to be
>
> $$\frac{1}{b-a} \int_a^b f(x) \, dx$$

Geometrically, the mean value of a function is the height of a rectangle of width $b - a$ (assuming $b > a$) which is equal in area to that under the curve $y = f(x)$, between the limits $x = a$ and $x = b$ (assuming the function is non negative over the range). The mean value is an 'average' value of the function.

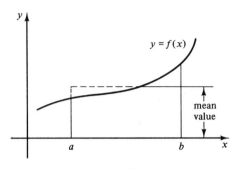

Fig. 10.18

If $f(x)$ is expressed as a function of another variable u (say), then the mean value with respect to u may be different. So it is important to stress which variable is being considered.

Example 10.7

A particle moves from rest in a straight line under an acceleration of $8 \, \mathrm{m \, s^{-2}}$. It may be shown that its velocity v is given by $v = 8t$ when expressed in terms of the time t s, or $v = 4\sqrt{s}$ when expressed in terms of the distance s m from its original position.

When the particle has travelled for 1 sec. then $v = 8 \, \mathrm{m \, s^{-1}}$ and it can readily be seen that $s = 4$ m.

Find the mean value of the velocity (a) with respect to time, (b) with respect to the distance covered in 1 second.

(a) The mean value of $v = 8t$ between the limits $t = 0$ and $t = 1$ is defined as

$$\frac{1}{1 - 0} \int_0^1 8t \, dt = [4t^2]_0^1 = 4.$$

(b) The mean value of $v = 4\sqrt{s}$ between the limits $s = 0$ and $s = 4$ is defined as

$$\frac{1}{4 - 0} \int_0^4 4s^{1/2} \, ds = \frac{1}{4} \left[\frac{8}{3} s^{3/2} \right]_0^4 = \frac{16}{3}.$$

Exercise 10.3

1. Find the finite area enclosed by the curve $y = 4x - x^2$ and the x axis. This area is rotated about the x axis; find the volume of revolution so formed.

2.   The area bounded by the x axis and the curve $y^2 = 1 - \cos 2x$ is rotated about the x axis. Obtain the volume of revolution.

3.   A solid is formed by rotating the area bounded by the curve $y = \sec x$ about the x axis between the limits $x = 0$ and $x = \dfrac{\pi}{3}$. Find the volume of this solid.

4.   The area enclosed by the x axis and the lines $y = \dfrac{r}{h}x$ and $x = h$ is rotated once about the x axis. Find the volume of revolution. Show that this represents the volume of a cone with base radius r and height h.

5.   Find the volume of a cylinder with base radius r and length h.

6.   Find the volume swept out when the area bounded by the parabola $y = x^2 + 1$ and the lines $x = 1$, $x = 2$ and $y = 0$ is rotated through 2π radians about the x axis.

7.   Find the mean value of the functions as indicated.

(a) $f(x) \equiv x^2$ between $x = 1$ and $x = 4$
(b) $f(\theta) \equiv \sin \theta$ in the range 0 to π
(c) $f(\theta) \equiv \cos \theta$ in the range 0 to π
(d) $f(\theta) \equiv \sin^2 \theta$ in the range 0 to π
(e) $f(\theta) \equiv \cos^2 \theta$ in the range 0 to π

8.   Find the mean value of the expression $(1 + \sin t)^2$ for values of t in the interval $0 \leqslant t \leqslant \pi$.

9.   Given that

$$f(x + a) \equiv f(x) \qquad x \in \mathbb{R}$$

for some constant a, show that

$$\int_{ka}^{(k+1)a} f(x)\,dx = \int_0^a f(x)\,dx$$

where k is any integer.

10.   Given the curve whose equation is

$$y = x^{-1/2},$$

find

(a) the mean value of $\dfrac{1}{y}$, with respect to x, in the interval $1 \leqslant x \leqslant 4$;
(b) the area of the region R bounded by the curve, the x-axis and the lines $x = 1$ and $x = 4$;

(c) the volume of the solid generated when the region R is rotated through an angle of 2π radians about the x-axis.

JMB 1983

11. The function f is defined on the interval $0 \leqslant x \leqslant \pi$ by

$$f(x) = 2 + \sin x.$$

(a) Find the mean value of $f(x)$ over its domain, giving your answer in terms of π.

(b) A finite region is bounded by the curve $y = 2 + \sin x$, the x-axis from $x = 0$ to $x = \pi$ and the lines $x = 0$, $x = \pi$. This region is rotated through four right angles about the x-axis. Find the volume generated, giving your answer in terms of π.

JMB 1982

12. The area enclosed by the curve $y = \cos x + \sin x$, the x-axis and the lines $x = 0$ and $x = \pi/2$, is rotated through four right angles about the x-axis. Find the volume generated.

W 1983

13. (a) Find the maximum point on the curve $y = \dfrac{1}{x^2 + 2x + 4}$ and show that there are points of inflexion at $(0, \frac{1}{4})$ and $(-2, \frac{1}{4})$.

(b) The area bounded by the curve $y = \tan x$, the x axis and the ordinate $x = \pi/3$ is rotated about the x axis. Calculate the volume of the solid of revolution so formed. (Give your answer to 3 significant figures.)

SU 1982

14. The curve defined by $y = a \cos \theta$, $x = a \sin^2 \theta$, is rotated once about the x axis. Find the volume of revolution between the limits where $\theta = 0$ and $\theta = \pi/2$.

15. A hemispherical bowl of internal radius a is fixed with its rim horizontal, and contains liquid to a depth h. Show by integration that the volume of liquid in the bowl is

$$\tfrac{1}{3}\pi h^2 (3a - h).$$

CAMB 1983

16. Find the equation of the chord which joins the points $A(-2, 3)$ and $B(0, 15)$ on the curve $y = 15 - 3x^2$.

(a) Show that the finite area enclosed by the curve and the chord AB is 4 square units.

(b) Find the volume generated when this area is rotated through $360°$ about the x-axis, leaving your answer in terms of π.

AEB 1983

17. A curve is given by the equation

$$y = \sin x + \tfrac{1}{2} \sin 2x, \qquad 0 \leqslant x \leqslant 2\pi.$$

Find the values of x for which y is zero.

Find the exact coordinates of the stationary points on the curve and sketch the curve.

Find the area of the region bounded by the curve and the x-axis for $0 \leqslant x \leqslant \pi$. Deduce the mean value of y over the interval $0 \leqslant x \leqslant \pi$.

JMB 1984

18. The arc of the curve $y = 1 + \cos x$ between $x = 0$ and $x = \pi/2$ is rotated about the x axis. Prove that the volume of revolution is approximately 13.7 ·cubic units.

19. The point P_1 has coordinates $(a, \tfrac{1}{2}a)$ and lies on the curve $y^2 = a(x - \tfrac{3}{4}a)$. The line OP_1, where O is the origin, cuts the curve again at P_2; the point A has coordinates $(\tfrac{3}{4}a, 0)$. The (finite) region bounded by the straight lines OA, OP_1 and the arc AP_1 of the curve is denoted by R_1. The region bounded by the portions of the curve and straight line lying between P_1 and P_2 is denoted by R_2.

(a) Show that the areas of R_1 and R_2 are equal.
(b) Find the volumes of the solids of revolution generated when R_1 and R_2 are rotated through 2π radians about the x-axis.

O & C 1983

20. Determine the maximum and the minimum values of the function given by

$$y = \sqrt{3} \sin 2x + \cos 2x$$

and find the values of x in the interval $0 \leqslant x \leqslant \pi$ for which they occur.

Find also the points in the interval $0 \leqslant x \leqslant \pi$ at which $y = 0$, and sketch the graph of y in this interval.

Calculate, in terms of π, the volume swept out when the region bounded by the curve

$$y = \sqrt{3} \sin 2x + \cos 2x,$$

the line $x = \pi/4$ and the coordinate axes is rotated about the x-axis through an angle of 2π radians.

JMB 1982

10.5 Integration of inverse trigonometric functions

Recall that

$$\frac{d}{dx}(\sin^{-1} x) = \frac{1}{\sqrt{1 - x^2}}$$

and

$$\frac{d}{dx}(\tan^{-1} x) = \frac{1}{1 + x^2}$$

as shown on page 237. It follows immediately that:

Statement

$$\int \frac{1}{\sqrt{1 - x^2}} dx = \sin^{-1} x + c$$

$$\int \frac{1}{1 + x^2} dx = \tan^{-1} x + c.$$

Example 10.8

Find

(a) $\displaystyle\int \frac{3}{\sqrt{1 - x^2}} dx$ (b) $\displaystyle\int \frac{1}{1 + 4x^2} dx$ (c) $\displaystyle\int \frac{1}{\sqrt{9 - x^2}} dx$

(a) $\displaystyle\int \frac{3}{\sqrt{1 - x^2}} dx = 3 \int \frac{1}{\sqrt{1 - x^2}} dx$

$$= 3 \sin^{-1} x + c$$

(b) We might guess the integral to be $\tan^{-1} 2x$. Now differentiating,

$$y = \tan^{-1} 2x$$

$$\Rightarrow \frac{dy}{dx} = \frac{1}{1 + 4x^2} \cdot 2.$$

Thus

$$\int \frac{1}{1 + 4x^2} dx = \tfrac{1}{2} \tan^{-1} 2x + c$$

(c) $\displaystyle\int \frac{1}{\sqrt{9 - x^2}} dx = \tfrac{1}{3} \int \frac{1}{\sqrt{1 - \frac{1}{9}x^2}} dx$

A guessed answer might be $\sin^{-1} \dfrac{x}{3}$.

Differentiating

$$y = \sin^{-1} \frac{x}{3}$$

$$\Rightarrow \frac{dy}{dx} = \frac{1}{\sqrt{1 - \frac{1}{9}x^2}} \cdot \frac{1}{3}$$

Thus

$$\int \frac{1}{\sqrt{9 - x^2}} dx = \sin^{-1} \frac{x}{3} + c.$$

Exercise 10.4

1. Find

(a) $\displaystyle\int \frac{1}{1 + 9x^2} dx$

(b) $\displaystyle\int \frac{1}{\sqrt{1 - 4x^2}} dx$

(c) $\displaystyle\int \frac{1}{\sqrt{1 - 16x^2}} dx$

(d) $\displaystyle\int \frac{4}{\sqrt{1 - \frac{1}{4}x^2}} dx$

(e) $\displaystyle\int \frac{1}{4 + x^2} dx$

(f) $\displaystyle\int \frac{2}{\sqrt{4 - x^2}} dx$

(g) $\displaystyle\int \frac{1}{\sqrt{16 - x^2}} dx$

(h) $\displaystyle\int \frac{1}{\sqrt{4 - 9x^2}} dx$

(i) $\displaystyle\int \frac{1}{\sqrt{16 - 25x^2}} dx$

(j) $\displaystyle\int \frac{3}{9 + x^2} dx.$

10.6 Integration by substitution

Suppose

$$I = \int f(x) \, dx,$$

then

$$\frac{dI}{dx} = f(x).$$

Now, if x is a function of u (say) then

$$\frac{dI}{du} = \frac{dI}{dx} \cdot \frac{dx}{du}$$

$$\Rightarrow \frac{dI}{du} = f(x) \frac{dx}{du}$$

Integrating w.r.t. u, we obtain

$$I = \int f(x) \frac{dx}{du} \, du$$

and so finally:

Statement

$$\int f(x)\,dx = \int f(x)\frac{dx}{du}\,du.$$

An example may illuminate this result.

Example 10.9

Use the method of substitution to obtain the following

(a) $\displaystyle\int (2x + 1)^3\,dx$ (b) $\displaystyle\int x(2x + 1)^{1/2}\,dx$

(a) This integral could be dealt with by expanding and then integrating or by simply guessing the answer.

However, suppose we set

$$u = 2x + 1$$

$$\Rightarrow \frac{du}{dx} = 2 \qquad \Rightarrow \frac{dx}{du} = \tfrac{1}{2}$$

so

$$\int (2x + 1)^3\,dx = \int (2x + 1)^3\frac{dx}{du}\,du$$

$$= \int u^3 \cdot \tfrac{1}{2}\,du$$

$$= \frac{u^4}{8} + c$$

$$= \frac{(2x + 1)^4}{8} + c$$

a fact which may be checked by differentiating!

(b) Consider now

$$I = \int x(2x + 1)^{1/2}\,dx.$$

It is unlikely that the reader will be able to guess the answer in this case. Suppose, however, we again set

$$u = 2x + 1$$

$$\Rightarrow \frac{dx}{du} = \tfrac{1}{2} \quad \text{as before}$$

$$\text{and} \quad x = \tfrac{1}{2}(u - 1).$$

So, on substituting

$$I = \int \tfrac{1}{2}(u - 1)u^{1/2} \cdot \tfrac{1}{2} \, du$$

$$= \tfrac{1}{4} \int u^{3/2} - u^{1/2} \, du$$

$$= \tfrac{1}{4} \left(\frac{2u^{5/2}}{5} - \frac{2u^{3/2}}{3} \right) + c$$

$$= \frac{u^{5/2}}{10} - \frac{u^{3/2}}{6} + c$$

$$= \frac{(2x + 1)^{5/2}}{10} - \frac{(2x + 1)^{3/2}}{6} + c.$$

Again this result may be checked by differentiating.

Change of limits

Example 10.10

Find $I = \displaystyle\int_0^4 \sqrt{4 - x} \, dx.$

Set

$$4 - x = u \quad \Rightarrow \frac{du}{dx} = -1 \quad \text{and so} \quad \frac{dx}{du} = -1,$$

so

$$I = \int_{x=0}^{x=4} u^{1/2} \cdot -1 \cdot du = \int_{x=0}^{x=4} -u^{1/2} \, du.$$

It is necessary to stress that the limits are those of x, for when $x = 0$, $u = 4$ and when $x = 4$, $u = 0$. This may be summarised as in Table 10.1.

4 − x = u	
x	u
4	0
0	4

Table 10.1

Continuing, we have

$$I = \left[\frac{-2u^{3/2}}{3} \right]_{x=0}^{x=4}$$

$$= \left[-2\frac{(4-x)^{3/2}}{3} \right]_{0}^{4} \qquad \text{(there is no ambiguity now)}$$

$$= \left[\frac{-2.0}{3} - \frac{-2.4^{3/2}}{3} \right]$$

$$= \frac{16}{3}.$$

However, rather than remain with the limits for x, which necessitates expressing the integral in terms of x, we may change the limits to those of u. In which case.

$$I = \left[\frac{-2u^{3/2}}{3} \right]_{4}^{0}$$

$$= 0 - \frac{-2(4)^{3/2}}{3}$$

$$= \frac{16}{3}.$$

Example 10.11
Find

$$I = \int_{0}^{2} (4 - x^2)^{1/2}\, dx$$

A suitable substitution is to let

$$x = 2\sin\theta \qquad \Rightarrow \frac{dx}{d\theta} = 2\cos\theta$$

$$x = 2\sin\theta$$

x	θ
2	$\dfrac{\pi}{2}$
0	0

Table 10.2

So

$$I = \int_{0}^{\pi/2} (4 - (2\sin\theta)^2)^{1/2} . 2\cos\theta\, d\theta \qquad \text{having changed the limits}$$

$$\Rightarrow I = \int_0^{\pi/2} 2(1 - \sin^2 \theta)^{1/2} \, 2 \cos \theta \, d\theta$$

$$= 4 \int_0^{\pi/2} \cos^2 \theta \, d\theta \qquad \text{since } (1 + \sin^2 \theta)^{1/2} \equiv \cos \theta$$

Now since $\cos^2 \theta = \frac{1}{2}(\cos 2\theta + 1)$,

$$I = 2 \int_0^{\pi/2} (\cos 2\theta + 1) \, d\theta$$

$$= 2 \left[\frac{\sin 2\theta}{2} + \theta \right]_0^{\pi/2}$$

$$= 2 \left[\left(0 + \frac{\pi}{2} \right) - (0 + 0) \right]$$

$$= \pi.$$

Changing the limits in this way is very convenient.

Example 10.12
Find

(a) $\displaystyle \int \frac{3}{4 + x^2} \, dx$ (b) $\displaystyle \int \frac{c}{\sqrt{a^2 - b^2 x^2}} \, dx$

(a) Recall that

$$\int \frac{1}{1 + x^2} \, dx = \tan^{-1} x + c$$

We set

$$x = 2u \qquad \Rightarrow \frac{dx}{du} = 2.$$

Now

$$\int \frac{3}{4 + x^2} \, dx = 3 \int \frac{1}{4 + x^2} \frac{dx}{du} \, du$$

$$= \tfrac{3}{4} \int \frac{1}{1 + u^2} \cdot 2 \, du$$

$$= \tfrac{3}{2} \tan^{-1} u + c$$

$$= \tfrac{3}{2} \tan^{-1} \frac{x}{2} + c \qquad \text{since } u = \frac{x}{2}.$$

(b) For

$$I = \int \frac{c}{\sqrt{a^2 - b^2 x^2}} \, dx$$

we set

$$bx = au \qquad \Rightarrow \frac{dx}{du} = \frac{a}{b}.$$

So

$$I = c \int \frac{1}{\sqrt{a^2 - a^2 u^2}} \cdot \frac{a}{b} \cdot du$$

$$= \frac{c}{b} \int \frac{1}{\sqrt{1 - u^2}} \, du$$

$$= \frac{c}{b} \sin^{-1} u + k$$

$$= \frac{c}{b} \sin^{-1} \frac{bx}{a} + k$$

Exercise 10.5

1. Integrate the following using the suggested substitutions.

(a) $\int (2x + 1)^3 \, dx$ Let $2x + 1 = u$.

(b) $\int \sqrt{3x + 2} \, dx$ Let $3x + 2 = u$.

(c) $\int x(1 + x)^4 \, dx$ Let $1 + x = u$.

(d) $\int x(1 + x^2)^3 \, dx$ Let $1 + x^2 = u$.

(e) $\int x\sqrt{1 - x^2} \, dx$ Let $1 - x^2 = u$.

(f) $\int (9 - x^2)^{1/2} \, dx$ Let $x = 3 \sin u$.

(g) $\int \frac{1}{9 + x^2} \, dx$ Let $x = 3u$.

(h) $\int \frac{1}{\sqrt{9 - 4x^2}} \, dx$ Let $2x = 3u$.

2. Integrate each of the following.

(a) $\int (3x - 1)^3 \, dx$

(b) $\int \left(1 - \frac{x}{2}\right)^5 \, dx$

(c) $\int \frac{1}{2}(1 - 5x)^2 \, dx$

(d) $\int \sqrt{ax + b} \, dx$, where a and b are constants.

(e) $\int x\sqrt{x^2 + 1} \, dx$

(f) $\int x\sqrt{x + 1} \, dx$

(g) $\int_0^1 (2x + 1)^3(x - 2) \, dx$

(h) $\int \left(1 - \frac{x^2}{4}\right)^{1/2} dx$

(i) $\int \frac{1}{1 + 4x^2} \, dx$

(j) $\int \frac{2}{1 + 2x^2} \, dx$

(k) $\int_0^1 \frac{1}{1 + (1 - x)^2} \, dx$

(l) $\int \frac{1}{9 + (2 + x)^2} \, dx$

(m) $\int \frac{1}{x^2 + 2x + 3} \, dx$

(n) $\int \frac{4}{x^2 + 4x + 6} \, dx$

[*Hint:* Complete the square.]

(o) $\int \frac{2}{5 + 2x + x^2} \, dx$

(p) $\int \frac{2}{\sqrt{4 - x^2}} \, dx$

(q) $\int \frac{3}{\sqrt{1 - 4x^3}} \, dx.$

10.7 Integration of e^x

In general,

$$\int f(x) \, dx = F(x) + c \Leftrightarrow F'(x) \equiv f(x).$$

It follows that

$$\int e^x \, dx = e^x + c \qquad \text{since } \frac{d}{dx}(e^x) = e^x.$$

More generally,

$$\int f'(x) e^{f(x)} \, dx = e^{f(x)} + c.$$

This may be checked by differentiating.

Example 10.13

Find

(a) $\int e^{3x}\,dx$ (b) $\int e^{x-1}\,dx$ (c) $\int 3x\,e^{x^2}\,dx$ (d) $\int_0^{\pi/2} \cos x\,e^{\sin x}\,dx.$

(a) $\int e^{3x}\,dx = \frac{1}{3}\int 3\,e^{3x}\,dx$ $\left[\text{rewriting and noting that } \frac{d(3x)}{dx} = 3\right]$

$\qquad = \frac{1}{3}e^{3x} + c.$

(b) $\int e^{x-1}\,dx = e^{x-1} + c$ $\left[\frac{d(x-1)}{dx} = 1\right].$

(c) $\int 3x\,e^{x^2}\,dx = \frac{3}{2}\int 2x\,e^{x^2}\,dx$ $\left[\frac{d}{dx}(x^2) = 2x\right]$

$\qquad = \frac{3}{2}e^{x^2} + c.$

(d) $\displaystyle\int_0^{\pi/2} \cos x\,e^{\sin x}\,dx$ $\left[\frac{d\sin x}{dx} = \cos x\right]$

$\qquad = [e^{\sin x}]_0^{\pi/2}$

$\qquad = e^{\sin(\pi/2)} - e^{\sin 0}$

$\qquad = e^1 - e^0$

$\qquad = e - 1.$

Example 10.14

Find $\displaystyle\int_0^\infty e^{-x}\,dx.$

Now

$$\int_0^n e^{-x}\,dx = [-e^{-x}]_0^n$$
$$= e^0 - e^{-n}$$
$$= 1 - e^{-n}.$$

So

$$\int_0^\infty e^{-x}\,dx = \lim_{n\to\infty} \int_0^n e^{-x}\,dx$$
$$= \lim_{n\to\infty} (1 - e^{-n})$$
$$= 1, \quad \text{since } e^{-n} \to 0 \quad \text{as} \quad n \to \infty.$$

Exercise 10.6

1. Find

   (a) $\displaystyle\int e^{-2x}\,dx$

   (b) $\displaystyle\int e^{x/2}\,dx$

   (c) $\displaystyle\int 3e^x\,dx$

   (d) $\displaystyle\int 5e^{5x}\,dx$

   (e) $\displaystyle\int 2x\,e^{x^2}\,dx$

   (f) $\displaystyle\int \sec^2 x\,e^{\tan x}\,dx$

   (g) $\displaystyle\int e^{nx}\,dx$

   (h) $\displaystyle\int a\,e^{nx}\,dx$

   (i) $\displaystyle\int 3e^{2x}\,dx$

   (j) $\displaystyle\int (1 + e^x)^2\,dx$

   (k) $\displaystyle\int (3x^2 + 2x)\,e^{x^3 + x^2}\,dx$

   (l) $\displaystyle\int (e^x - e^{-x})^2$

   (m) $\displaystyle\int \frac{1 + 3e^x}{e^{2x}}\,dx$

   (n) $\displaystyle\int_0^1 (1 + e^{x^2})2x\,dx$

   (o) $\displaystyle\int_0^a e^{x\ln a}\,dx$

   (p) $\displaystyle\int_{-\infty}^0 e^x\,dx$

   (q) $\displaystyle\int x\,e^{x^2}\,dx$

2. By writing a^x in the form $e^{x\ln a}$, find $\displaystyle\int a^x\,dx$.

10.8 The integral $\displaystyle\int\frac{1}{x}dx$

We cannot apply the result

$$\int x^n\,dx = \frac{x^{n+1}}{n+1} + c \qquad \text{when } n = -1,$$

but it was shown on page 253 that

$$\frac{d}{dx}(\ln x) = \frac{1}{x}.$$

It follows that

$$\boxed{\int \frac{1}{x}\,dx = \ln x + c.}$$

The definite integral $\displaystyle\int_a^b \frac{1}{x}\,dx$

Assuming $0 < a < b$,

$$\int_a^b \frac{1}{x}\,dx = \ln b - \ln a.$$

Now as $a \to 0$, $\ln a = \log_e a \to -\infty$ and so the integral is undefined if a and b have opposite signs.

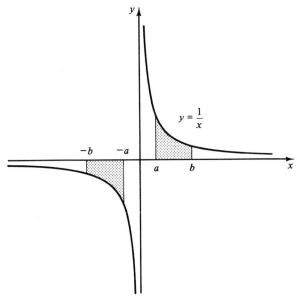

Fig. 10.19

Assuming again that $0 < a \leqslant b$, we see from Fig. 10.19 that by comparing the shaded areas

$$\ln b - \ln a = \int_a^b \frac{1}{x}\,dx$$

$$= -\int_{-b}^{-a} \frac{1}{x}\,dx \qquad \text{since the area is below the } x \text{ axis,}$$

$$= \int_{-a}^{-b} \frac{1}{x}\,dx \qquad \text{(exchanging the limits, reverses the sign).}$$

This may be summarised by stating

$$\int \frac{1}{x}\,dx = \ln|x| + c \qquad \text{for } x \neq 0$$

and the definite integral may be evaluated provided the limits are both positive or both negative.

We have, for example,

$$\int_{-3}^{-2} \frac{1}{x} dx = [\ln |x|]_{-3}^{-2}$$

$$= \ln 2 - \ln 3$$

$$= -0.405. \qquad \boxed{C}$$

Logarithmic integration

Example 10.15

Integrate each of the following.

(a) $\displaystyle\int \frac{1}{1+x} dx$ (b) $\displaystyle\int \frac{x}{1+x^2} dx$ (c) $\displaystyle\int \tan x\, dx$

(a) For $\displaystyle\int \frac{1}{1+x} dx$, set

$$1 + x = u \qquad \Rightarrow \frac{du}{dx} = 1$$

$$\Rightarrow \frac{dx}{du} = 1$$

$$\int \frac{1}{1+x} dx = \int \frac{1}{u} . 1 . du$$

$$= \ln |u| + c$$

$$= \ln |1 + x| + c$$

(b) For $\displaystyle\int \frac{x}{1+x^2} dx$, set

$$1 + x^2 = u \qquad \Rightarrow \frac{du}{dx} = 2x, \quad \frac{dx}{du} = \frac{1}{2x}$$

Then

$$\int \frac{x}{1+x^2} dx = \int \frac{x}{u} . \frac{1}{2x} du$$

$$= \tfrac{1}{2} \int \frac{1}{u} du$$

$$= \tfrac{1}{2} \ln |u| + c$$

$$= \tfrac{1}{2} \ln |1 + x^2| + c$$

(c) For $\displaystyle\int \tan x\, dx = \int \frac{\sin x}{\cos x}\, dx$, set

$$\cos x = u \qquad \Rightarrow \frac{du}{dx} = -\sin x$$

$$\Rightarrow \frac{dx}{du} = -\frac{1}{\sin x}.$$

$$\int \tan x\, dx = \int \frac{\sin x}{u} \cdot \frac{-1}{\sin x}\, du$$

$$= -\int \frac{1}{u}\, du$$

$$= -\ln|u| + c$$

$$= -\ln|\cos x| + c$$

$$= \ln|\sec x| + c$$

using the normal rules of logarithms to eliminate the negative sign.

Statement

$$\boxed{\int \frac{f'(x)}{f(x)}\, dx = \ln|f(x)| + c}$$

This important result may be used to good effect in the previous example.

For $\displaystyle\int \frac{f'(x)}{f(x)}\, dx$, set

$$f(x) = u \qquad \Rightarrow \frac{du}{dx} = f'(x)$$

$$\Rightarrow \frac{dx}{du} = \frac{1}{f'(x)}.$$

Then

$$\int \frac{f'(x)}{f(x)}\, dx = \int \frac{f'(x)}{u} \cdot \frac{1}{f'(x)}\, du$$

$$= \int \frac{1}{u}\, du$$

$$= \ln|u| + c$$

$$= \ln|f(x)| + c$$

Exercise 10.7

1. Integrate each of the following w.r.t. x.

(a) $\dfrac{1}{3x}$ $\quad x \neq 0$ (b) $\dfrac{3}{x}$ $\quad x \neq 0$ (c) $\dfrac{1}{x+3}$ $\quad x \neq -3$

(d) $\dfrac{1}{3x+1}$ $\quad x \neq -\frac{1}{3}$ (e) $\dfrac{3}{1-3x}$ $\quad x \neq \frac{1}{3}$ (f) $\dfrac{x}{1+x^2}$

(g) $\dfrac{x^2}{x^3-1}$ $\quad x \neq 1$ (h) $\dfrac{2x^2}{3x^3-2}$ $\quad 3x^3-2 \neq 0$ (i) $\cot x,$ $\quad \sin x \neq 0$

(j) $\dfrac{\cos x}{1+\sin x}$ $\quad \sin x \neq -1$ (k) $\dfrac{\cos x}{1-2\sin x}$ $\quad \sin x \neq \frac{1}{2}$

(l) $\dfrac{\sin 2x}{1+\sin^2 x}$ (m) $\dfrac{1}{x \ln x}$ $\quad x \neq 0$ $\left[Hint: \dfrac{1/x}{\ln x} \right]$

2. Find

(a) $\displaystyle\int_1^3 \dfrac{x}{1+x^2}\,dx$ (b) $\displaystyle\int_0^{\pi/4} \tan x$ (c) $\displaystyle\int_{-2}^{-1} \dfrac{x^2}{x^3-1}\,dx$

(d) $\displaystyle\int_1^2 \dfrac{3}{x+2}\,dx$ (e) $\displaystyle\int \dfrac{x+1}{x^2+1}\,dx$ $\quad$ [Hint: Separate.]

(f) $\displaystyle\int \dfrac{2x-1}{x^2+4}\,dx$ (g) $\displaystyle\int \dfrac{1}{(x+1)(x+2)}\,dx$ $\quad$ [Hint: Use partial fractions.]

Chapter 11 Sketch graphs

11.1 Polynomials

If a polynomial of degree n is differentiated, then the resulting polynomial will be of degree $n - 1$. Setting this equal to zero we find at most $n - 1$ distinct real roots. Thus a polynomial of degree n will have at most $n - 1$ stationary values. We may use the methods of calculus to distinguish between them.

If, in addition, we can locate the points where the graph crosses the axes, then it is a simple matter to sketch the graph of the polynomial.

Example 11.1

Sketch the graphs of each of the following equations.

(a) $y = (x - 1)(x + 2)(x - 3)$
(b) $y = x^3 + x^2 - x - 1$

(a) In this form it is easy to establish that the curve crosses the x axis at $x = 1$, $x = -2$ and $x = 3$, and the y axis at $y = 6$. This may be established either by inspection or by setting $y = 0$ and $x = 0$ respectively, and solving the resulting equations. The graph must pass through the points marked in Fig. 11.1. All polynomials are differentiable for all values of x—so the graph is continuous and smooth. (See page 166.)

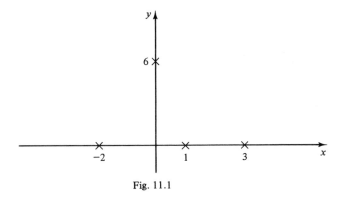

Fig. 11.1

The graph must have something like one of the forms indicated in Fig. 11.2. In all cases the graph will have at most two stationary values. In this case there must be a minimum in the interval $[1, 3]$, and a maximum in the interval $[-2, 1]$. (Recall that $[-2, 1]$ represents the set $\{x : -2 \leqslant x \leqslant 1\}$. See page 58). The variety of possibilities can be considerably reduced if we were to investigate further, to locate the turning points.

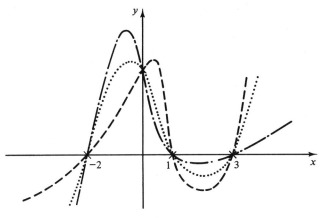

Fig. 11.2

Notice also that for large positive values of x, y is large and positive, and for large negative values of x, y is large and negative. This may be expressed as

$$x \to \pm \infty \Rightarrow y \to \pm \infty.$$

(b) Consider now the curve given by

$$y = x^3 + x^2 - x - 1.$$

Factorising we find

$$x^3 + x^2 - x - 1 \equiv (x + 1)^2 (x - 1),$$

so the graph of $y = (x + 1)^2 (x - 1)$ passes through the points $(-1, 0)$, $(1, 0)$ and $(0, -1)$ as indicated in Fig. 11.3.

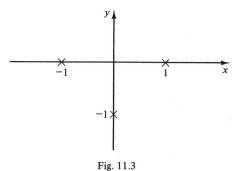

Fig. 11.3

It is again clear that

$$x \to \pm \infty \Rightarrow y \to \pm \infty$$

[If it is not clear, notice that for large values of x, positive or negative, $x + n \simeq x$ for $n \in \mathbb{R}$, and so $y \simeq x^3$ for large x. More generally, y is approximately equal to the term involving the highest power of x in a polynomial].

However, it is not yet clear what the general shape of the graph might be. Now

$$y = x^3 + x^2 - x - 1$$

$$\Rightarrow \frac{dy}{dx} = 3x^2 + 2x - 1$$

$$= (3x - 1)(x + 1)$$

For a stationary value $dy/dx = 0$, so

$$(3x - 1)(x + 1) = 0$$

$$\Rightarrow x = \tfrac{1}{3} \quad \text{or} \quad x = -1.$$

When $x = \tfrac{1}{3}$,

$$y = (\tfrac{1}{3})^3 + (\tfrac{1}{3})^2 - \tfrac{1}{3} - 1$$

$$= -1\tfrac{5}{27}.$$

When $x = -1$,

$$y = -1 + 1 + 1 - 1$$

$$= 0.$$

And since $\dfrac{d^2y}{dx^2} = 6x + 2,$

$$x = \tfrac{1}{3} \quad \Rightarrow \frac{d^2y}{dx^2} > 0 \Rightarrow \text{minimum}$$

$$x = -1 \Rightarrow \frac{d^2y}{dx^2} < 0 \Rightarrow \text{maximum.}$$

The situation is as now indicated in Fig. 11.4.

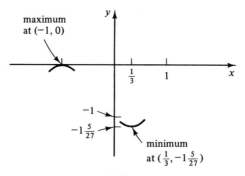

Fig. 11.4

We may now sketch the curve with some confidence to obtain the graph in Fig. 11.5.

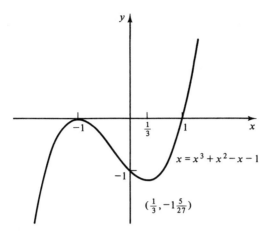

Fig. 11.5

Double roots

The graph in Fig. 11.5 touches the x axis at $(-1, 0)$, and also has a repeated root at $x = -1$. It is natural to wonder whether there is a connection.

If the polynomial $P(x)$ is to *touch* the x axis at $x = \alpha$, then $P(\alpha) = 0$ and $P'(\alpha) = 0$, since the gradient of the curve is zero at $x = \alpha$.

Now if $P(\alpha) = 0$ then $x - \alpha$ is a factor of $P(x)$ (see page 33). So we may write

$$P(x) \equiv (x - \alpha)Q(x) \quad \text{①}$$

for some polynomial $Q(x)$. Differentiating w.r.t. x

$$P'(x) \equiv (x - \alpha)Q'(x) + Q(x)$$

using the rule of products.

Substituting for α

$$P'(\alpha) = 0 \,.\, Q'(\alpha) + Q(\alpha)$$
$$\Rightarrow P'(\alpha) = Q(\alpha)$$

and by assumption

$$P'(\alpha) = 0$$
$$\Rightarrow Q(\alpha) = 0.$$

So $(x - \alpha)$ is a factor of $Q(x)$ and it follows that $Q(x) \equiv (x - \alpha)S(x)$ for some polynomial $S(x)$.

Combining with ① we have

$$P(x) \equiv (x - \alpha)^2 S(x).$$

Conversely, if $P(x) \equiv (x - \alpha)^2 S(x)$ then the graph meets the x axis at $x = \alpha$ with zero gradient. There are four possibilities as indicated in the following figure.

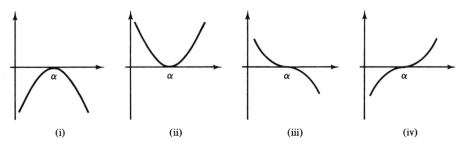

Fig. 11.6

Statement

> If the graph of $y = P(x)$, where $P(x)$ is a polynomial, touches the x axis at $x = \alpha$ then $P(x)$ has a factor of $(x - \alpha)^2$.

Exercise 11.1

1. Sketch the graphs of the following curves:

   (a) $y = x(x - 1)(x - 2)$
   (b) $y = (x - 1)^2(x - 2)$
   (c) $y = (x - 1)^3$
   (d) $y = x^3 - 3x$
   (e) $y = x^3 + x$
   (f) $y = x^4 + x^2 - 2$
   (g) $y = x^{2n}$ for $n \in \mathbb{N}$

2. Sketch the curves

   (a) $y = 1 + x^3$ (b) $y = 3(x^2 - 1)$

   and find the points of intersection.
       Hence find the area enclosed by the two curves.

3. Given that

   $$y = (1 - x^2)(4 - x^2),$$

   calculate the stationary values of y and sketch the graph of y.

   JMB 1982

4. Find the values of x for which the function $f(x) = 4x^3 + 6x^2 - 9x + 2$ has a maximum or minimum. Draw a rough sketch of the graph of the function $f(x)$, indicating clearly the positions of any maximum of minimum.

   OX 1983

5. Find the coordinates of each turning point on the graph of $y = 3x^4 - 16x^3 + 18x^2$ and determine in each case whether it is a maximum point or a minimum point.

   Sketch the graph of $y = 3x^4 - 16x^3 + 18x^2$, and state the set of values of k for which the equation $3x^4 - 16x^3 + 18x^2 = k$ has precisely two real roots for x.

   CAMB 1983

6. Sketch the curve $y = -x^2 + 4x$, between the origin O and the point A, where $x = 4$. The line $y = 3$ cuts the curve in points B and C. Find the area of $OBCA$.

   W 1982

7. The graph of

   $$y = x^3 + ax^2 + bx + c$$

   has stationary points where $x = -1$ and where $x = 3$, and passes through the point $(1, -2)$. Find the values of a, b and c.

   Show that the stationary point where $x = 3$ is a minimum point.

   Sketch the graph.

   CAMB 1982

8. Investigate the function f defined by

   $$f(x) = 3x^5 - 10x^3 + 10$$

   for local maxima, minima and points of inflexion. Sketch the graph of the function, showing clearly the stationary points and any points of inflexion.

   W 1981

9. Sketch separately the graphs of

   (a) $y = 2(x + 2)(x - 1)(x - 2)$;   (b) $y = (x - 1)(x - 3)^2$

   showing the coordinates of the points where they cross or touch the axes.

   Find a cubic function $f(x)$ such that the graph of $y = f(x)$ crosses the x-axis at $x = 1$, touches the x-axis at $x = 3$ and crosses the y-axis at $y = 18$. From consideration of the sketch graph of $y = f(x)$ show that if the roots of the equation $f(x) = k$ are all real then k must be negative. Give the sign of each root in this case. Show, also, that if $k > 0$ then the equation $f(x) = k$ has just one real root and give the range of values of k for which this root is negative.

   SU 1983

10. The curve with equation

    $$y = ax^3 + bx^2 + cx + d,$$

    where a, b, c and d are constants, passes through the points $(0, 3)$ and $(1, 0)$. At these points the curve has gradients -7 and 0, respectively.

    (a) Find the values of a, b, c and d.

(b) Show that the curve crosses the x-axis at the point $(3, 0)$.
(c) Find the x-coordinate of the maximum point on the curve.
(d) Sketch the curve.

CAMB 1983

11. It is given that

$$f(x) = (x - 2)^2 - \lambda(x + 1)(x + 2).$$

(a) Find the values of λ for which the equation $f(x) = 0$ has two equal roots.
(b) Show that, when $\lambda = 2, f(x)$ has a maximum value of 25.
(c) Given that the curve $y = f(x)$ has a turning point when $x = \frac{1}{4}$, find the value of λ and sketch the curve for this value of λ.

CAMB 1983

11.2 The exponential function

Example 11.2
Sketch the graph of $y = x e^{-x}$

Notice that

$$x = 0 \Rightarrow y = 0$$

and

$$y = 0 \Rightarrow x = 0$$

since $e^{-x} \neq 0$ for $x \in \mathbb{R}$. So the graph crosses the axes only at the origin.
Now

$$y = x e^{-x}$$

$$\Rightarrow \frac{dy}{dx} = -x e^{-x} + e^{-x} \qquad \text{(differentiating as a product.)}$$

$$\Rightarrow \frac{dy}{dx} = e^{-x}(1 - x). \qquad ①$$

For a turning value $dy/dx = 0$,

$$\Rightarrow \quad 0 = e^{-x}(1 - x)$$

$$\Rightarrow \quad x = 1 \qquad \text{since } e^{-x} \neq 0 \text{ for } x \in \mathbb{R}$$

When $x = 1, y = 1 . e^{-1} = e^{-1}$ and

$$\frac{d^2 y}{dx^2} = -e^{-x} - e^{-x}(1 - x)$$

$$= e^{-x}(x - 2).$$

When $x = 1$,

$$\frac{d^2y}{dx^2} < 0 \Rightarrow \text{maximum}.$$

The facts so far established are indicated in Fig. 11.7.

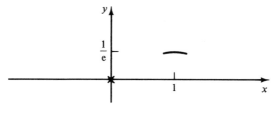

Fig. 11.7

Now

$$x \to -\infty \Rightarrow e^{-x} \to \infty$$

and so

$$y[=xe^{-x}] \to -\infty.$$

However, as $x \to \infty$ so $e^{-x} \to 0$ and it is not immediately clear what y tends to.

Statement

$$\lim_{x \to \infty} (x^n e^{-x}) = 0 \qquad \text{for all } n \in \mathbb{N}^+$$

No proof is given.

Using this result when $n = 1$, we have

$$x \to \infty \Rightarrow y[=xe^{-x}] \to 0$$

The graph is sketched in Fig. 11.8

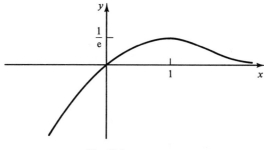

Fig. 11.8

Example 11.3

Sketch the graph of $y = e^{-x^2}$.

When $x = 0$, $y = 1$, and since $e^{-x^2} \neq 0$ for $x \in \mathbb{R}$, the graph does not cross the x axis.

$$y = e^{-x^2}$$

$$\Rightarrow \frac{dy}{dx} = -2x e^{-x^2}$$

$$\Rightarrow \frac{d^2y}{dx^2} = -2e^{-x^2} + 4x^2 e^{-x^2}$$

$$= 2e^{-x^2}(2x^2 - 1)$$

For a turning value, $dy/dx = 0$,

$$\Rightarrow 0 = -2x e^{-x^2}$$

$$\Rightarrow x = 0$$

$$x = 0 \Rightarrow \frac{d^2y}{dx^2} < 0 \Rightarrow \text{maximum.}$$

So there is a maximum at $(0, 1)$. As $x \to \pm\infty$, $y \to 0$.

Notice also that the graph (Fig. 11.9) is symmetrical about the y axis, for if (α, β) lies on the curve so too does $(-\alpha, \beta)$. This function is of enormous importance in the study of statistics.

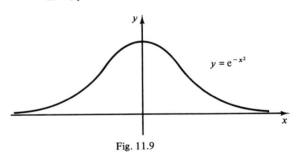

Fig. 11.9

The following points are worth remembering.

Statement

> If an equation contains no odd powers of x then the curve is symmetrical about the y axis.
>
> If a curve contains no odd powers of y then the curve is symmetrical about the x axis.
>
> So if the curve contains only even powers of x and y then the curve is symmetrical about both axes.
>
> If x and y may be interchanged without altering the relationship then the curve is symmetrical about the line $y = x$.

11.3 Asymptotes

If a curve approaches a straight line as either x or y tend to $\pm\infty$ then the line is called an *asymptote* to the curve. In the previous two examples the x axis is an asymptote to the curves.

Horizontal asymptotes may be found by letting $x \to \pm\infty$. For example, if

$$y = \frac{2x + 1}{x - 1}$$

then

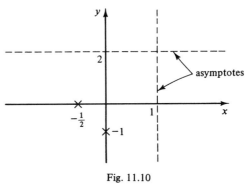

$$y = x\left(2 + \frac{1}{x}\right)\Big/x\left(1 - \frac{1}{x}\right)$$

$$= \left(2 + \frac{1}{x}\right)\Big/\left(1 - \frac{1}{x}\right)$$

Fig. 11.10

and so as $x \to \pm\infty$, $y \to 2$. It follows that $y = 2$ is an asymptote to the above curve.

It is possible that the curve cuts the asymptote for particular values of x. This may be checked by setting $y = 2$ in the equation of the curve and solving the resulting equation for x. In this example no solutions will be found. (However, when sketching a curve such a check is often worthwhile).

A vertical asymptote usually occurs when the function is undefined for a particular value of x. In this example $x = 1$ is a vertical asymptote.

Noting that when $x = 0$, $y = -1$ and when $y = 0$, $x = -\frac{1}{2}$ we now have the information summarised in Fig. 11.10.

The function

$$f(x) \equiv \frac{2x + 1}{x - 1}$$

is defined for all $x \neq 1$ and does not cross the axes except at the point indicated. So we may sketch the curve with some confidence, as shown in Fig. 11.11.

If additional information were required we could establish the value of y for specific values of x. Calculus may also be used to find stationary values.

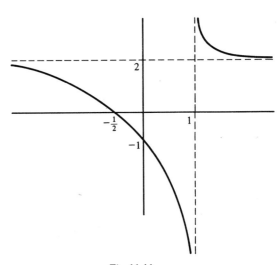

Fig. 11.11

11.4   Rational functions

For certain rational functions there is an algebraic method of establishing maximum and minimum values, as indicated in the following example.

Example 11.4

Sketch the graph of $y = \dfrac{4}{(x-1)(x-3)}$.

(i) $x = 0 \Rightarrow y = \frac{4}{3}$. There are no solutions when $y = 0$.
(ii) Asymptotes occur when $x = 1$ or $x = 3$ and letting $x \to \pm\infty$ we note that $y \to 0$.

An attempt to sketch the graph at this stage will show that there must be a turning value. This could be established using calculus. Alternatively:

$$y = \frac{4}{(x-1)(x-3)}$$
$$\Rightarrow y(x-1)(x-3) = 4$$
$$\Rightarrow yx^2 - 4yx + 3y - 4 = 0$$

writing as a quadratic in x (assuming $y \neq 0$).

For any particular value k of y, we may or may not obtain a corresponding x value (indicating that the curve crosses the line $y = k$ or not). We shall be *successful* if the discriminant is positive! That is if:

$$(-4y)^2 - 4y(3y - 4) \geqslant 0$$
$$\Rightarrow \quad 16y^2 - 12y^2 + 16y \geqslant 0$$
$$\Rightarrow \qquad\qquad y^2 + 4y \geqslant 0$$
$$\Rightarrow \qquad\qquad y(y + 4) \geqslant 0.$$

We now sketch the graph of $z = y(y + 4)$ as shown in Fig. 11.12. Notice that $z[= y(y + 4)]$ is greater than or equal to zero if $y \leqslant -4$ or $y \geqslant 0$.

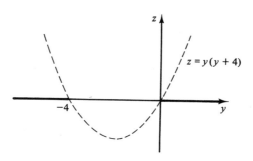

Fig. 11.12

So for values of y such that $y \leqslant -4$ or $y \geqslant 0$ a corresponding value of x can be found. However, in the range $-4 < y < 0$ the graph does not exist. We have also excluded the possibility that $y = 0$ and so the graph does not exist in the range $-4 < y \leqslant 0$. When $y = -4$ the equation reduces to

$$(x - 1)(x - 3) = -1$$
$$\Rightarrow \quad x^2 - 4x + 4 = 0$$
$$\Rightarrow (x - 2)(x - 2) = 0$$
$$\Rightarrow \quad\quad\quad x = 2.$$

Collecting all this information together we may sketch the graph as shown in Fig. 11.13.

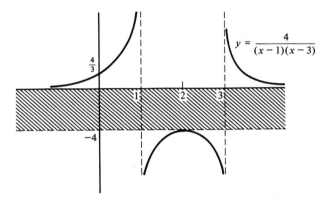

Fig. 11.13

Example 11.5

Sketch the graph of $y = \dfrac{1}{x^2 + 1}$.

(i) $x = 0 \Rightarrow y = 1$ and there are no solutions when $y = 0$.
(ii) The graph is defined for all values of x, and as $x \to \pm\infty$, $y \to 0$.
(iii) Notice that the graph is symmetrical about the y axis since only even powers of x appear.
(iv) Notice also that $y > 0$ for all $x \in \mathbb{R}$.

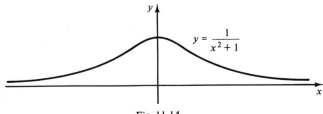

Fig. 11.14

The curve is only superficially similar to that of $y = e^{-x^2}$. However, in Fig. 11.15 the graph of $y = x^2 + 1$ is indicated, and methods for sketching $y = 1/f(x)$ were mentioned on page 144. These methods should be used in conjunction with the present methods.

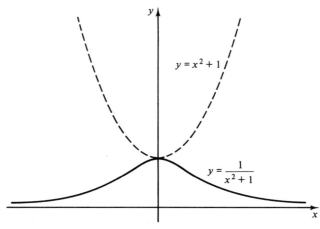

Fig. 11.15

Example 11.6

Sketch the graph of $y = \left| \dfrac{1}{x^2 - 2x - 3} \right|$.

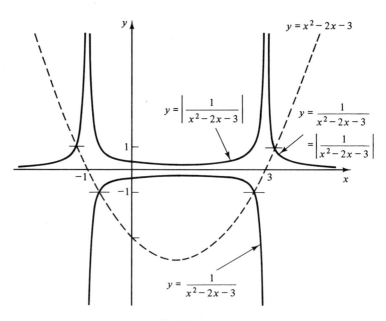

Fig. 11.16

This may be built up in stages. Sketch the graphs of

(i) $y = x^2 - 2x - 3 [\equiv (x - 3)(x + 1)]$

(ii) $y = \dfrac{1}{x^2 - 2x - 3}$

(iii) $y = \left| \dfrac{1}{x^2 - 2x - 3} \right|.$

These stages are shown in Fig. 11.16.

11.5 Some additional examples

Example 11.7

Sketch the curves

(a) $x^2 - y^2 = 1$ (b) $y^2 = x^3$ (c) $(y - 1)^2 = 2(x - 3)^3.$

(a) Consider $x^2 - y^2 = 1$. The graph is symmetrical about both axes. If $y = 0$ then $x = \pm 1$.
   Notice also that since

$$y^2 = x^2 - 1$$
$$y = \pm \sqrt{x^2 - 1}$$

and so the graph does not exist for x in the range $-1 < x < 1$. (For in such a case we are attempting to find the square root of a negative number.)
   Also when $x \to \pm \infty$, $y \simeq \pm x$ and so $y = x$ and $y = -x$ are asymptotes.
   Differentiating

$$2x - 2y \frac{dy}{dx} = 0$$

$$\Rightarrow \qquad \frac{dy}{dx} = \frac{x}{y}$$

So when $y = 0$ the gradient is undefined.
   The graph is as indicated in Fig. 11.17.

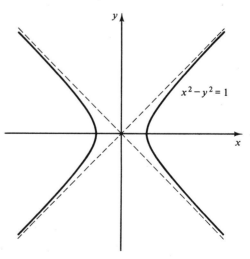

Fig. 11.17

(b) Consider $y^2 = x^3$.

The graph is symmetrical about the x axis and since

$$y = \pm\sqrt{x^3}$$

the graph is not defined for $x < 0$.

Differentiating

$$2y\frac{dy}{dx} = 3x^2$$

$$\Rightarrow \quad \frac{dy}{dx} = \frac{3x^2}{2y}$$

$$= \pm\tfrac{3}{2}x^{1/2} \qquad \text{since } y^2 = x^3.$$

In particular when $x = 0$, $y = 0$ and $dy/dx = 0$. As $x \to \infty$, $y \to \pm\infty$.
The graph is drawn in Fig. 11.18.

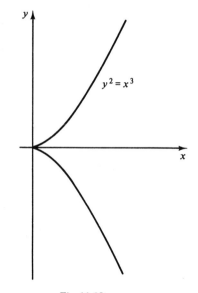

$$y^2 = x^3$$

Fig. 11.18

(c) The equation $(y - 1)^2 = 2(x - 3)^3$ is best dealt with by changing the origin.
Let $Y = y - 1$ and $X = x - 3$, then

$$Y^2 = 2X^3.$$

It is now a simple matter to sketch this graph in the X–Y plane. The origin of the
X–Y plane relative to the x–y plane is found by solving $0 = y - 1$ and $0 = x - 3$.

Thus relative to the x and y axes the origin [of the X–Y plane] is located at
$(3, 1)$. The graph crosses the x axis when $y = 0$ and so

$$(0 - 1)^2 = 2(x - 3)^3$$

$$\Rightarrow \qquad x = 3 + (\tfrac{1}{2})^{1/3}$$

$$\simeq 3.79.$$

The graph is drawn in Fig. 11.19.

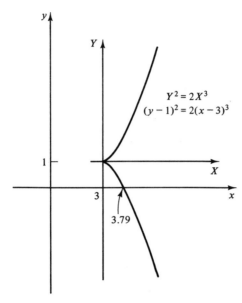

Fig. 11.19

Exercise 11.2

It may be necessary to be selective in this exercise.

1. Assuming the curve in Fig. 11.20 represents the graph of $y = f(x)$, sketch the graph of

    (a) $y = \dfrac{1}{f(x)}$

    (b) $y^2 = f(x)$

    (c) $y - 1 = f(x)$

    (d) $y = -f(x)$

    (e) $y = f(x + 2)$.

    [*Hint*: it may be helpful to return to Chapter 5.]

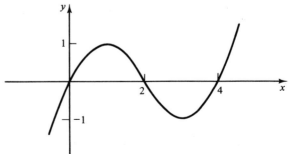

Fig. 11.20

2. Sketch the graphs of

    (a) $y^2 = x$ (b) $y = x^2$ (c) $y^3 = x^2$ (d) $y^3 = (x - 1)^2$.

3. Sketch the graphs of

   (a) $x^2 e^{-x}$ (b) $(1 - x) e^x$ (c) $y = x \ln x$ (d) $y = \dfrac{\ln x}{x}$

   (e) $y = e^{-x} \sin x$ (f) $y = \dfrac{e^{-x}}{x}$.

4. Sketch the graphs of each of the following

   (a) $y = \dfrac{x}{1 + x}$ (b) $y = \dfrac{x - 1}{x + 2}$ (c) $y = \dfrac{2x + 1}{x - 2}$

   (d) $y = \dfrac{1 - x}{1 - 2x}$ (e) $y = \dfrac{1 + x}{x^2}$ (f) $y = \dfrac{x^2}{1 + x}$

   (g) $y = \dfrac{x + 1}{x(x - 1)}$ (h) $y^2 = \dfrac{1}{x(x + 1)}$.

5. In each case in Fig. 11.21 the graph $y = f(x)$ has been drawn. Use this graph to sketch the general form of $y = |f(x)|$.

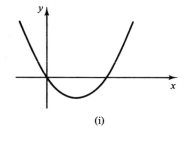

(i)

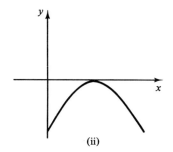

(ii)

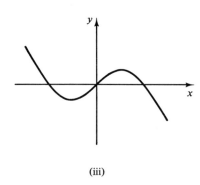

(iii)

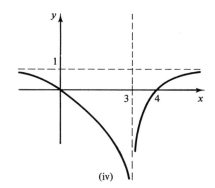

(iv)

Fig. 11.21

6. Find the range of possible values for y and hence sketch the curves

   (a) $y = \dfrac{3x}{x^2 + 1}$ (b) $y = \dfrac{x^2 - 3}{x + 2}$ (c) $y = \dfrac{x - 2}{x^2 - 2x + 2}$

7. Sketch the graphs of

   (a) $y^2 = (x - 1)(x + 1)$ (b) $y = \dfrac{1}{(x - 1)(x + 1)}$.

8. Sketch the graphs of the following equations.

   (a) $y = |1 - x|$ (b) $y = |x^2 + 5x + 4|$

   (c) $y = \left| \dfrac{1}{x^2 + 5x + 4} \right|$ (d) $y = |\cos x|$ $0 \leqslant x \leqslant 2\pi$

   (e) $y = \left| \dfrac{x + 2}{1 - x} \right|$

9. Sketch the curve whose equation is

$$y = \frac{2x + 3}{x + 1}.$$

   The curve cuts the x-axis at A and the y-axis at B, and the asymptotes to the curve intersect at C. Find the perpendicular distance from C to the line AB.

   JMB 1983

10. Sketch a graph of the function f given by

$$f(x) = 2|x| + |x - 1|, \qquad x \text{ real.}$$

    Solve the equation $f(x) = 4$.

    JMB 1982

11. Sketch (on separate diagrams) the graphs of

    (a) $y = x^2 - x^3$,   (b) $y = 1 - e^x$,   (c) $y = 1/(1 - e^x)$.

    You are only asked for rough sketches; details of maxima and minima are not required but you should indicate the behaviour of the curves for numerically large and small values of x.

    O&C 1982

12. Given that $y = \dfrac{x^2 + 1}{x^2 + x + 1}$,

    (a) state the limiting value of y
        (i) as x tends to ∞,
        (ii) as x tends to $-\infty$,
    (b) show that y is finite for all real values of x,
    (c) determine the set of values which y must take for real values of x,
    (d) determine the values of x for which y has a stationary value.
        Sketch the curve which represents the given equation.

    AEB 1982

13. The function f is defined by $f(x) = \dfrac{4x + 5}{x^2 - 1}$.

Show that $f(x)$ cannot take values between -4 and -1. Sketch the graph of the function showing clearly the behaviour of f as $x \to \pm 1$ and as $x \to \pm \infty$.

W 1981

14. Examine the function given by

$$h(x) = \frac{(x - 1)^2}{(x + 1)^3} \qquad (x \neq -1)$$

for maximum and minimum points and sketch its graph.

W 1983

15. Sketch (on separate diagrams) the graphs of

(a) $y = \ln\left(\dfrac{1}{x}\right)$,   (b) $y = \dfrac{x^2 - 1}{(x + 2)(x + 3)}$.

You are only asked for rough sketches; details of maxima and minima are not required, but you should indicate the behaviour of the graphs for numerically large values of x and y.

O&C 1983

16. Show that if $f(x) = \dfrac{(x - 4)^2}{x - 3}$ then $f'(x) = \dfrac{(x - 4)(x - 2)}{(x - 3)^2}$.

Find the local maxima or minima of the function f. Sketch the graph of f.

W 1982

17. A curve is defined by the equations

$$x = at^2, \quad y = at^3,$$

where a is a positive constant and t is a parameter. Sketch the curve and note on your sketch the set of values of t corresponding to each branch of the curve.

LOND 1983

18. The functions f and g are defined by

$$f(x) = x + 4, \quad x > -4,$$
$$g(x) = \ln x, \quad x > 0.$$

Denoting by h composite function gf, write down $h(x)$ and state the domain and range of h. Find $h^{-1}(x)$.

Sketch the graphs of $h(x)$ and $h^{-1}(x)$ on one diagram, labelling each graph clearly and indicating on the graphs the coordinates of any intersections with the axes.

JMB 1984

19. Sketch on separate diagrams the graphs corresponding to the following equations, showing the turning points and any asymptotes parallel to the coordinate axes:

(a) $y = (x - 1)(x - 3)$,

(b) $y = \dfrac{1}{(x - 1)(x - 3)}$ $(x \neq 1, x \neq 3)$.

The line $y = -\frac{4}{3}$ meets the graph of $y = \dfrac{1}{(x - 1)(x - 3)}$ at the points A and B. Calculate the area of the finite region enclosed by the line AB and the arc of the graph between A and B, giving three significant figures in your answer.

CAMB 1982

20. Show that the curve

$$y = \frac{x^3}{(1 + x^2)^{1/2}}$$

has a positive gradient at all points (other than the origin) and sketch the curve. Prove that the area of the region enclosed by the curve, the line $x = 2$ and the x-axis from $x = 0$ to $x = 2$, is $\frac{2}{3}(1 + \sqrt{5})$.

[The substitution $z = 1 + x^2$ is suggested for the integration.]

CAMB 1982

21.. Sketch the graphs of

(a) $y = x - 1$ (b) $y = \sqrt{x - 1}$ (c) $y = \dfrac{1}{\sqrt{x - 1}}$.

22. Sketch the curve defined by the equation

$$3ay^2 = x(a - x)^2 \qquad \text{where } a > 0.$$

Find the area of the loop.

11.6 Inequalities

Example 11.8
Find the range of values of x for which

$$(x - 2)(x + 1)(x + 3) \leqslant 0.$$

Consider the equation

$$y = (x - 2)(x + 1)(x + 3),$$

$$x = 0 \qquad \Rightarrow y = -6,$$

$$y = 0 \qquad \Rightarrow x = 2 \quad \text{or} \quad x = -1 \quad \text{or} \quad x = -3.$$

A very rough sketch may now be obtained of the graph of this relation, as in Fig. 11.22.

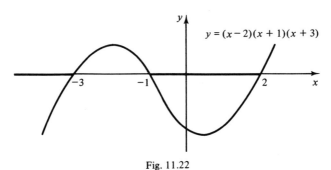

Fig. 11.22

It is immediately clear that y is negative or zero whenever x lies in the marked regions; $(-\infty, -3]$ or $[-1, 2]$. Since $y = (x - 2)(x + 1)(x + 3)$, it follows that

$$(x - 2)(x + 1)(x + 3) \leqslant 0$$

whenever

$$x \leqslant -3 \quad \text{or} \quad -1 \leqslant x \leqslant 2.$$

A sketch graph may be used to solve many inequalities.

Example 11.9

Find the range of values of x for which

$$\frac{4 - x}{x + 2} > 3.$$

As a first step, we obtain a zero on one side of the inequality:

$$\frac{4 - x}{x + 2} - 3 > 0$$

$$\Leftrightarrow \qquad \frac{(4 - x) - 3(x + 2)}{x + 2} > 0$$

$$\Leftrightarrow \qquad \frac{-2(2x + 1)}{x + 2} > 0$$

$$\Leftrightarrow \qquad \frac{2x + 1}{x + 2} < 0$$

(reversing the inequality since we divided by a negative quantity).

We now sketch the graph of

$$y = \frac{2x + 1}{x + 2}$$

$$y = 0 \Rightarrow x = -\tfrac{1}{2}$$

$$x = 0 \Rightarrow y = \tfrac{1}{2}$$

There is a vertical asymptote when $x = -2$

and as $x \to \pm \infty$ so $y \left[= \frac{2 + (1/x)}{1 + (2/x)} \right] \to 2.$

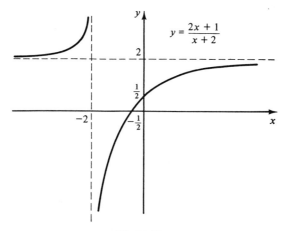

Fig. 11.23

It follows immediately from an inspection of the graph that

$$\frac{4 - x}{x + 2} > 3$$

$$\Leftrightarrow \frac{2x + 1}{x + 2} < 0$$

$$\Leftrightarrow x \text{ in the range } (-2, -\tfrac{1}{2})$$

An alternative approach

In order to solve

$$\frac{4 - x}{x + 2} > 3$$

we multiply both sides by $(x + 2)^2$, a positive quantity. Then

$$\Leftrightarrow (4 - x)(x + 2) > 3(x + 2)^2$$

$$\Leftrightarrow 4x + 8 - x^2 - 2x > 3x^2 + 12x + 12$$

$$\Leftrightarrow 4x^2 + 10x + 4 < 0$$

$$\Leftrightarrow 2x^2 + 5x + 2 < 0$$

$$\Leftrightarrow (2x + 1)(x + 2) < 0.$$

From a sketch of the graph of $y = (2x + 1)(x + 2)$ we see that the inequality holds if $-2 < x < -\frac{1}{2}$, as before.

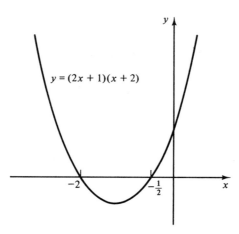

$y = (2x + 1)(x + 2)$

Fig. 11.24

Exercise 11.3

1. Find the range of values of x for which the following inequalities hold.

    (a) $(x + 2)(x - 1)(x + 3) < 0$

    (b) $(x - 1)^2(x + 2) \leqslant 0$

    (c) $x^3 + 3x^2 - 4 \geqslant 0$

    (d) $2x^3 + 3x^2 - 3x < 2$

    (e) $\dfrac{x + 1}{2 - x} > 1$

    (c) $\dfrac{1}{x^2 + 3x - 4} < 0$

    (g) $\dfrac{5}{x^2 + 2x + 1} \geqslant 0$

    (h) $\dfrac{3x - 1}{2x - 3} \leqslant \frac{1}{2}$

2. Solve $\dfrac{x + 5}{x - 3} > x$.

3. Solve the inequality $\dfrac{x + 2}{x - 1} < 3$.

    JMB 1984

4. Solve the inequality $\dfrac{1}{3 - x} < \dfrac{1}{x - 2}$.

5. Find the set of values of k for which, for all real values of x,

$$3x^2 + 3x + k > 0 \quad \text{and} \quad 3x^2 + kx + 3 > 0.$$

CAMB 1982

6. Using the same axes sketch the curves

$$y = \frac{1}{x - 1}, \qquad y = \frac{x}{x + 3},$$

giving the equations of the asymptotes. Hence, or otherwise, find the set of values of x for which

$$\frac{1}{x - 1} > \frac{x}{x + 3}.$$

LOND 1983

7. Solve the inequalities

(a) $x(x + 2) > 3$, (b) $\dfrac{x}{(x + 1)} \leqslant \dfrac{1}{6}.$

O&C 1983

8. Find, in each case, the set of values of x for which

(a) $x(x - 2) > x + 4$, (b) $x - 2 > \dfrac{x + 4}{x}.$

CAMB 1983

9. Show that the equation $x^3 - 8x - 8 = 0$ has a root which is an integer. Find this root and hence solve the equation completely.
     Find the ranges of values of x for which $x^3 - 8x - 8 < 0$.

11.7 Polar co-ordinates

There are many ways to locate the position of a point in a plane. The most common is to use Cartesian co-ordinates. An alternative is to locate a point which respect to a fixed point on a given line. The line is called *the initial line* and the point, O, is called the *pole* or *origin*. A point P is then fixed by noting the angle OP makes with the initial line, θ, and the displacement of P from the pole, r, say. The co-ordinates (r, θ) are called *polar co-ordinates*.

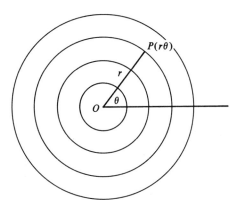

Fig. 11.25

The angle θ is positive if measured in the anticlockwise sense and negative if measured in the clockwise sense. Once the angle θ has been fixed, then r is positive along the arm OP and negative along the extension of PO (see Fig. 11.26).

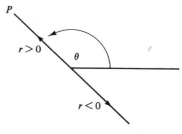

Fig. 11.26

Example 11.10

Plot the following polar co-ordinates.

$$A(2, \pi/4), \ B(-2, \pi/4), \ C(2, -\pi/4), \ D(-2, -\pi/4).$$

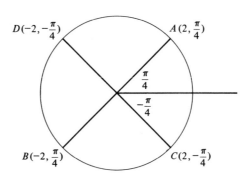

Fig. 11.27

Notice, for example, that the location of B could be given as $(2, -3\pi/4)$, C as $(-2, 3\pi/4)$ and D as $(2, 3\pi/4)$.

11.8 Polar equations

Some curves may be expressed more simply in polar form compared with their Cartesian form.

For example, the equation $r = 2$ is satisfied by any point of the form $(2, \theta)$ and so represents the circle with centre O and radius 2. Now the equation $\theta = \pi/4$ is satisfied by all points of the form $(r, \pi/4)$ and so represents the line through the pole which intersects the initial line at an angle $\pi/4$ (see Fig. 11.28 on opposite page).

Consider the circle passing through the origin with diameter OX of length a on the initial line (see Fig. 11.29). Since $\angle OPX$ is a right angle $\cos \theta = r/a$, so $r = a \cos \theta$ is the equation of this circle.

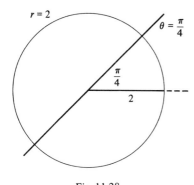

Fig. 11.28

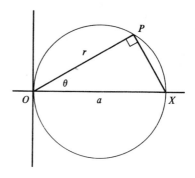

Fig. 11.29

Consider the straight line distance $d(>0)$ from the pole, where the perpendicular ON is at an angle α to the initial line.

In $\triangle ONP$

$$\cos(\theta - \alpha) = \frac{d}{r}$$

$$\Rightarrow r = \frac{d}{\cos(\theta - \alpha)} = d\sec(\theta - \alpha)$$

which is the polar equation of this line.

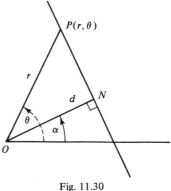

Fig. 11.30

The connection between polar and Cartesian equations

A point P can be located in the plane using Cartesian co-ordinates (x, y) or polar co-ordinates (r, θ). Assuming a common origin and that the initial line is taken to be the positive x axis then the co-ordinates are related as follows

$$x = r\cos\theta, \qquad r^2 = x^2 + y^2,$$

$$y = r\sin\theta, \qquad \tan\theta = \frac{y}{x}.$$

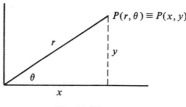

Fig. 11.31

These relationships may be used to convert the equation of a curve from Cartesian form to polar form and *vice versa*.

For example, if

$$\frac{1}{r} = 1 + \cos\theta$$

then

$$1 = r + r\cos\theta$$

$$\Rightarrow \qquad r^2 = (1 - r\cos\theta)^2$$

$$\Rightarrow \qquad x^2 + y^2 = (1 - x)^2$$

$$\Rightarrow \qquad y^2 = 1 - 2x$$

However, if $xy = c^2$ then

$$r\cos\theta\, r\sin\theta = c^2$$

$$\Rightarrow \qquad r^2 \frac{\sin 2\theta}{2} = c^2$$

$$\Rightarrow \qquad r^2 \sin 2\theta = 2c^2.$$

Curve sketching: polar equations

The graph of a polar equation may be sketched by tabulating values of θ and the corresponding values of r.

Some calculations may be eliminated by noting the various symmetries.

Symmetries

> If r is a function of $\cos\theta$, the curve is symmetrical about the line $\theta = 0$.
> If r is a function of $\sin\theta$, the curve is symmetrical about the line $\theta = \pi/2$.
> If only even powers of r appear the curve is symmetrical about the origin.

So if r is a function of $\cos\theta$ we need only tabulate values of θ from 0 to π, the values of θ from 0 to $-\pi$ being found by symmetry.

Example 11.11

Sketch the graphs of

(a) the *lemniscate* $r^2 = a^2 \cos 2\theta$,

(b) the *cardioid* $r = a(1 + \cos\theta)$.

(a) $$r^2 = a^2 \cos 2\theta = a^2(1 - 2\sin^2\theta)$$

$$= a^2(2\cos^2\theta - 1)$$

The curve is symmetrical about the line $\theta = 0$, the line $\theta = \pi/2$ and the origin. So it suffices to tabulate values of θ from 0 to $\pi/2$. However, when $\pi/4 < \theta < \pi/2$ then $\pi/2 < 2\theta < \pi$ and so $\cos 2\theta < 0$. So there is no real r corresponding to these values of θ and we need only consider values of θ from 0 to $\pi/4$.

θ	$0°$	$5°$	$10°$	$15°$	$20°$	$25°$	$30°$	$35°$	$40°$	$45°$
$r = \sqrt{a^2 \cos 2\theta}$	1	.99a	.97a	.93a	.88a	.80a	.71a	.58a	.42a	0

The values of θ are tabulated in degrees for ease of calculation, the corresponding r values are found using a calculator, expressed to 2 d.p. [1 d.p. is often sufficient for a rough sketch.] These values are marked in Fig. 11.32.

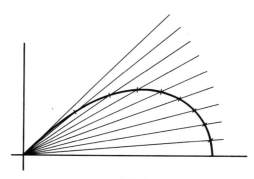

Fig. 11.32

Adopting the symmetries previously mentioned we have the sketch shown in Fig. 11.33. The lines $\theta = \pi/4$ and $\theta = -\pi/4$ are tangents to the curve at the origin.

$r^2 = a^2 \cos 2\theta$

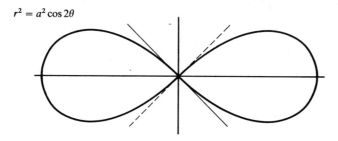

Fig. 11.33

In general

> If the curve $r = f(\theta)$ passes through the pole, the tangents to the curve at the pole may be found by solving the equation $f(\theta) = 0$.

Thus since $r^2 = a^2 \cos 2\theta$ setting $r = 0$ we have

$$\cos 2\theta = 0$$

$$\Rightarrow \quad 2\theta = \pi/2 \quad \text{or} \quad 3\pi/2$$

$$\Rightarrow \quad \theta = \pi/4 \quad \text{and} \quad 3\pi/4$$

giving the tangents as previously stated.

(b) $r = a(1 + \cos\theta)$. The graph is *symmetrical about the line $\theta = 0$*, and when $r = 0$

$$a(1 + \cos\theta) = 0$$

$$\Rightarrow \quad \cos\theta = -1$$

$$\Rightarrow \quad \theta = \pi$$

The tangent to the curve at the pole is the line $\theta = \pi$

θ	0°	20°	40°	60°	80°	100°	120°	140°	160°	180°
r	$2a$	$1.94a$	$1.77a$	$1.5a$	$1.17a$	$0.83a$	$0.5a$	$0.23a$	$0.06a$	0

The graph of $r = a(1 + \cos\theta)$ is given in Fig. 11.34.

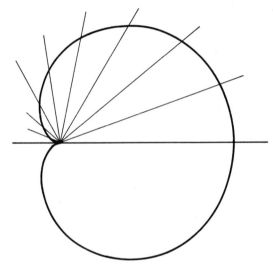

Fig. 11.34

Exercise 11.4

1. Plot on a diagram the points with polar co-ordinates:

   $A(1, 0)$, $B(2, \pi/2)$, $C(3, \pi)$, $D(-2, \pi)$, $E(-1, \pi/2)$, $F(2, -\pi/2)$, $G(2, -3\pi/2)$, $H(2, 2\pi/3)$.

2. The polar co-ordinates of the points P, Q and R are $(2, \pi/4)$, $(1, \pi)$ and $(2, -\pi/2)$ respectively and O is the origin.

   (a) Find the distance (i) QR, (ii) PR, (iii) PQ.

(b) Find the area of the triangle (i) $\triangle ROQ$, (ii) $\triangle PRO$, (iii) $\triangle PQR$.

3. Express each of the following polar co-ordinates in Cartesian form, assuming a common origin and taking the positive x axis as the initial line.

   (a) $(3, \pi/3)$,   (b) $(-2, -\pi/4)$,   (c) $(2, -\pi/4)$,   (d) $(4, -3\pi/4)$.

4. Adopting the same assumptions as in question 3 express the following Cartesian co-ordinates in polar form.

   (a) $(1, \sqrt{3})$,   (b) $(4, 0)$,   (c) $(0, 3)$,   (d) $(-1, -1)$.

5. Find the polar equations of the curves with given Cartesian equations

   (a) $xy = 4$,   (b) $x^2 + y^2 = 9$,
   (c) $a^2x^2 + b^2y^2 = a^2b^2$,   (d) $x^2 + (y - 1)^2 = 4$.

6. Find the Cartesian equations of the curves with given polar equations

   (a) $r = 2$,   (b) $\theta = -\pi/3$,   (c) $r = \sin \theta$.
   (d) $r \cos \theta = 4$,   (e) $r^2 = a^2 \sin \theta$,   (f) $r = \cos 2\theta$,
   (g) $r^2 \cos 2\theta = 4$, (h) $r = (1 - \sin \theta)$.

7. Sketch the curves corresponding to the following equations

   (a) $r = 3$,   (b) $r = 2 \cos \theta$,   (c) $r = -\sin \theta$,
   (d) $r = 2 \sin 2\theta$, (e) $r = 2(1 + \sin \theta)$,   (f) $r^2 = 9 \cos 2\theta$,
   (g) $r = 2 \cos^2 \theta$,(h) $r = 2\theta$, $0 \leqslant \theta \leqslant 2\pi$, (i) $r \cos \theta = 2$,
   (j) $r \sin \theta = 3$, (k) $r = \sqrt{2} \sec (\theta - \pi/4)$, (l) $r = \sqrt{3} \sec (\theta - \pi/3)$.

Revision exercise C

1. The arithmetic mean of two numbers p and q is 39 and their geometric mean is 15. Write down a quadratic equation whose roots are p and q.
   Hence, or otherwise, find the values of p and q.

   LOND 1984

2. (a) Evaluate $\displaystyle\int_0^{2\pi} (1 + \sin x)^2 \, dx$.

   (b) Transform the integral

   $$\int_1^e \frac{dt}{t(2 + \log_e t)^2}$$

   by the substitution $u = 2 + \log_e t$ and hence evaluate it.

(c) Give a rough sketch of the graph of $y = e^x - 1$. The area enclosed by the graph, the x-axis and the line $x = 1$ is rotated about the x-axis through 2π radians. Find the volume of the solid formed. (You may leave your answer in terms of e and π.)

SU 1983

3. The equation

$$2x^3 - 5x^2 - 2x - 3 = 0$$

has a root which is a positive integer. Find this root and show that the equation has no other real root.

CAMB 1982

4. Given that α and β are the roots of the equation $x^2 - bx + c = 0$,

(a) show that $(\alpha^2 + 1)(\beta^2 + 1) = (c - 1)^2 + b^2$,
(b) find, in terms of b and c, a quadratic equation whose roots are

$$\frac{\alpha}{\alpha^2 + 1} \quad \text{and} \quad \frac{\beta}{\beta^2 + 1}.$$

AEB 1983

5. (a) Find the values of x which satisfy the equation

$$4 \log_3 x = 9 \log_x 3.$$

(b) By taking $\log_{10} 5 \approx 0.7$, obtain an estimate of the root of the equation
$$10^{y-5} = 5^{y+2},$$

giving your answer to the nearest integer.

AEB 1983

6. Using the expressions for $\sin(A + B)$ and $\cos(A + B)$ derive the following results:

(a) $\sin 2A = 2 \sin A \cos A$,
    $\cos 2A = \cos^2 A - \sin^2 A$;
(b) $\sin 3A = 3 \sin A \cos^2 A - \sin^3 A$,
    $\cos 3A = \cos^3 A - 3 \cos A \sin^2 A$.

From the results in (b) above, obtain an expression for $\tan 3A$ in terms of $\tan A$. Find all values of x in the interval $0° \leqslant x \leqslant 180°$ for which

$$\tan 3x + 2 \tan x = 0,$$

giving your answers to the nearest $0.1°$ where necessary.

CAMB 1982

7. Three landmarks P, Q and R are on the same horizontal level. Landmark Q is 3 km and on a bearing of 328° from P, landmark R is 6 km and on a bearing of 191° from Q. Calculate the distance and the bearing of R from P, giving your answers in km to one decimal place and in degrees to the nearest degree.

LOND 1984

8. Determine the range of values of x for which the infinite geometric series

$$2x + \frac{2x}{1-x} + \frac{2x}{(1-x)^2} + \frac{2x}{(1-x)^3} + \ldots$$

converges. If x has a value such that the series converges, express the sum to infinity of the series in terms of x.

AEB 1983

9. (a) Show that for an arithmetic progression with first term a, common difference d and sum S, the number of terms, n, must satisfy the quadratic equation

$$n^2 + \left(\frac{2a}{d} - 1\right)n - \frac{2S}{d} = 0.$$

Find n when $a = 3$, $d = \frac{1}{2}$ and $S = 2828$.

(b) Find the least value of n for which the sum to n terms of the geometric series

$$1 + 0.99 + (0.99)^2 + (0.99)^3 + \ldots$$

is greater than half the sum to infinity.

CAMB 1983

10. Use induction to prove that the sum of the series

$$\frac{1}{1 \times 2 \times 3} + \frac{1}{2 \times 3 \times 4} + \frac{1}{3 \times 4 \times 5} + \ldots + \frac{1}{n(n+1)(n+2)}$$

is

$$\frac{n(n+3)}{4(n+1)(n+2)}.$$

O&C 1983

11. The tangent to the curve $4ay = x^2$ at the point $P(2at, at^2)$ meets the x-axis at the point Q. The point S is $(0, a)$.

(a) Prove that PQ is perpendicular to SQ.

(b) Find a Cartesian equation for the locus of the point, M, the mid-point of PS.

AEB 1983

12. (a) Differentiate, with respect to x, $\tan\left(\dfrac{1}{x}\right)$.

(b) Given that $y = A\cos(\log_e x) + B\sin(\log_e x)$, where A and B are constants and $x > 0$, show that

$$x^2\frac{d^2y}{dx^2} + x\frac{dy}{dx} + y = 0.$$

AEB 1983

13. An even function f, of period $\pi/2$, is defined by

$$f(x) = 4x^2 \qquad \text{for } 0 \leqslant x \leqslant \pi/4,$$
$$f(x) = \pi^2/4 \qquad \text{for } \pi/4 < x \leqslant \pi/2.$$

Sketch the graph of f for $-\pi \leqslant x \leqslant \pi$.

LOND 1984

14. The expression $12\cos\theta - 5\sin\theta$ is denoted by $f(\theta)$. Find the values of R and α, where R is positive and the angle α is acute, such that, for all values of θ,

$$f(\theta) = R\cos(\theta + \alpha),$$

giving the value of α in degrees to the nearest $0.1°$.
(a) Obtain the greatest and least values of $f(\theta)$ and give, to the nearest $0.1°$, the values of θ in the interval $0° < \theta < 360°$ for which these greatest and least values occur.
(b) Sketch the graph of $f(\theta)$ for $0° \leqslant \theta \leqslant 360°$.
(c) Calculate the values of θ, where $0° < \theta < 360°$, such that

$$12\cos\theta - 5\sin\theta + 4 = 0,$$

giving your answer to the nearest $0.1°$.

CAMB 1982

15. The function $g(x)$, with domain $\{x : x > -2\}$ is defined by

$$g(x) \equiv \ln(x+2) + 2$$

Find g^{-1}, stating its domain and range.

16. (a) Show that $y + 4x = 18$ is the normal to the curve $y^2 = x$ at a point P, and find the coordinates of P.
(b) A right circular cylinder of height h is inscribed in a sphere of radius R. Show that the volume of the cylinder, V, is given by the formula

$$V = \pi R^2 h - \frac{\pi h^3}{4}.$$

Prove that the maximum volume of a right circular cylinder which can be inscribed in a sphere of fixed radius R is

$$\frac{4\pi R^3}{3\sqrt{3}}.$$

(N.B. You must prove that it is the *maximum* volume.)

SU 1981

17. A plane figure $ABCD$ is such that $\angle ABC = \angle ACD = 90°$ and $\angle DAC = \angle CAB = \theta$. Given that Y denotes

$$\frac{AC + CD}{CB},$$

show that

$$Y = \sec\theta + \operatorname{cosec}\theta$$

and find its minimum value, showing clearly that the value you have found is the minimum value.

W 1981

18. (a) Differentiate with respect to x (i) $\dfrac{1}{1 + \sin x}$, (ii) $\log_e \sec 2x$.

(b) A body starts from O and moves in a straight line. Its distance from O after t seconds is x metres where $x = 10t\,e^{-1/2t^2}$. Find, correct to 3 decimal places, the distance from O where the body first comes to rest and show that it subsequently moves towards O without ever reaching it.

(c) A curve is given parametrically by $x = t^2 + 2t$, $y = t^2 - 2t$. Find

$$\frac{dy}{dx} \quad \text{and} \quad \frac{d^2y}{dx^2}$$

in terms of t and hence obtain the values of x and y at the stationary point of the curve, determining its nature.

SU 1980

19. A circular hollow cone of height h and semi-vertical angle α stands on a horizontal table. A circular cylinder placed in an upright position just fits inside the space between the cone and the table, as shown in the diagram. Prove that the maximum possible volume of the cylinder is

$$\frac{4\pi h^2 \tan^2 \alpha}{27}$$

JMB 1982

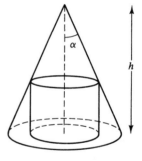

Fig. 11.35

20. The region enclosed by the x-axis, between the limits

$$x = 0 \quad \text{and} \quad x = \frac{\pi}{3},$$

and the curve $y = 2 \sin x + \tan x$ is rotated completely about the x-axis. Prove that the volume of the solid so formed is

$$\frac{\pi}{6} [2\pi + 24 \log_e(2 + \sqrt{3}) - 9\sqrt{3}].$$

AEB 1982

21. The curve C in the x–y plane is defined parametrically by

$$x = e^{3t}, \, y = t^2, \quad \text{where} \quad t \geqslant 0.$$

(a) Find $\dfrac{dy}{dx}$ in terms of t.

(b) Prove that $\dfrac{dy}{dx}$ has a maximum value when $t = \frac{1}{3}$.

(c) Sketch the curve C.

CAMB 1983

22. The domain of the function f defined by

$$f(x) = (x^2 - 3)e^{-x}$$

is the set of all real values of x. Determine, in terms of e, the maximum value and the minimum value of $f(x)$ and sketch the graph of $y = f(x)$. *Without* finding the coordinates of the points of inflexion, indicate the approximate positions of these points on your graph.

Calculate the area of the region bounded by the curve, the x-axis and the lines $x = 2$ and $x = 3$, giving your answer in the form $pe^{-2} + qe^{-3}$, where p and q are integers.

JMB 1983

23. Given that $y = \sqrt{9 - x}$ find, for values of x in the range $0 \leqslant x \leqslant 9$,

(a) the mean value of y w.r.t. x.
(b) the mean value of x w.r.t. y.

24. (a) Given that $\log_9 x = p$ and $\log_{\sqrt{3}} y = q$, express xy and x^2/y as powers of 3.
(b) Solve for x the equation

$$e^{2x} + e^x - 6 = 0.$$

AEB 1982

25. A truncated cone of height h has circular ends of radii $2r$ and r. In this cone is inserted a circular cylinder having its axis along the axis of the cone. One end of the cylinder lies in that face of the cone which is of radius $2r$ and the circumference of the other end lies in the curved surface of the cone (see Fig. 11.36). Given that the radius of the base of the cylinder is x, show that the volume of the cylinder is

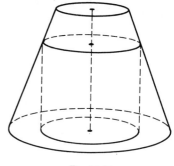

$$\frac{\pi h x^2 (2r - x)}{r}.$$

Fig. 11.36

If x is allowed to vary, find, in terms of h and r, the maximum volume of the cylinder.

CAMB 1982

26. The function f is defined by

$$f(x) = \frac{1 + e^x}{1 - e^x},$$

where x is real and non-zero. Show that f is an odd function.

JMB 1983

27. Given that

$$y = \frac{1}{(x + 2)(x + 3)},$$

express y in partial fractions and hence find the mean value of y over the interval $0 \leqslant x \leqslant 6$.

By reference to a sketch of the graph of y over this interval, show that your value for the mean value is a reasonable one.

AEB 1983

28. Given that a, b, c and d are four consecutive odd numbers and $a < b < c < d$, express each of b, c and d in terms of a and find the value of $a^2 + d^2 - b^2 - c^2$.

Find the constants A, B and k so that $(abcd + k)$ is always equal to

$$(a^2 + Aa + B)^2.$$

AEB 1982

Chapter 12 Further integration

12.1 Integration by parts

The product rule for differentiation states that

$$\frac{d}{dx}(u.v) = v.\frac{du}{dx} + u.\frac{dv}{dx}$$

where u and v are functions of x.

Integrating both sides w.r.t. x gives

$$\int \frac{d}{dx}(u.v)\,dx = \int v\frac{du}{dx}\,dx + \int u\frac{dv}{dx}\,dx$$

$$\Rightarrow \qquad uv = \int v\frac{du}{dx}\,dx + \int u\frac{dv}{dx}\,dx$$

which may be written as

$$\int v\frac{du}{dx}\,dx = uv - \int u\frac{dv}{dx}\,dx.$$

The advantage of this formula is that it enables us to express one product of the form $v\dfrac{du}{dx}$ in terms of another product $u\dfrac{dv}{dx}$; which may be easier to integrate.

Integrating in this way is called:

Integration by parts

$$\boxed{\int v.\frac{du}{dx}\,dx = uv - \int u.\frac{dv}{dx}\,dx.}$$

An example will indicate the power of this method.

Example 12.1

Find $\displaystyle\int \ln x\,dx$.

This integral may be thought of as

$$\int 1.\ln x\,dx$$

where

$$\frac{du}{dx} = 1 \quad \Rightarrow \quad u = x \qquad \text{(ignoring the constant of integration)}$$

$$v = \ln x \quad \Rightarrow \frac{dv}{dx} = \frac{1}{x}.$$

Using integration by parts

$$\int \ln x \, dx = \int 1 \cdot \ln x \, dx = x \cdot \ln x - \int x \cdot \frac{1}{x} \, dx$$

$$= x \cdot \ln x - \int 1 \, dx$$

$$= x \cdot \ln x - x + c,$$

a result which may be checked by differentiation.

Example 12.2

Find $\int x \sin x \, dx$.

When integrating a product by parts, one factor must be easy to differentiate — this is taken as v — and the other must be easy to integrate.

In this example both factors are readily integrated or differentiated. However, another condition is that the final integral must be 'simpler' than the original.

Suppose we take

$$v = \sin x \Rightarrow \frac{dv}{dx} = \cos x$$

and

$$\frac{du}{dx} = x \quad \Rightarrow \quad u = \frac{x^2}{2},$$

then we have

$$\int x \sin x \, dx = \frac{x^2}{2} \cdot \sin x - \int \frac{x^2}{2} \cos x \, dx.$$

In this case we have not improved matters. However, setting

$$v = x \quad \Rightarrow \frac{dv}{dx} = 1$$

$$\frac{du}{dx} = \sin x \Rightarrow \quad u = -\cos x$$

then

$$\int x \sin x \, dx = -x \cos x - \int (-\cos x) \, dx$$

$$= -x \cos x + \int \cos x \, dx$$

$$= -x \cos x + \sin x + c.$$

It may happen that one application of the method is insufficient.

Example 12.3

Find $\int x^2 e^x \, dx$.

Setting

$$v = x^2 \Rightarrow \frac{dv}{dx} = 2x$$

$$\frac{du}{dx} = e^x \Rightarrow u = e^x$$

then

$$\int x^2 e^x \, dx = x^2 e^x - \int 2x e^x \, dx.$$

Now to find $\int 2x e^x \, dx$, let

$$v = 2x \Rightarrow \frac{dv}{dx} = 2$$

$$\frac{du}{dx} = e^x \Rightarrow u = e^x.$$

So

$$\int 2x e^x \, dx = 2x e^x - \int 2e^x \, dx$$

$$= 2x e^x - 2e^x + c.$$

Substituting,

$$\int x^2 e^x \, dx = x^2 e^x - [2x e^x - 2e^x + c]$$

$$= (x^2 - 2x + 2)e^x + a \qquad \text{(where } a = -c\text{)}$$

Example 12.4

Find $\displaystyle\int_0^{\pi/2} e^x \cos x \, dx$.

Set

$$v = e^x \quad \Rightarrow \frac{dv}{dx} = e^x$$

$$\frac{du}{dx} = \cos x \Rightarrow u = \sin x.$$

So

$$\int_0^{\pi/2} e^x \cos x \, dx = \left[e^x \sin x \right]_0^{\pi/2} - \int_0^{\pi/2} e^x \sin x \, dx$$

$$= e^{\pi/2} - \int_0^{\pi/2} e^x \sin x \, dx.$$

Now for

$$\int_0^{\pi/2} e^x \sin x \, dx$$

set

$$v = e^x \quad \Rightarrow \frac{dv}{dx} = e^x$$

$$\frac{du}{dx} = \sin x \Rightarrow u = -\cos x.$$

Then

$$\int_0^{\pi/2} e^x \sin x \, dx = \left[-e^x \cos x \right]_0^{\pi/2} - \int_0^{\pi/2} e^x . (-\cos x) \, dx$$

$$= 1 + \int_0^{\pi/2} e^x \cos x \, dx.$$

Substituting we find

$$\int_0^{\pi/2} e^x \cos x \, dx = e^{\pi/2} - \left[1 + \int_0^{\pi/2} e^x \cos x \, dx \right]$$

$$\Rightarrow 2 \int_0^{\pi/2} e^x \cos x \, dx = e^{\pi/2} - 1$$

$$\Rightarrow \int_0^{\pi/2} e^x \cos x \, dx = \tfrac{1}{2}(e^{\pi/2} - 1).$$

Exercise 12.1

1. Using integration by parts find

   (a) $\int \tan^{-1} x \, dx$ (b) $\int \sin^{-1} x \, dx$ (c) $\int x \cos x \, dx$

   (d) $\int x . \ln x \, dx$ (e) $\int x^2 e^{-x} \, dx$ (f) $\int \dfrac{\ln x}{x} \, dx$

   (g) $\int e^x \sin x \, dx$ (h) $\int_0^1 x e^{2x} \, dx$ (i) $\int_0^{\pi} e^{2x} \sin x \, dx$

   (j) $\int x \tan^{-1} x \, dx$ (k) $\int x \sec^2 x \, dx$ (l) $\int_0^{\pi/2} x^2 \sin x \, dx$

   (m) $\int e^{-x} \cos x \, dx$ (n) $\int_1^e \ln x \, dx$

2. Prove that

$$\int e^{ax} \cos x \, dx = \frac{e^{ax}}{1 + a^2} (a \cos x + \sin x) + c.$$

3. Evaluate

$$\int_0^1 x e^{-3x} \, dx,$$

   giving the answer in its simplest form in terms of e.

   JMB 1984

12.2 Integration of rational functions

A revision of partial fractions may be of value at this stage.
   Most rational functions may be reduced to the sum of particularly simple forms. Methods of dealing with these are given by example.

The form $\dfrac{p}{ax + b}$

Example 12.5

Find $\int \dfrac{3}{4x - 1} \, dx.$

   This should be recognised as a logarithm.

$$\int \frac{3}{4x-1} \, dx = \tfrac{3}{4} \int \frac{4}{4x-1} \, dx$$

$$= \tfrac{3}{4} \ln |4x - 1| + c.$$

Writing c as $\tfrac{3}{4} \ln A$, then

$$\int \frac{3}{4x-1} \, dx = \tfrac{3}{4} \ln (A \,.\, |4x - 1|).$$

The form $\dfrac{px + q}{ax + b}$

This is an improper rational and division is required. The following example shows a quick method for division in such simple situations.

Example 12.6

Find $\displaystyle\int \frac{x+1}{2x-1} \, dx.$

Suppose $x + 1 \equiv A(2x - 1) + B$, then $A = \tfrac{1}{2}$ and so $B = \tfrac{3}{2}$ by inspection.
Thus

$$\int \frac{x+1}{2x-1} \, dx = \int \frac{\tfrac{1}{2}(2x-1) + \tfrac{3}{2}}{2x-1} \, dx$$

$$= \int \tfrac{1}{2} + \frac{\tfrac{3}{2}}{2x-1} \, dx$$

$$= \int \tfrac{1}{2} \, dx + \tfrac{3}{2} \int \frac{1}{2x-1} \, dx$$

$$= \tfrac{1}{2}x + \tfrac{3}{4} \int \frac{2}{2x-1} \, dx$$

$$= \tfrac{1}{2}x + \tfrac{3}{4} \ln |2x - 1| + c$$

The form $\dfrac{p}{(ax + b)^n}$

Example 12.7

Find $\displaystyle\int \frac{2}{(1 - 3x)^2}.$

This may be tackled by substitution. Alternatively,

$$\int \frac{2}{(1-3x)^2}\,dx = 2\int (1-3x)^{-2}\,dx$$

$$= \tfrac{2}{3}(1-3x)^{-1} + c$$

$$= \frac{2}{3(1-3x)} + c$$

which may be checked by differentiating.

The form $\dfrac{p}{ax^2 + bx + c}$

If the denominator factorises the rational should be split into partial fractions. Otherwise one should complete the square of the denominator.

Example 12.8

Find (a) $\displaystyle\int \frac{3x-1}{1-x^2}\,dx,$ (b) $\displaystyle\int \frac{3}{x^2+2x+5}\,dx.$

(a) Supposing

$$\frac{3x-1}{1-x^2} \equiv \frac{A}{1-x} + \frac{B}{1+x} \qquad \text{for } x \neq \pm 1$$

it suffices that

$$3x - 1 \equiv A(1+x) + B(1-x).$$

Setting $x = 1$,

$$2 = 2A \qquad \Rightarrow A = 1$$

Setting $x = -1$,

$$-4 = -2B \qquad \Rightarrow B = 2$$

So

$$\frac{3x-1}{1-x^2} \equiv \frac{1}{1-x} + \frac{2}{1+x}$$

and

$$\int \frac{3x-1}{1-x^2}\,dx = \int \frac{1}{1-x}\,dx + \int \frac{2}{1+x}\,dx$$

$$= -\ln|1-x| + 2\ln|1+x| + c$$

$$= \ln\left|\frac{A(1+x)^2}{1-x}\right| \qquad \text{where } \ln A = c.$$

(b) Since the denominator does not factorise we assume

$$x^2 + 2x + 5 \equiv (x + A)^2 + B$$

It then follows that $A = 1$ and $B = 4$. So

$$\int \frac{3}{x^2 + 2x + 5} \, dx = \int \frac{3}{(x + 1)^2 + 4} \, dx$$

$$= \tfrac{3}{4} \int \frac{1}{1 + \left(\dfrac{x + 1}{2}\right)^2} \, dx$$

Let $u = \dfrac{x + 1}{2}$, so

$$\frac{du}{dx} = \tfrac{1}{2} \Rightarrow \frac{dx}{du} = 2$$

and

$$\int \frac{3}{x^2 + 2x + 5} \, dx = \tfrac{3}{4} \int \frac{1}{1 + u^2} \cdot 2 \, du$$

$$= \tfrac{3}{2} \tan^{-1} u + c$$

$$= \tfrac{3}{2} \tan^{-1} \left(\frac{x + 1}{2}\right) + c.$$

The form $\dfrac{px + q}{ax^2 + bx + c}$

In this case the numerator may be expressed as the sum of a constant and a multiple of the derivative of the denominator.

Example 12.9

Find $\displaystyle\int \frac{x + 1}{x^2 - 2x + 1} \, dx$.

Suppose $x + 1 \equiv A(2x - 2) + B$ then $A = \tfrac{1}{2}$ and $B = 2$.

$$\int \frac{x + 1}{x^2 - 2x + 1} \, dx = \int \frac{\tfrac{1}{2}(2x - 2) + 2}{x^2 - 2x + 1} \, dx$$

$$= \tfrac{1}{2} \int \frac{2x - 2}{x^2 - 2x + 1} \, dx + 2 \int (x - 1)^{-2} \, dx$$

$$\text{since } x^2 - 2x + 1 \equiv (x - 1)^2$$

$$= \tfrac{1}{2} \ln |x^2 - 2x + 1| - 2(x - 1)^{-1} + c.$$

Of course the second integral may reduce to one of the other forms, depending on whether the denominator will factorise or not.

Reminder

> An improper rational should first be divided before attempting to integrate.

Exercise 12.2

1. Find

(a) $\int \dfrac{2}{1-x} \, dx,$

(b) $\int \dfrac{3}{2x-3} \, dx,$

(c) $\int \dfrac{2}{(1-x)^2} \, dx,$

(d) $\int \dfrac{3}{(2x-1)^2} \, dx,$

(e) $\int \dfrac{3}{(1+3x)^3} \, dx,$

(f) $\int \dfrac{4x}{1-x} \, dx,$

(g) $\int \dfrac{2x}{3x-2} \, dx,$

(h) $\int \dfrac{2x+1}{2x-1} \, dx,$

(i) $\int \dfrac{2x+1}{x+2} \, dx,$

(j) $\int \dfrac{1}{(1+x)(1+2x)} \, dx,$

(k) $\int \dfrac{1}{x^2+2x+2} \, dx,$

(l) $\int \dfrac{1}{x(1-x)^2} \, dx,$

(m) $\int \dfrac{1}{x^2(1+x)} \, dx,$

(n) $\int \dfrac{1}{x^2+4x+5} \, dx,$

(o) $\int \dfrac{1}{x^2+2x+10} \, dx,$

(p) $\int \dfrac{3}{1+(2x+1)^2} \, dx,$

(q) $\int \dfrac{1}{4x^2-4x+5} \, dx,$

(r) $\int \dfrac{3}{16+(3x-4)^2} \, dx,$

(s) $\int \dfrac{x+1}{x^2+2x-2} \, dx,$

(t) $\int \dfrac{3x+2}{x^2+2x-3} \, dx,$

(u) $\int \dfrac{4x+7}{x^2+6x+18} \, dx,$

(v) $\int \dfrac{x^2+1}{x^2-1} \, dx,$

(w) $\int \dfrac{x^2+1}{x+1} \, dx$

(x) $\int \dfrac{x^3-1}{x+1} \, dx.$

12.3 General integration

The following should be remembered —

Standard integrals   *page reference*

$$\int ax^n \, dx = \frac{ax^{n+1}}{n+1} + c \qquad n \neq 1$$ 201

$$\int \frac{1}{x} \, dx = \ln|x| + c$$ 344

$$\int e^{ax} \, dx = \frac{e^{ax}}{a} + c$$ 341

$$\int \cos x + \sin x + \sec^2 x \, dx = \sin x - \cos x + \tan x + c \qquad\qquad 240$$

$$\int \frac{1}{\sqrt{1-x^2}} \, dx = \sin^{-1} x + c \qquad\qquad 334$$

$$\int \frac{1}{1+x^2} \, dx = \tan^{-1} x + c \qquad\qquad 334$$

$$\int f'(x) e^{f(x)} \, dx = e^{f(x)} + c \qquad\qquad 341$$

$$\int \frac{f'(x)}{f(x)} \, dx = \ln |f(x)| + c \qquad\qquad 346$$

$$\int f'(x) g'(f(x)) \, dx = g(f(x)) + c$$

[which follows immediately upon differentiating $g(f(x))$]

together with the following general methods

(i) guessing the answer   203
(ii) changing the variable (substitution)   335
(iii) integration by parts   384

These will be found to be adequate for most functions likely to be met. A major exception is $\int e^{-x^2} \, dx$ which cannot be expressed in terms of standard functions.

Exercise 12.3

If a hint is required refer to the page indicated after each question.

1. Evaluate

(a) $\displaystyle\int \frac{dx}{\sqrt{x-4}}$ (335)   (b) $\displaystyle\int_6^8 \frac{dx}{x-4}$ (346)   (c) $\displaystyle\int_6^8 \frac{2x}{x-4} \, dx$ (389).

2. Find $\displaystyle\int_0^{\pi/2} \sin^3 x \, dx$ (241).

3. Use the $t = \tan \frac12 x$ substitution to show that

$$\int_0^\pi \frac{1}{1+\sin x} \, dx = 2 \quad (210)$$

4. Find

$$\int_0^1 (2x^2 + 1)(2x^3 + 3x + 4)^{1/2} \, dx \quad (203)$$

5. Evaluate

(a) $\displaystyle\int_1^e \frac{\ln x}{x^2}\,dx$ (384)   (b) $\displaystyle\int_0^{15} \frac{e^x}{e^x + 1}\,dx$ (335).

6. Show that

$$\int_0^{\pi/2} \frac{\cos\theta - \sin\theta}{\sin\theta + \cos\theta}\,d\theta = 0 \quad (346)$$

7. Show that

$$\int_{\pi/8}^{\pi/4} \sin 3x \sin x\,dx = \frac{3 - \sqrt{2}}{8} \quad (243)$$

8. Evaluate

(a) $\displaystyle\int_0^1 \frac{1}{(x + 1)(2x + 1)}\,dx$ (390)   (b) $\displaystyle\int_0^1 \frac{1}{(x + 1)(x^2 + 1)}\,dx$ (390)

(c) $\displaystyle\int_0^{\pi/2} \frac{\cos x}{(1 + \sin x)^2}\,dx$ (335)   (d) $\displaystyle\int_0^{\pi/2} \frac{\cos x}{4 + \sin x}\,dx$ (346).

9. Evaluate

(a) $\displaystyle\int_0^3 \frac{1}{\sqrt{9 - x^2}}\,dx$ (334)   (b) $\displaystyle\int_0^3 \frac{x}{\sqrt{9 - x^2}}\,dx$ (203)

10. Find

$$\int \frac{x - 2}{(x + 4)(x + 1)}\,dx. \quad (391)$$

CAMB 1982

11. (a) Prove that

$$\int_1^2 \frac{6}{(3 + x)(3 - x)}\,dx = \ln\tfrac{5}{2}. \quad (390)$$

(b) Evaluate, in terms of π,

$$\int_0^2 \frac{x^2}{x^2 + 4}\,dx. \quad (392)$$

JMB 1983

12. By using a substitution, or otherwise, find the exact value of

$$\int_0^{\pi/2} \frac{\cos x}{(4 + \sin x)^2}\,dx. \quad (335)$$

JMB 1984

Exercise 12.4 (Integration)

1. Find a if $\displaystyle\int_{2a}^{3a} (x - a)^2 \, dx = 63$.

W 1983

2. Find $\displaystyle\int \sin x(1 + \cos^2 x) \, dx$.

CAMB 1982

3. Find the exact value of $\displaystyle\int_0^1 x\sqrt{(4 - 3x^2)} \, dx$.

CAMB 1983

4. Using the substitution $x = 2 \sin \theta$, or otherwise, evaluate

$$\int_{-1}^{2} \sqrt{(4 - x^2)} \, dx.$$

W 1983

5. Using the change of variable $\tan \frac{1}{2}x = t$, or otherwise, show that

$$\int_0^{\pi/2} \frac{dx}{2 + \cos x} = \frac{\pi}{3\sqrt{3}}.$$

W 1981

6. Use the substitution $\tan \frac{1}{2}x = t$ or otherwise to find

$$\int \frac{dx}{3 + \cos x}.$$

W 1983

7. The function g is defined by $g(x) = \sqrt{(1 - x^2)}$.
   Find the area of the region bounded by the graph of g, the lines $x = 0$, $x = \frac{1}{2}$ and the x-axis, by using the substitution $x = \sin \theta$, or otherwise, to evaluate the integral.

W 1981

8. Using the substitution $t = \sin x$, evaluate to two decimal places the integral

$$\int_{\pi/6}^{\pi/2} \frac{4 \cos x}{3 + \cos^2 x} \, dx.$$

AEB 1983

9. Using the substitution $x = \sin \theta$, or otherwise, show that

$$\int_0^{1/2} \frac{x^2}{\sqrt{(1 - x^2)}} \, dx = \frac{\pi}{12} - \frac{\sqrt{3}}{8}.$$

W 1982

10. Find the exact value of $\int_{-2}^{1} x\sqrt{(x + 3)} \, dx$.

CAMB 1983

11. Use integration by parts to find $\int (x + 1)e^{2x} \, dx$.

W 1983

12. Using integration by parts, or otherwise, find $\int xe^{-x} \, dx$.

W 1981

13. Use integration by parts to find $\int x \log_e x \, dx$.

W 1982

14. (a) Show that

$$\int x \log_e x \, dx = \frac{x^2}{2} \log_e x - \frac{x^2}{4} + c.$$

Hence, or otherwise, evaluate $\int_1^3 x(\log_e x)^2 \, dx$, giving your answer to three significant figures.

   (b) By means of the substitution $x = \tan \theta$, show that

$$\int_0^1 \frac{dx}{(1 + x^2)^2} = \frac{\pi}{8} + \frac{1}{4}.$$

SU 1982

15. Express $1 - x^4$ as a product of linear and quadratic factors. Hence or otherwise find

$$\int \frac{x + 2}{1 - x^4} \, dx.$$

W 1983

16. Find $\int \dfrac{3x^2 + 1}{x^2(x^2 + 1)} dx$.

W 1982

17. Find $\int \dfrac{dx}{(x + 1)^2(x^2 + 1)}$.

W 1981

18. Evaluate the integrals

   (a) $\displaystyle\int_0^1 \dfrac{2x - 1}{(x - 3)^3} dx$ (by substitution or otherwise),

   (b) $\displaystyle\int_0^{\pi/4} \sin^2 3x \, dx$.

OX 1983

19. Find the following integrals:

   (a) $\displaystyle\int x e^{2x} dx$,   (b) $\displaystyle\int x \sin(x^2) dx$.

CAMB 1983

20. (a) Find $\displaystyle\int \dfrac{2 + \cos x}{\sin^2 x} dx$.

   (b) Using the substitution $x = 2 \sin \theta$, or otherwise, evaluate

$$\int_0^1 \dfrac{x^2}{\sqrt{(4 - x^2)}} dx,$$

   giving two significant figures in your answer.

CAMB 1982

21. Evaluate

   (a) $\displaystyle\int_1^4 \dfrac{dx}{x + \sqrt{x}}$,   (b) $\displaystyle\int_0^{1/2} 4x\sqrt{(1 - x^2)} dx$.

AEB 1982

22. Evaluate

   (a) $\displaystyle\int_1^2 \dfrac{1}{x^2 + 4x} dx$,   (b) $\displaystyle\int_0^{\pi/2} \sin^3 2x \, dx$.

OX 1982

23. Integrate the following with respect to x:

    (a) $x^3(x^4 - 3)^5$ (b) $\tan x$ (c) xe^{x^2}.

SU 1982

24. (a) Evaluate

    (i) $\displaystyle\int_2^3 \frac{dx}{2x^2 - x}$,   (ii) $\displaystyle\int_0^{\pi/4} \sin 2x \cos 3x \, dx$,

    giving your answers to three decimal places.

    (b) Find $\displaystyle\int x^2 e^{3x} \, dx$.

    (c) Using the substitution $x = \dfrac{1}{y}$, or otherwise, find $\displaystyle\int \frac{1}{x\sqrt{x^2 - a^2}} dx$.

O&C 1983

25. (a) Find the indefinite integral $\displaystyle\int (\sin 2x + \cos 2x)^2 \, dx$.

    (b) Evaluate $\displaystyle\int_0^{\pi/2} x \cos x \, dx$ correct to 3 significant figures.

    (c) Use the substitution $u = e^x$ to transform the integral $\displaystyle\int_0^1 \frac{dx}{1 + e^x}$ and obtain its value correct to 3 significant figures.

SU 1980

26. Express the function

$$f(x) = \frac{x + 2}{(x^2 + 1)(2x - 1)}$$

as the sum of partial fractions.
Hence find

$$\int f(x) \, dx.$$

JMB 1982

27. Find

    (a) $\displaystyle\int_0^{\pi/3} \tan 2\theta \, d\theta$,   (b) $\displaystyle\int_0^1 \frac{x^{1/2}}{1 + x} dx$.

28. Differentiate $x \sin^{-1} mx$, where m is a constant.
    Hence, or otherwise, integrate $\sin^{-1} mx$.

JMB 1982

29. Prove that $\dfrac{d}{dx}(\sin 2x - 2x\cos 2x) = 4x\sin 2x$.

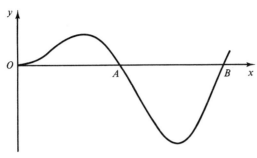

Fig. 12.1

The curve shown in the figure is part of the graph of the function $y = x\sin 2x$. Write down the co-ordinates of the points A and B.

Show that the area of the region bounded by the curve and the line segment AB is three times that of the area of the region bounded by the curve and the line segment OA.

AEB 1983

30. Find the co-ordinates of the turning points of the curve $y = x + 2\cos x$ for values of x in the range $0 \leqslant x \leqslant \pi$. Sketch the curve in this range, showing its salient features.

The region bounded by the curve, the ordinates $x = 0$ and $x = \pi$, and the x-axis is denoted by R. Find the area of R.

Evaluate

$$\int_0^\pi x\cos x\,dx \quad \text{and} \quad \int_0^\pi \cos^2 x\,dx$$

and use these results to evaluate the volume V of the solid of revolution obtained by rotating R through four right angles about the x-axis.

[You may leave your answers in terms of π.]

O&C 1982

Chapter 13 Co-ordinate geometry

13.1 Dividing a line in a given ratio

If a line segment PQ is divided into three equal parts as indicated in Fig. 13.1, then R divides the line in the ratio $1:2$ and S divides the line in the ratio $2:1$.

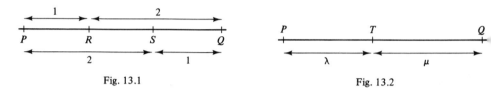

Fig. 13.1   Fig. 13.2

Similarly, if T divides the line segment PQ in the ratio $\lambda:\mu$ (see Fig. 13.2) then

$$PT = \frac{\lambda}{\lambda + \mu}PQ \quad \text{and} \quad TQ = \frac{\mu}{\lambda + \mu}PQ.$$

Example 13.1

Find the co-ordinates of the point R which divides the line joining $P(4, 3)$ to $Q(8, 11)$ in the ratio $3:1$.

By similar triangles S divides PT in the ratio $3:1$ (see Fig. 13.3), so

$$PS = \tfrac{3}{4}PT$$

$$= 3 \quad \text{since } PT \text{ is 4 units.}$$

The x co-ordinate of S, and hence R, is therefore $4 + 3 = 7$. Similarly the y co-ordinate of R is $3 + \tfrac{3}{4}.8 = 9$. The co-ordinates of R are therefore $(7, 9)$.

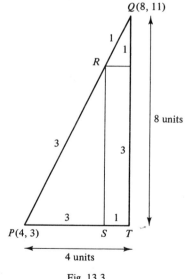

Fig. 13.3

More generally, if $R(x, y)$ divides the line joining $P(x_1, y_1)$ to $Q(x_2, y_2)$ in the ratio $\lambda : \mu$, then

$$x = x_1 + \frac{\lambda}{\lambda + \mu}(x_2 - x_1)$$

$$= \frac{\mu x_1 + \lambda x_2}{\lambda + \mu}$$

and

$$y = \frac{\mu y_1 + \lambda y_2}{\lambda + \mu}.$$

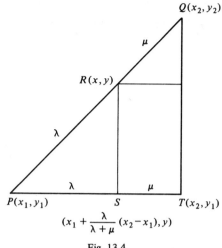

Fig. 13.4

Statement

If $R(x, y)$ divides the line joining $P(x_1, y_1)$ to $Q(x_2, y_2)$ in the ratio $\lambda : \mu$, then

$$x = \frac{\mu x_1 + \lambda x_2}{\lambda + \mu}$$

$$y = \frac{\mu y_1 + \lambda y_2}{\lambda + \mu}.$$

Example 13.2

Find the co-ordinate of the point $R(x, y)$ which divides the line joining $A(-1, 4)$ to $B(1, -2)$ in the ratio $2:3$.

Using the formula when $\lambda = 2$ and $\mu = 3$,

$$x = \frac{3.(-1) + 2.1}{5} = \frac{-1}{5}$$

$$y = \frac{3.4 + 2.(-2)}{5} = \frac{8}{5}$$

giving $(-1/5, 8/5)$ as the co-ordinates of R.

Example 13.3

Find the centroid G of a triangle whose vertices are $A(a_1, a_2)$, $B(b_1, b_2)$ and $C(c_1, c_2)$.

The centroid G divides the median AM in the ratio $2:1$, where M is the mid point of BC. In fact G is the point of intersection of the medians of the triangle.

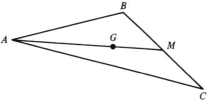

Fig. 13.5

The co-ordinates of M are

$$\left(\frac{b_1 + c_1}{2}, \frac{b_2 + c_2}{2}\right)$$

To find G we use the formula with $\lambda:\mu = 2:1$.
The x co-ordinate of G is

$$x = \frac{1 \cdot a_1 + 2 \cdot \frac{1}{2}(b_1 + c_1)}{1 + 2} = \frac{a_1 + b_1 + c_1}{3}$$

Similarly,

$$y = \frac{a_2 + b_2 + c_2}{3}$$

Statement

> The co-ordinates of the centroid of a triangle with verticles $A(a_1, a_2)$, $B(b_1, b_2)$ and $C(c_1, c_2)$ respectively, are
> $$\left(\frac{a_1 + b_1 + c_1}{3}, \frac{a_2 + b_2 + c_2}{3}\right).$$

13.2   The angle between two given lines

Given two straight lines,

$$y = m_1 x + c_1 \quad \text{and} \quad y = m_2 x + c_2,$$

suppose that the angle between these lines is θ. Suppose further that the lines cut the x axis in angles α and β respectively, as indicated in Fig. 13.6.

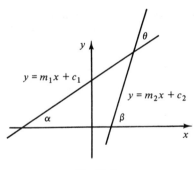

Fig. 13.6

Since $\tan \alpha = m_1$ and $\tan \beta = m_2$ and also $\theta = \beta - \alpha$, it follows that

$$\tan \theta = \frac{\tan \beta - \tan \alpha}{1 + \tan \beta \tan \alpha} \quad \text{①}$$

$$\Rightarrow \tan \theta = \frac{m_2 - m_1}{1 + m_1 m_2}.$$

There are generally two angles between two straight lines, one acute and the other obtuse. *The angle* between two straight lines is normally taken to be the acute angle, and so a negative sign for ① is ignored. If the lines are perpendicular then the formula breaks down since the r.h.s. is undefined. But this indicates that $\tan \theta$ is undefined, giving $\theta = \pi/2$ as expected.

Statement

> The *acute* angle θ between two straight lines with gradients m_1 and m_2 respectively is given by
>
> $$\tan \theta = \left| \frac{m_1 - m_2}{1 + m_1 m_2} \right|$$

Example 13.4

Find the angle between the lines

$$y = 3x + 4 \quad \text{and} \quad 2y - x = 1.$$

Taking m_1 as 3 and m_2 as $\frac{1}{2}$, then θ, the angle between the lines, is given by

$$\tan \theta = \left| \frac{3 - \frac{1}{2}}{1 + 3 \cdot \frac{1}{2}} \right|$$

$$= 1$$

$$\Rightarrow \quad \theta = \frac{\pi}{4}.$$

13.3   The distance of a point from a line

Statement

> The distance d of the point $P(h, k)$ from the line $ax + by + c = 0$ is given by
> $$d = \frac{|ah + bk + c|}{\sqrt{a^2 + b^2}}.$$

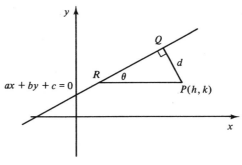

Fig. 13.7

Notice that if the line cuts the x axis at an angle θ then $\tan \theta = -a/b$ (the gradient of the line). If follows that

$$\sin \theta = \frac{|a|}{\sqrt{a^2 + b^2}}.$$

However, $\sin \theta = d/PR$, where d is the perpendicular distance of P from the line. To find the x coordinates of R we substitute $y = k$ into $ax + by + c = 0$ and find x given as

$$\frac{-bk - c}{a}.$$

So the *distance PR* is given by

$$PR = \left| h - \frac{-(bk + c)}{a} \right|$$

$$\Rightarrow PR = \left| \frac{ah + bk + c}{a} \right|$$

The modulus sign is necessary here since we are dealing with a distance, which is non negative.

Since $d = PR \cdot \sin \theta$, substituting

$$d = \left| \frac{ah + bk + c}{a} \right| \cdot \frac{|a|}{\sqrt{a^2 + b^2}}$$

$$d = \frac{|ah + bk + c|}{\sqrt{a^2 + b^2}} \qquad \text{as required.}$$

Example 13.5

Find the distance of the point $P(-2, -1)$ from the line $3x + 4y + 5 = 0$.

Using the formula

$$d = \frac{|3 . -2 + 4 . -1 + 5|}{\sqrt{3^2 + 4^2}}$$

$$= \frac{|-5|}{5}$$

$$= 1 \text{ unit.}$$

It is sometimes useful to note that the expression within the modulus will have the same sign (positive or negative) for points on the same side of the line.

Example 13.6

Establish which of the points $P(\frac{1}{2}, -2)$ and $Q(\frac{1}{2}, -\frac{1}{2})$ is to the left of the line $x + y + 1 = 0$.

For the above line the expression $ah + bk + c$ becomes $h + k + 1$ since $a = b = c = 1$.

For $P(\frac{1}{2}, -2)$ the value of the expression is $-\frac{1}{2}$, since $h = \frac{1}{2}$ and $k = -2$. For $Q(\frac{1}{2}, -\frac{1}{2})$ the value is 1, since $h = \frac{1}{2}$ and $k = -\frac{1}{2}$. As the signs differ, the points are on opposite sides of the line.

Now the origin $O(0,0)$ gives the value 1 and so is on the same side as Q (having the same sign.) But the line crosses the x axis at $(-1, 0)$ so the origin is to the right of the line.

It follows that P is the point to the left of the line.

Example 13.7

Given three points, $A(3, -1)$, $B(1,0)$ and $C(0, -2)$, show that

(a) $AB = BC$,   (b) $\angle ABC = 90°$.

Find the co-ordinates of D so that $ABCD$ forms a square.

(a) $AB^2 = (3 - 1)^2 + (-1 - 0)^2 = 5$

$BC^2 = (1 - 0)^2 + (0 - -2)^2 = 5$

It follows immediately that $AB = BC$

(b) To show $\angle ABC = 90°$ it suffices to show that the line AB is perpendicular to the line BC. Now

$$\text{grad } AB = \frac{-1 - 0}{3 - 1} = -\tfrac{1}{2}$$

and

$$\text{grad } BC = \frac{0 - -2}{1 - 0} = 2.$$

Since the product is -1 it follows that the lines are perpendicular.

To find the co-ordinates of D, we find the equations of two lines on which D must lie, and solve simultaneously.

D will lie on the line which passes through $A(3, -1)$ which is parallel to BC, and so have gradient 2. The equation of this line is

$$y - -1 = 2(x - 3)$$
$$\Rightarrow y - 2x + 7 = 0.$$

Similarly D will lie on the line passing through $C(0, -2)$ with gradient $-\tfrac{1}{2}$. The equation of this line is

$$y - -2 = -\tfrac{1}{2}(x - 0)$$
$$\Rightarrow 2y + x + 4 = 0.$$

Solving simultaneously

$$y - 2x + 7 = 0 \qquad ①$$
$$2y + x + 4 = 0 \qquad ②$$

$① + 2②$

$$5y + 15 = 0$$
$$\Rightarrow \qquad y = -3$$

$2① - ②$

$$-5x + 10 = 0$$
$$\Rightarrow \qquad x = 2$$

So the co-ordinates of D are $(2, -3)$.

Exercise 13.1

1. Find the co-ordinates of the two points which trisect the line joining the points A and B where the co-ordinates of A and B are given respectively as:

    (a) $(-2, 6), (4, -6)$ (b) $(0, 5), (-3, 8)$
    (c) $(1, -2), (4, 3)$ (d) $(a, b), (c, d)$.

2. Find the co-ordinates of the point which divides AB in the given ratio.

   (a) $A(5, 7)$, $B(0, 2)$ in the ratio $2:3$
   (b) $A(0, 4)$, $B(8, 0)$ in the ratio $3:1$
   (c) $A(a, -b)$, $B(6a, 4b)$ in the ratio $2:3$

3. Find the centroid of the triangle $\triangle ABC$ with vertices

   (a) $(4, 1)$, $(0, 1)$, $(-2, 4)$,   (b) $(2, 5)$, $(4, -3)$, $(5, 3)$.

4. Find the acute angle between the following pairs of lines.

   (a) $x + y + 3 = 0$, $x - y + 2 = 0$
   (b) $3x + y - 2 = 0$, $2x + 2y + 5 = 0$
   (c) $2x - 3y = 0$, $x + 2y = 0$
   (d) $y = 3x - 1$, $y = \frac{1}{4}x + 2$

5. Find the equation of the line with positive gradient which is inclined at an angle of $45°$ to the line $3y - x + 1 = 0$ passing through the point $(2, 0)$.

6. Find the distance of the origin from the line

   (a) $-4x + 3y + 4 = 0$,   (b) $4x - 3 = 3y$,   (c) $y = x + 1$.

7. Find the distance of the point $P(2, 3)$ from the line

   (a) $5x - 12y + 1 = 0$,   (b) $3x = 5y + 3$.

8. Show that the points $P(1, 4)$, $Q(-2, -2)$ are on the same side of the line $2x - y + 3 = 0$.

9. Find a relationship connecting m and c if the distance of the origin from the line $y = mx + c$ is 1 unit.

10. M is the mid-point of BC in the triangle $\triangle ABC$ where $A(8, 12)$, $B(11, 4)$ and $C(-5, -4)$ are the co-ordinates of the vertices. Find the distance AM.

11. R divides the line joining $A(0, -2)$ to $B(8, -6)$ in the ratio $1:3$. Find the co-ordinates of R and the distance of R from the line through the origin parallel to AB.

12. The point $R(X, Y)$ is equidistant from the line $x + y = 0$ and the point $(2, 2)$. Find an equation relating X and Y.

13. The point $P(X, Y)$ is equidistant from the lines $y = x$ and $y = 2x$. Find the equation of the line on which P must lie.

14. The point (x_2, y_2) is the image of the point (x_1, y_1) in the line $ax + by + c = 0$. Show that

$$a(x_1 + x_2) + b(y_1 + y_2) + 2c = 0.$$

15. Given the three points $A(4, 0)$, $B(0, 2)$ and $C(-2, -2)$, show that

    (a) $AB = BC$ (b) AB is perpendicular to BC.

    A square $ABCD$ is formed. Calculate the co-ordinates of D.
    [A diagram will be found helpful, but a solution using measurements from an accurate drawing is not acceptable.]

    CAMB 1982

16. The points $C(6, 6)$, $O(0, 0)$ and $A(4, 3)$ are three of the vertices of a parallelogram $COAB$.

    (a) Calculate the co-ordinates of the point B.
    (b) Show that the tangent of the acute angle AOC is $\frac{1}{7}$ and hence write down the exact value of the sine of the angle AOC.
    (c) Calculate the area of the parallelogram $COAB$.

    AEB 1982

17. Prove, by calculation, that the triangle formed by the lines $x - 2y + 1 = 0$, $9x + 2y - 11 = 0$ and $7x + 6y - 53 = 0$ is isosceles.
    Calculate, to the nearest degree, the smallest angle of the triangle.

    AEB 1982

The locus of a point

As a point moves over a plane surface it describes a curve or path called the locus of the point. If the point is constrained to move along a particular curve, the equation of the curve can often be found by making use of the constraints.

13.4   The circle

If a point P is free to move at a fixed distance r from the point (a, b) then P will describe part of a circle (see Fig. 13.8). The equation of the circle can be found using this property.

For suppose $P(x, y)$ is a point on the locus, then using Pythagoras' theorem

$$r^2 = (x - a)^2 + (y - b)^2.$$

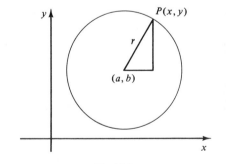

Fig. 13.8

Statement

> The graph of the equation
> $$(x - a)^2 + (y - b)^2 = r^2$$
> is a circle with centre (a, b) and radius r.

The equation of a circle with centre the origin has the form

$$x^2 + y^2 = r^2$$

This was established on page 329 using the same technique.

Example 13.8

Show that the graph of

$$x^2 + y^2 + 2(4y - 3x) + 9 = 0$$

is a circle and find its centre and radius.

Rewriting

$$x^2 - 6x + y^2 + 8y + 9 = 0$$
$$\Rightarrow (x - 3)^2 - 9 + (y + 4)^2 - 16 + 9 = 0 \qquad \text{(completing the square)}$$
$$\Rightarrow \qquad (x - 3)^2 + (y + 4)^2 = 16$$
$$\Rightarrow \qquad (x - 3)^2 + (y - -4)^2 = 16,$$

which is the equation of a circle with radius 4 and centre the point with co-ordinates $(3, -4)$.

Example 13.9

Find the equation of the tangent to the circle

$$x^2 + y^2 + 2y = 1$$

at the point $(1, -2)$. (The equation represents the circle with radius $\sqrt{2}$ and centre $(0, -1)$).

Differentiating

$$2x + 2y\frac{dy}{dx} + 2\frac{dy}{dx} = 0.$$

Substituting for $x = 1$ and $y = -2$

$$2 - 4\frac{dy}{dx} + 2\frac{dy}{dx} = 0$$

$$\Rightarrow \frac{dy}{dx} = 1$$

The equation of the tangent is

$$y + 2 = 1(x - 1)$$

$$\Rightarrow \qquad y = x - 3$$

[An alternation approach would be to note that the radius to $(1, -2)$ is perpendicular to the required tangent.]

Example 13.10

Show that the locus of the point P with co-ordinates

$$(1 + 2\cos\theta, 2 + 2\sin\theta)$$

is a circle and find its radius and centre.

Letting the point P have co-ordinates (x, y) we have

$$x = 1 + 2\cos\theta \quad \text{and} \quad y = 2 + 2\sin\theta.$$

If we eliminate θ we obtain the locus of P in terms of x and y rather than the parameter θ. Now

$$x - 1 = 2\cos\theta \qquad \text{and} \qquad y - 2 = 2\sin\theta$$

$$\Rightarrow (x - 1)^2 = 4\cos^2\theta \qquad \text{and} \quad (y - 2)^2 = 4\sin^2\theta$$

$$\Rightarrow (x - 1)^2 + (y - 2)^2 = 4\cos^2\theta + 4\sin^2\theta$$

$$\Rightarrow (x - 1)^2 + (y - 2)^2 = 4$$

Thus the locus of P is a circle with centre $(1, 2)$ and radius 2.

Exercise 13.2

1. Find the co-ordinates of the centre and radius of the circle given by

   (a) $(x - 2)^2 + (y - 5)^2 = 16$,
   (b) $(x + 1)^2 + (y - 2)^2 = 9$,
   (c) $(2x + 1)^2 + (2y - 3)^2 = 25$,
   (d) $x^2 + y^2 + 2(x + 2y) = 0$,
   (e) $x^2 + y^2 + 2fx + 2gy = c$.

   In the last example, state the condition which must hold for the circle to exist.

2. Find the gradient to the circle

   $$(x - 2)^2 + (y - 3)^2 = 4$$

   at the points (a) $A(2, 1)$,   (b) $B(2, 5)$,   (c) $C(0, 3)$.
   Obtain the co-ordinates of the points where the gradient is -1.

3. Obtain the equations of the tangent to the circle

$$(x - 1)^2 + (y + 2)^2 = 9$$

at the points where the circle crosses the x axis.

4. Find the centre and radius of the circle given by $x^2 + y^2 + 6x - 4y - 3 = 0$.

5. Obtain the equation of the circle with the points $A(1, 3)$ and $B(3, -1)$ at opposite ends of a diameter.

6. Use Pythagoras' theorem to find the length of a tangent from the origin to the circle

$$(x - 3)^2 + (y - 4)^2 = 4.$$

Hint: The radius to the point of contact of the tangent cuts the tangent in a right-angle.

7. Find the equation of the circle passing through the points $A(3, 2)$, $B(-1, 0)$ and $C(5, -2)$.

8. Find the equations of the tangent to the circle $x^2 + y^2 = 9$ with gradient 2.

9. Find the equation of the circle with origin as centre for which the line $2x + y - 2 = 0$ is a tangent. [*Hint*: The radius is equal to the distance of the centre from the tangent.]

10. Show that the point $P(2 + 3\cos\theta, 3(1 + \sin\theta))$ lies on a circle, and find the centre and radius of this circle. Find also the tangent to the circle at the point where $\theta = \pi/3$.

11. Show that the line $(x - a)\cos\theta + y\sin\theta = p$ is a tangent to the circle $(x - a)^2 + y^2 = p^2$. [*Hint*: Find the distance from the centre of the circle to the line.]

12. The circumcircle of a triangle is a circle which passes through the vertices of the triangle. Show that the circle

$$(x - 2)^2 + (y + 1)^2 = 5$$

is the circumcircle of the triangle $\triangle ABC$, where the co-ordinates of A, B and C are $(3, 1)$, $(1, 1)$ and $(0, 0)$, respectively.

Find the circumcircle of the triangle whose vertices are located at $(2, 3)$, $(3, 5)$ and $(5, 3)$.

13. The point R moves so that the ratio $PR : QR = 3 : 2$, where $P(1, -2)$ and $Q(6, 8)$ are the co-ordinates of P and Q, respectively.

Show that the locus of R is a circle.

14. Find the radius and the co-ordinates of the centre of the circle S given by the equation

$$x^2 + y^2 + 10x - 6y - 2 = 0.$$

A chord AB of S is a tangent to the circle with the same centre but with radius equal to one half that of S. The tangents to S at A and B intersect at C. Prove that

(a) the triangle ABC is equilateral,
(b) the point C is on the circle whose equation is

$$x^2 + y^2 + 10x - 6y - 110 = 0.$$

JMB 1982

15. The points A and B have co-ordinates $(2a, 0)$ and $(-a, 0)$ respectively. A point P moves in the x-y plane in such a way that $AP = 2PB$. Prove that the locus of P is a circle and find its centre and radius.

CAMB 1982

16. Find the centre and radius of the circle $x^2 + y^2 - 4x - 6y - 12 = 0$.
    Find the points of intersection of the line $y = 2x + 4$ and the given circle and prove that the length of the chord cut off is $4\sqrt{5}$.
    Show that the circle which has the same centre as the given circle and which touches the given line passes through the point $(0, 2)$.
    What is the equation of the tangent to this second circle at $(0, 2)$?

SU 1981

17. Prove that the line $y = x$ does not meet the circle

$$(x - 2)^2 + (y - 10)^2 = 18.$$

Calculate the coordinates of the points N and F on this circle which are respectively nearest to and furthest from the line $y = x$.

AEB 1982

18. Prove that the circle with equation $x^2 + y^2 - 2ax - 2by + a^2 = 0$ touches the x axis.
    Find the equations of the two circles which pass through the points $(2, 1)$ and $(3, 2)$ and touch the x-axis.

AEB 1983

19. (a) The points A and B have co-ordinates $(\frac{18}{5}, \frac{21}{5})$ and $(\frac{2}{5}, \frac{9}{5})$ respectively. Show that the circle having AB as diameter has equation

$$x^2 + y^2 - 4x - 6y + 9 = 0.$$

Sketch the circle.
    Show also that the distance of the centre C of the circle from the line $3x + 2y - 4 = 0$ is greater than the radius of the circle.

(b) Given that the line $y = mx + c$ is a tangent to the curve $4x^2 + 3y^2 = 12$, show that $c^2 = 3m^2 + 4$.

Given that a tangent to the curve passes through the point $(1, 2)$ show that $c = 2$ or 4.

W 1981

20. The point A has co-ordinates $(3, 9)$ and the point B has co-ordinates $(10, 8)$. Find the equation of the perpendicular bisector of AB.

A circle S has the following properties:

(i) it passes through A and B,
(ii) it has the line $4x + 3y = 64$ as tangent at the point B.

State why the centre of S must be on the line through B with gradient $\frac{3}{4}$.

Calculate the coordinates of the centre of S. Show that the equation of S is

$$x^2 + y^2 - 12x - 10y + 36 = 0.$$

Verify that the x-axis is a tangent to S.

CAMB 1982

21. Verify that the circle with equation

$$x^2 + y^2 - 2rx - 2ry + r^2 = 0$$

touches both the co-ordinate axes.

Find the radii of the two circles which pass through the point $(16, 2)$ and touch both the co-ordinate axes.

CAMB 1983

22. The points O, A, B, C have co-ordinates $(0, 0)$, $(1, 0)$, $(1, 1)$, $(0, 1)$, respectively, and P is the point (x, y). Write down, in terms of x and y, an expression for d^2, where

$$d^2 = PO^2 + PA^2 + PB^2 + PC^2,$$

and show that

$$d^2 = 4\left[\left(x - \frac{1}{2}\right)^2 + \left(y - \frac{1}{2}\right)^2 + \frac{1}{2}\right].$$

(a) Deduce that no point P exists for which $d^2 < 2$.
(b) Given that $d^2 = 2$, find the coordinates of P.
(c) Given that $d^2 = 6$, show that the possible positions of P lie on a circle and give the centre and radius of this circle.
(d) Given that P does not lie outside the square $OABC$, find the greatest value of d.

JMB 1983

23. Find the area of the largest square contained within the circle

$$x^2 + y^2 - 2x + 4y + 1 = 0.$$

24. Two coplanar circles, each of radius 5 cm, have their centres 6 cm apart. Calculate the area of the region common to the interiors of both the circles, giving your answer in cm² correct to two significant figures.

CAMB 1982

13.5   The parabola

Suppose a point P is constrained to move so that the distance from P to a fixed point is equal to the distance from P to a fixed line (see Fig. 13.9).

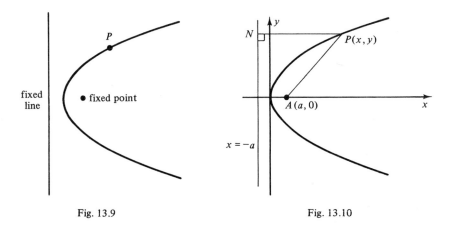

Fig. 13.9   Fig. 13.10

The curve described by P is called a parabola. Its equation depends on the choice of the fixed line and point.

If axes are chosen so that the point has co-ordinate $(a, 0)$ and the line has equation $x = -a$ then we can obtain the equation of this parabola, which has a particularly simple form.

Suppose that $P(x, y)$ is a point on the locus, N is the foot of the perpendicular from P to the line and A is the fixed point $A(a, 0)$. There we require

$$PA = PN$$

as equivalently and more conveniently

$$PA^2 = PN^2$$

$$\Rightarrow \qquad (x - a)^2 + y^2 = (a + x)^2$$

$$\Rightarrow x^2 - 2ax + a^2 + y^2 = a^2 + 2ax + x^2$$

$$\Rightarrow \qquad\qquad\qquad y^2 = 4ax.$$

Statement

> The general equation of a parabola is $y^2 = 4ax$

Some terminology associated with the parabola is indicated in Fig. 13.11.

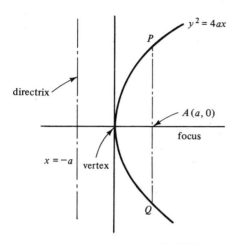

The distance PQ is called the latus rectum. $PQ = 4a$ for the curve with equation $y^2 = 4ax$.

A chord joins two points on the curve.

A focal chord is a chord which passes through the the focus.

Fig. 3.11

Example 13.11

Show that $y^2 = 2(x + 2y)$ is the equation of a parabola and find the co-ordinates of the focus.

$$y^2 = 2(x + 2y)$$
$$\Rightarrow y^2 - 4y = 2x$$
$$\Rightarrow (y - 2)^2 = 2x + 4$$
$$\Rightarrow (y - 2)^2 = 2(x + 2)$$

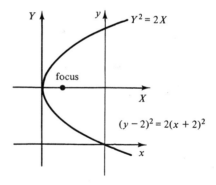

Fig. 13.12

Setting $Y = y - 2$ and $X = x + 2$ we have $Y^2 = 2X$.

This agrees with the general form with $a = \frac{1}{2}$. Relative to the X–Y axes the focus is at $(\frac{1}{2}, 0)$. That is

$$X = \tfrac{1}{2}, \qquad Y = 0$$
$$\Rightarrow x + 2 = \tfrac{1}{2}, \quad y - 2 = 0$$
$$\Rightarrow \qquad x = -\tfrac{3}{2}, \qquad y = 2$$

So the co-ordinates of the focus are $(-\frac{3}{2}, 2)$.

Exercise 13.3

1. Sketch the parabolas represented by each of the following.

    (a) $y^2 + 3x = 0$ (b) $x + 4y^2 = 0$ (c) $4y + x^2 = 0$ (d) $x^2 = 8y$

2. Find the co-ordinates of the vertex and focus of each of the parabolas

    (a) $y^2 = 4(x - 3)$ (b) $(y + 2)^2 = 2(x - 1)$
    (c) $y^2 - 2y = 8x + 7$ (d) $y^2 = 6y + x$

3. Obtain the equation of the parabola with
    (a) focus at $(3, 0)$ and directrix $x = -3$,
    (b) focus $(4, 0)$ directrix $x = 2$,
    (c) focus $(0, 2)$ directrix $y = 0$.

4. By first drawing a sketch, find the equation of the locus of the point which is equidistant from the points $A(1, 0)$, $B(3, 0)$.

5. Find the locus of the point equidistant from the points $A(3, 1)$, $B(5, 2)$.

6. A point $P(x, y)$ moves so that it is equidistant from the line $3x + 4y + 1 = 0$ and the point $A(3, 4)$. Find the equation of its locus.

7. A point $P(x, y)$ moves so that the distance from P to the line $x + y = 0$ is 2 units. Show by sketching that P lies on one of two lines, and find the equations of these lines.

8. The co-ordinates of the point P are $(2t^2, 4t)$ where $t \in \mathbb{R}$. Obtain the Cartesian equation of the locus of P.
    [*Hint*: let $x = 2t^2$, $y = 4t$ and eliminate t.]

9. Find the locus of the point

    (a) $P(t^2, 2t)$,   (b) $Q(at^2, 2at)$.

10. Find the locus of the point $P(2 + \sin \theta, 2 - \cos \theta)$.

11. Find the locus of the point $P(a \cos \theta, b \sin \theta)$.

12. The points P and Q move on the x and y axes respectively so that $OP = 2OQ$ where O is the origin.
    Find the locus of the mid point of PQ.

13. A point P lies on the line $x + y = 1$. Show that the co-ordinate of P will be of the form $(a, 1 - a)$ for some $a \in \mathbb{R}$. If P moves along the line, find the locus of the mid point of PQ where the co-ordinates of Q are $(2, 4)$.

14. Find the equation of the tangent to the curve $y^2 = 4ax$ at the point $P(at^2, 2at)$.

   The line through the origin O parallel to this tangent meets the curve again at Q. Find the co-ordinates of R, the midpoint of OQ. Show that the line through P parallel to the x axis passes through R.

W 1983

15. Find the equation of the normal to the parabola $y^2 = 4ax$ at the point $(at^2, 2at)$. The straight line $4x - 9y + 8a = 0$ meets the parabola at the points P and Q; the normals to the parabola at the points P and Q meet at R. Find the coordinates of R, and verify that it lies on the parabola.

OX 1983

13.6 The conic sections

The conic sections are so called because they may all be obtained from a cone by cutting the cone along a plane. The cone is imagined to extend indefinitely in both directions. The angle at which the intersecting plane cuts the cone determines the shape of the section.

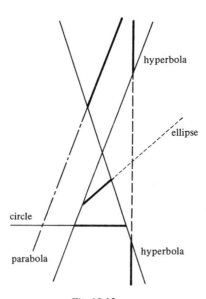

Fig. 13.13

General equations and graphs of the conic sections are shown in Fig. 13.14.

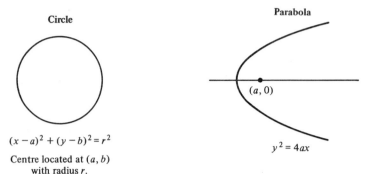

Circle

$(x - a)^2 + (y - b)^2 = r^2$

Centre located at (a, b)
with radius r.

Parabola

$(a, 0)$

$y^2 = 4ax$

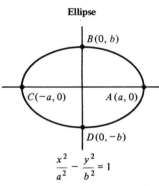

Ellipse

$B(0, b)$

$C(-a, 0)$ $A(a, 0)$

$D(0, -b)$

$$\frac{x^2}{a^2} - \frac{y^2}{b^2} = 1$$

AC is the major axis, of length $2a$.
BD is the minor axis of length $2b$.

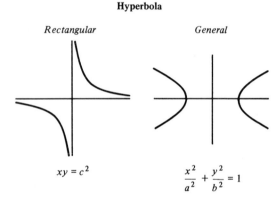

Hyperbola

Rectangular

$xy = c^2$

General

$$\frac{x^2}{a^2} + \frac{y^2}{b^2} = 1$$

Fig. 13.14

Example 13.12

Show that the equation

$$x^2 - 4x + 4y^2 - 8y + 4 = 0$$

represents the equation of an ellipse and sketch its graph.

One immediately recognises that the equation does not represent a circle as the coefficient of x^2 is not equal to the coefficient of y^2. However, the equation is equivalent to

$$(x - 2)^2 - 4 + 4(y - 1)^2 - 4 + 4 = 0 \qquad \text{(completing the square)}$$

$$\Leftrightarrow \frac{(x - 2)^2}{4} + (y - 1)^2 = 1$$

$$\Leftrightarrow \frac{(x - 2)^2}{4} + \frac{(y - 1)^2}{1} = 1.$$

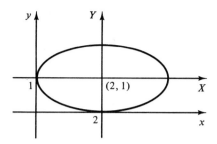

Fig. 13.15

Now setting $X = x - 2$ and $Y = y - 1$ we obtain

$$\frac{X^2}{4} + \frac{Y^2}{1} = 1,$$

an ellipse in standard form relative to the X–Y axes with origin located at $(2, 1)$ (see Fig. 13.15).

The sketch was obtained by noting that the major axis was of length 4 ($= 2a$ where $a = 2$) and the minor axis of length 2 ($= 2b$ where $b = 1$).

Exercise 13.4

1. Show that the equations

   (a) $x^2 + 2y^2 = 4$
   (b) $x^2 + 3y^2 + 2(x - 6y) + 4 = 0$

   each represent an ellipse. In each case state the length of the major and minor axis, and hence sketch the curves.

2. Show that the locus of the point $P(3t, 3/t)$ is a rectangular hyperbola. Find the equation of the tangent at the point where

   (a) $t = 1$,   (b) $t = -2$.

   Sketch the locus and the tangents on the same axes.

3. Show that the point P with co-ordinates $(a \cos \theta, b \sin \theta)$ lies on the ellipse

   $$\frac{x^2}{a^2} + \frac{y^2}{b^2} = 1.$$

4. Show that $y^2 + 8x - 2y - 15 = 0$ represents the equation of a parabola.

5. The points $P(8, 3)$ and $Q(6, 4)$ lie on an ellipse whose equation is

   $$\frac{x^2}{a^2} + \frac{y^2}{b^2} = 1.$$

Find a and b.

The ellipse intersects the positive x-axis at A, and O is the origin. Prove that AP is perpendicular to OQ.

JMB 1983

6. (a) A is the point $(2, 1)$ and C is the point $(5, 2)$. Show that the line bisecting AC at right angles is given by $y = -3x + 12$. This line meets the y-axis at B. Find the area of the triangle ABC. If D is the point such that $ABCD$ is a parallelogram, find the co-ordinates of D and the angles of the parallelogram.

   (b) O is the origin and A and B are the points $(a, 0)$ and $(-a, 0)$ respectively. The point P moves such that $PA \cdot PB = OP^2$. Show that the coordinates (x, y) of P satisfy the equation $x^2 - y^2 = a^2/2$. Show also that the slope of this curve at the point $(a, a/\sqrt{2})$ is $\sqrt{2}$.

W 1983

7. (a) Show that the circle having the line joining the points $(1, 1)$ and $(2, 4)$ as diameter is given by the equation

$$x^2 + y^2 - 3x - 5y + 6 = 0.$$

   Find the equation satisfied by m if the line $y = mx$ is a tangent to this circle. If $y = m_1 x$, and $y = m_2 x$ are the tangents, find values for $(m_1 + m_2)$ and $m_1 m_2$.

   (b) Show that the equation to the tangent to the curve $y = x^3$ at the point (c, c^3) is

$$y = 3c^2 x - 2c^3.$$

   A tangent to the curve intercepts the Ox and Oy axes at P and Q respectively. Given that the triangle OPQ has area $\frac{32}{3}$, find the possible forms of the equations to the tangent.

W 1982

8. (a) The points $(0, 1)$, $(2, 6)$, $(6, 0)$ are denoted by A, B, and C respectively. Show that the equation of AC is

$$6y = -x + 6$$

   and that the line through B perpendicular to AC is given by

$$y = 6x - 6.$$

   This second line cuts AC in E. Given that F is the point $(\frac{36}{37}, \frac{15}{37})$ show that FE is parallel to AB.

   BE produced meets the x-axis at D. Show that the area of triangle ADC is $2\frac{1}{2}$ square units.

   (b) Show that the equation to the tangent of the curve $xy = a^2$ at the point $T(at, a/t)$ is given by

$$t^2 y = -x + 2at.$$

Tangents at point T and the point $S(as, a/s)$ intersect at the point R. Given that R lies on the line $y = x$ and $s + t \neq 0$, show that $st = 1$.

W 1982

9. Find the values of m such that $y = mx$ is a tangent from the origin $O(0, 0)$ to the circle whose equation is $(x - 3)^2 + (y - 4)^2 = 1$.
   Find the cosine of the acute angle between these tangents.

AEB 1983

13.7 Tangents and normals to the conic sections

Example 13.13
Find the equation of the tangent to the parabola $y^2 = 2x$ at the points

(a) $P(18, -6)$,   (b) $Q(2t^2, 2t)$.

(a) $y^2 = 2x$

$$\Rightarrow 2y\frac{dy}{dx} = 2$$

$$\Rightarrow \frac{dy}{dx} = -\frac{1}{6} \quad \text{at } P(18, -6).$$

So the equation of the tangent at P is

$$y - -6 = -\frac{1}{6}(x - 18)$$

$$\Rightarrow 6y + x + 18 = 0 \qquad \text{①}$$

(b) $y^2 = 2x$

$$\Rightarrow 2y\frac{dy}{dx} = 2$$

$$\Rightarrow \frac{dy}{dx} = \frac{1}{2t} \quad \text{at } Q(2t^2, 2t)$$

So the equation of the tangent is

$$y - 2t = \frac{1}{2t}(x - 2t^2)$$

$$\Rightarrow 2ty - x - 2t^2 = 0 \qquad \text{②}$$

Notice that Q is at P if $t = -3$ and it can be checked that equation ② is identical with ① when $t = -3$ is substituted.

Equation ② can be thought of as the general equation of the tangent to the parabola $y^2 = 2x$. For each choice of t, equation ② will be the equation of the tangent at the point Q.

Example 13.14

Show that the equation of the normal at the point $P(a\cos\theta, b\sin\theta)$ on the ellipse

$$\frac{x^2}{a^2} + \frac{y^2}{b^2} = 1$$

is

$$xa\sin\theta - yb\cos\theta = (a^2 - b^2)\sin\theta\cos\theta.$$

The normal cuts the x axis at Q and the y axis at R. Show that, as P varies, the locus of the mid point of QR is also an ellipse.

To find the gradient of the tangent we may either: observe that

$$\frac{x^2}{a^2} + \frac{y^2}{b^2} = 1$$

$$\Rightarrow \frac{2x}{a^2} + \frac{2y}{b^2}\frac{dy}{dx} = 0 \qquad \text{(differentiating implicitly)}$$

$$\Rightarrow \frac{dy}{dx} = -\frac{2x}{a^2}\bigg/\frac{2y}{b^2} \qquad y \neq 0$$

$$= -\frac{xb^2}{ya^2} \qquad y \neq 0$$

$$= \frac{-b^2\, a\cos\theta}{a^2\, b\sin\theta} \qquad \text{at } P \qquad \theta \neq \frac{\pi}{2}$$

$$= -\frac{b}{a}\cot\theta \qquad\qquad\qquad \theta \neq \frac{\pi}{2}$$

or note that

$$x = a\cos\theta \qquad \text{and} \qquad y = b\sin\theta$$

$$\Rightarrow \frac{dx}{d\theta} = -a\sin\theta \quad \text{and} \quad \frac{dy}{d\theta} = b\cos\theta$$

$$\Rightarrow \frac{dy}{dx} = \frac{b\cos\theta}{-a\sin\theta} = \frac{-b}{a}\cot\theta \qquad \text{by the chain rule.}$$

In either case the gradient of the normal at P is $\dfrac{a}{b}\tan\theta$. So the equation of the normal is

$$y - b\sin\theta = \frac{a}{b}\tan\theta\,(x - a\cos\theta)$$

$$\Rightarrow by\cos\theta - b^2\sin\theta\cos\theta = a\sin\theta\, x - a^2\cos\theta\sin\theta$$

$$\Rightarrow ax\sin\theta - by\cos\theta = (a^2 - b^2)\sin\theta\cos\theta \qquad \text{as required}$$

The co-ordinates of Q and R may be found by setting $y = 0$ and $x = 0$ respectively, in the equation of the normal. Giving the co-ordinates as

$$Q\left(\frac{(a^2 - b^2)}{a}\cos\theta, 0\right) \quad \text{and} \quad R\left(0, -\frac{(a^2 - b^2)}{b}\sin\theta\right).$$

The mid point $M(x, y)$ is such that

$$x = \frac{(a^2 - b^2)\cos\theta + 0}{2a} \quad \text{and} \quad y = \frac{0 - (a^2 - b^2)\sin\theta}{2b}.$$

We need to eliminate θ to find the Cartesian equation of the locus of M. Now

$$\cos\theta = \frac{2ax}{a^2 - b^2}, \qquad \sin\theta = \frac{-2by}{a^2 - b^2}.$$

Squaring and adding

$$\frac{4a^2x^2}{(a^2 - b^2)^2} + \frac{4b^2y^2}{(a^2 - b^2)^2} = 1$$

$$\Rightarrow \frac{x^2}{[(a^2 - b^2)/2a]^2} + \frac{y^2}{[(a^2 - b^2)/2b]^2} = 1,$$

the equation of an ellipse.

Exercise 13.5

1. The normal to the parabola $y^2 = 4x$ at the point $P(2, 1)$ cuts the curve again at Q. Find the co-ordinates of Q.

2. Show that the equation of the tangent to the parabola $y^2 = 4ax$ at the point $P(x_1, y_1)$ on the curve is

$$y y_1 = 2a(x + x_1).$$

   Hence write down the equation of the tangent at the point $(a, 2a)$.

3. Find the equation of the tangent to the curve $y^2 = 4ax$ at the point $(at^2, 2at)$. Find in terms of t and a the distance of the origin from this tangent.

4. Obtain the equation of the normal to the curve $y^2 = 4x$ at the point $P(t^2, 2t)$. If this normal passes through the point $(2, 1)$ find the co-ordinates of P.

5. The point P moves on the parabola $y^2 = 8x$. Find the locus of the mid-point of the line PR when R has co-ordinates $(4, 0)$.
   [Hint: Express the co-ordinates of P in terms of a parameter t.]

6. Find the equations of the tangents to the curve $y^2 = 16x$ at the points $P(1, 4)$ and $Q(4, -8)$ and find the co-ordinates of the point of intersection of these tangents.

7. The normal at the point $P(a\cos\theta, b\sin\theta)$ to the ellipse

$$\frac{x^2}{a^2} + \frac{y^2}{b^2} = 1$$

meets the line $y = x\tan\theta$ at Q. Show that $OQ = a + b$ where O is the origin.
Find the equation of the tangent to the ellipse

$$\frac{x^2}{a^2} + \frac{y^2}{b^2} = 1$$

at the point $(a\cos\theta, b\sin\theta)$.

8. Find the equation of the normal to the rectangular hyperbola $xy = c^2$ at the point $P(ct, c/t)$. Show that the normal meets the curve again at $Q(-c/t^3, -ct^3)$ and hence find the locus of the mid-point of OQ where O is the origin.

13.8   The use of parametric co-ordinates

A focal chord of a parabola $y^2 = 4ax$ is a chord which passes through the focus located at $(a, 0)$. The equation of the focal chord with gradient 2 is

$$y - 0 = 2(x - a)$$

$$\Rightarrow \quad y = 2(x - a)$$

We could find the points of intersection of this focal chord with the parabola by solving simultaneously. An alternative approach which reduces the algebraic manipulation is preferable and available.

The point $P(at^2, 2at)$ necessarily lies on the parabola for all values of t. Substituting $x = at^2$ and $y = 2at$ into the equation of the focal chord gives

$$2at = 2(at^2 - a)$$

a quadratic in t, which once solved may be used to obtain the co-ordinates of the points of intersection.

Thus

$$2at^2 - 2at - 2a = 0$$

$$\Rightarrow \quad t^2 - t - 1 = 0$$

$$\Rightarrow \quad t = \frac{1 \pm \sqrt{5}}{2}$$

and since

$$t^2 = \left(\frac{1 \pm \sqrt{5}}{2}\right)^2 = \frac{1 \pm 2\sqrt{5} + 5}{4} = \frac{3}{2} \pm \frac{\sqrt{5}}{2},$$

the co-ordinates of the point of intersection are

$$\left(\frac{a}{2}(3 + \sqrt{5}), a(1 + \sqrt{5})\right) \quad \text{and} \quad \left(\frac{a}{2}(3 - \sqrt{5}), a(1 - \sqrt{5})\right)$$

[compare with $(at^2, 2at)$].

Another technique for avoiding algebraic manipulation is illustrated in the following example.

Example 13.15

The line $2x - y - 4a = 0$ cuts the parabola $y^2 = 4ax$ at the points $P(at_1^2, 2at_1)$ and $Q(at_2^2, 2at_2)$. Find the values of $t_1 + t_2$ and $t_1 . t_2$ and hence find the mid point of PQ.

The general point $(at^2, 2at)$ on the parabola $y^2 = 4ax$ is on the line $2x - y - 4a = 0$ if the equation

$$2at^2 - 2at - 4a = 0 \quad \text{is satisfied.}$$

$$\Rightarrow \quad t^2 - t - 2 = 0.$$

We expect the solutions to this quadratic to be t_1 and t_2 and so the sum of the roots $t_1 + t_2 = 1$ and the product $t_1 . t_2 = -2$.

We could of course solve for t to find t_1 and t_2 directly, then obtain the points P and Q and then the co-ordinates of the mid-point. However, the co-ordinates of the mid-point are

$$\left(\frac{at_1^2 + at_2^2}{2}, \frac{2at_1 + 2at_2}{2}\right)$$

which are

$$\left(\frac{a}{2}[(t_1 + t_2)^2 - 2(t_1 . t_2)], a(t_1 + t_2)\right)$$

that is

$$\left(\frac{a}{2}[1 + 4], a\right) \quad \text{(since } t_1 + t_2 = 1 \text{ and } t_1 . t_2 = -2\text{)}$$

and finally

$$\left(\frac{5a}{2}, a\right).$$

This technique will prove to be of great value.

Example 13.16

Find the locus of the mid-points of focal chords of the parabola $y^2 = 4ax$.

Since the focus is at $(a, 0)$ a focal chord will have form

$$y - 0 = m(x - a)$$

$$\Rightarrow y = m(x - a) \quad \text{as } m \text{ varies.}$$

The point $P(at^2, 2at)$ on the parabola is on this focal chord if

$$2at = m(at^2 - a)$$
$$\Rightarrow mt^2 - 2t - m = 0.$$

The end points may be obtained from the roots t_1, t_2 of this equation, where

$$t_1 + t_2 = \frac{2}{m} \quad \text{and} \quad t_1 . t_2 = -1$$

The mid-point M of $Q(at_1^2, 2at_1)$ and $R(at_2^2, 2at_2)$ is

$$\left(a\frac{(t_1^2 + t_2^2)}{2}, 2a\frac{(t_1 + t_2)}{2} \right)$$

$$\equiv \left(\frac{a}{2}[(t_1 + t_2)^2 - (2t_1 . t_2)], a(t_1 + t_2) \right)$$

$$\equiv \left(\frac{a}{2}\left[\frac{4}{m^2} + 2\right], a\frac{2}{m} \right)$$

$$\equiv \left(\frac{2a}{m^2} + a, \frac{2a}{m} \right) \equiv \left(\frac{a}{m^2}(2 + m^2), \frac{2a}{m} \right)$$

To find the locus of M we set

$$x = \frac{a}{m^2}(2 + m^2) \quad \text{and} \quad y = \frac{2a}{m}.$$

$$\Rightarrow m = \frac{2a}{y}$$

and

$$x = \frac{y^2}{4a}\left(2 + \frac{4a^2}{y^2}\right)$$

$$\Rightarrow 4ax = 2y^2 + 4a^2$$

$$\Rightarrow y^2 = 2a(x - a) \qquad \text{a parabola}$$

Statement

> *The parametric forms for the conics*
>
> The circle $(x - a)^2 + (y - b)^2 = c^2$:
>
> $x = a + c\cos\theta, \quad y = b + c\sin\theta.$
>
> The parabola, $y^2 = 4ax$:
>
> $x = at^2, \quad y = 2at.$

The ellipse $\dfrac{x^2}{a^2} + \dfrac{y^2}{b^2} = 1$:

$$x = a\cos\theta, \quad y = b\sin\theta.$$

The rectangular hyperbola $xy = c^2$:

$$x = ct, \quad y = \dfrac{c}{t}.$$

The hyperbola $\dfrac{x^2}{a^2} - \dfrac{y^2}{b^2} = 1$:

$$x = a\sec\theta, \quad y = b\tan\theta.$$

Example 13.17

P is the point $(ap^2, 2ap)$ on the parabola $y^2 = 4ax$. QSR is the chord which passes through the focus S of the parabola and is parallel to the tangent at P.

If Q has co-ordinates $(aq^2, 2aq)$, show, by considering the gradient of the line QS or otherwise, that $q^2 - 2pq - 1 = 0$ and hence that

$$q = p \pm \sqrt{p^2 + 1}.$$

If W is the mid point of QR, prove that PW is parallel to the axis of the parabola and that the length of PW is $a(1 + p^2)$.

SU 1981

Since $y^2 = 4ax$,

$$\Rightarrow 2y\frac{dy}{dx} = 4a.$$

At $P(ap^2, 2ap)$ the gradient is

$$\frac{dy}{dx} = \frac{1}{p}$$

The gradient of QS is

$$\frac{2aq - 0}{aq^2 - a} = \frac{2q}{q^2 - 1}$$

Equating

$$\frac{2q}{q^2 - 1} = \frac{1}{p}$$

$$\Rightarrow 2pq = q^2 - 1$$

$$\Rightarrow q^2 - 2pq - 1 = 0$$

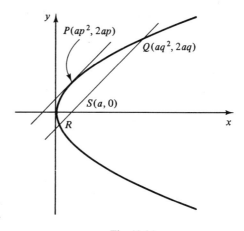

Fig. 13.16

Solving this quadratic in q

$$q = \frac{2p \pm \sqrt{4p^2 + 4}}{2}$$

$$\Rightarrow q = p \pm \sqrt{p^2 + 1}.$$

The two solutions q_1, q_2 for q correspond to the points Q and R, and the mid-point of QR, W, has co-ordinates

$$\left(\frac{aq_1^2 + aq_2^2}{2}, \frac{2aq_1 + 2aq_2}{2} \right)$$

$$\equiv \left(\frac{a}{2}(q_1^2 + q_2^2), a(q_1 + q_2) \right)$$

where

$$q_1 + q_2 = p + \sqrt{p^2 + 1} + p - \sqrt{p^2 + 1} = 2p$$

It follows that the y co-ordinate of W is $2ap$, the same as that for P. Thus PW is parallel to the x axis – the axis of the parabola.

Now

$$q_1^2 + q_2^2 = (p + \sqrt{p^2 + 1})^2 + (p - \sqrt{p^2 + 1})^2$$

$$= p^2 + 2p\sqrt{p^2 + 1} + p^2 + 1 + p^2 - 2p\sqrt{p^2 + 1} + p^2 + 1$$

$$= 4p^2 + 2.$$

So the x co-ordinate of W is

$$2ap^2 + a,$$

the length PW is

$$2ap^2 + a - ap^2$$

$$= ap^2 + a$$

$$= a(p^2 + 1).$$

Exercise 13.6

1. Find the range of possible values of m, if the line $y = mx$ is to cut the curve $y^2 = x - 4$ in two points. Hence find the values of m so that the line is a tangent to the curve.

2. Find the equations of the tangent to the curve $y = x(x - 4)$ which passes through the point (α, β) on the curve.

3. Find the equations of the tangent from the point $(0, -2)$ to the curve $y = 2x^2$.

4. The chord PQ of the curve given by $y = x^2$ is constrained so that it always passes through the point $(0, 2)$. Find the locus of the mid-point M of PQ.

5. Obtain the equation of the chord joining the points $P(at_1^2, 2at_1)$ and $Q(at_2^2, 2at_2)$ on the parabola $y^2 = 4ax$, and find the co-ordinates of the point of intersection of the tangents at P and Q. Find the locus of the point of intersection if PQ is constrained so that it passes through the point (a, a).

6. The line $y = mx + c$ cuts the parabola $y^2 = 4ax$ in two points P and Q. Show that the co-ordinates of the mid point M of PQ are

$$\left(\frac{2a - mc}{m^2}, \frac{2a}{m} \right).$$

Find the locus of M if m varies and $c = 1$.

7. The points $P(at_1^2, 2at_1)$ and $Q(at_2^2, 2at_2)$ lie on the parabola $y^2 = 4ax$. M is the mid point of PQ and the tangents at P and Q intersect at R. Show that the co-ordinates of R are $(at_1 . t_2, a(t_1 + t_2))$ and find the locus of M if R is constrained to lie on the line $x = b$.

8. Show that the equation of the chord joining the points $P(at_1^2, 2at_1)$ and $Q(at_2^2, 2at_2)$ which lie on the parabola $y^2 = 4ax$ is

$$(t_1 + t_2)y = 2x + 2at_1t_2.$$

The tangents to the parabola at P and Q meet at the point T. When P and Q move on the parabola such that PQ passes through the fixed point $(4a, a)$, show that the locus of T is a straight line.

AEB 1982

9. Show that the equation of the chord of the parabola $y^2 = 4x$ joining the points $P(p^2, 2p)$ and $Q(q^2, 2q)$ is $(p + q)y - 2x = 2pq$ and find the condition that this chord should pass through the point $A(3, 0)$. If, in addition, the line PO meets the line $x = -3$ at the point L, show that LQ is parallel to OA, O being the origin. If PQ is also normal to the parabola at P show that $p^2 + pq + 2 = 0$ and hence find the possible values of p and the length of PQ.

SU 1980

10. $P(ap^2, 2ap)$ and $Q(aq^2, 2aq)$ are points on the parabola $y^2 = 4ax$. Prove that the equation of the chord PQ can be written in the form $(p + q)y - 2x = 2apq$ and that the gradient of the normal to the parabola at P is $-p$.

The gradient of the normal at P is six times the gradient of the tangent at Q. Prove (a) that PQ passes through a fixed point on the axis of the parabola; (b) that the mid-point of PQ lies on the parabola $y^2 = 2a(x - 6a)$. Sketch the graphs of the two parabolae on the same axes and give the co-ordinates of the focus of each.

SU 1983

11. Obtain the equation of the normal to the curve

$$y^2 = x^3$$

at the point (t^2, t^3). Show that the equation of the normal at the point where $t = \frac{1}{2}$ is

$$32x + 24y - 11 = 0.$$

Find the perpendicular distance from the point $(-1, 2)$ to this normal.

JMB 1982

Exercise 13.7 (miscellaneous)

1. Find the equation of the tangent at the point $P(3at^2, 2at^3)$ on the curve $4x^3 = 27ay^2$.
   The tangent meets the curve again at $Q(3at_1^2, 2at_1^3)$. Show that $2t_1 = -t$. If the tangents at P and Q are perpendicular find the co-ordinates of P.

2. Find the Cartesian equation of the curve given by

$$x = t + \frac{1}{t} \qquad y = t - \frac{1}{t}$$

and sketch the curve.

3. Find the Cartesian equation of the curve given parametrically by

$$x = 2(t - 1) \qquad y = 1 + 3t^2.$$

Sketch the curve by first changing the origin to the point $(-2, 1)$.

4. PQ is a chord of gradient $m (\neq 0)$ through the focus $(1, 0)$ of the parabola

$$y^2 = 4x.$$

Find the co-ordinates of the mid-point M of PQ in terms of m. Deduce that M lies on the curve whose equation is

$$y^2 = 2(x - 1).$$

Given that P lies in the first quadrant, find the equation of the tangent at P to the parabola $y^2 = 4x$ in the case when $m = 2\sqrt{2}$. Prove that, in this case, the acute angle between PQ and the tangent at P is $\tan^{-1}(1/\sqrt{2})$.

JMB 1982

5. Sketch the curve whose equation is

$$y = \frac{x - 2}{x + 2},$$

and state the equations of its asymptotes.

On the same diagram sketch the curve whose equation is

$$x^2 + 4y^2 = 4.$$

Hence, or otherwise, find all real solutions of the equation

$$4(x - 2)^2 = (4 - x^2)(x + 2)^2.$$

JMB 1982

6. Show that the tangent to the parabola $y^2 = 4ax$ at the point $P(ap^2, 2ap)$ has the equation

$$py = x + ap^2.$$

This tangent meets the line $x + a = 0$ at T. Find, in terms of a and p, the co-ordinates of the mid-point of PT.

Prove that, for all values of p, this mid-point lies on the curve

$$y^2(2x + a) = a(3x + a)^2.$$

CAMB 1982

7. Find the centre and radius of each of the circles C_1 and C_2 whose equations are $x^2 + y^2 - 16y + 32 = 0$ and $x^2 + y^2 - 18x + 2y + 32 = 0$ respectively and show that the circles touch externally. Find the co-ordinates of their point of contact and show that the common tangent at that point passes through the origin. The other tangents from the origin, one to each circle, are drawn. Find, correct to the nearest degree, the angle between these tangents.

SU 1980

8. Points $P(ap^2, 2ap)$ and $Q(aq^2, 2aq)$ lie on the parabola $y^2 = 4ax$.

(a) Find the equation of the tangent to this parabola at the point P.
(b) Verify that the point $R(apq, a(p + q))$ lies on the tangent at P (and so also on the tangent at Q).

If PQ passes through the point $S(a, 0)$,

(c) prove that $pq = -1$,
(d) prove that RS is perpendicular to PQ.

O&C 1982

9. Prove that the normal to the parabola $y^2 = 4ax$ at the point $(at^2, 2at)$ has equation $y + tx = 2at + at^3$. This normal meets the parabola again at the point $(aT^2, 2aT)$. Prove that $T = -t - 2/t$.

Points $P(ap^2, 2ap)$ and $Q(aq^2, 2aq)$ of the parabola are such that the normals to the parabola at P and Q intersect on the parabola. Find the equation satisfied by the co-ordinates of the mid-point R of PQ as P and Q vary, showing that it also respresents a parabola.

Draw on one diagram a rough sketch to show both parabolas.

O&C 1983

10. Find the equations of the two circles which pass through the point $(2, 0)$ and have both the y-axis and the line $y - 1 = 0$ as tangents.
    Calculate the co-ordinates of the second point at which the circles intersect.

    AEB 198

11. The triangle ABC has vertices $A(0, 8)$, $B(6, 0)$ and $C(-6, 0)$.
    (a) Calculate the co-ordinates of the point P, lying inside the triangle ABC which is such that the perpendicular distances of P from each of the three sides of the triangle are equal.
    (b) Calculate the co-ordinates of the point Q which is such that $QA = QB = QC$. Hence, or otherwise, find the equation of the circle which passes through A, B and C.

    CAMB 198

12. Show that the equation of the normal to the rectangular hyperbola $xy = c^2$ at the point $P(cp, c/p)$ is

    $$py - p^3 x = c(1 - p^4).$$

    This normal cuts the rectangular hyperbola again at the point $Q(cq, c/q)$ Prove that $q = -\dfrac{1}{p^3}$.

    Given that $p^2 \neq 1$ and that the line joining the origin to P cuts the hyperbola again at the point R, show that RP and RQ are perpendicular.

    CAMB 198

13. (a) The points $(1, 2)$ and $(2, 6)$ are denoted by A and B respectively. Show that the equation of the line AB is $y = 4x - 2$.
        The line AB meets the x-axis at C. D is the point $(0, \frac{13}{2})$. Show that CB is perpendicular to DB and find the angle between DC and DB.
    (b) Show that the equation of the normal to the curve $y^2 = 4ax$ at the point P having co-ordinates $(ap^2, 2ap)$, is $y = -px + ap^3 + 2ap$.
        The normal meets the line $y = 2x$ at the point Q. Given the point $S(0, a)$ and that SQ is parallel to the tangent at P show that p satisfies the equation

        $$2p^3 - p^2 + 3p - 4 = 0.$$

    W 198

14. A point P moves in the x-y plane so that its distance from the origin, O, is twice its distance from the point with co-ordinates $(3a, 0)$. Show that the locus of P is a circle and obtain the co-ordinates of its centre and its radius. If the circle meets the x-axis in A and B, where $OA < OB$, find the coordinates of A and B. If the tangents from O to the circle are OL and OM, find the angle LOM and the equations of OL and OM. Calculate the area of the triangle enclosed by the lines OL, OM and the tangent to the circle at A.

    SU 198

15. The tangent to the parabola $y^2 = 4ax$ at the point $P(at^2, 2at)$ meets the y-axis at G. The normal at the point P meets the x-axis at the point H.

(a) Find the equation of the normal and show that the mid-point M of HG has co-ordinates

$$\left(a + \frac{at^2}{2}, \frac{at}{2}\right).$$

(b) Find the cartesian equation of the locus of M as P moves on the parabola.

AEB 1983

16. A curve is defined by the parametric equations

$$x = t, \quad y = \frac{1}{t}, \quad t \neq 0.$$

Sketch the curve, and find the equation of the tangent to the curve at the point

$$\left(t, \frac{1}{t}\right).$$

The points P and Q on the curve are given by $t = p$ and $t = q$, respectively, and $p \neq q$. The tangents at P and Q meet at R. Show that the co-ordinates of R are

$$\left(\frac{2pq}{p + q}, \frac{2}{p + q}\right).$$

(a) Given that p and q vary so that $pq = 2$, state the equation of the line on which R moves.

(b) Given that p and q vary so that $p^2 + q^2 = 1$, show that the point R moves on the curve

$$y(2x + y) = k,$$

where k is a constant. State the value of k.

JMB 1984

17. The tangents to the parabola $y^2 = x$ at the variable points $P(p^2, p)$ and $Q(q^2, q)$ intersect at N. Given that M is the mid-point of PQ, prove that

$$MN = \tfrac{1}{2}(p - q)^2.$$

Given also that PN is always perpendicular to QN,

(a) show that the locus of N is a straight line, and write down its equation,

(b) find the Cartesian equation of the locus of M.

JMB 1983

Chapter 14 Power series and differential equations

14.1 Power series

We already know that some functions may be expressed as an infinite series of ascending powers of x.

For example

$$(1 - x)^{-1} = 1 + x + x^2 + x^3 + \ldots + x^n + \ldots$$

provided $|x| < 1$. The expansion is not valid if $|x| \geqslant 1$ since the series does not converge [the series does not have a finite sum].

It turns out that many functions can be expressed as a *power series*, that is an infinite series composed of terms of the form of ax^n.

Supposing that

$$f(x) \equiv a_0 + a_1 x + a_2 x^2 + a_3 x^3 + \ldots$$

at least for a range of x values including zero, then setting $x = 0$ we see that

$$f(0) = a_0.$$

Now differentiating w.r.t. x

$$f'(x) \equiv a_1 + 2a_2 x + 3a_3 x^2 + \ldots$$

assuming it to be valid to differentiate term by term: and assuming $f'(x)$ exists.

Setting $x = 0$ again, we find

$$f'(0) = a_1.$$

Repeating this process of differentiating:

$$f''(x) \equiv 2a_2 + 3.2a_3 x + 4.3a_4 x^2 + \ldots$$

and setting $x = 0$

$$f''(0) = 2a_2$$

Continuing this process we find

$$f^3(0) = f'''(0) = 3.2a_3$$
$$f^4(0) = 4.3.2a_4 \qquad \text{where } f^r(x) \text{ is the } r\text{th derivative.}$$
$$f^5(0) = 5.4.3.2a_5$$
$$\vdots$$
$$f^r(0) = r!\, a_r$$

It follows that $a_r = \dfrac{f^r(0)}{r!}$, which leads to:

Maclaurin's theorem

$$f(x) \equiv f(0) + f^1(0)x + \frac{f^2(0)x^2}{2!} + \frac{f^3(0)x^3}{3!} + \ldots \frac{f^r(0)x^r}{r!} + \ldots$$

This procedure is valid, at least for those functions introduced here, though a formal proof is beyond the scope of this book.

Statement

$$e^x \equiv 1 + x + \frac{x^2}{2!} + \frac{x^3}{3!} + \frac{x^4}{4!} + \ldots \qquad \text{for } x \in \mathbb{R}$$

The following sets out the details required to check the above result.

$f(x) \equiv e^x$	$f(0) = e^0 = 1$	$a_0 = f(0) = 1$
$f'(x) \equiv e^x$	$f'(0) = e^0 = 1$	$a_1 = f'(0) = 1$
$f''(x) \equiv e^x$	$f''(0) = e^0 = 1$	$a_2 = \dfrac{f''(0)}{2!} = \dfrac{1}{2!}$
$f^3(x) \equiv e^x$	$f^3(0) = e^0 = 1$	$a_3 = \dfrac{f^3(0)}{3!} = \dfrac{1}{3!}$
$\vdots$	$\vdots$	$\vdots$
$f^r(x) \equiv e^x$	$f^r(0) = e^0 = 1$	$a_r = \dfrac{f^r(0)}{r!} = \dfrac{1}{r!}$

The result then follows by substitution using Maclaurin's theorem. The fact that the expansion is valid for all $x \in \mathbb{R}$ is beyond the scope of this book.

Statement

$$\cos x \equiv 1 - \frac{x^2}{2!} + \frac{x^4}{4!} - \frac{x^6}{6!} + \frac{x^8}{8!} - \ldots \qquad x \in \mathbb{R}$$

$$\sin x \equiv x - \frac{x^3}{3!} + \frac{x^5}{5!} - \frac{x^7}{7!} + \ldots \qquad x \in \mathbb{R}$$

$$\ln(1 + x) \equiv x - \frac{x^2}{2} + \frac{x^3}{3} - \frac{x^4}{4} + \ldots \qquad -1 < x \leqslant 1$$

The first two assertions are left as an exercise (see Exercise 14.1, question 3). For the third, notice that no attempt could be made to find the *Maclaurin expansion* for $\ln x$ since the function is not defined for $x = 0$.

Now

$$f(x) \equiv \ln(1 + x) \qquad \Rightarrow f(0) = \ln 1 = 0 \qquad \Rightarrow a_0 = 0$$

$$f'(x) \equiv \frac{1}{1 + x} \equiv (1 + x)^{-1} \qquad f'(0) = 1 \qquad a_1 = 1$$

$$f''(x) \equiv -(1 + x)^{-2} \qquad f''(0) = -1 \qquad a_2 = \frac{-1}{2}$$

$$f^3(x) \equiv 2(1 + x)^{-3} \qquad f^3(0) = 2! \qquad a_3 = \frac{1}{3}$$

$$f^4(x) \equiv -3.2(1 + x)^{-4} \qquad f^4(0) = -3! \qquad a_4 = \frac{-1}{4}$$

$$f^5(x) \equiv 4.3.2.(1 + x)^{-5} \qquad f^5(0) = 4! \qquad a_5 = \frac{1}{5}$$

$$\vdots$$

which upon substitution gives us our result.

An alternative approach runs as follows — using the binomial theorem.

$$\frac{1}{1 + x} \equiv (1 + x)^{-1} \equiv 1 - x + x^2 - x^3 + x^4 - x^5 + \ldots$$

Now integrating both sides w.r.t. x

$$\ln(1 + x) \equiv x - \frac{x^2}{2} + \frac{x^3}{3} - \frac{x^4}{4} + \frac{x^5}{5} - \frac{x^6}{6} \ldots$$

assuming such integration to be valid. (Setting $x = 0$ shows that the constant of integration is zero.)

The use of the binomial theorem reminds us that $|x| < 1$ for the expansion to be valid, though this particular series also converges when $x = 1$.

Further expansion of series

Example 14.1

Expand

(a) $\ln\left(\dfrac{1 + x}{1 - x}\right)$ (b) $\frac{1}{2}(e^{2x} - e^{-2x})$ (c) $\dfrac{1 + x^2}{1 - x^2}$

as a series of ascending powers of x, up to and including the term in x^5. State the range of values of x for which the expansion is valid.

(a) $\ln\left(\dfrac{1 + x}{1 - x}\right) \equiv \ln(1 + x) - \ln(1 - x)$

Now by Maclaurin's theorem

$$\ln(1 + x) \equiv x - \frac{x^2}{2} + \frac{x^3}{3} - \frac{x^4}{4} + \frac{x^5}{5} \ldots \qquad \text{for } -1 < x \leqslant 1.$$

Substituting $-x$ for x, we have

$$\ln(1 - x) \equiv -x - \frac{x^2}{2} - \frac{x^3}{3} - \frac{x^4}{4} - \frac{x^5}{5} - \ldots \qquad \text{for } -1 < -x \leqslant 1$$
$$\Rightarrow -1 \leqslant x < 1.$$

For *both* expansions to be valid we have $-1 < x < 1$ and then

$$\ln\frac{1 + x}{1 - x} \equiv \ln(1 + x) - \ln(1 - x)$$

$$\equiv x - \frac{x^2}{2} + \frac{x^3}{3} - \frac{x^4}{4} + \frac{x^5}{5} \ldots$$

$$- \left(x + \frac{x^2}{2} + \frac{x^3}{3} + \frac{x^4}{4} + \frac{x^5}{5} + \ldots \right)$$

$$\equiv 2x + \frac{2x^3}{3} + \frac{2x^5}{5} + \ldots.$$

So

$$\ln\frac{1 + x}{1 - x} \equiv 2x + \frac{2x^3}{3} + \frac{2x^5}{5} + \ldots \qquad |x| < 1$$

(b) Recall that

$$e^x \equiv 1 + x + \frac{x^2}{2!} + \frac{x^3}{3!} + \frac{x^4}{4!} + \frac{x^5}{5!} + \ldots \qquad x \in \mathbb{R}$$

So

$$e^{2x} \equiv 1 + 2x + \frac{4x^2}{2!} + \frac{8x^3}{3!} + \frac{16x^4}{4!} + \frac{32x^5}{5!} + \ldots \qquad x \in \mathbb{R}$$

and

$$e^{-2x} \equiv 1 - 2x + \frac{4x^2}{2!} - \frac{8x^3}{3!} + \frac{16x^4}{4!} - \frac{32x^5}{5!} + \ldots \qquad x \in \mathbb{R}$$

$$(x \in \mathbb{R} \Leftrightarrow -2x \in \mathbb{R})$$

Thus

$$\tfrac{1}{2}(e^{2x} - e^{-2x}) \equiv \tfrac{1}{2}\left(4x + \frac{16x^3}{3!} + \frac{64x^5}{5!} + \ldots \right)$$

$$\equiv 2\left(x + \frac{4x^3}{3!} + \frac{16x^5}{5!} + \ldots \right) \qquad x \in \mathbb{R}$$

(c) $\dfrac{1 + x^2}{1 - x^2} \equiv (1 + x^2)(1 - x^2)^{-1}$

$$\equiv (1 - x^2)^{-1} + x^2(1 - x^2)^{-1} \qquad \textcircled{1}$$

Now using the binomial theorem

$$(1 - x^2)^{-1} \equiv 1 + x^2 + x^4 + x^6 + \dots \qquad |x^2| < 1$$

continuing with $\textcircled{1}$

$$\dfrac{1 + x^2}{1 - x^2} \equiv 1 + x^2 + x^4 + \dots$$
$$+ x^2(1 + x^2 + x^4 + \dots$$
$$\equiv 1 + 2x^2 + 2x^4 + \dots \qquad |x| < 1$$

$$(\text{since } |x^2| < 1 \Leftrightarrow |x| < 1)$$

14.2   A proof of the binomial theorem

The binomial theorem states that

$$(1 + x)^p = 1 + px + \frac{p(p - 1)x^2}{2!} + \frac{p(p - 1)(p - 2)x^3}{3!} + \dots$$

where p is a constant for $|x| < 1$.

We may use Maclaurin's theorem to show that this result is reasonable. Suppose $f(x) \equiv (1 + x)^p$, then:

$f(x) \equiv (1 + x)^p$ $\qquad \Rightarrow f(0) = 1$ $\qquad \Rightarrow a_0 = f(0) = 1$

$f'(x) \equiv p(1 + x)^{p-1}$ $\qquad f'(0) = p$ $\qquad a_1 = f'(0) = p$

$f''(x) \equiv p(p - 1)(1 + x)^{p-2}$ $\qquad f''(0) = p(p - 1)$ $\qquad a_2 = \dfrac{f''(0)}{2!} = \dfrac{p(p - 1)}{2!}$

$f^3(x) \equiv p(p - 1)(p - 2)(1 + x)^{p-3}$

$\qquad\qquad\qquad f^3(0) = p(p - 1)(p - 2) \qquad a_3 = \dfrac{f^3(0)}{3!}$

$$= \dfrac{p(p - 1)(p - 2)}{3!}$$

from which the result follows using Maclaurin's theorem.

Example 14.2

Show that $\displaystyle \lim_{h \to 0} \left(\frac{e^h - 1}{h} \right) = 1.$

[This result was required on page 250.]

Now

$$e^h \equiv 1 + h + \frac{h^2}{2!} + \frac{h^3}{3!} + \frac{h^4}{4!} + \frac{h^5}{5!} + \dots$$

So

$$\frac{e^h - 1}{h} \equiv 1 + \frac{h}{2!} + \frac{h^2}{3!} + \frac{h^3}{4!} + \frac{h^4}{5!} + \dots$$

It follows immediately that

$$\lim_{h \to 0} \left(\frac{e^h - 1}{h} \right) = 1.$$

Example 14.3

Show that

$$\ln(1 + x - 2x^2) \simeq x - \frac{5}{2}x^2 + \frac{7}{3}x^3 \dots$$

and establish the range of values for which the expansion is valid.

Now

$$\ln(1 + X) \equiv X - \frac{X^2}{2} + \frac{X^3}{3} \dots \qquad \text{for } -1 < X \leqslant 1.$$

To obtain the expansion of $\ln(1 + x - 2x^2)$ set $X = (x - 2x^2)$ in the above.

$$\ln(1 + x - 2x) = (x - 2x^2) - \frac{(x - 2x^2)^2}{2} + \frac{(x - 2x^2)^3}{3} \dots$$

$$\equiv x - 2x^2 - \tfrac{1}{2}[x^2 - 4x^3 + \dots] + \tfrac{1}{3}[x^3 + \dots] + \dots$$

$$\equiv x - 2x^2 - \frac{x^2}{2} + 2x^3 + \frac{x^3}{3} + \dots$$

$$\equiv x - \frac{5}{2}x^2 + \frac{7}{3}x^3 \dots$$

The expansion is valid if

$$-1 < x - 2x^2 \leqslant 1.$$

A sketch of the curve $y = x - 2x^2$ (see Fig. 14.1 on the next page) shows that for all x

$$x - 2x^2 \leqslant 1.$$

Now

$$-1 < x - 2x^2$$

provided

$$-\tfrac{1}{2} < x < 1.$$

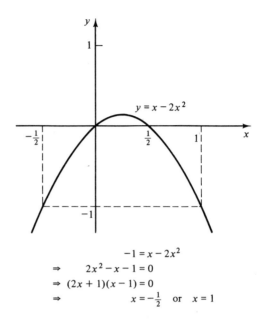

$$-1 = x - 2x^2$$
$$\Rightarrow \quad 2x^2 - x - 1 = 0$$
$$\Rightarrow \quad (2x + 1)(x - 1) = 0$$
$$\Rightarrow \quad \qquad x = -\tfrac{1}{2} \quad \text{or} \quad x = 1$$

Fig. 14.1

It happens that the above quadratic $1 + x - 2x^2$ factorises so that

$$\ln(1 + x - 2x^2) \equiv \ln[(1 - x)(1 + 2x)]$$
$$\equiv \ln(1 - x) + \ln(1 + 2x)$$

It is much easier to find the expansion in this form for:

$$\ln(1 - x) \equiv -x - \frac{(-x)^2}{2} + \frac{(-x)^3}{3} + \dots$$

$$\equiv -x - \frac{x^2}{2} - \frac{x^3}{3},$$

and

$$\ln(1 + 2x) \equiv 2x - \frac{(2x)^2}{2} + \frac{(2x)^3}{3} + \dots$$

$$\equiv 2x - 2x^2 + \frac{8x^3}{3} + \dots$$

so that

$$\ln(1 + x - 2x^2) \equiv \ln(1 - x) + \ln(1 + 2x)$$

$$\equiv -x - \frac{x^2}{2} - \frac{x^3}{3} + \dots$$

$$+2x - 2x^2 + \frac{8x^3}{3} + \dots$$

$$\equiv x - \frac{5x^2}{2} + \frac{7x^3}{3} + \dots$$

as before.

Exercise 14.1

1. Write down the first four terms of the expansion of the following in ascending powers of x.

   (a) e^{2x} (b) e^{-x} (c) $e^{x/2}$ (d) e^{-x^2}

   (e) $(1 + x)e^x$ (f) $\frac{(1 - x)}{e^x}$ (g) $(1 - x^2)e^{2x}$ (h) $e^{(1-x)}$

2. Use the Maclaurin expansion of e^x to find

   (a) e (b) $e^{1/2}$ (c) $e^{1/10}$

   correct to 3 d.p.

3. Use Maclaurin's theorem to obtain the expansion of (a) $\sin x$, (b) $\cos x$ as a power series [given on page 435].

4. Obtain the first four terms in the expansion of

   (a) $\ln(1 + 2x)$ (b) $\ln(1 + x^2)$ (c) $\log_{10}(1 + x)$

   [*Hint*: change to base e.]

   In each case, state the range of x, for which the expansion is valid.

5. Find the first three terms in the expansion of $e^x \sin x$ in ascending powers of x.

6. Determine the constants a and b if the coefficient of x^3 and x^4 are zero in the expansion of

   $$(1 + ax + bx^2)e^{-x}.$$

7. By expanding $(1 - x)^{-1/2}$ and using the fact that

   $$\sin^{-1} x = \int (1 - x)^{-1/2} dx,$$

   find the first four terms in the expansion of $\sin^{-1} x$ in ascending powers of x.

8. Determine the possible values of the constants a and b in the expansion of

$$\frac{1+x}{1+ax} - e^{bx},$$

so that the coefficients of x and x^2 are both zero.

9. Find the first three terms in the expansion of

(a) $e^{\sin x}$ (b) a^x [Hint: $a^x \equiv e^{x\ln a}$]

10. Show that

$$e^x - \ln\sqrt{\frac{1+x}{1-x}} \simeq 1 + \frac{x^2}{2} - \frac{x^3}{6}$$

provided x is sufficiently small so that powers of x^4 and above may be neglected.

11. Use the fact that

$$\tan^{-1} x = \int \frac{1}{1+x^2}\,dx$$

to obtain the first four terms in the expansion of $\tan^{-1} x$ in ascending powers of x.

12. If x is known to be small, find a quadratic approximation to the function

$$f(x) \equiv \frac{e^x}{(1+x)^2} \qquad x \neq -1.$$

13. The hyperbolic sine of x is written $\sinh(x)$ [pronounced shine x or occasionally cinch x] and is defined by

$$\sinh x \equiv \tfrac{1}{2}(e^x - e^{-x}).$$

Prove that $\dfrac{d}{dx}(\sinh x) = \cosh x$ [pronounced cosh x], where the hyperbolic cosine is defined by

$$\cosh x \equiv \tfrac{1}{2}(e^x + e^{-x})$$

Expand $\sinh x$ in ascending powers of x using

(a) Maclaurin's theorem
(b) the definitions in terms of e^x.

14. Find y explicitly in terms of x if $e^y(1+x) = 1 - 2x$. Hence find a cubic approximation to y in terms of x.

15. Obtain the expansion in ascending powers of x, up to and including the term in x^4, of the function $y = e^{-x}\cos^2 x$.

O & C 1983

16. Find the first three non-zero terms of the expansion of $e^{x/2} \log_e (1 + 2x)$ in ascending powers of x. Use the first two terms to give a value for $e^{0.01} \log_e 1.04$ and the third term to estimate the error in your value. Hence correct your first answer to an appropriate number of decimal places.

SU 1983

17. Expand $e^{x/2} \log_e (1 + x)$ in ascending powers of x as far as the term in x^4 and hence show that, for certain values of x to be stated,

$$e^{x/2} \log_e (1 + x) + e^{-x/2} \log_e(1 - x) = ax^4 + \ldots$$

and give the value of a.

SU 1980

18. (a) Write down the first 4 terms and the term in x^n in the expansion of e^x.
    Prove that the coefficient of x^n in the expansion of $(x^2 + x)e^x$ is

$$\frac{n}{(n - 1)!} \qquad \text{when } n \geqslant 2.$$

    Deduce that $(x^2 + x)e^x - x$ is always positive when x is positive.

    (b) Express $\dfrac{\sqrt{1 - x}}{1 + x}$ in ascending powers of x up to and including the term in x^2.

    By substituting $x = \dfrac{1}{9}$, show that $\sqrt{2} \approx \dfrac{2755}{1944}$.

SU 1981

19. Prove the identity

$$1 + 3x + 6x^2 + 4x^3 \equiv \frac{(1 + x)(1 - 8x^3)}{1 - 2x}.$$

Hence, or otherwise, expand $\log_e (1 + 3x + 6x^2 + 4x^3)$ in ascending powers of x as far as, and including, the term in x^6.
For what values of x is this expansion valid?

SU 1982

20. Express $E = \dfrac{2x^2 - x}{(2 + x)(1 + x^2)}$ in partial fractions.

Expand E in ascending powers of x as far as, and including, the term in x^3.
For what values of x is this expansion valid?

SU 1982

21. Obtain the expansion in ascending powers of x of the function

$$f(x) = \ln \left\{ \frac{(1 - 3x)^2}{(1 + 2x)} \right\}.$$

Give the values of the coefficients of the powers of x up to and including x^3, and give an expression for the coefficient of x^n. For what range of values of x is this expansion valid?

OX 1983

22. Given that $x > 2$, use the binomial expansion to express $\left(\dfrac{x+2}{x}\right)^{-1/2}$ in the form $a + \dfrac{b}{x} + \dfrac{c}{x^2} + \dfrac{d}{x^3} + \ldots$, evaluating the constants a, b, c and d. Taking $x = 100$, use your series to find an approximation for $\left(\dfrac{450}{51}\right)^{1/2}$ giving your answer to 4 decimal places.

AEB 1983

23. Show that, if x is so small that x^6 and higher powers of x may be neglected,
$$\ln(1+x) = x - \tfrac{1}{2}x^2 + \tfrac{1}{3}x^3 - \tfrac{1}{4}x^4 + \tfrac{1}{5}x^5.$$

Deduce that, for such values of x,
$$\ln\left(\frac{1+x}{1-x}\right) = 2(x + \tfrac{1}{3}x^3 + \tfrac{1}{5}x^5).$$

By giving x a suitable value in this last result, prove that, for large N,
$$\ln\left(\frac{N+1}{N-1}\right) \approx \frac{2}{N} + \frac{2}{3N^3} + \frac{2}{5N^5}.$$

Hence show that, for large N,
$$\left(\frac{N+1}{N-1}\right)^N \approx e^2.$$

CAMB 1982

24. The function f is defined for x real and not equal to an odd multiple of $\pi/2$ by
$$f(x) = \frac{1}{2}\log_e\left(\frac{1+\sin x}{1-\sin x}\right)$$

Show that $f'(x) = \sec x$.

Write down the expansion of $f(x)$ and of
$$\frac{1}{1-\sin^2 x}$$

in ascending powers of $\sin x$, giving in each case the first three non-zero terms. State the values of x in the interval $0 \leqslant x < 2\pi$ for which each expansion is valid.

Show that $\sec x$ can be expanded in the form
$$\cos x(1 + \sin^2 x + \sin^4 x + \ldots),$$

and verify that differentiating the first three terms in the expansion of $f(x)$ gives these same three terms.

JMB 1982

25. Given that

$$f(x) \equiv e^{-3x} \quad \text{and} \quad g(x) \equiv (1-x)^{-3} \qquad x \neq 1,$$

find the first three non-zero terms in the expansions in ascending powers of x of

(a) $f(x) - g(x)$,
(b) $\log_e f(x) - \log_e g(x)$,
(c) $\dfrac{f(x)}{g(x)}$.

In each of the expansions (a) and (b) obtain the coefficient of x^n for $n > 1$.

JMB 1982

26. The sum of the first three terms in the expansion in ascending powers of x of

$$(1 + ax) e^{bx + x^2}$$

has the form $(1 + cx)^2$. Show that

$$(a - b)^2 = 4 + 2b^2.$$

Find a pair of positive integers a and b satisfying the above condition choosing b to be as small as possible.

JMB 1982

14.3 Differential equations

Any equation relating y with x which contains one or more of the derivatives dy/dx, d^2y/dx^2, ... is called a *differential equation*.

As examples,

(a) $(x + y)\dfrac{dy}{dx} = 1 + x$

(b) $\dfrac{d^2y}{dx^2} + \dfrac{dy}{dx} + 2 = 0.$

(a) is an example of a *first order* equation since no derivative higher than dy/dx appears and (b) is an example of a *second order* equation since no derivative higher than d^2y/dx^2 appears.

Such equations occur in many of the sciences as well as mathematics. They are said to be solved if y can be expressed in terms of x, without involving derivatives. Integration is used to solve these equations and so arbitrary constants of integration are introduced.

For example, if

$$\frac{dy}{dx} = x$$

then

$$y = \frac{x^2}{2} + A$$

where A is an arbitrary constant. This is called the *general solution* of the differential equation. If it is known that $y = 3$ when $x = 2$ then we find $y = \frac{1}{2}x^2 + 1$. This solution which does not involve arbitrary constants, is called a *particular solution*.

Example 14.4

Show that $y = \sin x$ is a particular solution of the second order equation

$$\frac{d^2y}{dx^2} = -y.$$

If

$$y = \sin x$$

$$\Rightarrow \quad \frac{dy}{dx} = \cos x$$

$$\Rightarrow \frac{d^2y}{dx^2} = -\sin x$$

$$\Rightarrow \frac{d^2y}{dx^2} = -y \qquad \text{since } \sin x = y.$$

Thus $y = \sin x$ is a solution of the differential equation and since it contains no arbitrary constants it is a particular solution.

Finding solutions of differential equations can be difficult. One of the simpler types is dealt with as follows.

Variable separable

Suppose a differential equation can be expressed in the form

$$f(y)\frac{dy}{dx} = g(x)$$

then

$$\int f(y)\frac{dy}{dx}dx = \int g(x)\,dx.$$

So

$$\int f(y)\,dy = \int g(x)\,dx \quad \left(\text{since} \int \ldots \frac{dy}{dx}\,dx \equiv \int \ldots dx \right).$$

In this way the variables have been separated.

Example 14.5

Solve the differential equations.

(a) $\dfrac{dy}{dx} = 1 + y^2$ (b) $x^2 \dfrac{dy}{dx} = y$

(a) $$\frac{dy}{dx} = 1 + y^2$$

$$\Rightarrow \frac{1}{1+y^2}\frac{dy}{dx} = 1$$

$$\Rightarrow \int \frac{1}{1+y^2}\,dy = \int dx$$

$$\Rightarrow \quad \tan^{-1} y = x + A$$

$$\Rightarrow \qquad y = \tan(x + A)$$

(b) $$x^2 \frac{dy}{dx} = y$$

$$\Rightarrow \quad \frac{1}{y}\frac{dy}{dx} = \frac{1}{x^2}$$

$$\Rightarrow \int \frac{1}{y}\,dy = \int \frac{1}{x^2}\,dx$$

$$\Rightarrow \quad \ln y = \frac{-1}{x} + A$$

$$\Rightarrow \qquad y = e^{(-1/x + A)}$$

$$\Rightarrow \qquad y = B e^{-1/x}, \quad \text{where } B = e^A.$$

If the value of y is given for some value of x then the value of the arbitrary constant may be found and so we obtain a particular solution. Such values of x and y are called *initial conditions* or *boundary values*.

Example 14.6

Solve the equation

$$e^{(x+y)}\frac{dy}{dx} = 1$$

given that $y = \ln 2$ when $x = 0$.

$$e^{(x+y)}\frac{dy}{dx} = 1$$

$$\Rightarrow \quad e^y\frac{dy}{dx} = e^{-x} \qquad [\text{since } e^{x+y} = e^x e^y]$$

$$\Rightarrow \quad \int e^y\,dy = \int e^{-x}\,dx$$

$$\Rightarrow \qquad e^y = -e^{-x} + A$$

When $x = 0$, $y = \ln 2$ and so

$$2 = -1 + A \qquad \Rightarrow A = 3$$

$$\Rightarrow e^y = 3 - e^{-x}$$

$$\Rightarrow y = \ln(3 - e^{-x}).$$

Exercise 14.2

1. Use the method of separating the variable to solve the following differential equations.

   (a) $\dfrac{dy}{dx} = -\dfrac{x}{y}$

   (b) $\dfrac{dy}{dx} = \dfrac{1 + x^2}{y}$

   (c) $\dfrac{dy}{dx} = y e^x$

   (d) $e^x\dfrac{dy}{dx} = (1 + y)^2$

   (e) $x\dfrac{dy}{dx} = \tan y$

   (f) $(1 + x^2)\dfrac{dy}{dx} = \dfrac{x}{y}$

   (g) $y^2\dfrac{dy}{dx} + y e^x + \dfrac{dy}{dx} = 0$

   (h) $x\dfrac{dy}{dx} = y + x^2 y$

   (i) $y\dfrac{dy}{dx} = \cos x$

   (j) $\dfrac{dy}{dx} = e^y \sin x$

2. Find the particular solutions of the following differential equations.

   (a) $\sqrt{1 - x^2}\,\dfrac{dy}{dx} = 1$, if $y = 0$ when $x = 1$

   (b) $\tan x\dfrac{dy}{dx} = 1$, if $y = 1$ when $x = \pi/3$.

3. Solve the differential equation $\dfrac{d\theta}{dt} - 2\theta + 1 = 0$ given that $\theta = \pi$ when $t = 0$.

4. Given that

$$\sqrt{(9 - x^2)}\frac{dy}{dx} = (1 - y)^2$$

and that $y = 0$ when $x = 0$, calculate the exact value of y when $x = 3/2$.

JMB 1982

5. The gradient of the curve $y = f(x)$ is given by the differential equation

$$(x - 1)^3\frac{dy}{dx} + 8y^2 = 0$$

and the curve passes through the point $(2, -\frac{1}{3})$. By solving this differential equation show that

$$f(x) \equiv \frac{x^2 - 2x + 1}{x^2 - 2x - 3}.$$

Find the equations of the asymptotes to the curve and the coordinates and nature of its stationary point. Sketch the curve.
    Sketch also the curve whose equation is

$$y = \frac{1}{f(x)}.$$

JMB 1983

6. Solve the differential equation

$$xy\frac{dy}{dx} = 1 - x^2, \qquad x > 0,$$

given that $y = 2$ when $x = 1$.

LOND 1984

7. Solve the differential equation

$$\frac{dy}{dx} = y^2 e^{-2x},$$

given that $y = 1$ when $x = 0$. Give your answer in a form expressing y in terms of x.

CAMB 1982

8. Find the solution of the differential equation $2\frac{dy}{dx} = 2x e^{-2y} + e^{-2y}$ for which $y = 0$ when $x = 0$.

SU 1980

9. By first substituting $y = ux$, where u is a function of x, or otherwise, solve the differential equation

$$x\frac{dy}{dx} - y = x,$$

subject to the initial condition that $y = 1$ when $x = 1$.

O & C 1983

10. Find the equation of the curve which passes through the point $(1, 2)$ and is such that all points of the curve satisfy $x\frac{dy}{dx} + 2y = xy$, giving your answer in the form $y = f(x)$.

SU 1982

11. Given that $y > \frac{1}{2}$, find the general solution of the differential equation

$$\frac{dy}{dx} + 2xy = x.$$

Given that $y = 1$ when $x = 1$, express y in terms of x.

AEB 1982

12. Find the general solution of the differential equation

$$\frac{dy}{dx} = \frac{y^2}{(x^2 - x - 2)}$$

in the region $x > 2$. Find also the particular solution which satisfies $y = 1$ when $x = 5$.

OX 1983

13. For all positive values of x the gradient of a curve at the point (x, y) is $\dfrac{y}{x^2 + x}$.

The point $A(3, 6)$ lies on this curve.

(a) Calculate the equation of the normal to the curve at A.
(b) Find the equation of the curve in the form $y = f(x)$.

AEB 1982

14.4   Obtaining differential equations

If x and y are such that the rate of change of y w.r.t. x is proportional to x, then we may write

$$\frac{dy}{dx} = kx$$

where k is the constant of proportionality.

Though, if we know that the rate of *decrease* of y w.r.t. x is proportional to x then setting

$$\frac{dy}{dx} = -kx$$

is more helpful, the negative sign indicating that dy/dx is decreasing.

Example 14.7

A radioactive substance decays so that the rate of decrease of mass is proportional to the mass present at that time. Setting x to represent the mass remaining at time t, express x as a function of time.

Given that one quarter of the mass is lost in 10 days, find to the nearest day, the time required for the mass to be reduced to half its initial value.

We are given that

$$\frac{dx}{dt} = -kx$$

where k is a constant. Notice that when a rate of change of x is mentioned without specifying a variable, then x is changing w.r.t. time. So

$$\frac{1}{x}\frac{dx}{dt} = -k$$

$$\Rightarrow \int \frac{1}{x}\,dx = -k \int dt$$

$$\Rightarrow \quad \ln x = -kt + A.$$

When $t = 0$, $x = M$, say, the original mass

$$\Rightarrow \quad A = \ln M$$

$$\Rightarrow \ln x = -kt + \ln M$$

$$\Rightarrow \quad x = e^{-kt + \ln M}$$

$$\Rightarrow \quad x = e^{\ln M} e^{-kt}$$

$$\Rightarrow \quad x = M e^{-kt}$$

$t = 10$ when $x = \frac{3}{4}M$

$$\Rightarrow \quad \tfrac{3}{4} = e^{-k.10}$$

$$\Rightarrow \ln \tfrac{3}{4} = -k.10$$

$$\Rightarrow \quad k = \tfrac{1}{10}\ln \tfrac{4}{3}$$

To find t when $x = M/2$, we have

$$\frac{M}{2} = M e^{-kt}, \quad \text{where } k = \tfrac{1}{10} \ln \tfrac{4}{3}$$

$$\Rightarrow \quad \tfrac{1}{2} = e^{-kt}$$

$$\Rightarrow \quad -kt = \ln \tfrac{1}{2}$$

$$\Rightarrow \quad kt = \ln 2$$

$$\Rightarrow \quad t = \frac{10 \ln 2}{\ln \tfrac{4}{3}}$$

$$\Rightarrow \quad t \simeq 24 \text{ days.} \quad \boxed{\text{C}}$$

Exercise 14.3

1. A solution contains x bacteria at time t hours, where x increases at a rate proportional to x. Initially, there are 100 bacteria, which increase to 300 after 1 hour. How many bacteria are there after 2 hours? Find also the time, to the nearest minute when there are 600 bacteria.

2. The amount x of a particular chemical present at time t mins in a reaction is given by

$$\frac{dx}{dt} = 4k + x$$

   where k is a constant. Initially, there are $2k$ units of the substance present. Find the time taken for x to equal k.

3. A body of mass m falls from rest under gravity in a medium which offers a resistance of kmv^2, where v is the velocity and k is a constant.
   It can be shown in such a circumstance that

$$v \frac{dv}{dx} = k(u^2 - v^2)$$

   where u is a constant.
   Obtain v^2 in terms of x and the constants and hence express x in terms of v and the constants.

4. Newton's Law of cooling states that the rate of change of temperature of a substance is proportional to the difference between the temperature of the substance θ and the room temperature θ_0, assumed constant.

In fact,

$$\frac{d\theta}{dt} = -k(\theta - \theta_0)$$

Initially the substance has temperature

(a) $2\theta_0$,   (b) $\dfrac{\theta_0}{2}$.

In each case express the temperature of the substance in terms of t and the constant θ_0.

5. Two variables x and t are connected by the differential equation

$$\frac{dx}{dt} = \frac{kx}{10 - x},$$

where $0 < x < 10$ and where k is a constant. It is given that $x = 1$ when $t = 0$ and that $x = 2$ when $t = 1$. Find the value of t when $x = 5$, giving three significant figures in your answer.

CAMB 1983

6. Express

$$\frac{2}{(1 + x)(1 + 3x)}$$

in partial fractions.
    Hence, or otherwise, solve the differential equation

$$\frac{dy}{dx} = \frac{2(y + 2)}{(1 + x)(1 + 3x)}.$$

given that $y = -1$ when $x = 0$.

LOND 1983

7. The current, i, flowing in a certain circuit at any time t, is increasing at a rate proportional to $I - i$, where I is a constant.
    Given that $i = i_0$ when $t = 0$, prove that

$$I - i = (I - i_0)e^{-kt},$$

where k is a constant.
    Given that $i_0 = 1.5$ amp, that $i = 3.5$ amp when $t = 3$ s, and that $i = 5$ amp when $t = 6$ s, determine I.

AEB 1983

8. During a chemical reaction two substances A and B decompose. The number of grams, x, of substance A present at time t is given by

$$x = \frac{10}{(1+t)^3}.$$

There are y grams of B present at time t and $\dfrac{dy}{dt}$ is directly proportional to the product of x and y. Given that $y = 20$ and $\dfrac{dy}{dt} = -40$ when $t = 0$, show that

$$\frac{dy}{dt} = \frac{-2y}{(1+t)^3}.$$

Hence determine y as a function of t.

Determine the amount of substance B remaining when the reaction is essentially complete.

AEB 1982

9. For the curve $y = f(x)$ the second derivative is proportional to x. At the point $(-1, -4)$ on the curve the gradient is 8. Find the equation of this curve if it passes through the origin.

10. A curve C in the x-y plane has the property that the gradient of the tangent at the point $P(x, y)$ is three times the gradient of the line joining the point $(2, 1)$ to P. Express this property in the form of a differential equation. Given that $x > 2$ and $y > 1$ at all points on C, and that C passes through the point $(3, 2)$, find the equation of C in the form $y = f(x)$.

The curve C may be obtained by a translation of part of the curve $y = x^3$. Describe this translation.

JMB 1984

Chapter 15   Vectors

15.1   Vectors

Many quantities such as temperature or distance can be specified by a single number (in appropriate units). Such quantities are called scalars [perhaps because they are often measured on a scale!].

However, forces, velocities and many other quantities cannot be completely specified in this way. In order to describe these quantities uniquely both the magnitude and *direction* needs to be specified. One way to do this is to draw a *directed line segment* (Fig. 15.1).

Fig. 15.1   Fig. 15.2

In this way the direction may be specified uniquely as the direction of the line and the magnitude may be represented by the length of the line segment.

Such directed line segments are also representations of mathematical entities called *vectors*.

For convenience we shall often refer to the line segment as though it were the vector. If the end points of a vector are A and B respectively then the vector may be referred to as $\vec{AB}$ (or sometimes $\mathbf{AB}$). Then $\vec{BA}$ would be the vector with the same magnitude and opposite direction (Fig. 15.2).

Alternatively, a vector may be given a name $\mathbf{a}$ or $\mathbf{b}$ where $-\mathbf{a}$ would have the same magnitude but opposite direction to $\mathbf{a}$. The vector $\lambda\mathbf{a}$ where $\lambda \in \mathbb{R}$ is represented by a line segment parallel to $\mathbf{a}$ but λ times as long (see Fig. 15.3).

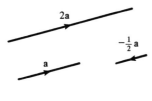

Fig. 15.3

Thus if $\mathbf{a}$ is parallel to $\mathbf{b}$ then there is a $\lambda \in \mathbb{R}$ such that $\mathbf{a} = \lambda\mathbf{b}$.

If $\lambda > 0$ then $\mathbf{a}$ and $\mathbf{b}$ are in the same direction and if $\lambda < 0$ then $\mathbf{a}$ and $\mathbf{b}$ are in opposite directions. If $\lambda = 0$ then $\mathbf{a} = \mathbf{0}$, the zero vector. The zero vector has zero magnitude and no defined direction.

It should be clear that two vectors are equal if and only if they are represented by line segments which have the same length and direction.

15.2 Vector addition

Suppose $\vec{AB}$ and $\vec{BC}$ are two vectors, we define $\vec{AB} + \vec{BC} = \vec{AC}$.

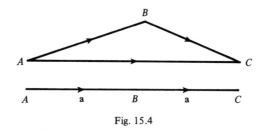

Fig. 15.4

Notice that it follows from the definition that if $\vec{AB} = \mathbf{a}$ and $\vec{BC} = \mathbf{a}$ then since

$$\vec{AB} + \vec{BC} = \vec{AC},$$

$$\Rightarrow \quad \mathbf{a} + \mathbf{a} = 2\mathbf{a},$$

which is as we would wish.

More generally, if $\mathbf{a}$ and $\mathbf{b}$ are any two vectors, then $\mathbf{a} + \mathbf{b}$ is equal to that vector $\mathbf{c}$ which relates to $\mathbf{a}$ and $\mathbf{b}$ as suggested in Fig. 15.5. Notice that the direction of the arrows is important, and $\mathbf{a}$ and $\mathbf{b}$ must be 'moved' to form two sides of a triangle with the arrows pointed in the same sense round the triangle. The sum is that vector represented by the third side of the triangle with its arrow pointed in the opposite sense round the triangle.

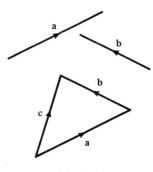

Fig. 15.5

This definition of addition is sometimes referred to as the triangle law of addition.

The link between vectors and 'vector quantities' such as force and velocity is more than of just passing interest, since the application of two forces is in many respects equivalent to the application of a single force. The magnitude and direction of this single force bears the same relationship to the two forces as $\mathbf{a} + \mathbf{b}$ does to $\mathbf{a}$ and $\mathbf{b}$! It is because of this link that the study of vectors has assumed so much importance.

Example 15.1

Show that for any vectors $\mathbf{a}$ and $\mathbf{b}$

$$\mathbf{a} + \mathbf{b} = \mathbf{b} + \mathbf{a}.$$

We may feel intuitively that this is obvious, but in order to prove it we must return to the definition of addition.

Suppose $\vec{AB} = \mathbf{a}$ and $\vec{BC} = \mathbf{b}$. Assuming $\vec{AB}$ is not parallel to $\vec{BC}$ then a point D can be found so that $ABCD$ forms a parallelogram (see Fig. 15.6).

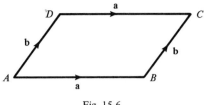

Fig. 15.6

Clearly AD is parallel to BC with the same magnitude so $\vec{AD} = \mathbf{b}$. Similarly $\vec{DC} = \mathbf{a}$.
  Now

$$\mathbf{a} + \mathbf{b} = \vec{AB} + \vec{BC} = \vec{AC} \quad \text{by definition}$$

but also

$$\mathbf{b} + \mathbf{a} = \vec{AD} + \vec{DC} = \vec{AC} \quad \text{by definition.}$$

It follows immediately that

$$\mathbf{a} + \mathbf{b} = \mathbf{b} + \mathbf{a}.$$

If $\mathbf{a}$ is parallel to $\mathbf{b}$ so that $\mathbf{b} = \lambda\mathbf{a}$ then

$$\mathbf{a} + \mathbf{b} = \mathbf{a} + \lambda\mathbf{a} = (1 + \lambda)\mathbf{a}$$

and

$$\mathbf{b} + \mathbf{a} = \lambda\mathbf{a} + \mathbf{a} = (\lambda + 1)\mathbf{a}.$$

But $1 + \lambda = \lambda + 1$, since we are dealing with real numbers, and so again the required result follows. A parallelogram is useful when vectors are displayed as in Fig. 15.7 where $\mathbf{a}$ and $\mathbf{b}$ are represented by two adjacent sides of a parallelogram. Here the directions are not in the same sense round the parallelogram and $\mathbf{a} + \mathbf{b} = \mathbf{c}$.

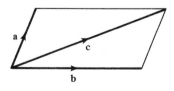

Fig. 15.7

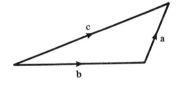

Fig. 15.8

This is referred to as the *parallelogram law of vector addition*. Compare this with the triangle law of vector addition (Fig. 15.8). Both these laws describe the same rule of addition of vectors.

Example 15.2

In $\triangle ABC$, M is the midpoint of AB (Fig. 15.9).

If $\vec{AM} = \mathbf{a}$ and $\vec{BC} = \mathbf{b}$, describe each of the following vectors in terms of $\mathbf{a}$ and $\mathbf{b}$.

(a) $\vec{MB}$ (b) $\vec{CM}$ (c) $\vec{AC}$

(a) $\vec{MB} = \mathbf{a}$ since MB has the same length
   as AM and is clearly parallel to AM.
(b) Notice that $\vec{MB} + \vec{BC} = \vec{MC}$, by
   definition.

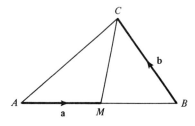

$$\Rightarrow \mathbf{a} + \mathbf{b} = \vec{MC}$$

$$\Rightarrow \vec{CM} = -(\mathbf{a} + \mathbf{b})$$

(c) Notice that $\vec{AB} + \vec{BC} = \vec{AC}$ where
   $\vec{AB} = 2\mathbf{a}$ and $\vec{BC} = \mathbf{b}$, so $\vec{AC} = 2\mathbf{a} + \mathbf{b}$.

Fig. 15.9

Example 15.3

In the tetrahedron $ABCD$, $\vec{AB} = \mathbf{a}$, $\vec{AC} = \mathbf{b}$, and $\vec{BD} = \mathbf{c}$. Find, in terms of $\mathbf{a}$, $\mathbf{b}$ and $\mathbf{c}$.

(a) $\vec{AD}$,   (b) $\vec{DC}$.

(a) Consider triangle $\triangle ABD$.

$$\vec{AD} = \vec{AB} + \vec{BD}$$

$$\Rightarrow \vec{AD} = \mathbf{a} + \mathbf{c}.$$

(b) Consider triangle $\triangle ACD$.

$$\vec{DC} = \vec{DA} + \vec{AC}$$

$$\Rightarrow \vec{DC} = -(\mathbf{a} + \mathbf{c}) + \mathbf{b}$$

$$= \mathbf{b} - (\mathbf{a} + \mathbf{c})$$

Exercise 15.1

1. In the parallelogram $ABCD$ (Fig. 15.10), find $\vec{DB}$ in terms of $\mathbf{a}$ and $\mathbf{b}$ if

    (a) $\vec{DA} = \mathbf{a}$ and $\vec{AB} = \mathbf{b}$
    (b) $\vec{BC} = \mathbf{b}$ and $\vec{AB} = \mathbf{a}$
    (c) $\vec{BA} = -\mathbf{a}$ and $\vec{DA} = \mathbf{b}$

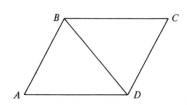

Fig. 15.10

2. In $\triangle ABC$ (Fig. 15.11), M is the mid-
   point of AB and N is the midpoint of
   BC.

   By letting $\vec{BM} = \mathbf{a}$ and $\vec{BN} = \mathbf{b}$
   express both $\vec{MN}$ and $\vec{AC}$ in terms of
   a and **b** and hence deduce that MN is
   parallel to AC.

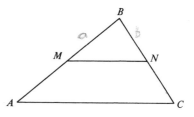

Fig. 15.11

3. Vectors are not restricted to two dimensions.
   In the cuboid shown in Fig. 15.12, suppose

   $$\vec{EH} = \mathbf{a}$$

   $$\vec{EF} = \mathbf{b}$$

   $$\vec{EA} = \mathbf{c}.$$

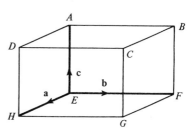

Fig. 15.12

   Express the following vectors in terms of **a**, **b** and **c**.

   (a) $\vec{FB}$ (b) $\vec{EB}$ (c) $\vec{BC}$ (d) $\vec{EC}$
   (e) $\vec{HC}$ (f) $\vec{HB}$ (g) $\vec{AG}$ (h) $\vec{FD}$

4. Which of the following are vectors and which are scalars?

   (a) volume,   (b) speed,   (c) mass,
   (d) acceleration,   (e) velocity

5. What can be deduced if $\vec{AB} = \mathbf{a}$, $\vec{CD} = \mathbf{b}$ and $\mathbf{a} = \lambda\mathbf{b}$ for some $\lambda \in \mathbb{R}$?

6. In the diagram (Fig. 15.13), AC is
   parallel to DE, BC is 3 cm long and
   DE is 6 cm long. Given that

   $$\vec{BF} = \mathbf{a} - \mathbf{b}$$

   $$\vec{DF} = 2\mathbf{b}$$

   $$\vec{DA} = 3\mathbf{b} - 2\mathbf{a},$$

   find in terms of **a** and **b**:

   (a) $\vec{FC}$,   (b) $\vec{BC}$,   (c) $\vec{DE}$.

   Show also that AB is of length 3 cm.

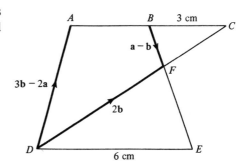

Fig. 15.13

7. The regular hexagon $ABCDEF$ with centre O is such that $\vec{AB} = \mathbf{a}$ and $\vec{BC} = \mathbf{b}$. Find in terms of $\mathbf{a}$ and $\mathbf{b}$

    (a) $\vec{OC}$,   (b) $\vec{FC}$,   (c) $\vec{OB}$,   (d) $\vec{FA}$,   (e) $\vec{AE}$.

15.3 Some definitions

Definition

> The modulus of the vector $\mathbf{a}$, written $|\mathbf{a}|$ or simply a, is the magnitude (or length) of the line segment corresponding to $\mathbf{a}$.
> A *unit vector* is a vector which has modulus 1 unit.

If $\mathbf{a}$ is a vector then $\hat{\mathbf{a}}$ is used to denote the unit vector in the same direction as $\mathbf{a}$. It follows that for a vector $\mathbf{a}$

$$\mathbf{a} = a\hat{\mathbf{a}}$$

and so

$$\hat{\mathbf{a}} = \frac{\mathbf{a}}{a}.$$

Definition

> Two or more vectors are said to be coplanar if the line segments corresponding to the vectors may be drawn in the same plane.

It is interesting to note that any *two* vectors are coplanar, and in fact if they are not parallel they determine uniquely the orientation of the plane.

Statement

> If $\mathbf{a}$, $\mathbf{b}$ and $\mathbf{c}$ are coplanar, and $\mathbf{a}$ and $\mathbf{b}$ are not parallel, then there are reals k and l such that
>
> $$\mathbf{c} = k\mathbf{a} + l\mathbf{b}.$$

In other words, $\mathbf{c}$ can be expressed in terms of $\mathbf{a}$ and $\mathbf{b}$.

To see that this is so (Fig. 15.14), let $\vec{AC} = \mathbf{c}$, where AC is the diagonal of a parallelogram, with one side AB parallel to $\mathbf{a}$ and an adjacent side AD parallel to $\mathbf{b}$. For some k, $\vec{AB} = k\mathbf{a}$ and for some l, $\vec{AD} = l\mathbf{b}$.

It then follows from the parallelogram law of vector addition, that $k\mathbf{a} + l\mathbf{b} = \mathbf{c}$.

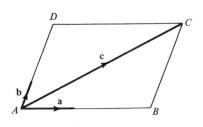

Fig. 15.14

15.4   Cartesian components

In the x-y plane it is usual to consider two specific unit vectors: **i** in the direction of the positive x axis and **j** in the direction of the positive y axis (Fig. 15.15). [Exceptionally a circumflex is not used to denote these unit vectors.]

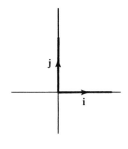

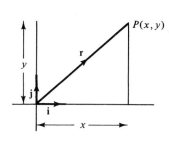

Fig. 15.15   Fig. 15.16

By what has been said above, any vector **r** in the xy plane may be expressed in terms of **i** and **j**. Indeed if $\mathbf{r} = \vec{OP}$, where O is the origin and P has co-ordinates (x, y) then

$$\mathbf{r} = x\mathbf{i} + y\mathbf{j}$$

(see Fig. 15.16). The vector **r** is sometimes expressed in the form $\begin{pmatrix} x \\ y \end{pmatrix}$, which is referred to as a *column vector*.

Example 15.4
If $\mathbf{r} = 3\mathbf{i} + 4\mathbf{j}$ find

(a) $|\mathbf{r}|$ (b) $\hat{\mathbf{r}}$.

(a) The modulus of **r** is found using Pythagoras' Theorem

$$r = |\mathbf{r}| = \sqrt{3^2 + 4^2}$$
$$= 5.$$

(b) The unit vector in the direction of **r** is given by

$$\hat{\mathbf{r}} = \frac{\mathbf{r}}{r}$$
$$= \frac{1}{5}(3\mathbf{i} + 4\mathbf{j})$$
$$= \frac{3}{5}\mathbf{i} + \frac{4}{5}\mathbf{j}.$$

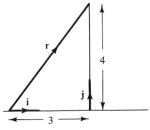

Fig. 15.17

Statement

$$\text{If } \mathbf{r} = x\mathbf{i} + y\mathbf{j} \quad \text{then} \quad |\mathbf{r}| = r = \sqrt{x^2 + y^2}$$

The extension to three dimensions

In three dimensions, the axes are traditionally called x, y and z. Unit vectors in the direction of the axes are called $\mathbf{i}$, $\mathbf{j}$ and $\mathbf{k}$ respectively, as indicated in Fig. 15.18. If any vector $\mathbf{r}$ is represented by a line segment $\vec{OR}$ where O is the origin and R has coordinates (a, b, c) [a units along the x axis, b units along the y axis and c units along the z axis], then

$$\mathbf{r} = a\mathbf{i} + b\mathbf{j} + c\mathbf{k}.$$

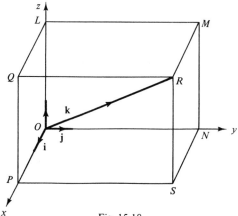

Fig. 15.18

[This is merely an extension to three dimensions of the corresponding situation in two dimensions — the plane.]

To check this result, notice that

$$\vec{OP} = a\mathbf{i}, \quad \vec{PS} = \vec{ON} = b\mathbf{j} \quad \text{and} \quad \vec{SR} = \vec{OL} = c\mathbf{k}.$$

Then, by the definition of vector addition,

$$\vec{OP} + \vec{PS} = \vec{OS} \quad \text{and} \quad \vec{OS} + \vec{SR} = \vec{OR}.$$

In other words,

$$\vec{OP} + \vec{PS} + \vec{SR} = \vec{OR}.$$

Substituting

$$a\mathbf{i} + b\mathbf{j} + c\mathbf{k} = \mathbf{r},$$

It is a simple matter to check that

$$r = \sqrt{a^2 + b^2 + c^2}$$

Example 15.5

Find a unit vector in the direction of the vector

$$4\mathbf{i} - 2\mathbf{j} - 4\mathbf{k}.$$

If $\mathbf{r} = 4\mathbf{i} - 2\mathbf{j} - 4\mathbf{k}$,

$$r = |\mathbf{r}| = \sqrt{16 + 4 + 16}$$
$$= 6.$$

So

$$\hat{\mathbf{r}} = \frac{1}{6}(4\mathbf{i} - 2\mathbf{j} - 4\mathbf{k})$$
$$= \frac{2\mathbf{i}}{3} - \frac{\mathbf{j}}{3} - \frac{2\mathbf{k}}{3}$$

Statement

> (i) If $\mathbf{r} = a\mathbf{i} + b\mathbf{j} + c\mathbf{k}$ and $\mathbf{s} = d\mathbf{i} + e\mathbf{j} + f\mathbf{k}$
> then $\mathbf{r} + \mathbf{s} = (a + d)\mathbf{i} + (b + e)\mathbf{j} + (c + f)\mathbf{k}$.
> (ii) If $\mathbf{r} = \mathbf{s}$ then $a = d$, $b = e$ and $c = f$.
> (iii) $\lambda\mathbf{r} = \lambda a\mathbf{i} + \lambda b\mathbf{j} + \lambda c\mathbf{k}$.

So, for example,

$$3(2\mathbf{i} - \mathbf{j} + \mathbf{k}) - (4\mathbf{i} - \mathbf{j} - 2\mathbf{k})$$
$$= 2\mathbf{i} - 2\mathbf{j} + 5\mathbf{k}.$$

The proofs are omitted but depend only on the fact that for any vectors $\mathbf{a}$, $\mathbf{b}$ and $\mathbf{c}$

$$(\mathbf{a} + \mathbf{b}) + \mathbf{c} = \mathbf{a} + (\mathbf{b} + \mathbf{c}) \qquad \text{(the \textit{associative law}),}$$

and

$$\mathbf{a} + \mathbf{b} = \mathbf{b} + \mathbf{a} \qquad \text{(proved earlier — the \textit{commutative law}).}$$

Example 15.6

Show that $\mathbf{c} = \mathbf{i} - 5\mathbf{j}$ may be expressed in terms of $\mathbf{a}$ and $\mathbf{b}$ where

$$\mathbf{a} = 2\mathbf{i} - \mathbf{j} + 3\mathbf{k} \quad \text{and} \quad \mathbf{b} = \mathbf{i} + \mathbf{j} + 2\mathbf{k}.$$

Suppose there are reals l and m so that

$$\mathbf{c} = l\mathbf{a} + m\mathbf{b}$$

then

$$\mathbf{i} - 5\mathbf{j} + 0\mathbf{k} = l(2\mathbf{i} - \mathbf{j} + 3\mathbf{k}) + m(\mathbf{i} + \mathbf{j} + 2\mathbf{k})$$
$$= (2l + m)\mathbf{i} + (m - l)\mathbf{j} + (3l + 2m)\mathbf{k}$$

It follows that

$$2l + m = 1 \qquad ①$$
$$m - l = -5 \qquad ②$$

$$3l + 2m = 0 \qquad ③$$

① − ② gives

$$3l = 6 \qquad \Rightarrow l = 2.$$

Substitute into ①:

$$4 + m = 1 \qquad \Rightarrow m = -3.$$

It must be stressed that we need to check that all three equations hold, which they do
in this case. Hence

$$c = 2a - 3b.$$

Notice that this proves that **a**, **b** and **c** are coplanar. If they were not coplanar we
would not have found solutions for l and m which satisfy all *three* equations.

Exercise 15.2

1. Find the magnitude of each of the following vectors
   (a) $4i - 3j$,   (b) $5i + 12j$,   (c) $i + j$.

2. Find the unit vector in the direction of the following vectors:
   (a) $-5i + 12j$,   (b) $i - j$,   (c) $3i - 2j$.

3. If $r = ai + bj$ and $s = ci + dj$ draw suitable diagrams to show that
   $$r + s = (a + c)i + (b + d)j.$$
   Hence simplify the following:
   (a) $(2i + j) + (3i - j)$,   (b) $(i - 2j) + (3i - 5j)$,
   (c) $\begin{pmatrix} 4 \\ 1 \end{pmatrix} + \begin{pmatrix} 2 \\ 3 \end{pmatrix}$,   (d) $(4i + j) - (3i + 2j)$,
   (e) $(2i + 3j) - (3i - j)$,   (f) $2(3i + 4j) - 3(4i + j)$.

4. If $a = 2i - j$ and $b = i + 2j$, express each of the following vectors in terms of **a**
   and **b**:
   (a) $3i + j$,   (b) $5i$,   (c) $-5j$,   (d) $7i + 4j$.

5. If $r = 2i - 3j + k$ and $s = 4i + j + 2k$ find the following in terms of **i**, **j** and **k**:
   (a) $r + s$,   (b) $r - s$,   (c) $2r$,   (d) $2r - s$,   (e) $3r - 2s$.

6. Find the modulus of the following vectors
   (a) r,   (b) $p + q$,   (c) $2p - q$,

given that

$$\mathbf{r} = x\mathbf{i} + y\mathbf{j} + \mathbf{k},$$
$$\mathbf{p} = 2\mathbf{i} - 3\mathbf{j} + \mathbf{k},$$
$$\mathbf{q} = \mathbf{i} + 2\mathbf{j} - \mathbf{k}.$$

7. Find the unit vector in the direction of $\mathbf{r}$ where

    (a) $\mathbf{r} = 2\mathbf{i} + \mathbf{j} - 3\mathbf{k}$ (b) $\mathbf{r} = 4\mathbf{i} - \mathbf{j} + 2\mathbf{k}$

8. If $\mathbf{a} = \mathbf{i} + \mathbf{j} + 2\mathbf{k}$ and $\mathbf{b} = 3\mathbf{i} - 2\mathbf{j} + \mathbf{k}$, express the vector $\mathbf{c} = -\mathbf{i} + 4\mathbf{j} + 3\mathbf{k}$ in terms of $\mathbf{a}$ and $\mathbf{b}$.

9. If

    $$\mathbf{a} = \mathbf{i} + \mathbf{j} + \mathbf{k} \qquad \mathbf{b} = 2\mathbf{i} - 3\mathbf{j} + \mathbf{k}$$
    $$\mathbf{c} = -\mathbf{i} + 9\mathbf{j} + \mathbf{k} \qquad \mathbf{d} = 3\mathbf{i} - 3\mathbf{j} + 4\mathbf{k}$$

    show that $\mathbf{a}$, $\mathbf{b}$ and $\mathbf{c}$ are coplanar, but that $\mathbf{a}$, $\mathbf{b}$ and $\mathbf{d}$ are not.

10. Show that $\mathbf{a} = 2\mathbf{i} - 3\mathbf{j} + \mathbf{k}$ and $\mathbf{b} = 4\mathbf{i} - 6\mathbf{j} + 2\mathbf{k}$ are parallel. If $\mathbf{c} = -\mathbf{i} + m\mathbf{j} - \frac{1}{2}\mathbf{k}$ is parallel to $\mathbf{a}$ and $\mathbf{b}$ find the value of m.

15.5 The scalar (or dot) product

If $\mathbf{a}$ and $\mathbf{b}$ are any vectors we define $\mathbf{a} . \mathbf{b}$ to mean $ab \cos \theta$ where θ is the angle between the two vectors. (Recall that any two vectors can be represented by line segments in a plane.)

Definition

For any vectors $\mathbf{a}$ and $\mathbf{b}$

$$\mathbf{a} . \mathbf{b} = ab \cos \theta$$

where θ is the angle between the vectors.

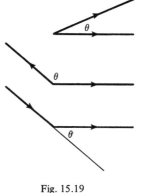

Fig. 15.19

Notice that $\mathbf{a} \cdot \mathbf{b}$ is a *scalar quantity*, a number, not a vector. The following facts are immediate observations from the definition.

Statement

> (i) $\mathbf{a} \cdot \mathbf{a} = a^2$ the angle θ between parallel vectors is zero, so $\cos \theta = 1$
> (ii) $\mathbf{a} \cdot \mathbf{b} = \mathbf{b} \cdot \mathbf{a}$
> (iii) $\mathbf{a} \cdot \mathbf{b} = 0 \Rightarrow \mathbf{a} = \mathbf{0}$ or $\mathbf{b} = \mathbf{0}$ or $\cos \theta = 0$.
>   (In the latter case $\theta = \pi/2$ and the vectors are said to be *perpendicular*.)

It should be remembered that to check that two non zero vectors are perpendicular it suffices to show that the dot product is zero.

In the case of the unit vectors $\mathbf{i}$, $\mathbf{j}$ and $\mathbf{k}$ it follows that

$$\mathbf{i} \cdot \mathbf{j} = 0 = \mathbf{j} \cdot \mathbf{i}$$
$$\mathbf{i} \cdot \mathbf{k} = 0 = \mathbf{k} \cdot \mathbf{i}$$
$$\mathbf{j} \cdot \mathbf{k} = 0 = \mathbf{k} \cdot \mathbf{j}$$
and
$$\mathbf{i} \cdot \mathbf{i} = \mathbf{j} \cdot \mathbf{j} = \mathbf{k} \cdot \mathbf{k} = 1$$

A geometrical interpretation

Suppose $\vec{OA} = \mathbf{a}$ and $\vec{OB} = \mathbf{b}$. Now

$$\mathbf{a} \cdot \mathbf{b} = ab \cos \theta$$
$$= \mathbf{a} \cdot b \cos \theta$$

where $b \cos \theta$ is the 'projection' of OB onto OA.
Similarly,

$$\mathbf{a} \cdot \mathbf{b} = ab \cos \theta$$
$$= \mathbf{b} \cdot a \cos \theta$$

where $a \cos \theta$ is the 'projection' of OA onto OB.

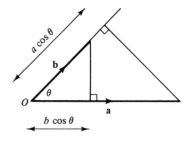

Fig. 15.20

$b \cos \theta$ is called the component (or resolved part) of $\mathbf{b}$ in the direction of $\mathbf{a}$ and $a \cos \theta$ is the component (or resolved part) of $\mathbf{a}$ in the direction of $\mathbf{b}$.

It is interesting to note that if $\mathbf{r} = x\mathbf{i} + y\mathbf{j}$ then

$$\mathbf{r} \cdot \mathbf{i} = |\mathbf{i}| \, r \cos \theta$$
$$= x$$

and

$$\mathbf{r}.\mathbf{j} = |\mathbf{j}|\, r \cos\left(\frac{\pi}{2} - \theta\right)$$

$$= r \sin \theta$$

$$= y.$$

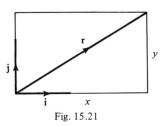

Fig. 15.21

So x and y are the components (or resolved part) of $\mathbf{r}$ in the direction of $\mathbf{i}$ and $\mathbf{j}$ respectively.

Statement

> If $\mathbf{a}$, $\mathbf{b}$ and $\mathbf{c}$ are any vectors then
>
> (i) $(\lambda\mathbf{a}).\mathbf{b} = \lambda(\mathbf{a}.\mathbf{b})$ for $\lambda \in \mathbb{R}$
> (ii) $\mathbf{a}.(\mathbf{b} + \mathbf{c}) = \mathbf{a}.\mathbf{b} + \mathbf{a}.\mathbf{c}$

The first result is easy to prove since $|\lambda\mathbf{a}| = \lambda a$, so:

$$(\lambda\mathbf{a}).\mathbf{b} = |\lambda\mathbf{a}|\, b \cos \theta$$

$$= \lambda ab \cos \theta$$

$$= \lambda(\mathbf{a}.\mathbf{b})$$

where θ is the angle between $\mathbf{a}$ and $\mathbf{b}$.

For the second result, suppose $\vec{OA} = \mathbf{a}$, $\vec{OB} = \mathbf{b}$ and $\vec{BC} = \mathbf{c}$, then

$$\mathbf{a}.\mathbf{b} = a.|OD|$$

$$\mathbf{a}.\mathbf{c} = a.|DE|$$

and

$$\mathbf{a}.(\mathbf{b} + \mathbf{c}) = a.|OE|$$

but $|OE| = |OD| + |DE|$, so

$$\mathbf{a}.\mathbf{b} + \mathbf{a}.\mathbf{c} = \mathbf{a}.(\mathbf{b} + \mathbf{c}).$$

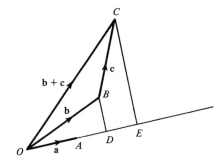

Fig. 15.22

Example 15.7

Show that $a\mathbf{i}.(b\mathbf{i} + c\mathbf{j} + d\mathbf{k}) = ab$.

We have

$$a\mathbf{i}.(b\mathbf{i} + c\mathbf{j} + d\mathbf{k}) = ab\mathbf{i}.\mathbf{i} + ac\mathbf{i}.\mathbf{j} + ad\mathbf{i}.\mathbf{k}$$

$$= ab$$

since $\mathbf{i}.\mathbf{j} = 0 = \mathbf{i}.\mathbf{k}$ and $\mathbf{i}.\mathbf{i} = 1$.

Statement

> If $\mathbf{r} = c\mathbf{i} + b\mathbf{j} + c\mathbf{k}$ and $\mathbf{s} = d\mathbf{i} + e\mathbf{j} + f\mathbf{k}$, then
>
> $$\mathbf{r} \cdot \mathbf{s} = ad + be + cf.$$

We shall take the above result on trust, though it can be proved in a way similar to that shown in the previous example.

Notice that if

$$\mathbf{r} = x\mathbf{i} + y\mathbf{j} + z\mathbf{k}$$

then from the above result

$$\mathbf{r} \cdot \mathbf{r} = x^2 + y^2 + z^2$$

and from the definition

$$\mathbf{r} \cdot \mathbf{r} = rr\cos\theta$$
$$= r \cdot r$$
$$= \sqrt{x^2 + y^2 + z^2} \cdot \sqrt{x^2 + y^2 + z^2}$$
$$= x^2 + y^2 + z^2$$

which is what we would expect.

Example 15.8

Show that the vectors $2\mathbf{i} - 2\mathbf{j} + \mathbf{k}$ and $3\mathbf{i} + 4\mathbf{j} + 2\mathbf{k}$ are perpendicular.

Now

$$(2\mathbf{i} - 2\mathbf{j} + \mathbf{k}) \cdot (3\mathbf{i} + 4\mathbf{j} + 2\mathbf{k}) = 2.3 - 2.4 + 1.2$$
$$= 0$$

and since neither vector has zero length we may deduce that they are perpendicular.

Example 15.9

Find the cosine of the angle between the vectors $3\mathbf{i} + \mathbf{j} - \mathbf{k}$ and $2\mathbf{i} + 3\mathbf{j} - \mathbf{k}$.

$$(3\mathbf{i} + \mathbf{j} - \mathbf{k}) \cdot (2\mathbf{i} + 3\mathbf{j} - \mathbf{k}) = 3.2 + 1.3 - 1.(-1)$$
$$= 10$$

but also, since

$$|3\mathbf{i} + \mathbf{j} - \mathbf{k}| = \sqrt{3^2 + 1^2 + (-1)^2} = \sqrt{11}$$

and

$$|2\mathbf{i} + 3\mathbf{j} - \mathbf{k}| = \sqrt{2^2 + 3^2 + (-1)^2} = \sqrt{14}$$

we have

$$(3\mathbf{i} + \mathbf{j} - \mathbf{k}) \cdot (2\mathbf{i} + 3\mathbf{j} - \mathbf{k}) = \sqrt{11} \cdot \sqrt{14} \cos \theta$$

from the definition, where θ is the angle between the vectors. It follows that

$$\sqrt{11} \cdot \sqrt{14} \cdot \cos \theta = 10$$

$$\Rightarrow \qquad \cos \theta = \frac{10}{\sqrt{11}\sqrt{14}}$$

$$= \frac{10}{\sqrt{154}}.$$

We could then obtain the angle θ, if required.

If $\cos \theta$ is negative, then we shall obtain an obtuse angle.

Direction cosines

If $\mathbf{r}$ is any vector in three dimensions, then the cosine of the angle that $\mathbf{r}$ makes with each of the vectors $\mathbf{i}$, $\mathbf{j}$ and $\mathbf{k}$ are called the *direction cosines l, m* and *n* respectively.

Now

$$\mathbf{r} \cdot \mathbf{i} = r \cos \theta_1 \qquad \text{where } \cos \theta_1 = l$$

$$\mathbf{r} \cdot \mathbf{j} = r \cos \theta_2 \qquad \text{where } \cos \theta_2 = m$$

and

$$\mathbf{r} \cdot \mathbf{k} = r \cos \theta_3 \qquad \text{where } \cos \theta_3 = n$$

But if $\mathbf{r} = x\mathbf{i} + y\mathbf{j} + z\mathbf{k}$, then

$$\mathbf{r} \cdot \mathbf{i} = x, \quad \mathbf{r} \cdot \mathbf{j} = y \quad \text{and} \quad \mathbf{r} \cdot \mathbf{k} = z.$$

It follows that

Statement

> If $\mathbf{r} = x\mathbf{i} + y\mathbf{j} + z\mathbf{k}$ and l, m and n are the direction cosines then
>
> $$l = \frac{x}{r}, \quad m = \frac{y}{r} \quad \text{and} \quad n = \frac{z}{r}.$$
>
> Further
>
> $$l^2 + m^2 + n^2 = 1.$$

This latter result is almost immediate since

$$l^2 + m^2 + n^2 = \frac{x^2}{r^2} + \frac{y^2}{r^2} + \frac{z^2}{r^2}$$

$$= \frac{x^2 + y^2 + z^2}{r^2}$$

but $r^2 = x^2 + y^2 + x^2$!

Example 15.10

Find the direction cosines of $\mathbf{r} = 2\mathbf{i} - \mathbf{j} + 2\mathbf{k}$.

Now

$$\mathbf{r} = \sqrt{4 + 1 + 4}$$
$$= 3$$

So $l = 2/3$, $m = -1/3$ and $n = 2/3$.

Example 15.11

If $|\mathbf{a}| = |\mathbf{b}|$ show that $\mathbf{a} + \mathbf{b}$ and $\mathbf{a} - \mathbf{b}$ are perpendicular, assuming neither is the zero vector.

It suffices to show that the dot product of $\mathbf{a} + \mathbf{b}$ with $\mathbf{a} - \mathbf{b}$ is zero.
Now

$$(\mathbf{a} + \mathbf{b}) . (\mathbf{a} - \mathbf{b}) = \mathbf{a} . \mathbf{a} - \mathbf{a} . \mathbf{b} + \mathbf{b} . \mathbf{a} - \mathbf{b} . \mathbf{b}$$
$$= a^2 - \mathbf{a} . \mathbf{b} + \mathbf{a} . \mathbf{b} - b^2$$
$$= 0, \quad \text{since } a = b. \ [\text{NB } a = |\mathbf{a}|]$$

The proof of the cosine rule

Consider the triangle $\triangle ABC$ (Fig. 15.23) where

$$\vec{CB} = \mathbf{a}$$
$$\vec{AC} = \mathbf{b}$$
$$\vec{AB} = \mathbf{c}.$$

Then $\mathbf{a} . \mathbf{a} = a^2$, and also $\mathbf{a} = \mathbf{c} - \mathbf{b}$. So

$$\mathbf{a} . \mathbf{a} = (\mathbf{c} - \mathbf{b}) . (\mathbf{c} - \mathbf{b})$$
$$= \mathbf{c} . \mathbf{c} - \mathbf{c} . \mathbf{b} - \mathbf{b} . \mathbf{c} + \mathbf{b} . \mathbf{b}$$
$$= c^2 - 2\mathbf{b} . \mathbf{c} + b^2$$
$$= b^2 + c^2 - 2bc \cos A.$$

So

$$a^2 = b^2 + c^2 - 2bc \cos A. \quad !$$

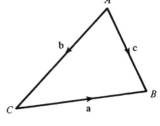

Fig. 15.23

The proof of the compound formulae in trigonometry

Any vector of the form

$$\hat{\mathbf{r}} = \cos A\mathbf{i} + \sin A\mathbf{j}$$

is a unit vector:

$$|\hat{\mathbf{r}}|^2 = \hat{\mathbf{r}} \cdot \hat{\mathbf{r}} = (\cos A\mathbf{i} + \sin A\mathbf{j}) \cdot (\cos A\mathbf{i} + \sin A\mathbf{j})$$
$$= \cos^2 A + \sin^2 A$$
$$= 1$$

The vector, $\hat{\mathbf{s}} = -\sin A\mathbf{i} + \cos A\mathbf{j}$, is also a unit vector—readily checked as for $\hat{\mathbf{r}}$, which is $\pi/2$ 'ahead' of $\hat{\mathbf{r}}$ in the sense depicted in Fig. 15.24. It is as though $\hat{\mathbf{r}}$ and $\hat{\mathbf{s}}$ were obtained by rotating $\mathbf{i}$ and $\mathbf{j}$ clockwise through an angle A.

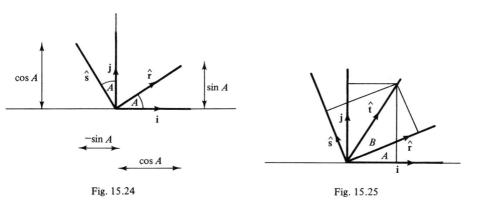

Fig. 15.24   Fig. 15.25

Any unit vector $\hat{\mathbf{t}}$ in the plane determined by $\mathbf{i}$ and $\mathbf{j}$ (and hence $\hat{\mathbf{r}}$ and $\hat{\mathbf{s}}$) may be expressed in terms of $\hat{\mathbf{r}}$ and $\hat{\mathbf{s}}$ as (say)

$$\hat{\mathbf{t}} = \cos B\hat{\mathbf{r}} + \sin B\hat{\mathbf{s}}$$

but then

$$\hat{\mathbf{t}} = \cos(A + B)\mathbf{i} + \sin(A + B)\mathbf{j}$$

as indicated in Fig. 15.25.

Now since

$$\hat{\mathbf{t}} = \cos B\hat{\mathbf{r}} + \sin B\hat{\mathbf{s}}$$

where

$$\hat{\mathbf{r}} = \cos A\mathbf{i} + \sin A\mathbf{j} \quad \text{and} \quad \hat{\mathbf{s}} = -\sin A\mathbf{i} + \cos A\mathbf{j}$$

we have

$$\hat{\mathbf{t}} = \cos B(\cos A\mathbf{i} + \sin A\mathbf{j}) + \sin B(-\sin A\mathbf{i} + \cos A\mathbf{j}).$$
$$\Rightarrow \hat{\mathbf{t}} = (\cos A \cos B - \sin A \sin B)\mathbf{i} + (\sin A \cos B + \cos A \sin B)\mathbf{j}$$

and since $\hat{\mathbf{t}}$ is also represented as

$$\hat{\mathbf{t}} = \cos(A + B)\mathbf{i} + \sin(A + B)\mathbf{j}$$

it follows that

$$\cos(A + B) = \cos A \cos B - \sin A \sin B$$

$$\sin(A + B) = \sin A \cos B + \cos A \sin B.$$

Notice that this proof does not depend on A or B being acute!

Exercise 15.3

1. If $\mathbf{a} = 4\mathbf{i} - 3\mathbf{j} + \mathbf{k}$ and $\mathbf{b} = 2\mathbf{i} + 3\mathbf{j} + \mathbf{k}$ show that $\mathbf{a}$ and $\mathbf{b}$ are perpendicular.

2. Find the component of $\mathbf{b} = 2\mathbf{i} - 3\mathbf{j} + \mathbf{k}$ in the direction of $\mathbf{a} = 4\mathbf{i} + 3\mathbf{k}$. [*Hint*: show that the required component is $\mathbf{a} \cdot \mathbf{b}/|\mathbf{a}|$.]

3. Find the cosine of the angle between the following pairs of vectors:

   (a) $\mathbf{a} = 2\mathbf{i} + \mathbf{j} - 3\mathbf{k}$, $\mathbf{b} = -\mathbf{i} - \mathbf{j} + \mathbf{k}$;
   (b) $\mathbf{a} = \mathbf{i} - \mathbf{j}$, $\mathbf{b} = \mathbf{j} - 2\mathbf{k}$;
   (c) $\mathbf{a} = 4\mathbf{i} - \mathbf{j} - \mathbf{k}$, $\mathbf{b} = 2\mathbf{i} - 3\mathbf{j} + \mathbf{k}$.

4. Find the angle between the following pairs of vectors:

   (a) $\mathbf{a} = 3\mathbf{i} + \mathbf{j}$, $\mathbf{b} = 5\mathbf{i} - 12\mathbf{j}$;
   (b) $\mathbf{a} = 2\mathbf{i} + \mathbf{j} - \mathbf{k}$, $\mathbf{b} = 3\mathbf{i} - \mathbf{j} + \mathbf{k}$.

5. Show that $\mathbf{a}$ is parallel to $\mathbf{b}$ but $\mathbf{a}$ is perpendicular to $\mathbf{c}$ where

   $\mathbf{a} = 3\mathbf{i} - 6\mathbf{j} + \mathbf{k}$,
   $\mathbf{b} = -6\mathbf{i} + 12\mathbf{j} - 2\mathbf{k}$,
   $\mathbf{c} = 9\mathbf{i} + 5\mathbf{j} + 3\mathbf{k}$.

6. Find the direction cosines l, m and n of the following vectors and check directly that $l^2 + m^2 + n^2 = 1$:

   (a) $2\mathbf{i} + 3\mathbf{j} + 6\mathbf{k}$,   (b) $2\mathbf{i} - 3\mathbf{j} - 6\mathbf{k}$,   (c) $\mathbf{i} + \mathbf{j} + \mathbf{k}$.

7. The vectors $\mathbf{i}$, $\mathbf{j}$ and $\mathbf{k}$ are drawn along adjacent edges of a unit cube, in the standard manner. Express each of the vectors corresponding to the *four* diagonals of the cube, in terms of $\mathbf{i}$, $\mathbf{j}$ and $\mathbf{k}$.

8. In the tetrahedron $ABCD$, $\vec{DB} = \mathbf{a}$, $\vec{DA} = \mathbf{c}$ and $\vec{CA} = \mathbf{b}$.
   If M is the midpoint of AB, express in terms of $\mathbf{a}$, $\mathbf{b}$ and $\mathbf{c}$ each of the following vectors.

   (a) $\vec{AB}$,   (b) $\vec{BC}$,   (c) $\vec{DM}$.

9. Given that $\mathbf{a} = 2\mathbf{i} + \mathbf{j} - \mathbf{k}$ and $\mathbf{b} = \mathbf{i} - \mathbf{j} + 2\mathbf{k}$, find

   (a) $(\mathbf{a} . \mathbf{b})\mathbf{a}$ [Notice that $(\mathbf{a} . \mathbf{b}) . \mathbf{a}$ is not defined!]
   (b) $(\mathbf{a} - \mathbf{b}) . \mathbf{a}$ (c) $\mathbf{a} . \mathbf{b} - \mathbf{b} . \mathbf{a}$ (d) $(\mathbf{a} + \mathbf{b}) . (\mathbf{b} - \mathbf{a})$

10. If $\mathbf{a} = 13\mathbf{i} + p\mathbf{j} + q\mathbf{k}$ is perpendicular to both $\mathbf{b} = 3\mathbf{i} + 4\mathbf{j} - \mathbf{k}$ and $\mathbf{c} = 2\mathbf{i} + \mathbf{j} + 3\mathbf{k}$ find the values of p and q.

11. If $\mathbf{a}$ and $\mathbf{b}$ are perpendicular, show that

$$|\mathbf{a} - \mathbf{b}| = |\mathbf{a} + \mathbf{b}|.$$

12. The magnitudes of the vectors $\mathbf{a}$ and $\mathbf{b}$ are 8 and 3, respectively, and the angle between the vectors is 60°. Sketch a diagram showing these vectors and the vector $\mathbf{a} - \mathbf{b}$. Calculate

   (a) the magnitude of $\mathbf{a} - \mathbf{b}$,
   (b) the resolved part of $\mathbf{a}$ in the direction of $\mathbf{b}$.

JMB 1984

13. The resolved part of the vector $2a\mathbf{i} - 2a\mathbf{j} + a\mathbf{k}$ in the direction $\mathbf{i} - 2\mathbf{j} - 2\mathbf{k}$ is found to be 6. Find the value of a.

15.6   Position vectors

For every point P in space relative to Cartesian co-ordinates with origin O, there corresponds the vector $\vec{OP}$, called a position vector. A *position vector* is simply a vector which is represented by a line segment $\vec{OP}$ with O the origin.

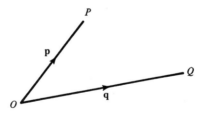

p is the position vector of P and q is the position vector of Q.

Fig. 15.26

Example 15.12

Show that the points with position vectors $\mathbf{i} + \mathbf{j} + \mathbf{k}$, $4\mathbf{i} + 3\mathbf{j}$ and $-5\mathbf{i} - 3\mathbf{j} + 3\mathbf{k}$ are collinear.

Suppose the three points are A, B and C with position vectors $\mathbf{a}$, $\mathbf{b}$ and $\mathbf{c}$ as given above then A, B and C are collinear if $\vec{AB} = \lambda \vec{BC}$ for some $\lambda \in \mathbb{R}$.

Now

$$\vec{AB} = \mathbf{b} - \mathbf{a}$$
$$= (4\mathbf{i} + 3\mathbf{j}) - (\mathbf{i} + \mathbf{j} + \mathbf{k})$$
$$= 3\mathbf{i} + 2\mathbf{j} - \mathbf{k}$$

and

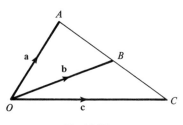

$$\vec{BC} = \mathbf{c} - \mathbf{b}$$
$$= (-5\mathbf{i} - 3\mathbf{j} + 3\mathbf{k}) - (4\mathbf{i} + 3\mathbf{j})$$
$$= -9\mathbf{i} - 6\mathbf{j} + 3\mathbf{k}.$$

Fig. 15.27

By inspection notice that

$$\vec{AB} = -\tfrac{1}{3}\vec{BC}$$

It follows that the points are collinear.

The ratio theorem

Suppose R divides the line PQ in the ratio $m:n$, where P and Q have position vectors $\mathbf{p}$ and $\mathbf{q}$ respectively relative to the origin O. Find the position vector of R in terms of $\mathbf{p}$ and $\mathbf{q}$.

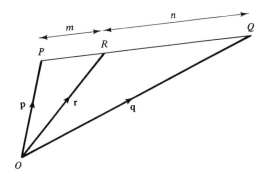

Fig. 15.28

By the triangle law of addition

$$\mathbf{p} + \vec{PQ} = \mathbf{q}$$
$$\vec{PQ} = \mathbf{q} - \mathbf{p}.$$

Now

$$|PR| = \frac{m}{m+n}|PQ|$$

Thus

$$\vec{PR} = \frac{m}{m+n}\vec{PQ}$$

$$= \frac{m}{m+n}(\mathbf{q} - \mathbf{p}).$$

But the position vector of R, $\mathbf{r}$, is such that

$$\mathbf{r} = \mathbf{p} + \vec{PR}$$

$$= \mathbf{p} + \frac{m}{(m+n)}(\mathbf{q} - \mathbf{p})$$

$$= \frac{n\mathbf{p} + m\mathbf{q}}{m+n}.$$

In particular, if R is the midpoint of PQ so that R divides PQ in the ratio $1:1$ then

$$\mathbf{r} = \frac{\mathbf{p} + \mathbf{q}}{2}$$

$$= \tfrac{1}{2}(\mathbf{p} + \mathbf{q}).$$

This result is not surprising when one considers the parallelogram law of vector addition.

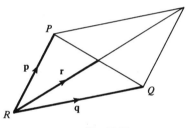

Fig. 15.29

15.7 The vector equation of the straight line

A straight line (in two or three dimensions) is uniquely determined (relative to an axis system) if the position vector of a point on the line is given, and a vector parallel to the line is known.

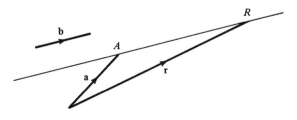

Fig. 15.30

Consider the line, with $\mathbf{a}$ the position vector of the point A on the line and which is parallel to the vector $\mathbf{b}$ (see Fig. 15.30). For *any* point R on the line there will be a $t \in \mathbb{R}$

such that

$$\vec{AR} = t\mathbf{b}.$$

(t may be positive or negative; if $t = 0$ then R coincides with A.)

Then by the triangle law of addition

$$\mathbf{r} = \mathbf{a} + t\mathbf{b}.$$

If $|AR_1| = |\mathbf{b}|$ and $|AR_2| = |2\mathbf{b}|$
then

$$\mathbf{r}_1 = \mathbf{a} + \mathbf{b}$$

and

$$\mathbf{r}_2 = \mathbf{a} - 2\mathbf{b}.$$

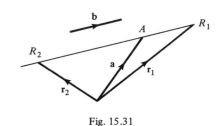

Fig. 15.31

Statement

> If $\mathbf{r}$ is the position vector of the point R and as R moves, $\mathbf{r}$ may be expressed in the form
>
> $$\mathbf{r} = \mathbf{a} + t\mathbf{b} \qquad \text{for some } t \in \mathbb{R},$$
>
> then the locus of R is the straight line parallel to the vector $\mathbf{b}$ passing through the point A with position vector $\mathbf{a}$.

The equation

$$\mathbf{r} = \mathbf{a} + t\mathbf{b}$$

is called a *vector equation of a straight line*.

Example 15.13

The points A and B with position vectors

$$\mathbf{i} + \mathbf{j} + \mathbf{k} \quad \text{and} \quad 2\mathbf{i} - \mathbf{j} + \mathbf{k}$$

respectively, relative to the origin O, determine a unique straight line.
Find a vector equation of the straight line.

Suppose

$$\mathbf{a} = \mathbf{i} + \mathbf{j} + \mathbf{k} \quad \text{and}$$

$$\mathbf{b} = 2\mathbf{i} - \mathbf{j} + \mathbf{k}$$

then, by the triangle law of vector addition,

$$\mathbf{a} + \vec{AB} = \mathbf{b}$$

$$\Rightarrow \qquad \vec{AB} = \mathbf{b} - \mathbf{a}$$

$$= (2\mathbf{i} - \mathbf{j} + \mathbf{k}) - (\mathbf{i} + \mathbf{j} + \mathbf{k})$$

$$= \mathbf{i} - 2\mathbf{j}.$$

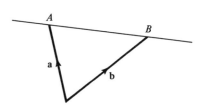

Fig. 15.32

Now $\vec{AB}$ is parallel to the line through A and B. So a vector equation of the line is

$$\mathbf{r} = \mathbf{a} + t\vec{AB}$$
$$\mathbf{r} = (\mathbf{i} + \mathbf{j} + \mathbf{k}) + t(\mathbf{i} - 2\mathbf{j})$$
$$\mathbf{r} = (1 + t)\mathbf{i} + (1 - 2t)\mathbf{j} + \mathbf{k}.$$

Since B is on the line we could easily have represented the line in the form

$$\mathbf{r} = \mathbf{b} + s\vec{AB} \qquad\qquad \text{where } s \in \mathbb{R}$$
$$\mathbf{r} = (2\mathbf{i} - \mathbf{j} + \mathbf{k}) + s(\mathbf{i} - 2\mathbf{j})$$
$$\mathbf{r} = (2 + s)\mathbf{i} - (1 + 2s)\mathbf{j} + \mathbf{k}.$$

It is not immediately obvious from an inspection of the equation that these represent the same line. But if we set $s = t - 1$ in the second equation we do, in fact, obtain the first.

Example 15.14

The position vector with respect to the origin O of points A, B, C, D are $\mathbf{a}$, $\mathbf{b}$, $3\mathbf{a}$, $2\mathbf{b}$ respectively, where $\mathbf{a}$ and $\mathbf{b}$ are not parallel. Find the position vector of the point of intersection E of the lines AB and CD in terms of $\mathbf{a}$ and $\mathbf{b}$.

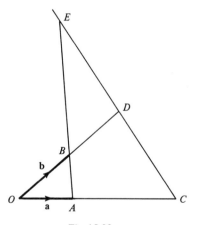

Fig. 15.33

Notice that $\vec{AB} = \mathbf{b} - \mathbf{a}$ and $\vec{CD} = 2\mathbf{b} - 3\mathbf{a}$.

If R with position vector $\mathbf{r}$ is a point on the line through A and B then

$$\mathbf{r} = \mathbf{a} + \mu(\mathbf{b} - \mathbf{a}) \qquad \text{for some } \mu \in \mathbb{R}.$$

If S, with position vector $\mathbf{s}$ is a point on the line through C and D then

$$\mathbf{s} = 3\mathbf{a} + \lambda(2\mathbf{b} - 3\mathbf{a}) \qquad \text{for some } \lambda \in \mathbb{R}$$

If E is the point of intesection of these two lines then

$$\vec{OE} = \mathbf{a} + \mu(\mathbf{b} - \mathbf{a}) \qquad \text{for some } \mu \in \mathbb{R}$$

and $O\vec{E} = 3\mathbf{a} + \lambda(2\mathbf{b} - 3\mathbf{a})$ for some $\lambda \in \mathbb{R}$.

In which case

$$\mathbf{a} + \mu(\mathbf{b} - \mathbf{a}) = 3\mathbf{a} + \lambda(2\mathbf{b} - 3\mathbf{a})$$
$$\Rightarrow (1 - \mu)\mathbf{a} + \mu\mathbf{b} = 3(1 - \lambda)\mathbf{a} + 2\lambda\mathbf{b}.$$

In which case

$$\mu = 2\lambda \qquad \text{①}$$

and

$$1 - \mu = 3(1 - \lambda) \qquad \text{②}$$

Substituting

$$1 - 2\lambda = 3 - 3\lambda$$
$$\Rightarrow \qquad \lambda = 2$$
$$\Rightarrow \qquad \mu = 4.$$

In either case we find

$$O\vec{E} = -3\mathbf{a} + 4\mathbf{b}.$$

Exercise 15.4

1. Show that the points with given position vectors are collinear.

    (a) $\mathbf{i} + 2\mathbf{j} + \mathbf{k}, \mathbf{i} - \mathbf{j} - 2\mathbf{k}, \mathbf{i} + 3\mathbf{j} + 2\mathbf{k}$.
    (b) $2\mathbf{i} - \mathbf{j}, 2\mathbf{j} + \mathbf{k}, -6\mathbf{i} + 11\mathbf{j} + 4\mathbf{k}$.
    (c) $\mathbf{a} - \mathbf{b}, \mathbf{a} + 2\mathbf{b}, \mathbf{a} - 4\mathbf{b}$.

2. Show that points with the following co-ordinates are not collinear.

    $$(3, 2, 1), (2, 4, 1), (1, 3, 2)$$

3. In the triangle $\triangle ABO$, A has position vector $\mathbf{a}$ and B has position vector $\mathbf{b}$, relative to O. Find the position vector of R if R divides AB in the ratio

    (a) $1:1$,   (b) $1:2$,   (c) $2:3$.

4. The position vectors, relative to the origin O, of A, B and C are $\mathbf{a}$, $\mathbf{b}$, $\mathbf{c}$ respectively. M is the midpoint of AB and G cuts CM in the ratio $2:1$. Find in terms of $\mathbf{a}$, $\mathbf{b}$ and $\mathbf{c}$ the following

    (a) $A\vec{B}$,   (b) $A\vec{M}$,   (c) $A\vec{C}$,   (d) $A\vec{G}$ (e) The position vector of G.

    Because the position vector of G is symmetrical with respect to $\mathbf{a}$, $\mathbf{b}$ and $\mathbf{c}$ deduce that the medians of the triangle intersect at G. This point is called the centroid of the triangle.

5. The position vectors of the points A and B are **a**, **b** respectively. A point P is chosen on the line OB (produced if necessary) so that AP is perpendicular to OB. By setting $\vec{OP} = \lambda\,\hat{\mathbf{b}}$, show that

$$\lambda = \frac{\mathbf{a} \cdot \mathbf{b}}{b}$$

and hence that

$$\vec{OP} = \frac{\mathbf{a} \cdot \mathbf{b}}{\mathbf{b} \cdot \mathbf{b}}\,\mathbf{b}.$$

Find $\vec{OP}$ if $\mathbf{a} = 3\mathbf{i} + 2\mathbf{j} + \mathbf{k}$ and $\mathbf{b} = 2\mathbf{i} - \mathbf{j} - \mathbf{k}$.
Check directly that $\vec{OP} \cdot \vec{AP} = 0$.

6. If $\mathbf{a} = \mathbf{i} + 2\mathbf{j} + 3\mathbf{k}$ and $\mathbf{b} = 2\mathbf{i} - \mathbf{j} + \mathbf{k}$, we can find a vector **c** which is perpendicular to both **a** and **b**. [Recall that two non parallel vectors determine a plane and **c** will be perpendicular to this plane.]
   One way to find a suitable vector is to set

$$\mathbf{c} = \mathbf{i} + l\mathbf{j} + m\mathbf{k}.$$

Then using the fact that $\mathbf{c} \cdot \mathbf{a}$ and $\mathbf{c} \cdot \mathbf{b}$ are both zero we obtain two equations in l and m, which may be solved. Any multiple of **c** will also be a suitable vector.

   (a) Find a vector perpendicular to the vectors **a** and **b** above.
   (b) Find a unit vector perpendicular to both **a** and **b**.

7. Any three non collinear points determine a unique plane! The points A, B and C have position vectors $2\mathbf{i}$, $3\mathbf{j} + \mathbf{k}$, $\mathbf{i} + \mathbf{j} - \mathbf{k}$, respectively. Find a unit vector perpendicular to the plane determined by A, B and C.

8. State a relation which exists between the vectors **a** and **b** if these vectors are

   (a) parallel,   (b) perpendicular.

   The position vectors relative to O of the points A and B in the parallelogram $OABC$ are $\mathbf{i} + \mathbf{j} - \mathbf{k}$ and $\mathbf{i} - \mathbf{j} - \mathbf{k}$, respectively. Find the position vector of C. Determine whether $OABC$ is a rhombus.

9. Show that the straight line with vector equation

$$\mathbf{r} = t\mathbf{i} + 2\mathbf{j} - (t + 1)\mathbf{k}$$

passes through the points A and B with position vectors $\mathbf{i} + 2\mathbf{j} - 2\mathbf{k}$ and $-2\mathbf{i} + 2\mathbf{j} + \mathbf{k}$, respectively, relative to O. Find the acute angle between the given line and OA.

10. Obtain a vector equation of the line parallel to **a** passing through the point with position vector **b** where

   (a) $\mathbf{a} = \mathbf{i} + 2\mathbf{j}$ \qquad\qquad $\mathbf{b} = \mathbf{i} - \mathbf{j} + \mathbf{k}$,

(b) $\mathbf{a} = 2\mathbf{i} - \mathbf{j} + \mathbf{k}$ $\mathbf{b} = \mathbf{i} - \mathbf{j} + \mathbf{k}$,
(c) $\mathbf{a} = \mathbf{i} + 2\mathbf{j} - \mathbf{k}$ $\mathbf{b} = \mathbf{0}$,
(d) $\mathbf{a} = \mathbf{b} = 3\mathbf{i} + \mathbf{j} - \mathbf{k}$.

11. Find the vector equation of the straight line passing through the points A and B with position vectors $\mathbf{a}$ and $\mathbf{b}$ respectively relative to the origin, where

(a) $\mathbf{a} = 3\mathbf{i} - \mathbf{j}$ $\mathbf{b} = 2\mathbf{i} + 4\mathbf{j}$
(b) $\mathbf{a} = 2\mathbf{i} + \mathbf{j} - \mathbf{k}$ $\mathbf{b} = \mathbf{i} + 2\mathbf{j} - 3\mathbf{k}$
(c) $\mathbf{a} = \mathbf{0}$ $\mathbf{b} = \mathbf{i} + \mathbf{j} + \mathbf{k}$.

12. Find the angle between the lines L_1 and L_2 where L_1 has vector equation

$$\mathbf{r} = 2\mathbf{i} - \mathbf{j} + \lambda(3\mathbf{i} - 4\mathbf{j}) \qquad \lambda \in \mathbb{R}$$

and L_2 has vector equation

$$\mathbf{s} = 3\mathbf{i} + \mathbf{j} + \mu(2\mathbf{i} - 3\mathbf{j}) \qquad \mu \in \mathbb{R}.$$

Determine which line is perpendicular to the vectors

(a) $8\mathbf{i} + 6\mathbf{j}$,   (b) $6\mathbf{i} + 4\mathbf{j}$.

Write down a unit vector perpendicular to both lines!

13. $ABCD$ is a parallelogram. The position vectors of B and C relative to A are $\mathbf{b}$, $\mathbf{c}$ respectively. Find the vector equations of the lines (a) CD, (b) BD, and *hence* find the position vector of D.

14. The vector $\mathbf{r} = x\mathbf{i} + y\mathbf{j} + z\mathbf{k}$ is often written as

$$\begin{pmatrix} x \\ y \\ z \end{pmatrix}$$

for convenience.
Find

(a)

$$\begin{pmatrix} 3 \\ 1 \\ 7 \end{pmatrix} \cdot \begin{pmatrix} 7 \\ 4 \\ 12 \end{pmatrix}$$

(b) the vector equation of the line passing through the points $(3, 1, 7)$ and $(7, 4, 12)$.
(c) the position vector of the point P on the above line such that OP is perpendicular to the vector

$$\begin{pmatrix} 4 \\ 3 \\ 5 \end{pmatrix}.$$

15. P is a point on the line

$$\mathbf{r} = \mathbf{i} + \mathbf{j} + \mathbf{k} + \lambda(2\mathbf{i} + 2\mathbf{j} - \mathbf{k}).$$

Find the value of $O\vec{P}.\ (2\mathbf{i} + 2\mathbf{j} - \mathbf{k})$ in terms of λ, where O is the origin. Hence find the co-ordinates of P if $O\vec{P}$ is perpendicular to the line, and hence obtain the distance of the origin O from the line.

16. Find the position vector of the point P on the line

$$r = \begin{pmatrix} 16 \\ -1 \\ -6 \end{pmatrix} + t \begin{pmatrix} -6 \\ 4 \\ 1 \end{pmatrix}$$

which is nearest the origin O. Hence find the coordinates of the reflection of O in the line.

17. The points A and B have position vectors $\mathbf{a}$ and $\mathbf{b}$, respectively, relative to an origin O. The point P lies on AB and $AP:PB = 1:5$; the point Q lies on AB produced and $AQ:BQ = 4:1$. Write down in terms of $\mathbf{a}$ and $\mathbf{b}$ the position vectors of

(a) P,   (b) Q,   (c) R, the mid-point of PQ.

State the ratio in which R divides AB.

JMB 1982

18. The lines L_1 and L_2 are given by the vector equations

$$\mathbf{r} = \begin{pmatrix} 0 \\ 1 \\ -1 \end{pmatrix} + t \begin{pmatrix} 2 \\ -1 \\ 2 \end{pmatrix} \quad \text{and} \quad \mathbf{r} = \begin{pmatrix} p \\ 3 \\ 0 \end{pmatrix} + s \begin{pmatrix} 2 \\ 2 \\ -1 \end{pmatrix},$$

respectively, where t and s are parameters. Given that L_1 and L_2 intersect, find the value of the constant p.

Find the distance from the point with co-ordinates $(p, 3, 0)$ to the point of intersection of L_1 and L_2.

JMB 1982

19. Find a unit vector which is in the opposite direction to the sum of the vectors $(3\mathbf{i} + 2\mathbf{j} + \mathbf{k})$ and $(-5\mathbf{i} - 3\mathbf{j} + 6\mathbf{k})$. Prove that this unit vector is perpendicular to the vector $(9\mathbf{i} - 4\mathbf{j} + 2\mathbf{k})$.

LOND 1984

20. The position vectors $\mathbf{a}$, $\mathbf{b}$, $\mathbf{c}$ of three points A, B, C respectively are given by

$$\mathbf{a} = \mathbf{i} + \mathbf{j} + \mathbf{k},$$
$$\mathbf{b} = \mathbf{i} + 2\mathbf{j} + 3\mathbf{k},$$
$$\mathbf{c} = \mathbf{i} - 3\mathbf{j} + 2\mathbf{k}.$$

Find

(a) a unit vector parallel to $\mathbf{a} + \mathbf{b} + \mathbf{c}$,
(b) the cosine of the angle between $\mathbf{a} + \mathbf{b} + \mathbf{c}$ and the vector $\mathbf{a}$,

(c) the vector of the form $\mathbf{i} + \lambda\mathbf{j} + \mu\mathbf{k}$ perpendicular to both $\mathbf{a}$ and $\mathbf{b}$,

(d) the position vector of the point D which is such that $ABCD$ is a parallelogram having BD as a diagonal.

CAMB 1982

21. The position vectors, relative to the origin O, of A, B, C are $\mathbf{a}$, $\mathbf{b}$, $\mathbf{c}$ respectively. Show that the point G with position vector

$$\mathbf{g} = \tfrac{1}{3}(\mathbf{a} + \mathbf{b} + \mathbf{c})$$

lies on the median through A (i.e. on the line joining A to the mid-point of BC) and deduce that the three medians of the triangle all pass through G, the centroid of the triangle.

The centroids of $\triangle OBC$, $\triangle OCA$, $\triangle OAB$ are respectively G_1, G_2, G_3 and the centroid of $\triangle G_1 G_2 G_3$ is G_4. Show that O, G_4, G are collinear and determined the ratio $OG_4 : G_4 G$.

O&C 1983

22. (a) Use the scalar product to find the angle (to the nearest degree) between $\mathbf{i} - 2\mathbf{j} + 3\mathbf{k}$ and $2\mathbf{i} + \mathbf{j} + \mathbf{k}$.

(b) The three vectors $2\mathbf{i} - 3\mathbf{j} + 4\mathbf{k}$, $a\mathbf{i} + \mathbf{j} + \mathbf{k}$ and $\mathbf{i} + b\mathbf{j} + c\mathbf{k}$ are mutually perpendicular; find a, b, c.

O&C 1982

23. The position vectors with respect to a point O of four distinct points A, B, D, E are $\mathbf{a}$, $\mathbf{b}$, $p\mathbf{a}$, $q\mathbf{b}$ respectively, where $\mathbf{a}$, $\mathbf{b}$ are non-parallel vectors and p, q are unequal scalars. Show that any point on AB has position vector given by $(1 - t)\mathbf{a} + t\mathbf{b}$, where t is a scalar.

The lines AB, DE intersect at L. Show that

$$\vec{OL} = \frac{1}{p - q}[p(1 - q)\mathbf{a} + q(p - 1)\mathbf{b}].$$

Two further points C, F have position vectors $\mathbf{c}$ and $r\mathbf{c}$ respectively, where $\mathbf{c}$ is not parallel to $\mathbf{a}$ or to $\mathbf{b}$ and r is a scalar not equal to p or to q. The lines CB, FE intersect at M, and CA, FD intersect at N. Write down $\vec{OM}$ and $\vec{ON}$ in terms of $\mathbf{a}$, $\mathbf{b}$, $\mathbf{c}$, p, q, r.

Taking the values $p = 5$, $q = 3$, $r = 2$,

(a) show that L, M, N are collinear points,

(b) if $\mathbf{a}$, $\mathbf{b}$ are the perpendicular unit vectors $\mathbf{i}$, $\mathbf{j}$ respectively and $\mathbf{c} = \mathbf{i} + 6\mathbf{j}$, find the acute angle between LMN and OAD.

O&C 1983

24. Referred to an origin O, the position vectors of points A and B are given respectively by

$$\vec{OA} = 3\mathbf{i} + \mathbf{j} + 3\mathbf{k},$$
$$\vec{OB} = 5\mathbf{i} - 4\mathbf{j} + 3\mathbf{k}.$$

Show that the cosine of angle AOB is equal to $4/\sqrt{(38)}$.

Hence, or otherwise, find the position vector of the point P on OB such that AP is perpendicular to OB.

The reflection of A in the line OB is A'. Find the position vector of A'.

CAMB 1983

25. The position vectors of the points A and B are $4\mathbf{i} - 8\mathbf{j}$ and $7\mathbf{i} - 2\mathbf{j} + 3\mathbf{k}$ respectively. Find a vector equation of the line AB.

The line CD has vector equation

$$\mathbf{r} = 8\mathbf{i} - 2\mathbf{j} + \mu(\mathbf{i} + \mathbf{j} - \mathbf{k}).$$

Find the position vector of P, the point of intersection of the lines AB and CD.

Given that O is the origin, show that OP is perpendicular to the plane $ABCD$.

AEB 1982

26. The position vectors of points A, B with respect to an origin O are $\mathbf{a}$, $\mathbf{b}$ respectively. Show that for all values of the real constant λ the point P with position vector $\mathbf{p}$ given by $\mathbf{p} = \lambda\mathbf{a} + (1 - \lambda)\mathbf{b}$ is collinear with A and B.

(a) Show that the point Q with position vector $\mathbf{q}$ given by $\mathbf{q} = 3\mathbf{a} - 2\mathbf{b}$ lies on BA produced, and find the ratio $QA : QB$.

(b) The point R has position vector $\mathbf{r}$ given by $\mathbf{r} = 3\mathbf{a} + 5\mathbf{b}$. Find the position vector of the point where OR meets AB (produced if necessary).

(c) Prove that the value of λ for which OP and AB are perpendicular is given by

$$\lambda = \frac{\mathbf{b} \cdot (\mathbf{b} - \mathbf{a})}{(\mathbf{b} - \mathbf{a}) \cdot (\mathbf{b} - \mathbf{a})}.$$

CAMB 1982

27. The position vectors, with respect to a fixed origin, of the points L, M and N are given by $\mathbf{l}$, $\mathbf{m}$ and $\mathbf{n}$ respectively, where

$$\mathbf{l} = a(\mathbf{i} + \mathbf{j} + \mathbf{k}), \quad \mathbf{m} = a(2\mathbf{i} + \mathbf{j}), \quad \mathbf{n} = a(\mathbf{j} + 4\mathbf{k})$$

and a is a non-zero constant. Show that the unit vector $\mathbf{j}$ is perpendicular to the plane of the triangle LMN.

Find a vector perpendicular to both $\mathbf{j}$ and $(\mathbf{m} - \mathbf{n})$, and hence, or otherwise, obtain a vector equation of the perpendicular bisector of MN which lies in the plane LMN.

Verify that the point K with position vector $a(5\mathbf{i} + \mathbf{j} + 4\mathbf{k})$ lies on this bisector and show that K is equidistant from L, M and N.

LOND 1983

28. Given the points $A(4, 2, 6)$ and $B(7, 8, 9)$, find a vector equation for the line AB in terms of a parameter t. The perpendicular to the line AB from the point $C(1, 8, 3)$ meets the line at N. Find the co-ordinates of N.

Obtain a vector equation for the line which is the reflection of the line AC in the line AB.

JMB 1983

29. The points A and B have position vectors $\mathbf{a} = 3\mathbf{i} + 6\mathbf{j} + 2\mathbf{k}$ and $\mathbf{b} = \mathbf{i} + 8\mathbf{j} + 4\mathbf{k}$ with respect to the origin O.

Express the cosine of the angle $\angle AOB$ as a surd and hence find the area of $\triangle AOB$.

30. The lines L_1 and L_2 have vector equations

$$\mathbf{r} = \begin{pmatrix} 1 \\ 4 \\ -3 \end{pmatrix} + s \begin{pmatrix} 2 \\ -2 \\ 5 \end{pmatrix} \quad \text{and} \quad \mathbf{r} = \begin{pmatrix} 8 \\ 2 \\ 5 \end{pmatrix} + t \begin{pmatrix} 3 \\ 2 \\ 6 \end{pmatrix},$$

respectively, where s and t are parameters. Show that L_1 and L_2 intersect and state the position vector of A, their point of intersection.

The plane π has the vector equation

$$\mathbf{r} . \begin{pmatrix} 1 \\ 3 \\ 4 \end{pmatrix} = d,$$

where d is a constant. Given that A lies on π, find the value of d. Show that the plane π contains the line L_1 but not the line L_2.

JMB 1984

15.8 The vector equation of a plane

Any vector $\mathbf{n}$ which is perpendicular to a plane is called a *normal* to the plane. A *unit normal* is a unit vector which is normal to the plane (Fig. 15.34). A plane is uniquely determined if the position vector of a point in the plane and a normal to the plane is given.

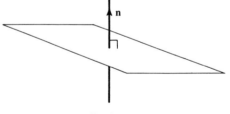

Fig. 15.34

Suppose $\mathbf{a}$ represents the position vector of the point A in a plane, $\mathbf{n}$ is a normal to the plane and $\mathbf{r}$ is the position vector of a point R. Then R is in the plane if and only if $\vec{AR}$ is perpendicular to $\mathbf{n}$ (see Fig. 15.35).

Now $\vec{AR} = \mathbf{r} - \mathbf{a}$ so the condition becomes

$$(\mathbf{r} - \mathbf{a}) . \mathbf{n} = 0$$

which may be rewritten

$$\mathbf{r} . \mathbf{n} = \mathbf{a} . \mathbf{n}$$

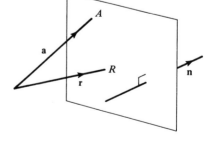

Fig. 15.35

that is

$$\boxed{\mathbf{r} \cdot \mathbf{n} = d}$$

where $d = \mathbf{a} \cdot \mathbf{n}$.

If

$$\mathbf{r} = x\mathbf{i} + y\mathbf{j} + z\mathbf{k}$$

and

$$\mathbf{n} = a\mathbf{i} + b\mathbf{j} + c\mathbf{k}$$

then by direct substitution we obtain the

Cartesian equation of the plane

$$\boxed{ax + by + cz = d}$$

Notice that a normal to the plane given in Cartesian form may be found by inspecting the coefficient of x, y and z, respectively.

Example 15.15

Obtain the Cartesian equation of the plane, given that $\mathbf{i} + 2\mathbf{j} + 3\mathbf{k}$ is normal to the plane and $A(2, -1, 3)$ is a point in the plane.

Recalling that $\mathbf{n} \cdot \mathbf{r} = \mathbf{a} \cdot \mathbf{n}$ where

$$\mathbf{n} = \mathbf{i} + 2\mathbf{j} + 3\mathbf{k}$$
$$\mathbf{a} = 2\mathbf{i} - \mathbf{j} + 3\mathbf{k},$$

in this case, and

$$\mathbf{r} = x\mathbf{i} + y\mathbf{j} + z\mathbf{k},$$

we have:

$$(\mathbf{i} + 2\mathbf{j} + 3\mathbf{k}) \cdot (x\mathbf{i} + y\mathbf{j} + z\mathbf{k}) = (2\mathbf{i} - \mathbf{j} + 3\mathbf{k}) \cdot (\mathbf{i} + 2\mathbf{j} + 3\mathbf{k})$$
$$\Rightarrow x + 2y + 3z = 2 - 2 + 9$$
$$\Rightarrow x + 2y + 3z = 9.$$

Example 15.16

Find a unit normal to the plane $x + 3y + 4z = 1$.

A normal to the plane is $\mathbf{i} + 3\mathbf{j} + 4\mathbf{k}$ with length $\sqrt{1^2 + 3^2 + 4^2} = \sqrt{26}$. A unit normal is therefore

$$\frac{1}{\sqrt{26}}(\mathbf{i} + 3\mathbf{j} + 4\mathbf{k}).$$

(The other unit normal is $-\dfrac{1}{\sqrt{26}}(\mathbf{i} + 3\mathbf{j} + 4\mathbf{k}).$)

Example 15.17

Find the point where the line, with vector equation

$$\mathbf{r} = (1 + t)\mathbf{i} + 2\mathbf{j} + (3 - 2t)\mathbf{k},$$

intersects the plane given by $3x - 2y + 3z = 5$.

Any point will lie on the line if it can be expressed in the form $(1 + t, 2, 3 - 2t)$ for some value of t. Such a point will lie in the plane if it satisfies the equation.

$$3x - 2y + 3z = 5.$$

That is, if

$$3(1 + t) - 2(2) + 3(3 - 2t) = 5$$
$$\Rightarrow 3 + 3t - 4 + 9 - 6t = 5$$
$$\Rightarrow -3t = -3$$
$$t = 1$$

Thus the line intersects the plane at the point $(2, 2, 1)$.

If the line is in the plane than the above method will result in the equation $0 = 0$, rather than give a specific value for t. In this case, of course any value for t gives rise to a point both on the line and in the plane. On the other hand, if the line does not intersect the plane, the above method will result in the equation $0 = 1$. As this is inconsistent, no value of t will be found.

The distance of the origin from the plane

If $\mathbf{n}$ is both a normal to the plane and the position vector of a point in the plane, then the vector equation of the plane is

$$\mathbf{n}.\mathbf{r} = n^2 \qquad ①$$

In this case the distance of the origin from the plane is simply n, the length of the vector $\mathbf{n}$.

If

$$ax + by + cz = d$$

then

$$kax + kby + kcz = kd \quad \text{(multiplying throughout by } k\text{)}.$$

Now this will be in the form of equation ① above, if we can take $\mathbf{n}$ to be

$$ka\mathbf{i} + kb\mathbf{j} + kc\mathbf{k}.$$

This will obviously be possible if $n^2 = kd$.

That is if

$$(ka\mathbf{i} + kb\mathbf{j} + kc\mathbf{k}) . (ka\mathbf{i} + kb\mathbf{j} + kc\mathbf{k}) = kd$$

$$\Rightarrow k^2(a^2 + b^2 + c^2) = kd$$

$$\Rightarrow \qquad\qquad k = \frac{d}{a^2 + b^2 + c^2}.$$

It follows that n^2 (in equation ①) is represented by

$$kd = \frac{d^2}{a^2 + b^2 + c^2}$$

So the distance of the origin from the plane given by $ax + by + cz = d$ is

$$\frac{|d|}{\sqrt{a^2 + b^2 + c^2}}.$$

Example 15.18
Find the distance of the origin from the plane given by

$$x - y + 2z = 5.$$

The distance is given, in standard notation, by

$$\frac{|d|}{\sqrt{a^2 + b^2 + c^2}} = \frac{5}{\sqrt{(1)^2 + (-1)^2 + (2)^2}} = \frac{5}{\sqrt{6}}.$$

Example 15.19
Find the distance of the point $(1, 2, -2)$ from the plane given by $x - 2y - 2z = 3$.

We first find the equation of the plane which passes through the point $(1, 2, -2)$ and is 'parallel' to the plane $x - 2y - 2z = 3$. In other words, both planes have common normals.

A normal to the required plane is $\mathbf{i} - 2\mathbf{j} - 2\mathbf{k}$ (this is a normal to the given plane), so the required plane has equation given by,

$$(\mathbf{i} - 2\mathbf{j} - 2\mathbf{k}) . (x\mathbf{i} + y\mathbf{j} + z\mathbf{k}) = (\mathbf{i} + 2\mathbf{j} - 2\mathbf{k}) . (\mathbf{i} - 2\mathbf{j} - 2\mathbf{k})$$

$$\Rightarrow x - 2y - 2z = 1 - 4 + 4$$

$$\Rightarrow x - 2y - 2z = 1$$

Now the distance between the plane

$$x - 2y - 2z = 3 \qquad ①$$

$$\text{and} \quad x - 2y - 2z = 1 \qquad ②$$

will be the same as the distance of the point from the plane!

The next step is to find in each case the distance of the origin from the plane. For equation ① the distance is

$$\frac{3}{\sqrt{1+4+4}} = 1$$

and for equation ② it is

$$\frac{1}{\sqrt{1+4+4}} = \frac{1}{3}.$$

The distance between these planes is given by

$$1 - \frac{1}{3} = \frac{2}{3}.$$

Had the planes been on opposite 'sides' of the origin then it would have been necessary to add the distances, (from the origin).

It is worth noting for this purpose that the two planes

$$ax + by + cz = d$$

$$\text{and} \quad ax + by + cz = d'$$

are on the same 'side' if d and d' are both positive or both negative. If they have opposite signs then they are on opposite 'sides'.

Angles involving planes

The angle between a line and a plane is defined to be the complement of the angle between the line and a normal. The angle between two planes is the same as the angle between the two normals.

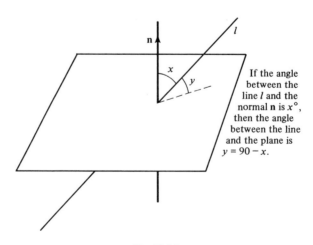

If the angle between the line l and the normal n is $x°$, then the angle between the line and the plane is $y = 90 - x$.

Fig. 15.36

Example 15.20

Find the angle between the line given by

$$\mathbf{r} = (1 - t)\mathbf{i} + (2 + t)\mathbf{j} + 3t\mathbf{k}$$

and the plane given by

$$x + y + z = 4.$$

A vector parallel to the line is

$$-\mathbf{i} + \mathbf{j} + 3\mathbf{k}$$

and a normal to the plane is

$$\mathbf{i} + \mathbf{j} + \mathbf{k}$$

If θ is the angle between these vectors we have

$$\cos \theta = \frac{(-\mathbf{i} + \mathbf{j} + 3\mathbf{k}).(\mathbf{i} + \mathbf{j} + \mathbf{k})}{\sqrt{1 + 1 + 9}.\sqrt{1 + 1 + 1}}$$

$$= \frac{-1 + 1 + 3}{\sqrt{11}.\sqrt{3}}$$

$$= \frac{3}{\sqrt{33}}$$

$$\Rightarrow \quad \theta = 58.5° \qquad \boxed{C}$$

Thus the angle between the line and the plane is

$$(90 - 58.5) = 31.5°.$$

Exercise 15.5

1. Obtain the equation of the plane with normal $2\mathbf{i} - 3\mathbf{j} + 6\mathbf{k}$ which passes through the point $(1, -5, 4)$.

2. Show that the point $(1, 3, -1)$ lies in the plane $x - y + 2z + 4 = 0$ and that the point $(-3, 1, 4)$ does not.

3. Find unit normals to the plane

   (a) $x + 2y + 2z = 1$
   (b) $3x - y + 4z = 2$

4. Find the point where the line l meets the plane π where l and π are given as:

   (a) $l : \mathbf{r} = (1 - t)\mathbf{i} + (1 + t)\mathbf{j} + \mathbf{k}$ $\pi : 2x - y + 3z = 1$
   (b) $l : \mathbf{r} = t\mathbf{i} + 2t\mathbf{j} + 3t\mathbf{k}$ $\pi : x + y + z = 12$
   (c) $l : \mathbf{r} = (2t - 1)\mathbf{i} + (1 - 2t)\mathbf{k}$ $\pi : 2x + y - z = 15$

5. Find the distance of the following planes to the origin.

    (a) $2x + y - 2z = 3$
    (b) $6x - 3y + 2z = 5$
    (c) $3x + 6y - 2z = 14$

6. Find the distance of the point $(1, 2, 1)$ to the plane $3x - 2y + 6z = 7$

7. Find the acute angle between the two planes

$$3x + 4y - z = 1 \quad \text{and} \quad x - y + z = 2.$$

8. Find the acute angle between the given line l and plane π

    (a) $l : \mathbf{r} = t\mathbf{i} - 2t\mathbf{j} + (1 - t)\mathbf{k}$
        $\pi : 3x + 4y - z = 1$

    (b) $l : \mathbf{r} = (1 - 2t)\mathbf{i} + \mathbf{j} - 2t\mathbf{k}$
        $\pi : x - y + z = 2$

9. Obtain a vector which is perpendicular to both $\mathbf{i} + 2\mathbf{j} - \mathbf{k}$ and $2\mathbf{i} - \mathbf{j}$. [*Hint*: let such a vector be $a\mathbf{i} + b\mathbf{j} + c\mathbf{k}$. Obtain two equations in a, b and c. Choose an arbitrary value for a (try $a = 1$) and solve simultaneously for b and c.]
    Hence find a vector which is normal to the plane which contains the points

    $A(2, 3, 0)$, $B(1, 1, 1)$ and $C(-1, 2, 1)$.

10. Use a method similar to that used in question 9 to find the equation of the plane passing through the points

    (a) $A(-1, 3, 2)$, $B(1, -3, 4)$ and $C(0, 3, 1)$,
    (b) $A(1, 4, -1)$, $B(0, 2, -2)$ and $C(-2, 1, -3)$.

11. The lines L and M have the equations

$$\mathbf{r} = \begin{pmatrix} 3 \\ 2 \\ 4 \end{pmatrix} + s \begin{pmatrix} 1 \\ 3 \\ -5 \end{pmatrix} \quad \text{and} \quad \mathbf{r} = \begin{pmatrix} -3 \\ 4 \\ 6 \end{pmatrix} + t \begin{pmatrix} 1 \\ -2 \\ 2 \end{pmatrix},$$

    respectively. The plane π has the equation

$$\mathbf{r} . \begin{pmatrix} 2 \\ 0 \\ -1 \end{pmatrix} = 16.$$

    (a) Verify that the point A with co-ordinates $(1, -4, 14)$ lies on L and on M but not on π.
    (b) Find the position vector of the point of intersection B of L and π.
    (c) Show that M and π have no common point.

(d) Find the cosine of the angle between the vectors

$$\begin{pmatrix} 1 \\ 3 \\ -5 \end{pmatrix} \text{ and } \begin{pmatrix} 2 \\ 0 \\ -1 \end{pmatrix}.$$

Hence find, to the nearest degree, the angle between L and π.

JMB 1985

12. The line L passes through the point $A(2, -9, 11)$ and is in the direction of the vector $\mathbf{n}$, where

$$\mathbf{n} = \begin{pmatrix} 3 \\ 6 \\ -2 \end{pmatrix}.$$

Write down a vector equation for L and show that L passes through the point $B(14, 15, 3)$.

The plane π passes through the point $C(4, 7, 13)$ and is at right angles to $\mathbf{n}$. Find an equation for π.

Verify that the resolved parts of $\vec{AC}$ and $\vec{BC}$ in the direction of $\mathbf{n}$ have the same magnitude but opposite signs. Give a geometrical interpretation of this result.

JMB 1986

Chapter 16   Complex numbers

16.1   Introduction

A sequence of extensions to the number system was introduced in Chapter 1: the *natural* numbers — the *integers* — the *rationals* — the *reals*. In a geometrical sense this process appears to be complete. Each real number corresponds to a unique point on the number line and vice versa. There does not appear to be any room left for the introduction of new numbers.

However, the extension to the number systems was not based entirely on geometrical considerations. Rather it was the need to solve problems which motivated the extensions. For example, the equation $x + 1 = 2$ only requires a knowledge of natural numbers to solve it. However, $x + 2 = 1$ cannot be solved using natural numbers, but can be solved using integers. Similarly, $2x = 1$ requires a knowledge of rational numbers for its solution, just as $x^2 = 2$ requires a knowledge of real numbers.

For a long time the equation $x^2 = -1$ was said to have no solution — more properly, no real solution. This means, of course, that there is no real number whose square is -1. It took a giant leap of imagination to suggest that $\sqrt{-1}$ should be accepted as a new number. The great mathematician, Euler, did just this in the 18th century when he introduced the symbol i for this square root. That is

$$i = \sqrt{-1}$$

and so

$$i^2 = -1$$

(some specialists use j instead of i).

It can be shown that i obeys all the usual rules of arithmetic. In other words, it behaves just like a real number, with the additional property that $i^2 = -1$. Given that this can be proved, as indeed it can, it seems churlish not to accept this extension to the number system. The more so since the extension has proved to be invaluable not only in the study of pure mathematics, but also in our understanding of the behaviour of electricity and other physical phenomena.

The acceptance of this one new number opens up a huge extension to the number system. For instance,

$$\sqrt{-4} = \sqrt{4} \cdot \sqrt{-1}, \qquad \text{using the ordinary rules of arithmetic,}$$
$$= 2i.$$

Generally,

$$\sqrt{-b^2} = \sqrt{b^2} \cdot \sqrt{-1}, \qquad \text{where } b \text{ is real,}$$
$$= bi.$$

The sum of a real number, a, and a number of the form bi is written $a + bi$, and cannot be simplified any further.

Definition

A number of the form $a + bi$ where a, $b \in \mathbb{R}$ is called a *complex number*. This set of complex numbers is denoted by $\mathbb{C}$

The variables z and w are usually reserved as members of $\mathbb{C}$. If $z \in \mathbb{C}$, that is z is a complex number, then there are reals a, b, such that

$$z = a + bi.$$

a is called the real part of z, written $\mathrm{Re}(z)$, and b is called the imaginary part of z written $\mathrm{Im}(z)$, so

$$z = \mathrm{Re}(z) + \mathrm{Im}(z)i.$$

If $b = 0$ then z is, in fact, a real number, which indicates that the set of real numbers is a subset of $\mathbb{C}$, i.e. $\mathbb{R} \subset \mathbb{C}$. If $a = 0$ then z is said to be *purely imaginary*.

Addition, subtraction, multiplication, division and all other operations on numbers can be extended to apply to complex numbers, giving rise to further complex numbers. The method is simple if we recall that such numbers obey the usual rules of arithmetic, with the additional property that

$$i^2 = -1.$$

Definition

If $z = x + iy$ and $w = u + iv$ where $x, y, u, v \in \mathbb{R}$, then

$$z + w = x + u + i(y + v)$$
$$z - w = x - u + i(y - v)$$

dealing with the real and imaginary parts separately.

Thus

$$(3 + 4i) + (2 + 3i) = 5 + 7i$$

$$\text{and} \quad (2 - 4i) - (5 + 3i) = -3 - 7i.$$

It is not necessary to recall a formula for the multiplication of complex numbers, we simply proceed as follows:

$$(3 + 4i)(2 - 3i) = 3 \cdot 2 - 3 \cdot 3i + 4i \cdot 2 - 4i \cdot 3i$$
$$= 6 - 9i + 8i - 12i^2$$
$$= 6 - i + 12, \quad \text{since } i^2 = -1,$$
$$= 18 - i.$$

As a special case

$$(a + ib)(a - ib) = a^2 - b^2 i^2$$

$$= a^2 + b^2, \quad \text{a real number.}$$

Definition

> If
>
> $$z = a + bi \qquad \text{where } a, b \in \mathbb{R}$$
>
> then the complex conjugate of z, written z^* is
>
> $$z^* = a - bi,$$
>
> so $\quad zz^* = a^2 + b^2.$

Sometimes $\bar{z}$ is used to denote the complex conjugate of z.

Division of complex numbers can be handled using the property that zz^* is always real. For example,

$$\frac{3}{2 + 3i} = \frac{3(2 - 3i)}{(2 + 3i)(2 - 3i)}$$

(multiplying the numerator and denominator by the complex conjugate of the denominator)

$$= \frac{6 - 9i}{4 + 9}$$

$$= \frac{6}{13} - \frac{9i}{13}$$

$$= \tfrac{3}{13}(2 - 3i).$$

Example 16.1

Find the imaginary part of z where

$$z = \frac{2 - 3i}{1 - i}$$

$$\frac{2 - 3i}{1 - i} = \frac{(2 - 3i)(1 + i)}{(1 - i)(1 + i)}$$

$$= \frac{2 + 2i - 3i - 3i^2}{1 + 1}$$

$$= \tfrac{1}{2}(5 - i), \quad \text{since } i^2 = -1.$$

So

$$\text{Im}(z) = -\tfrac{1}{2} \qquad \text{(Recall that Im}(z) \text{ is always real!)}$$

Exercise 16.1

1. Express each of the following in the form $a + bi$
   - (a) $3 + \sqrt{-9}$,
   - (b) $4 - \sqrt{-25}$,
   - (c) $1 - \sqrt{-3}$.

2. Simplify
   - (a) $(2 + 3i) + (4 - 3i)$
   - (b) $(3 - 7i) - (2 - 5i)$
   - (c) $(3 + 2i) + (5 - 3i)$
   - (d) $(3 - \sqrt{-9}) + (3 + \sqrt{-16})$

3. Find $a, b \in \mathbb{R}$ such that
   - (a) $(2 + 3i) + (a + bi) = 4 - 2i$,
   - (b) $2(a - 2i) - 2(3 - bi) = a(1 + i)$,
   - (c) $3(a + bi) - (5b + 2ai) = -1$.

4. Simplify

   $i^3, \quad i^4, \quad i^5, \quad i^6, \quad i^{19}, \quad i^{32}, \quad i^{4n+1}$.

5. Express in the form $a + bi$
   - (a) $i(3 - 2i)$
   - (b) $2i(1 + 2i)$
   - (c) $(3 + 2i)(4 - i)$
   - (d) $(5 - i)(3 + 2i)$
   - (e) $(4 - 3i)(4 + 3i)$
   - (f) $(3 + i)^2$
   - (g) $(1 + \sqrt{2}i)^2$
   - (h) $(1 + i)^3$
   - (i) $i(1 + i)(2 + i)$
   - (j) $(5 - 2i)(5 + 2i)$

6. Express the following in the form $a + bi$
   - (a) $\dfrac{1}{1 + i}$
   - (b) $\dfrac{i}{1 - 2i}$
   - (c) $\dfrac{3 + i}{3 - i}$
   - (d) $\dfrac{7 - i}{2 + i}$
   - (e) $\dfrac{1}{(1 + i)^2}$
   - (f) $\dfrac{1}{(1 + i)^3}$
   - (g) $\dfrac{1}{2 + i} + \dfrac{2}{3 - i}$
   - (h) $\dfrac{1 + i}{2 + i} - \dfrac{3 + i}{2 - i}$

7. Write down the complex conjugates of
   - (a) $3 - 2i$
   - (b) $4 + \sqrt{-25}$
   - (c) $(1 + i)(2 - i)$
   - (d) 6
   - (e) 0.

8. If $z = 1 - 2i$, find
   - (a) $z + z^*$
   - (b) zz^*
   - (c) $\dfrac{z}{z^*}$
   - (d) $(z^*)^2$
   - (e) $(z^2)^*$
   - (f) $\dfrac{1}{zz^*}$.

9. Given that $z_1 = a_1 + b_1 i$ and $z_2 = a_2 + b_2 i$ show that

(a) $(z^*)^* = z$ (b) $(z_1 + z_2)^* = (z_1)^* + (z_2)^*$

(c) $(z_1 z_2)^* = z_1^* z_2^*$

(d) $\left(\dfrac{z_1}{z_2}\right)^* = \dfrac{z_1^*}{z_2^*}.$

10. Show that

$z = z^*$ if $z \in \mathbb{R}.$

11. If $z = 3 - 4i$ and $w = 4 - 3i$ find

(a) $z + w$,   (b) $(z + w),^*$ (c) $2w$,   (d) $z^* w^*$,

(e) $\dfrac{z}{w}$,   (f) $\dfrac{z^*}{w^*}.$

12. Show that

$$(x + iy)(a + ib) = (xa - yb) + i(ya + xb).$$

13. If $z = x + iy$ find the possible values of x and y, if

$$zz^* + 2iz = 6(2 + i).$$

14. Simplify

(a) $\dfrac{1 - i}{(2 - i)^2}$,   (b) $(x + i)^4$,   where $x \in \mathbb{R}.$

15. Simplify

(a) $(\cos\theta + i\sin\theta)(\cos\theta - i\sin\theta)$,
(b) $(\cos\theta + i\sin\theta)^2$,

(c) $\dfrac{1}{\cos\theta + i\sin\theta}.$

16. Show that

$$(\cos\theta + i\sin\theta)^3 \equiv \cos 3\theta + i\sin 3\theta.$$

Prove by induction that

$$(\cos\theta + i\sin\theta)^n = \cos n\theta + i\sin n\theta$$

If $z = \cos\theta + i\sin\theta$, find the values of

$$z + \frac{1}{z} \quad \text{and} \quad z - \frac{1}{z}.$$

16.2 Equating real and imaginary parts

If $a + bi = 0$ then both a and b must be zero. In other words, if $z = 0$ then $\text{Re}(z) = 0$ and $\text{Im}(z) = 0$. An equation involving complex numbers is equivalent to two separate equations; obtained by '*equating real and imaginary parts*'.

Example 16.2

Find $\sqrt{i}$.

If we suppose

$$\sqrt{i} = a + ib \qquad \text{where } a, b \in \mathbb{R},$$

then

$$i = (a + ib)^2$$

$$\Rightarrow \quad i = a^2 - b^2 + 2abi.$$

Equating real and imaginary parts

$$a^2 - b^2 = 0$$

$$\text{and} \qquad 2ab = 1 \qquad \text{where } a \text{ and } b \text{ are real.}$$

By substitution

$$a^2 - \frac{1}{4a^2} = 0$$

$$\Rightarrow \qquad 4a^4 = 1$$

$$\Rightarrow \qquad a^4 = \frac{1}{4}$$

$$\Rightarrow \qquad a^2 = \frac{1}{2} \quad \text{or} \quad a^2 = -\frac{1}{2}$$

$$\Rightarrow \qquad a = \pm\frac{1}{\sqrt{2}} \quad \text{(the second equation gives no real solutions).}$$

Now $2ab = 1$, so

$$a = \frac{1}{\sqrt{2}} \quad \Rightarrow b = \frac{1}{\sqrt{2}}$$

$$\text{and} \quad a = -\frac{1}{\sqrt{2}} \Rightarrow b = -\frac{1}{\sqrt{2}}.$$

As is expected, there are two square roots of i, namely

$$\frac{1}{\sqrt{2}} + \frac{1}{\sqrt{2}}i \quad \text{and} \quad -\frac{1}{\sqrt{2}} - \frac{1}{\sqrt{2}}i.$$

So

$$\sqrt{i} = \pm\frac{\sqrt{2}}{2}(1 + i).$$

16.3   Quadratic equations

One justification for the introduction and study of complex numbers is the ability to solve more equations.

Example 16.3

Solve the equation $z^2 + z + 1 = 0$.

Notice that the use of the variable z reminds us that we are able to admit complex numbers as solutions.

We know, from the usual rules that

$$z = \frac{-1 \pm \sqrt{1 - 4}}{2}$$

$$= \frac{-1 \pm \sqrt{-3}}{2}$$

$$= \frac{-1 \pm i\sqrt{3}}{2}$$

So

$$z = -\frac{1}{2} + \frac{i\sqrt{3}}{2} \quad \text{and} \quad z = -\frac{1}{2} - \frac{i\sqrt{3}}{2}.$$

It is interesting to note that the solutions to the quadratic is a conjugate pair — that is each solution is the complex conjugate of the other.

This can be proved using some properties proved in the previous exercise. Namely:

Statement

> (i) $z_1 = z_1^*$ iff z_1 is real,
>
> (ii) $(z_1 + z_2)^* = z_1^* + z_2^*$,
>
> (iii) $(z_1 z_2)^* = z_1^* z_2^*$,
>
> for any complex numbers z_1, z_2.

Consider the quadratic

$$az^2 + bz + c = 0 \qquad \text{where } a, b, c \in \mathbb{R}$$

We now show that if z_1 is a solution then z_1* is also a solution. Suppose then

$$az_1^2 + bz_1 + c = 0$$

$\Rightarrow \quad (az_1^2 + bz_1 + c)* = (0)* \qquad$ taking the complex conjugate of both sides

$\Rightarrow (az_1^2)* + (bz_1)* + (c)* = (0)*$

$\Rightarrow \quad a*(z_1^2)* + b*z_1^* + c* = 0*$

but for $x \in \mathbb{R}$, $(x)* = x$ and so

$$a(z_1^*)^2 + b(z_1)* + c = 0 \qquad \text{noting also that } (z^2)* = (z*)^2,$$

thus z_1* is also a solution.

In fact this proof can be extended to show that if z is a solution of a polynomial with real coefficients, then $z*$ is also a solution.

Example 16.4

Write down the quadratic equation with roots

$$1 + 2i \quad \text{and} \quad 1 - 2i.$$

Method 1. The required quadratic is

$$(z - (1 + 2i))(z - (1 - 2i)) = 0$$
$$\Rightarrow z^2 - z(1 - 2i) - z(1 + 2i) + (1 + 2i)(1 - 2i) = 0$$
$$\Rightarrow \qquad\qquad\qquad\qquad z^2 - 2z + 5 = 0$$

Method 2. If $\alpha = z$ and $\beta = z*$ are the roots of the quadratic then the required quadratic is

$$z^2 - (\alpha + \beta)z + \alpha\beta = 0.$$

$\alpha + \beta$ and $\alpha\beta$ are readily found and may be substituted.

Example 16.5

Find the complex numbers p and q if the quadratic

$$z^2 + pz + q = 0$$

has roots $1 - i$ and $2 - i$.

If $\alpha = 1 - i$ and $\beta = 2 - i$, then

$$\alpha + \beta = -p$$
$$\alpha\beta = q.$$

But

$$\alpha + \beta = 3 - 2i$$
$$\alpha\beta = 1 - 3i,$$

so

$$p = -3 + 2i$$
$$q = 1 - 3i.$$

Notice that the roots are not complex conjugate of one another, but then the quadratic does not have real coefficients.

16.4   Cubic equations

A cubic equation will have three roots, possibly complex and possibly repeated. Since complex solutions occur in conjugate pairs at least one solution will be real, assuming real coefficients.

Example 16.6

Solve the cubic $z^3 + z - 2 = 0$.

Let

$$P(z) \equiv z^3 + z - 2$$

then

$$P(1) = 1 + 1 - 2 = 0$$

so $(z - 1)$ is a factor.
  Suppose

$$P(z) \equiv (z - 1)(az^2 + bz + c)$$

then by equating coefficients $a = 1$, $c = 2$ and $b = 1$. So

$$P(z) \equiv (z - 1)(z^2 + z + 2).$$

Now if $z^2 + z + 2 = 0$,

$$z = \frac{-1 \pm \sqrt{1 - 8}}{2}$$

$$= \frac{-1 \pm i\sqrt{7}}{2},$$

giving the solutions

$$z = 1 \quad \text{or} \quad z = -\frac{1}{2} + i\frac{\sqrt{7}}{2} \quad \text{or} \quad z = -\frac{1}{2} - i\frac{\sqrt{7}}{2}.$$

The cube root of unity

A cube root of 1 is a solution of the equation

$$z^3 = 1$$

$$\Leftrightarrow z^3 - 1 = 0.$$

Factorising

$$z^3 - 1 = (z - 1)(z^2 + z + 1)$$

so

$$(z - 1)(z^2 + z + 1) = 0$$

$$\Rightarrow z = 1 \quad \text{or} \quad z^2 + z + 1 = 0$$

$$\Rightarrow z = 1 \quad \text{or} \quad z = -\frac{1}{2} + i\frac{\sqrt{3}}{2} \quad \text{or} \quad z = -\frac{1}{2} - i\frac{\sqrt{3}}{2}.$$

(as found in Example 16.3).

Notice that

$$\left(-\frac{1}{2} + i\frac{\sqrt{3}}{2}\right)^2 = \frac{1}{4} - i\frac{\sqrt{3}}{2} + \frac{3}{4}i^2 = -\frac{1}{2} - i\frac{\sqrt{3}}{2}$$

and similarly

$$\left(-\frac{1}{2} - i\frac{\sqrt{3}}{2}\right)^2 = \frac{1}{4} + i\frac{\sqrt{3}}{2} + \frac{3}{4}i^2 = -\frac{1}{2} + i\frac{\sqrt{3}}{2}.$$

Thus if w is either complex cube root of 1, then the other is w^2. In other words, $w^* = w^2$.

For this reason the cube roots of 1 are written

$$1, w \text{ and } w^2$$

and since w is a solution of $z^2 + z + 1 = 0$

$$w^2 + w + 1 = 0.$$

Example 16.7

Simplify

(a) $\dfrac{1}{w + w^2}$,   (b) $w^4 + w^2$.

(a) Since

$$1 + w + w^2 = 0$$

$$w + w^2 = -1$$

thus

$$\frac{1}{w + w^2} = \frac{1}{-1} = -1$$

(b) $w^4 = w$ since $w^3 = 1$

so $w^4 + w^2 = w + w^2 = -1$ since $1 + w + w^2 = 0$

Example 16.8

Using the fact that $w^3 = 1$ and w is complex, prove that

$$1 + w + w^2 = 0.$$

Suppose

$$1 + w + w^2 = z$$

then

$$w(1 + w + w^2) = wz$$

but

$$w(1 + w + w^2) = w + w^2 + w^3$$
$$= 1 + w + w^2 = z$$

thus

$$wz = z$$
$$\Rightarrow z(w - 1) = 0$$

and since w is complex, $w - 1 \neq 0$.

$$\Rightarrow z = 0$$
$$\Rightarrow 1 + w + w^2 = 0 \qquad !$$

Statement

> If w is a complex cube root of unity, then
>
>   (i) $w^* = w^2$
>   (ii) $w^2 + w + 1 = 0$
>   (iii) the three cube roots of unity are
>
> $$1, w \text{ and } w^2$$

Exercise 16.2

1. Solve the quadratic equations

   (a) $z^2 + 2z + 2 = 0$ (b) $z^2 + z + 2 = 0$ (c) $z^2 + 3z + 3 = 0$
   (d) $2z^2 + 5z + 4 = 0$ (e) $z^2 + 3z + 4 = 0$

2. Write down the quadratic equation with roots

   (a) $1 + i, 1 - i$ (b) $2 - 5i, 2 + 5i$ (c) $3 - 4i, 3 + 4i$
   (d) $4 + i, 1 - 4i$!!   (e) $2 - i, 3 + i$!!

3. Given that $1 + i$ is a root of

$$z^2 - pz + p + i = 0$$

where p is real, find   (a) p,   (b) the other root.

4. Verify that $\frac{1}{2}(1 + i)$ is a root of the quadratic

$$2z^2 = 2z - 1$$

and write down the other root.

5. Solve the cubic equations

(a) $z^3 + 1 = 0$,   (b) $z^3 - 8 = 0$,   (c) $(z - 2)^3 = 8$,   (d) $(z - 1)^3 = 1$.

6. If w is a complex cube root of unity, simplify

(a) $w^3 + w^4 + w^5$,   (b) $1 + w + w^2 + w^3$,   (c) $(1 + 3w + w^2)^2$,

(d) $(2 + w + 2w^2)^3$,   (e) $\dfrac{1}{1 + w}$,   (f) $\dfrac{1}{(1 + w)^3}$,

(g) $\dfrac{1}{2 + w^2}$,   (h) $\dfrac{1}{(1 - w)(w - w^2)}$.

7. Obtain the square roots of

(a) $3 + 4i$ (b) $-2i$.

8. Find real numbers p and q so that

$$p(2 + i) - (p - 2qi) = 4.$$

9. Show that

(a) $1 + \dfrac{1}{w} + \dfrac{1}{w^2} = 0$,   (b) $1 + w^4 + w^8 = 0$,

where w is a complex cube root of unity.

10. If x and y are real, solve the equation

$$\frac{iy}{1 + iy} = \frac{3y + 4i}{3x + yi}.$$

11. If w is a complex cube root of unity and

$$a = \alpha + \beta$$
$$b = w\alpha + w^2\beta$$
$$c = w^2\alpha + w\beta,$$

show that $a + b + c = 0$ and find the value of $a^2 + b^2 + c^2$.

12. Show that $1 + w^n + w^{2n}$ may take one of two possible values depending on the value of the integer n.

13. Solve the equation
$$z^4 - 6z^2 + 25 = 0$$
for z^2, and hence find the four roots of the equation.

14. Find the values of A, B and C in the identity
$$\frac{A}{x-1} + \frac{B}{x-w} + \frac{C}{x-w^2} \equiv \frac{1}{x^3 - 1}$$
where w is a complex cube root of unity.

15. (a) Express $\dfrac{-1 + i\sqrt{3}}{-1 - i\sqrt{3}}$ in the form $a + ib$, where a and b are real numbers.

 (b) Find the quadratic equation whose roots are $-3 + 4i$ and $-3 - 4i$, expressing your answer in the form $x^2 + px + q = 0$, where p and q are real numbers.

CAMB 1982

16. (a) Find $(2 - i)^3$, expressing your answer in the form $a + ib$.
 (b) Verify that $2 + 3i$ is one of the square roots of $-5 + 12i$. Write down the other square root.

CAMB 1983

17. Given that $z = 1 + i$, show that $z^3 = -2 + 2i$. For this value of z, the real numbers p and q are such that
$$\frac{p}{1 + z} + \frac{q}{1 + z^3} = 2i.$$
Find the values of p and q.

JMB 1984

18. The complex numbers
$$z_1 = 1 + ia, \qquad z_2 = a + ib,$$
where a and b are real, are such that $z_1 - z_2 = 3i$. Find a and b, and show that
$$\frac{1}{z_1} + \frac{1}{z_2} = (7 - i)/10.$$
Hence, or otherwise, find
$$\frac{z_1^2 - z_2^2}{z_1 z_2}$$
in the form $x + iy$, where x and y are real.

JMB 1983

16.5   The Argand diagram

A complex number $a + ib$ can be represented
by the point (a, b) in a plane. This plane is
referred to as an Argand diagram, after the
French mathematician Argand who first
represented complex numbers in this way.

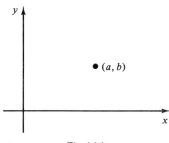

Fig. 16.1

Real numbers are represented on the horizontal axis and purely imaginary numbers on
the vertical axis. For this reason the horizontal axis is called the real axis and the
vertical axis is called the imaginary axis. The
point $P(x_1, y_1)$ representing $z_1 = x_1 + iy_1$, on
the Argand diagram will sometimes be called
P_{z_1} or simply P_1, in what follows.

The modulus and argument of a complex number

The *modulus* of the complex number $z = x + iy$
written $|z|$ is the distance between P_z and the
origin. This distance r (say) can be found using
Pythagoras' theorem.

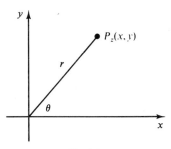

Fig. 16.2

$$|z| = r = \sqrt{x^2 + y^2}$$

The *argument* of a complex number $z = x + iy$ written arg z, is the angle θ which OP_z
makes with the positive real axis. This angle θ
measured in radians is taken to be between $-\pi$
and π. It follows that $\tan \theta = y/x$ (and $\sin \theta = y/r$, $\cos \theta = x/r$).

Example 16.9

Find the modulus and argument of the complex number

$$z = 3 - 4i$$

$$|z| = \sqrt{3^2 + (-4)^2} = 5. \quad \text{Notice that } |z| \geqslant 0.$$

If arg $z = \theta$ then

$$\tan \theta = -\frac{4}{3}$$

$$\Rightarrow \quad \theta = -0.927 \quad (-53.1° \text{ approx.})$$

The other solution $\pi - 0.927$ may be eliminated by noting that z is located in the
fourth quadrant.

All complex numbers possess a unique argument, except zero—for which no argument is defined. The modulus and argument of some complex numbers may be obtained immediately by inspection of its location as represented on the Argand diagram. Consider the examples in the Table 16.1.

z	P	$|z|$	$\arg z$
2	$P(2, 0)$	2	0
-2	$P(-2, 0)$	2	π
i	$P(0, 1)$	1	$\pi/2$
$1 + i$	$P(1, 1)$	$\sqrt{2}$	$\pi/4$

Table 16.1

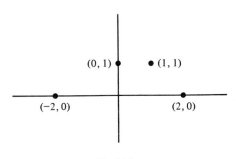

Fig. 16.3

If we are given the modulus and argument of a complex number then we can write it in the standard form immediately. For if

$$|z| = r \quad \text{and} \quad \arg z = \theta$$

then

$$z = r(\cos \theta + i \sin \theta).$$

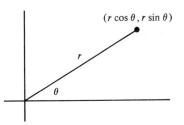

Fig. 16.4

Example 16.10

Find the modulus and argument of the complex numbers

(a) $z_1 = \dfrac{-4}{1 + i}$,

(b) $z_2 = 3\left(\cos \dfrac{3\pi}{2} + i \sin \dfrac{3\pi}{2}\right)$,

(c) $z_3 = -\left(\cos \dfrac{\pi}{3} + i \sin \dfrac{\pi}{3}\right)$,

(d) $z_4 = \left(\cos \dfrac{2\pi}{3} - i \sin \dfrac{2\pi}{3} \right).$

(a) $\quad z_1 = \dfrac{-4}{(1+i)} \dfrac{(1-i)}{(1-i)} = -2(1-i)$

$|z_1| = \sqrt{4+4} \qquad = \sqrt{8}$

$\arg z = \dfrac{3\pi}{4},$ since z is located in the second quadrant (see Fig. 16.5).

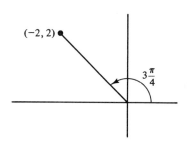

Fig. 16.5

(b) $|z_2| = 3,$ by inspection

but

$$\arg z_2 = -\frac{\pi}{2}.$$

Since $-\pi < \arg z_2 \leqslant \pi,$ all $z \neq 0.$

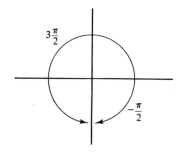

Fig. 16.6

(c) $|z_3| = 1$ since

$$\sqrt{\left(-\cos\frac{\pi}{3} \right)^2 + \left(-\sin\frac{\pi}{3} \right)^2} \equiv 1.$$

Now P_{z_3} is diametrically opposite $(\cos \pi/3, \sin \pi/3)$ as shown in Fig. 16.7. Thus

$\arg z_3 = -\dfrac{2\pi}{3}.$

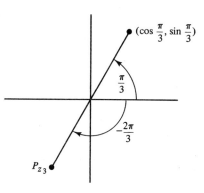

Fig. 16.7

(d) $|z_4| = 1.$

Now

$$\cos \theta = \cos(-\theta) \quad \text{and} \quad \sin \theta = -\sin(-\theta)$$

so z_4 may be written

$$\cos\left(-\frac{2\pi}{3}\right) + i\sin\left(-\frac{2\pi}{3}\right).$$

It follows that

$$\arg z_4 = \frac{-2\pi}{3}.$$

Alternatively, notice that z_4 is the conjugate of

$$\cos\frac{2\pi}{3} + i\sin\frac{2\pi}{3}$$

and so $\arg z_4 = \dfrac{-2\pi}{3}$.

Generally $\arg z^* = -\arg z$.

16.6 Multiplication and division of complex numbers

It is instructive to consider the geometrical effects of multiplying or dividing complex numbers.

Suppose

$$z_1 = r_1(\cos\theta_1 + i\sin\theta_1)$$
$$z_2 = r_2(\cos\theta_2 + i\sin\theta_2) \qquad \text{where} \quad r_1, r_2 > 0,$$

then

$$z_1 z_2 = r_1(\cos\theta_1 + i\sin\theta_1)r_2(\cos\theta_2 + i\sin\theta_2)$$
$$= r_1 r_2[(\cos\theta_1\cos\theta_2 - \sin\theta_1\sin\theta_2)$$
$$+ i(\cos\theta_1\sin\theta_2 + \sin\theta_1\cos\theta_2)]$$
$$= r_1 r_2[\cos(\theta_1 + \theta_2) + i\sin(\theta_1 + \theta_2)].$$

It follows immediately that

$$|z_1 z_2| = |z_1||z_2|.$$

The modulus of the product is obtained by multiplying the moduli of z_1, and z_2.

The argument of $z_1 z_2$ may be obtained by adding the argument of z_1 and z_2 — remembering to express this angle so that it lies between $-\pi$ and π.

Example 16.11

Given $z_1 = 2(\cos\pi/2 + i\sin\pi/2)$ and $z_2 = 3(\cos 5\pi/4 + i\sin 5\pi/4)$ write down the modulus and argument of $z_1 z_2$.

Now

$$z_1 z_2 = 6 \left(\cos \frac{7\pi}{4} + i \sin \frac{7\pi}{4} \right)$$

$$= 6 \left(\cos -\frac{\pi}{4} + i \sin -\frac{\pi}{4} \right).$$

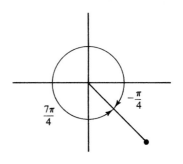

So

$$|z_1 z_2| = 6$$

$$\arg(z_1 z_2) = -\frac{\pi}{4}.$$

Fig. 16.8

If

$$z = r(\cos \theta + i \sin \theta) \qquad \text{where } r > 0$$

then

$$z^2 = r^2(\cos 2\theta + i \sin 2\theta).$$

Similarly, if

$$z = r(\cos \theta + i \sin \theta)$$

then

$$\sqrt{z} = \pm \sqrt{r} \left(\cos \frac{\theta}{2} + i \sin \frac{\theta}{2} \right).$$

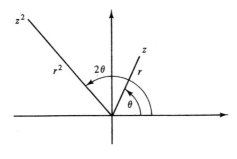

Fig. 16.9

Example 16.12

Write down the two square roots of

(a) i,   (b) $1 + \sqrt{3}i$.

(a) $i = 1 \left(\cos \frac{\pi}{2} + i \sin \frac{\pi}{2} \right)$ since its modulus is 1 and its argument is $\frac{\pi}{2}$.

It follows that

$$\sqrt{i} = \pm\sqrt{1}\left(\cos\frac{\pi}{4} + i\sin\frac{\pi}{4}\right)$$

$$= \pm\left(\frac{1}{\sqrt{2}} + i\frac{1}{\sqrt{2}}\right)$$

in agreement with example 16.2.

(b) $1 + \sqrt{3}i = 2\left(\cos\frac{\pi}{3} + i\sin\frac{\pi}{3}\right),$

so

$$\sqrt{1 + \sqrt{3}i} = \pm\sqrt{2}\left(\cos\frac{\pi}{6} + i\sin\frac{\pi}{6}\right)$$

$$= \pm\sqrt{2}\left(\frac{\sqrt{3}}{2} + i\frac{1}{2}\right)$$

$$= \pm\frac{\sqrt{2}}{2}(\sqrt{3} + i).$$

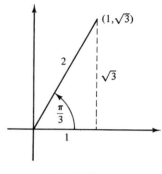

Fig. 16.10

Statement

If
$z_1 = r_1(\cos\theta_1 + i\sin\theta_1)$ and $z_2 = r_2(\cos\theta_2 + i\sin\theta_2)$
then
$z_1z_2 = r_1r_2[\cos(\theta_1 + \theta_2) + i\sin(\theta_1 + \theta_2)]$
$\dfrac{z_1}{z_2} = \dfrac{r_1}{r_2}\left[\cos(\theta_1 - \theta_2) + i\sin(\theta_1 - \theta_2)\right]$

The latter result is left as an exercise. It follows that

$$\left|\frac{z_1}{z_2}\right| = \frac{|z_1|}{|z_2|}$$

and the argument of z_1/z_2 may be obtained by subtracting the argument of z_2 from that of z_1—remembering to express this angle so that it lies between $-\pi$ and π.

Exercise 16.3

1. Plot the following complex numbers on an Argand diagram and thus obtain the modulus and argument in each case.

    (a) $z_1 = 2i$ (b) $z_2 = -3$ (c) $z_3 = -3 + 4i$

(d) $z_4 = -5 - 12i$ (e) $z_5 = \dfrac{2}{1-i}$ (f) $z_6 = (3 - 4i)*$

2. Find the modulus and argument of $z = 3 - 2i$ and hence write down the modulus and argument of $z*$.

3. Find the modulus and argument of

(a) $-5 + 12i$,   (b) $5 + 12i$,   (c) $i(1 - i)$,

(d) $(7 - 3i)(2 + i)$,   (e) $3i\left(\cos\dfrac{5\pi}{4} + i\sin\dfrac{5\pi}{4}\right)$,

(f) $1 + z$ where $z = 2 - 5i$.

4. Find the modulus and argument of

(a) $3\left(\cos\dfrac{\pi}{3} + i\sin\dfrac{\pi}{3}\right)$,

(b) $5(\cos 200° + i\sin 200°)$,

(c) $\cos\dfrac{3\pi}{4} - i\sin\dfrac{3\pi}{4}$,

(d) $\cos\theta - i\sin\theta$, where $-\pi < \theta \leqslant \pi$,

(e) $-(\cos\theta + i\sin\theta)$, where $-\pi < \theta \leqslant \pi$.

5. Find the two square roots of the complex number

(a) $z_1 = -1 + \sqrt{3}i$,   (b) $z_2 = -2i$,   (c) $z_3 = 3 + 4i$.

     In each case plot the complex number, together with its square roots, on an Argand diagram.

6. Show that if $z = \cos\theta + i\sin\theta$ then

(a) $z* = \dfrac{1}{z}$,   (b) $\mathrm{Re}\left(\dfrac{1-z}{1+z}\right) = 0$

7. Represent each of the following complex numbers on an Argand diagram.

$4 + 3i$, $4 - 3i$, $\dfrac{4 + 3i}{4 - 3i}$.

8. Given that

$$z = \dfrac{1}{\lambda + i} \quad \text{for} \quad \lambda \in \mathbb{R},$$

is represented on the Argand diagram by P, show that the locus of P as λ varies, is a circle. [*Hint*: equate real and imaginary parts and eliminate λ.]

9. If
$$z_1 = r_1(\cos\theta_1 + i\sin\theta_2) \quad \text{and} \quad z_2 = r_2(\cos\theta_2 + i\sin\theta_2)$$
prove that
$$\frac{z_1}{z_2} = \frac{r_1}{r_2}(\cos(\theta_1 - \theta_2) + i\sin(\theta_1 - \theta_2)).$$

10. Find the value of the constant a, such that, for all values of the real constant b, one root of the equation
$$2x^3 + ax + 4 = b(x - 2)$$
is equal to 2.

(a) When a has this value, find the set of values of b for which the given equation has three distinct real roots.

(b) When $a = -10$ and $b = -12$, find the roots, real and complex, of the given equation.

CAMB 1982

11. Given that $z_1 = 3 + 2i$ and $z_2 = 4 - 3i$,

(a) find $z_1 z_2$ and $\dfrac{z_1}{z_2}$, each in the form $a + ib$;

(b) verify that $|z_1 z_2| = |z_1||z_2|$.

12. You are given that $z = \cos\theta + i\sin\theta\,(0 < \theta < \frac{1}{2}\pi)$. Draw an Argand diagram to illustrate the relative positions of the points representing z, $z + 1$, $z - 1$.
Hence, or otherwise,

(a) determine the modulus and argument of each of these three complex numbers;

(b) prove that the real part of $\dfrac{z - 1}{z + 1}$ is zero.

O&C 1982

13. Find the modulus and argument of the complex numbers z_1, z_2 and z_3, where
$$z_1 = (1 - i), \quad z_2 = z_1^3, \quad z_3 = \frac{\sqrt{3} - i}{\sqrt{3} + i}.$$
Mark on an Argand diagram the points representing z_1, z_2 and z_3.

LOND 1983

14. Mark in an Argand diagram the points P_1 and P_2 which represent the two complex numbers z_1 and z_2, where $z_1 = 1 - i$ and $z_2 = 1 + i\sqrt{3}$.
On the same diagram, mark the points P_3 and P_4 which represent $(z_1 + z_2)$ and $(z_1 - z_2)$ respectively.

Find the modulus and argument of

(a) z_1,　　(b) z_2,　　(c) z_1z_2,　　(d) z_1/z_2.

LOND 1984

15. By use of the remainder theorem, or otherwise, factorise $(x^3 + 1)$ into factors with real coefficients. Hence, or otherwise, determine the three solutions of the equation $z^3 + 1 = 0$.

By use of the substitution $w = (1 + i)z$, find the three solutions of the equation

$$w^3 + 2(i - 1) = 0.$$

Plot the solutions of both equations in the same Argand diagram. Label the solutions of the first equation as z_1, z_2, z_3 and the solutions of the second equation as w_1, w_2, w_3 in such a way that

$$|z_1 - w_1| = |z_2 - w_2| = |z_3 - w_3| = k$$

and state the value of k.

O&C 1982

16. Prove that the non-real cube roots of unity are

$$-\frac{1}{2} \pm i\frac{\sqrt{3}}{2}.$$

These roots are represented in an Argand diagram by the points A, B and the number $z = -2$ is represented by the point C. Show that the area of the sector of the circle with centre C through A and B which is bounded by CA, CB and the minor arc AB is $\frac{1}{2}\pi$.

JMB 1983

17. Find the two possible pairs of values of real numbers x and y such that

$$(x + iy)^2 = -3 + 4i.$$

Mark in an Argand diagram the points representing your solutions and also the point representing $-3 + 4i$. Prove that the triangle formed by these points is right-angled.

JMB 1982

18. Two complex numbers are given by

$$z_1 = r_1(\cos\theta_1 + i\sin\theta_1)$$

and

$$z_2 = r_2(\cos\theta_2 + i\sin\theta_2).$$

Show that

$$z_1z_2 = r_1r_2[\cos(\theta_1 + \theta_2) + i\sin(\theta_1 + \theta_2)].$$

Express the complex number z, given by $z = -2 + 2\sqrt{3}i$, in the form $r(\cos\theta + i\sin\theta)$, where $r > 0$ and $-\pi < \theta \leqslant \pi$.

Hence, or otherwise, express z^3 in the form $a + ib$.

Find the modulus and argument of $1/z^4$.

JMB 1983

16.7 A geometrical representation of the addition of two complex numbers

Suppose $z_1 = x_1 + iy_1$ and $z_2 = x_2 + iy_2$ are represented by P_1 and P_2 in the Argand diagram (Fig. 16.11), then $P_3(x_1 + x_2, y_1 + y_2)$ represents $z_1 + z_2$ since $z_1 + z_2 = (x_1 + x_2) + i(y_1 + y_2)$.

It takes a simple geometrical argument to show that P_3 is located at one vertex of the parallelogram determined by the points O, P_1, P_2. (In vector notation: if a complex number z is represented as a vector $\vec{OP_z}$ rather than the point P_z, then complex addition obeys the laws of vector addition.)

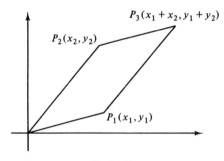

Fig. 16.11

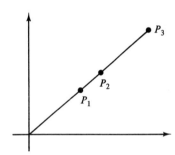

Fig. 16.12

If O, P_1 and P_2 are colinear then P_3 may readily be located along the line OP_1P_2 (Fig. 16.12).

One immediate observation is that

$$|z_1| + |z_2| \geqslant |z_1 + z_2|$$

This is the complex equivalent of noting that the shortest distance between two points is a straight line.

However, $|z_1| + |z_2| = |z_1 + z_2|$ if and only if z_1 and z_2 possess the same argument.

When two complex numbers have the same modulus then the parallelogram formed is a rhombus. The modulus and argument of the sum is then readily found (Fig. 16.3).

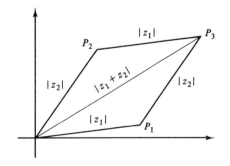

Fig. 16.13

Example 16.13

If $z = \cos\theta + i\sin\theta$, find the modulus and argument of $1 + z$, assuming $0 \leqslant \theta \leqslant \pi/2$.

It is clear from the diagram (Fig. 16.14) that

$$\arg(1 + z) = \frac{\theta}{2}$$

and

$$|1 + z| = 2\cos\frac{\theta}{2}.$$

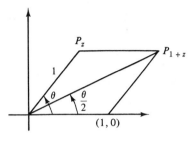

Fig. 16.14

Statement

> If z and w are represented by the points P_z, P_w in the Argand diagram then $|z - w|$ is equal to the distance between P_z and P_w.

The point P_{-w} representing $-w$ is diametrically opposite P_w. P_{z-w} is located at one vertex of the parallelogram determined by O, P_z and P_{-w}.

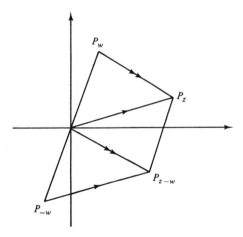

Fig. 16.15

Now

$$|z - w| = OP_{z-w}$$

$$= P_w P_z \qquad \text{as required.}$$

This result, as with many others, may be proved algebraically. Suppose $z = x + iy$ and $w = u + iv$, then

$$z - w = (x - u) + i(y - v)$$

so

$$|z - w| = \sqrt{(x - u)^2 + (y - v)^2}$$

and this is the distance between $P_z(x, y)$ and $P_w(u, v)$!

One useful interpretation of this result is to notice that $|z - (1 + 2i)|$ is equal to the distance of the point P_z to the point $(1, 2)$. It follows that if z is constrained so that $|z - (1 + 2i)| = 3$, then the point P_z, representing z, lies on a circle with centre $(1, 2)$ and radius 3 on the Argand diagram.

16.8   Loci

If z is constrained so that $|z - w| = r$ then the distance of z to w as represented on the Argand diagram is fixed as r units. In other words, z is constrained to move on a circle of radius r and with centre the point P_w. Similarly, if $\arg z = \pi/4$ then z is constrained to a half line meeting the origin with gradient 1.

Notice that the line cannot be extended into the third quadrant since the arguments of the complex numbers represented by such points are negative. It is also worth noticing that 0 is excluded since it does not possess an argument.

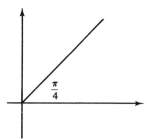

Fig. 16.16

Example 16.14.

Indicate on an Argand diagram the locus of z given that

$$\left| \frac{z - 1 - i}{z + 2} \right| = 1.$$

Recall that

$$\left| \frac{z_1}{z_2} \right| = \frac{|z_1|}{|z_2|},$$

so the above is equivalent to

$$|z - 1 - i| = |z + 2|,$$

which may be rewritten as

$$|z - (1 + i)| = |z - (-2)|,$$

indicating that the distance of P_z to $(1, 1)$ equals the distance of P_z to $(-2, 0)$.

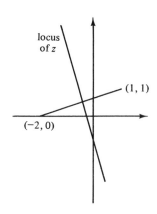

Fig. 16.17

The Cartesian equation of the line representing the locus of z may be found by letting $z = x + iy$ in the constraint

$$|z - 1 - i| = |z + 2|$$

giving

$$|(x - 1) + i(y - 1)| = |x + 2 + iy|$$

that is

$$(x - 1)^2 + (y - 1)^2 = (x + 2)^2 + y^2$$
$$\Rightarrow x^2 - 2x + 1 + y^2 - 2y + 1 = x^2 + 4x + 4 + y^2$$
$$\Rightarrow \qquad -2x - 2y + 2 = 4x + 4$$
$$\Rightarrow \qquad 6x + 2y + 2 = 0$$
$$\Rightarrow \qquad 3x + y + 1 = 0$$

Example 16.15

The complex number $z = x + iy$ is constrained so that

$$\frac{z + 3i}{z + 4} = ki \qquad \text{where} \quad k \in \mathbb{R}.$$

Show that, regardless of the value of k, a point P representing z on the Argand diagram must lie on a particular circle. Find the Cartesian equation of this circle.

$$\frac{z + 3i}{z + 4} = \frac{x + i(y + 3)}{(x + 4) + iy}$$

$$= \frac{(x + i(y + 3))((x + 4) - iy)}{(x + 4)^2 + y^2}$$

$$= \frac{[x(x + 4) + y(y + 3)] + i[(y + 3)(x + 4) - xy]}{(x + 4)^2 + y^2}$$

Since the real part of this expression is zero we have

$$x(x + 4) + y(y + 3) = 0$$

(providing $x = -4$ and $y = 0$ are not both true—since in that case the denominator is zero)

$$\Rightarrow (x + 2)^2 + \left(y + \frac{3}{2}\right)^2 = 4 + \frac{9}{4}$$

$$\Rightarrow (x + 2)^2 + \left(y + \frac{3}{2}\right)^2 = \frac{25}{4}$$

This is the Cartesian equation of a circle with radius 5/2 and centre $(-2, -3/2)$. As k varies so z will be represented by a point which moves on this circle.

Alternatively,

$$\frac{z + 3i}{z + 4} = ki$$

$$\Rightarrow \qquad z + 3i = ki(z + 4)$$

$$\Rightarrow x + i(3 + y) = -ky + ik(x + 4).$$

Equating real and imaginary parts,

$$x = -ky$$

$$(3 + y) = k(x + 4).$$

Eliminating k,

$$(3 + y) = -\frac{x}{y}(x + 4)$$

$$\Rightarrow y(3 + y) + x(x + 4) = 0.$$

We may then proceed as before.

Suppose P_z and P_w are points in the Argand diagram representing z and w. It can be seen from the diagram (Fig. 16.18) that OP_{z-w} is parallel to P_zP_w. If $\arg(z - w) = \alpha$ then the line P_wP_z is inclined at an angle α to the positive real axis.

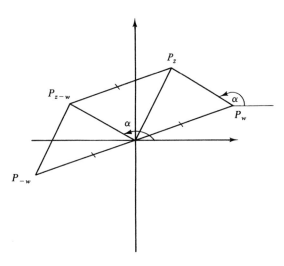

Fig. 16.18

Now if z is constrained so that $\arg(z - w) = \alpha$ it follows that P_z lies on the half line from P_w which is inclined at an angle α to the positive real axis.

On the other hand, if $\arg(w - z) = \alpha$ then $\arg(z - w) = \alpha \pm \pi$—the sign being chosen to ensure that the argument lies between $-\pi$ and π. Thus z is represented by a point on the half line drawn from w which makes an angle $\alpha \pm \pi$ with the positive real axis.

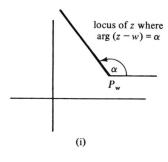

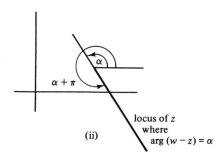

Fig. 16.19

Exercise 16.4

1. Sketch on an Argand diagram the locus of z if $|w| = 2$ and

    (a) $z = w$,   (b) $z = w + 2$,   (c) $z = iw$.

2. Show on an Argand diagram the locus of z if

    (a) $\dfrac{z - i}{z - 1}$ is purely imaginary,

    (b) $\dfrac{z - i}{z - 1}$ is real.

3. Sketch on the same Argand diagram the loci of z and w, where

    $$|z| = |z - 4i| \quad \text{and} \quad |w + 2| = 1.$$

    State

    (a) the minimum value of $|z - w|$,

    (b) the Cartesian form of z for which $\arg z = -\dfrac{\pi}{4}$.

    AEB 1982

4. Shade in an Argand diagram the region in which *both* of the following inequalities are satisfied:

    $$|z - 2i| \leqslant 2,$$

    $$\frac{\pi}{6} \leqslant \arg z \leqslant \frac{\pi}{2}.$$

    JMB 1984

5. Indicate on an Argand diagram the set of points representing the complex
    numbers z satisfying both

$$|z - 1| \leqslant 2 \quad \text{and} \quad \arg(z - i) = \frac{\pi}{4}.$$

JMB 1982

6. (a) Given that $(2 + 3i)z = 4 - i$, find the complex number z, giving your answer
        in the form $a + bi$.
    (b) Find the modulus and argument of the complex number $5 - 3i$.
        The complex number w is represented in an Argand diagram by the point
        W. Describe geometrically the locus of W if $|w| = |5 - 3i|$.

7. (a) Show that $(4 + 2i)$ is a root of $z^2 - (5 + 4i)z + 10i = 0$.
    (b) Find the quadratic equation whose roots are $2 + i$ and $2 - i$.
    (c) If $|z - 3| = |z - 1 - 2i|$, show clearly on an Argand diagram the locus of the
        point which represents z.
        Explain why there is no complex number z such the $\arg z = \pi/4$ and
        $|z - 3| = |z - 1 - 2i|$.

SU 1982

8. (a) If z_1 and z_2 are complex numbers, solve the simultaneous equations
        $4z_1 + 3z_2 = 23$, $z_1 + iz_2 = 6 + 8i$, giving both answers in the form $x + yi$.
    (b) If $(a + bi)^2 = -5 + 12i$, find a and b given that they are both real. Give the
        two square roots of $-5 + 12i$.
    (c) In each of the following cases define the locus of the point which represents z
        in the Argand diagram. Illustrate each statement by a sketch.

    (i) $|z - 2| = 3$,   (ii) $|z - 2| = |z - 3|$.

9. (a) If $-\pi < \arg z_1 + \arg z_2 \leqslant \pi$ show that $\arg(z_1 z_2) = \arg z_1 + \arg z_2$. The
        complex numbers $a = 4\sqrt{3} + 2i$ and $b = \sqrt{3} + 7i$ are represented in the
        Argand diagram by points A and B respectively. O is the origin. Show that
        triangle OAB is equilateral and find the complex number c which the point C
        represents where $OABC$ is a rhombus. Calculate $|c|$ and $\arg c$. (You may
        leave answers in surd form.)
    (b) z is a complex number such that

$$z = \frac{p}{2 - i} + \frac{q}{1 + 3i}$$

    where p and q are real. If $\arg z = \pi/2$ and $|z| = 7$ find the values of p and q.

SU 1980

10. (a) Given that $z_1 = 2 + i$ and that $z_2 = -2 + 4i$, find, in the form $a + bi$, the
        complex number z which is such that

$$\frac{1}{z} = \frac{1}{z_1} + \frac{1}{z_2}.$$

Find the modulus and arguments of z, giving your answer for $\arg z$ in the interval $-\pi < \arg z \leqslant \pi$, correct to three decimal places.

(b) Find, in a simplified form, the Cartesian equation of the locus of the point P representing the complex number z where

$$|z - 1| = 2|z + 1|.$$

Sketch the locus on an Argand diagram.

CAMB 1982

11. (a) $a = 2 + 5i$, $b = 5 + i$ and $c = 10 + 11i$ are complex numbers represented by the points A, B and C respectively in the Argand diagram. Express $(c - a)/(b - a)$ in the form $p + iq$ and hence, or otherwise show that angle BAC is a right angle. Calculate the angle ABC correct to the nearest degree.

If the number z (represented by Z) is such that $|z - b| = 12$ sketch the locus of Z in the Argand diagram and *prove* that all points of the triangle ABC lie inside the locus.

(b) If $z = x + iy$ and $z^* = x - iy$ solve (for z) the equation $z + 5z^* = 6 + 8i$.

SU 1983

12. The complex numbers z_1 and z_2 are given by

$$z_1 = \frac{1}{2}(1 + i\sqrt{3}), \qquad z_2 = i.$$

State the modulus and the argument of z_1 and of z_2. Represent the complex numbers z_1, z_2 and $z_1 + z_2$ on an Argand diagram. By using your diagram, or otherwise, show that

$$\tan\frac{5\pi}{12} = 2 + \sqrt{3}.$$

JMB 1982

Chapter 17 Numerical methods

17.1 Approximate solutions of equations

A graphical approach

A solution or root of the equation $f(x) = 0$ is a value of x, x_1 say, such that $f(x_1)$ is zero.

Geometrically this means that the curve

$$y = f(x)$$

crosses the x axis at $(x_1, 0)$. Conversely, if the graph of $y = f(x)$ crosses the x axis at $(x_1, 0)$ then x_1 is a root of the equation $f(x) = 0$.

Example 17.1

Find an approximation to the root of

$$\ln(1 + x) + 1 = 0.$$

As a first step it is useful to *sketch* a graph. This may be done in stages.

(i) $y = \ln x$
(ii) $y = \ln(1 + x)$ translate one unit to left
(iii) $y = \ln(1 + x) + 1$ translate one unit upwards

It can be seen from the sketch (Fig. 17.1) that there is just one root lying between -1 and 0.

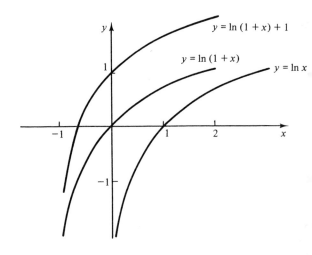

Fig. 17.1

The next stage involves drawing an accurate graph of the function $\ln(x+1)+1$ in the region of the root. In this case we draw the graph between -0.9 and 0 [The function is undefined when $x = -1$]. The following table of values is used.

x	-0.9	-0.8	-0.7	-0.6	-0.5	-0.4	-0.3	-0.2	-0.1	0
y	-1.30	-0.61	-0.20	$+0.08$	$+0.31$	$+0.49$	$+0.64$	$+0.78$	$+0.89$	$+1$

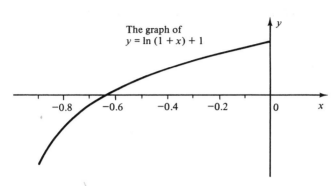

The graph of
$y = \ln(1+x) + 1$

Fig. 17.2

It may be seen from the graph (Fig. 17.2) that the root is approximately -0.64. If necessary an even more accurate graph may be drawn for values of x between -0.6 and -0.7. This is turn may be used to find a better approximation to the root.

Sometimes it is convenient to express the equation $f(x) = 0$ in the form $g(x) = h(x)$. This is particularly helpful when both $g(x)$ and $h(x)$ are standard functions. If we draw the graphs of $y = g(x)$ and $y = h(x)$ on the same axes, then if (x_1, y_1) is a point of intersection of the two graphs, we have $y_1 = g(x_1)$ and $y_1 = h(x_1)$.

It follows that

$$g(x_1) = h(x_1)$$

and so x_1 is a root of $f(x) = 0$.

Example 17.2

Find an approximation to the smallest positive root of the equation

$$x^3 - 3x + 1 = 0.$$

This may be rewritten as

$$x^3 = 3x - 1.$$

We now sketch the curves $y = x^3$ and $y = 3x - 1$ on the same axes. [See Fig. 17.3 on the next page.]

The graphs intersect for some value of x between 0 and 1 and for another value between 1 and 2. We could then proceed as before by drawing more accurate graphs between 0 and 2. However, once a reasonable approximation to the root has been found other methods may be employed to obtain a better approximation.

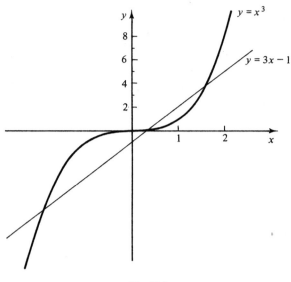

Fig. 17.3

Consider the root, which lies between 0 and 1, of the equation

$$x^3 - 3x + 1 = 0$$

where $f(x) \equiv x^3 - 3x + 1$.

Now $f(0) = 1$ and $f(1) = -1$. If we join the points $(0, f(0))$ and $(1, f(1))$ by a straight line, then the line is an approximation to the curve $y = f(x)$ and so the value of x where the line crosses the x axis is an approximation to the root of $f(x) = 0$. By symmetry we see that in this case the line crosses at $x = 0.5$.

Now $f(0.5) = -0.375$. A moments thought indicates that the actual root lies between 0 and 0.5. The above procedure may now be repeated, to find a better approximation to the root.

When a root is found in this way the process is called *linear interpolation*.

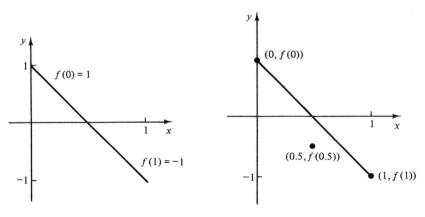

Fig. 17.4   Fig. 17.5

Linear interpolation

Suppose we have two values of x, a and b such that $f(a) > 0$ and $f(b) < 0$ then provided the function is continuous over this range of values there will be at least one root of $f(x) = 0$ between a and b. If the line segment joining $(a, f(a))$ to $(b, f(b))$ cuts the x axis at $R(r,0)$ then r will be an approximation to the root. In the diagram (Fig. 17.6) we have assumed $a < b$.

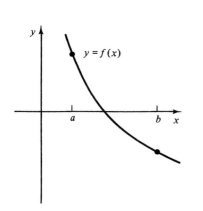

Fig. 17.6

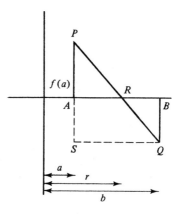

Fig. 17.7

By similar triangles we have

$$\frac{AR}{AP} = \frac{SQ}{SP},$$

that is

$$\frac{AR}{|f(a)|} = \frac{|b - a|}{|f(a)| + |f(b)|}.$$

The modulus signs are introduced to ensure that the quantities involved are positive.
Now

$$AR = \frac{|f(a)| \cdot |b - a|}{|f(a)| + |f(b)|}.$$

Finally, assuming $a < b$ we have

$$r = a + AR$$

$$= a + \frac{|f(a)| \cdot |b - a|}{|f(a)| + |f(b)|}$$

Statement

> If $a < b$ and $f(a)$ and $f(b)$ are of alternate signs (one positive the other negative) then r is an approximate root of $f(x) = 0$ where
> $$r = a + \frac{|f(a)| \cdot (b - a)}{|f(a)| + |f(b)|}$$

Example 17.3

Use linear interpolation to find an approximation to the negative root of
$$x^3 - 3x + 1 = 0.$$

Setting $P(x) \equiv x^3 - 3x + 1$, notice that
$$P(-1) = 3$$
$$P(-2) = -1$$

which shows that there is a root between -2 and -1. By the usual process of considering similar triangles
$$\frac{d}{3} = \frac{1}{4}$$
$$\Rightarrow d = 0.75$$

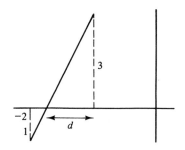

Fig. 17.8

Thus the root is approximately -1.75.
Now $P(-1.75) = 0.89$ and $P(-2) = -1$, which shows that the actual root is less than -1.75. Again
$$\frac{d}{0.89} = \frac{0.25}{1.89}$$
$$\Rightarrow \quad d = 0.118$$

The root is approximately -1.87. Now in this case
$$P(-1.87) = 0.07$$

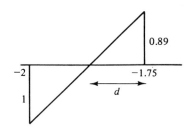

Fig. 17.9

This process may be repeated to find an even more accurate approximation.

Interval bisection

Suppose we wish to find an approximate solution of the equation
$$f(x) = 0$$

and it is known that $f(a) < 0$ and $f(b) > 0$ for some a and b. Suppose further that it is known that there is just one root between a and b.

We may select a value c (say) such that $a < c < b$. If $f(c) = 0$, we have solved our problem! However, if $f(c) < 0$ then our root lies in the interval (c, b) and if $f(c) > 0$ it is in the interval (a, c).

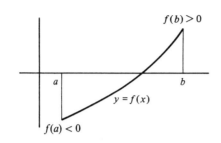

Fig. 17.10

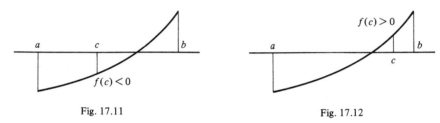

Fig. 17.11   Fig. 17.12

In either case we have reduced the length of the interval within which the root is known to lie. This procedure may be repeated as often as required.

Example 17.4

Find the root of

$$\sin x - \frac{x}{2} = 0$$

which lies between $x = 1$ and $x = 2$, correct to 1 d.p.

Now

$$f(1) = 0.34 \quad > 0$$
$$f(2) = -0.09 < 0$$

Clearly there is a root between 1 and 2, which is probably closer to 2 than 1.

Now

$$f(1.8) = 0.07 > 0,$$

so the root lies between 1.8 and 2.

$$f(1.9) = -0.004 < 0,$$

so the root lies between 1.8 and 1.9.

$$f(1.85) = 0.04 > 0,$$

so the root lies between 1.85 and 1.9.

It follows that the root is 1.9 correct to 1 d.p.

This process may be repeated to find an even more accurate approximation. Even this method can be tedious, and other methods which find good approximations quickly and efficiently are to be preferred.

17.2   Newton's method of approximation

Once an approximate root of $f(x) = 0$ has been found, perhaps by using one of the methods mentioned above, then an approach due to Newton may be employed. This approach is also known as the *Newton–Raphson method*.

Suppose we wish to solve the equation

$$f(x) = 0,$$

and $x = x_1$ is an *approximate* root. Then for some small h, $x_1 + h$ will be an *exact* solution.

Assuming f is differentiable,

$$f'(x_1) \simeq \frac{f(x_1 + h) - f(x_1)}{h} \quad \text{[see page 166]}$$

but we have assumed that $f(x_1 + h) = 0$ since $x_1 + h$ is the root of the equation.

It follows that

$$h \simeq -\frac{f(x_1)}{f'(x_1)}$$

and so letting

$$x_2 = x_1 - \frac{f(x_1)}{f'(x_1)}$$

then x_2 will be a better approximation.

The process may be repeated to find an improved approximation x_3 where

$$x_3 = x_2 - \frac{f(x_2)}{f'(x_2)} \quad \text{and so on.}$$

Newton's method of approximation.

If the function f is differentiable and x_n is an approximate root of $f(x) = 0$ then an improved approximation x_{n+1} may (usually) be obtained by setting

$$x_{n+1} = x_n - \frac{f(x_n)}{f'(x_n)}$$

The process of finding a good approximation by repeating the same procedure is often referred to as '*iteration*'.

Example 17.5

Find an approximate root of

$$\cos x - x = 0$$

using Newton's method of approximation.

Before we can use Newton's method we need an approximate solution. Rewriting the equation in the form

$$\cos x = x$$

suggests sketching the graphs of $y = \cos x$ and $y = x$ on the same axes (Fig. 17.13). This sketch shows that there is just one root; between 0 and $\pi/2$.

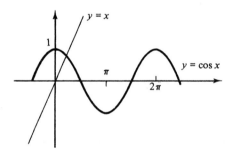

Fig. 17.13

Now setting

$$f(x) \equiv \cos x - x$$

notice that

$$f(0) = 1$$

$$f(\pi/2) = -\pi/2.$$

Using linear interpolation,

$$\frac{d}{1} = \frac{\pi}{2} \bigg/ \left(1 + \frac{\pi}{2}\right)$$

$$\Rightarrow d = 0.6 \qquad \boxed{C}$$

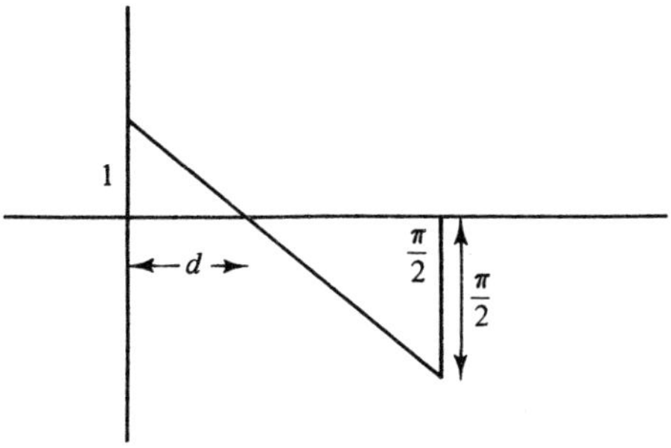

Fig. 17.14

Now $f(x) = \cos x - x$, so

$$f'(x) = -(1 + \sin x).$$

Using $x = 0.6$ as our starting point

$$f(x_1) = 0.22534 \qquad \boxed{C}$$

$$f'(x_1) = -1.56464 \qquad \boxed{C}$$

So

$$x_2 = x_1 - \frac{f(x_1)}{f'(x_1)}$$

$$= 0.74402. \qquad \boxed{C}$$

Repeating

$$f(x_2) = -0.00827 \qquad \boxed{C}$$
$$f'(x_2) = -1.67725 \qquad \boxed{C}$$

so

$$x_3 = x_2 - \frac{f(x_2)}{f'(x_2)}$$

$$= 0.73909. \qquad \boxed{C}$$

Repeating

$$f(x_3) = -0.00001.$$

This is so small that we may with confidence state that the root of $\cos x - x = 0$ is 0.7391 to 4 d.p.

Newton's method can be interpreted geometrically as follows. Suppose $x = a$ is an approximate root of $f(x) = 0$. The tangent at a meets the x axis at a point P which is generally closer to the actual root than is a (See Fig. 17.15).

So that $x - PA = a - h$ is a better approximation. Now it may be seen from the diagram that

$$f'(a) = \frac{f(a)}{h}$$

that is

$$h = \frac{f(a)}{f'(a)}$$

and so

$$a - \frac{f(a)}{f'(a)}$$

is a better approximation.

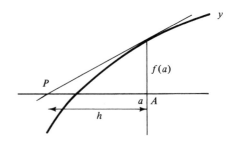

Fig. 17.15

Newton's method is not always successful, particularly when the first approximation is not sufficiently close to the actual root.

Fig. 17.16(i) shows how successive approximations converge to the root, and Fig. 17.16(ii) shows how successive approximations diverge from the root. When this happens a better first approximation can sometimes help.

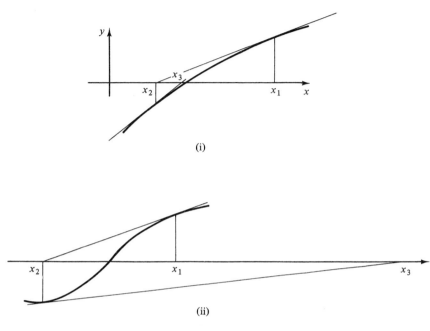

(i)

(ii)

Fig. 17.16

Exercise 17.1

1. Draw the graph of $y = \ln x$ for values of x such that $0.1 < x < 6$. Hence find a first approximation to the root of

   (a) $\ln x = 2 - x$ (b) $\ln x = 3 - 2x$

2. The points A and P are on the circumference of a circle centre O. The tangent to the circle at A meets OP produced at Q. The angle POA is θ radians and the length of the line segment PQ is the same as the length of the circular arc PA. Prove that $\sec \theta = \theta + 1$.

   By drawing accurate graphs for $0 < \theta < 1.2$, estimate a value of θ in this interval which satisfies this equation, giving your answer to two decimal places.

   AEB 1982

3. Show by means of a sketch, or otherwise, that $x - 1 - \sin x = 0$ has only one root and that it lies between 0 and π.

   Given an approximate value for the root, explain how an iterative method could be used to obtain a better approximation. (No calculation required.)

   W 1982

4. Sketch the curve

$$y = 2\cos\left(\tfrac{1}{2}\pi x\right)$$

for values of x such that $0 \leqslant x \leqslant 6$.

By drawing a suitable straight line on your diagram, show that the smallest positive root of the equation

$$4\cos\left(\tfrac{1}{2}\pi x\right) = 6 - x$$

lies between 3 and 4.

Taking 3.5 as the first approximation to this root, use the Newton–Raphson method to find a second approximation, giving two places of decimals in your answer.

CAMB 1983

5. A and B are points on the circumference of a circle centre O and the tangents at A and B meet at T. The area of the quadrilateral $OATB$ is equal to the area of the larger sector of the circle. If angle $AOB = 2\theta$ show that $\tan\theta = \pi - \theta$. Find, graphically, the relevant solution to this equation in radians correct to two places of decimals. Suggest one method by which a more accurate solution could be obtained. (You are not required to produce such a solution.)

SU 1983

6. Given that

$$f(x) = e^{-x} - 2x,$$

show by considering the sign of $f'(x)$, or otherwise, that the equation $f(x) = 0$ has just one real root. Show that this root lies between 0.3 and 0.4. Use linear interpolation once to obtain an approximate solution, giving two decimal places in your answer.

JMB 1982

7. Show that the equation $x^3 - 3x^2 - 2 = 0$ has a root between $x = 3$ and $x = 4$. Use Newton's method to find an approximation to this root using just two iterations.

8. Show that the equation $x^3 + x + 3 = 0$ has just one root, which lies between -2 and -1. Use Newton's method to find an approximation to this root correct to 2 d.p.

9. Draw the graphs of

$$y = \frac{1}{x} \quad \text{and} \quad y = x + 1$$

on the same axes for values of x and y such that $-3 \leqslant x \leqslant 3$ and $-3 \leqslant y \leqslant 4$.

Use the graphs to obtain both solutions to the equation

$$x^2 + x - 1 = 0 \quad \text{correct to 1 d.p.}$$

10. Given that $x = a$ is an approximation to a root of the equation $f(x) = 0$, show that, in general, $x = a - \dfrac{f(a)}{f'(a)}$ is a closer approximation, illustrating your argument with a clearly labelled diagram.

Illustrate in a diagram how this method for improving a numerical estimate for a root of an equation can fail.

By sketching the graphs of $y = 0.5 - x$ and $y = \log_e x$, show that the only real root of the equation

$$x + \log_e x = 0.5$$

lies in the interval $[\tfrac{1}{2}, 1]$.

Use the method of paragraph 1 to determine the value of the root correct to three decimal places.

AEB 1983

11. Given that $x = 2$ is an approximate root of the equation

$$x^3 - 2x^2 + x - 1 = 0$$

find the value of this root correct to two significant figures

W 1983

2. The equation $e^x + x = 2$ has a root between 0 and 1. Find this root correct to two places of decimals.

W 1981

3. Use Newton's method once, starting with the approximation $x = 2$, to obtain a second approximation x_1 for a root of the equation

$$x^5 = x^3 + 25.$$

What does it mean to claim that x_1 is correct to two decimal places? State briefly how you would prove this (no detailed working is expected).

O&C 1982

4. Show, by means of a sketch graph, or otherwise, that the equation

$$e^{2x} + 4x - 5 = 0$$

has only one real root, and that this root lies between 0 and 1.

Starting with the value 0.5 as a first approximation to this root, use the Newton–Raphson method to evaluate successive approximations, showing the stages of your work and ending when two successive approximations give answers which, when rounded to two decimal places, agree.

CAMB 1982

5. Given that $(1 + x)y = \ln x$. show that, when y is stationary,

$$\ln x = (1 + x)/x.$$

Show graphically, or otherwise, that this latter equation has only one real root and prove that this root lies between 3.5 and 3.8.

By taking 3.5 as a first approximation to this root and applying the Newton–Raphson process once to the equation $\ln x - (1 + x)/x = 0$, find a second approximation to this root, giving your answer to 3 significant figures.

Hence find an approximation to the corresponding stationary value of y.

LOND 1984

17.3   The use of the derivative in approximations

By definition

$$\frac{dy}{dx} = \lim_{x \to 0}\left(\frac{\delta y}{\delta x}\right).$$

It follows that

$$\frac{dy}{dx} \simeq \frac{\delta y}{\delta x}$$

with the approximation improving as $\delta x \to 0$. This may be written as

$$\delta y \simeq \frac{dy}{dx} \cdot \delta x.$$

This result may be used to determine the change in one variable resulting from a small change in another variable.

Example 17.6

If the radius of a circle increases from 5 cm to 5.1 cm, find the approximate increase in the area.

Now

$$A = \pi r^2$$

$$\Rightarrow \frac{dA}{dr} = 2\pi r$$

So

$$\delta A \simeq \frac{dA}{dr} \cdot \delta r$$

$$= 2\pi r \, \delta r$$

When $r = 5$ cm, $\delta r = 0.1$, so

$$\delta A \simeq 2 \times \pi \times 5 \times 0.1$$

$$= 3.142$$

Example 17.7

Use the method of approximations to evaluate tan (45.05°).

Suppose $y = \tan x$, then

$$y + \delta y = \tan(x + \delta x) \qquad ①$$

and

$$\frac{dy}{dx} = \sec^2 x$$

provided x is measured in radians. So

$$\delta y \simeq \frac{dy}{dx} \cdot \delta x$$

$$= \sec^2 x \, \delta x$$

when $x = 45°$, $\delta x = \dfrac{0.05\,\pi}{180}$ radians and $\sec^2 45° = 2$, so

$$\delta y \simeq 2 \cdot \frac{0.05\pi}{180}$$

$$\simeq 0.001745. \qquad \boxed{C}$$

Now

$$\tan(45.05°) = \tan 45° + \delta y \qquad [\text{from } ① \text{ since } \tan 45 = y]$$

$$\simeq 1.001745.$$

Compare this with the 1.001747 obtained directly from a calculator.

Exercise 17.2

1. Given that $\sin 45° = 0.7071$, obtain the value of $\sin(45.05°)$ correct to 4 d.p.

2. Obtain the value of $\cos(59.95°)$ correct to 4 d.p. using the method of approximations.

3. The radius of a circle increases from 3 cm to 3.2 cm. Find the approximate increase in area. Compare this with a direct calculation of the increase in area.

4. The radius of a sphere increases from 3 cm to 3.1 cm. Find the approximate increase in volume.

5. If $y = \sqrt{x}$, find the approximate increase in y as x increases from 4 to 4.1. Hence obtain $\sqrt{4.1}$ correct to 3 d.p., without using a calculator.

6. The volume of water in a spherical vessel is given by

$$v = \frac{\pi x^2}{3}(9 + x)$$

where x represents the depth of water. How much water must be added to increase the depth from 6 cm to 6.1 cm?

7. Prove from first principles that the derivative of $\sin x$ is $\cos x$.

[You may use $\lim\limits_{\theta \to 0} \dfrac{\sin \theta}{\theta} = 1$ without proof.]

When x increases from π to $\pi + \varepsilon$, where ε is small, the increment in

$$\frac{\sin x}{x}$$

is approximately equal to $p\varepsilon$. Find p in terms of π.

JMB 198.

8. Given that $y = \ln[(x + 1)(2x - 1)]$, $x \geqslant 1$, and that when x increases by a small amount δx, y increases by a small amount δy, show that

$$\delta y \approx \frac{4x + 1}{(x + 1)(2x - 1)}\delta x.$$

CAMB 198.

17.4   Numerical integration

The area bounded by the curve $y = f(x)$, the x axis and the lines $x = a$, $x = b$ may often be found by integration. Typically,

$$\text{Area} = \int_a^b f(x)\,dx$$

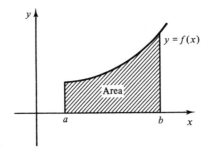

Fig. 17.17

It sometimes happens that the function $f(x)$ is unknown, and only specific values may be found. Even if $f(x)$ is known there may be no function whose derivative is $f(x)$. For example e^{x^2} is such a function. In either case the integration cannot be performed and so an approximate method must be used to find the area under a curve or the value of the definite integral.

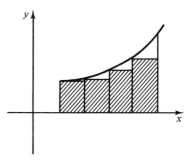

Fig. 17.18

One approach, mentioned in Section 9.1 is to plot an accurate graph and 'count squares'. Another is to approximate the area by a number of strips or rectangles (Fig. 17.18). The greater the number of strips the more accurate is the approximation.

The trapezium rule

A more accurate approach is to approximate the area by using trapeziums rather than rectangles.

In effect, the curve is reduced to a number of line segments. When this is done, even adopting relatively few strips can give quite accurate results.

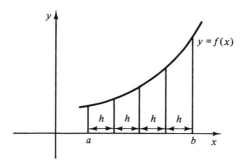

Fig. 17.19

Suppose then that the area representing

$$\int_a^b f(x)\,dx$$

is divided into strips of equal width h. Each strip is a trapezium bounded by ordinates y_1 and y_2 (say). Then the area of the trapezium is

$$\tfrac{1}{2}(y_1 + y_2)\,.\,h$$

Summing these areas gives us our approximation to the definite integral.

Quite generally, if there are n strips, bound by the ordinates $y_0, y_1, y_2 \ldots y_n$ of equal width h, then:

$$\int_a^b f(x)\,dx \simeq \tfrac{1}{2}(y_0 + y_1)h + \tfrac{1}{2}(y_1 + y_2) + \tfrac{1}{2}(y_2 + y_3)h + \ldots + \tfrac{1}{2}(y_{n-1} + y_n)h$$

$$= \tfrac{1}{2}h(y_0 + y_1 + y_1 + y_2 + y_2 + y_3 + \ldots + y_{n-1} + y_n)$$

$$= \tfrac{1}{2}h(y_0 + 2y_1 + 2y_2 + \ldots + 2y_{n-1} + y_n)$$

$$= \tfrac{1}{2}h(\text{first + last ordinate}) + h(\text{sum of other ordinates})$$

which is known as the *trapezium rule*.

Statement — The trapezium rule

$$\int f(x)\,dx = \tfrac{1}{2}h(\text{first and last ordinate}) + h(\text{sum of others}).$$

Example 17.8

Use the trapezium rule with ordinates at equal width $h = 0.2$ to find an approximate value for

$$\int_1^2 \frac{1}{x}\,dx.$$

We first construct the following table.

x	1	1.2	1.4	1.6	1.8	2.0	
$y = 1/x$	1	0.833	0.714	0.625	0.556	0.5	
	y_0	y_1	y_2	y_3	y_4	y_5	$h = 0.2$

$$\int_1^2 \frac{1}{x}\,dx \simeq \tfrac{1}{2}.0.2(1 + 0.5) + 0.2\,(0.833 + 0.714 + 0.625 + 0.556)$$

$$\simeq 0.696$$

By analytical methods

$$\int_1^2 \frac{1}{x}\,dx = [\ln x]_1^2 = \ln 2 = 0.693. \qquad \boxed{\text{C}}$$

Simpson's rule

The trapezium rule approximates the curve $y = f(x)$ by a series of line segments. A more sophisticated approach is to use a simple *curve* as an approximation.

For Simpson's rule a series of parabolas are used.

To begin with, suppose the area under the curve $y = f(x)$ is divided into *two* strips by the ordinates y_0, y_1 and y_2; set an equal distance h apart.

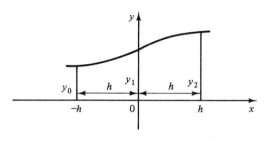

Fig. 17.20

Then

$$\int_{-h}^{h} f(x)\,dx \simeq \int_{-h}^{h} ax^2 + bx + c\,dx$$

(for some, as yet undetermined quadratic, $y = ax^2 + bx + c$)

$$= \left[\frac{ax^3}{3} + \frac{bx^2}{2} + cx\right]_{-h}^{h}$$

$$= \left(\frac{ah^3}{3} + \frac{bh^2}{2} + ch\right) - \left(-\frac{ah^3}{3} + \frac{bh^2}{2} - ch\right)$$

$$= \frac{2ah^3}{3} + 2ch$$

$$= \frac{h}{3}(2ah^2 + 6c).$$

Now the required quadratic will pass through the points $(-h, y_0)$, $(0, y_1)$ and (h, y_2) and so:

$$y_0 = ah^2 - bh + c$$

$$y_1 = c$$

and

$$y_2 = ah^2 + bh + c.$$

Notice that

$$y_0 + y_2 = 2ah^2 + 2c$$

and so

$$y_0 + 4y_1 + y_2 = 2ah^2 + 6c$$

It follows that

$$\int_{-h}^{h} f(x)\,dx \simeq \frac{h}{3}(2ah^2 + 6c)$$

$$= \frac{h}{3}(y_0 + 4y_1 + y_2)$$

For $\int_{a}^{b} f(x)\,dx$, we divide the area into an *even* number of strips (an *odd* number of ordinates) and apply the above result to each pair of strips in turn. For four strips we obtain

$$\int_{a}^{b} f(x)\,dx \simeq \frac{h}{3}(y_0 + 4y_1 + y_2) + \frac{h}{3}(y_2 + 4y_3 + y_4)$$

$$= \frac{h}{3}(y_0 + 4y_1 + 2y_2 + 4y_3 + y_4).$$

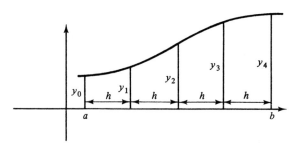

Fig. 17.21

Generalising to $2n$ strips ($2n + 1$ ordinates) we have, referring to the ordinates in the natural way:

Statement — Simpson's rule

$$\int_{a}^{b} f(x)\,dx = \frac{h}{3}\left[\text{first} + \text{last} + 4(\text{2nd} + \text{4th} + \ldots) + 2(\text{3rd} + \text{5th} + \ldots)\right]$$

Example 17.9

Find an approximation to

$$\int_{0}^{1} e^x\,dx,$$

using four strips and applying Simpson's rule.

Since the range is from 0 and 1 each strip is of width 0.25. We may set the work out as follows.

x	First and last ordinate	2nd, 4th, 6th, ... ordinate	3rd, 5th, 7th, ... ordinate
0	$e^0 = 1$		
0.25		$e^{0.25} = 1.284$	
0.5			$e^{0.5} = 1.649$
0.75		$e^{0.75} = 2.117$	
1.0	$e^1 = 2.718$		
Totals	3.718	3.401	1.649
		4	2
		13.604	3.298

Now

$$\int_0^1 e^x \, dx \simeq \frac{0.25}{3} (3.718 + 13.604 + 3.298)$$

$$= 1.718 \qquad \boxed{C}$$

Compare this with

$$\int_0^1 e^x \, dx = [e^x]_0^1 = e^1 - e^0 = 1.718. \qquad \boxed{C}$$

Exercise 17.3

1. Find an approximate value for each of the following definite integrals. Use the trapizoidal rule and ordinates spaced at equal intervals of width as indicated.

   (a) $\int_0^{\pi/2} \sin x \, dx$, width $\pi/8$

   (b) $\int_0^2 e^{-x} \, dx$, width $\frac{1}{3}$

   (c) $\int_0^1 \frac{1}{1+x} dx$, width 0.2

   (d) $\int_0^2 \tan^{-1} x \, dx$, width 0.25

2. Find an approximate value of the following definite integrals, using Simpson's rule and ordinates spaced at equal intervals of width as indicated.

   (a) $\int_0^{\pi/2} \sin x \, dx$, width $\pi/8$. Compare your result with the exact value.

   (b) $\int_0^\pi (1 + \cos \theta)^2 \, d\theta$, width $\pi/6$.

3. Find an approximation to the integral $\int_0^\pi \sin x \, dx$ by using Simpson's rule with five ordinates.
   Find the error in your result by comparing it with the exact value.

4. Use Simpson's rule with ten strips to approximate

$$\int_0^1 \frac{1}{1+x}\,dx$$

and hence estimate $\ln 2$.

5. Use Simpson's rule with five ordinates to find an approximation to

(a) $\displaystyle\int_0^1 \frac{dx}{1+x^3}$,   (b) $\displaystyle\int_1^5 \ln x\,dx$.

Use integration by parts to evaluate the latter integral.

6. Some values of a function f, are given in the table below. Use the trapezium rule to estimate $\displaystyle\int_0^{0.5} f(x)\,dx$ to 2 decimal places.

x	0	0.1	0.2	0.3	0.4	0.5
$f(x)$	0.10	0.23	0.45	0.52	0.44	0.21

LOND 1984

7. Evaluate

$$\int_2^4 \frac{\ln x}{x}\,dx$$

(a) by using the substitution $u = \ln x$,
(b) by using the trapezium rule with four strips,

giving each answer correct to three significant figures.

CAMB 1982

8. Evaluate $\displaystyle\int_0^\pi e^x \sin x\,dx$ approximately

(a) by using the trapezoidal rule with ordinates at $x = 0, \frac{1}{4}\pi, \frac{1}{2}\pi, \frac{3}{4}\pi, \pi$;
(b) by using Simpson's rule with the same five ordinates.

Sketch the graph of $y = e^x \sin x$, for $0 \leqslant x \leqslant \pi$.
Without further calculation, but with reference to your sketch, explain why both the trapezoidal rule and Simpson's rule (with three ordinates in each case) would give good approximations to $\displaystyle\int_0^{\pi/2} e^x \sin\,dx$. Explain also why the approximation to $\displaystyle\int_0^\pi e^x \sin\,dx$ calculated by Simpson's rule in (b) above is considerably more accurate than the trapezoidal rule approximation calculated in (a).

CAMB 1983

9. Using the same (co-ordinate) axes, sketch roughly the graphs of $y = \sin x$ and $y = \sqrt{\sin x}$ for values of x from 0 to $\frac{1}{2}\pi$, making it clear which curve is which. Deduce that, if

$$I = \int_0^{\pi/2} \sqrt{\sin x}\, dx,$$

then $1 < I < \frac{1}{2}\pi$.

Use Simpson's rule with three ordinates to obtain an approximation for I. (Give your answer to 3 significant figures.)

O & C 1983

10. Use Simpson's Rule, with six intervals, to find an approximate value for

$$\int_0^\pi \sqrt{1 + 2\sin x}\, dx,$$

giving your answer to three significant figures.

AEB 1982

11. Use Simpson's rule with seven equally spaced ordinates to find an estimate of the mean value of the function $\log_{10}(1 + x^3)$ between the values $x = 1$ and $x = 19$. Show your working in the form of a table and give your final answer to 3 significant figures.

AEB 1982

12. Use Simpson's rule with five ordinates to estimate

$$\int_0^{2\pi/9} \log_{10}(\cos x)\, dx,$$

giving your answer to 3 decimal places.

AEB 1982

13. The region enclosed by the curve $y = e^{-x^2/2}$, the x-axis and the lines $x = -3$ and $x = 3$ is rotated completely about the x-axis. Showing all your working in the form of a table, use Simpson's method with six intervals to find, in terms of e and π, an estimate of the volume of the solid so formed.

AEB 1983

17.5 Determination of laws

When two variables are related, values of one variable and corresponding values of the other variable may be obtained by experiment or observation. The relationship between the variables may be investigated by plotting the points found on a graph. If the graph shows that the points lie on a straight line—allowing for experimental error, then there is a linear relationship between the variables of the form

$$y = mx + c.$$

In such cases the gradient of the line may be used to estimate the value of m. The intercept can be used as an estimate of c. Alternatively particular (x and y) values (of points) on the line may be substituted into the equation to determine the value of c.

Example 17.10

The table below represents the observed values of two quantities x and y. Show that they appear to be linearly related, and estimate the precise form of the relationship.

x	10	20	30	40	50
y	4.8	7.1	9.0	10.9	13.2

We first plot a graph of the given values.

The points appear to lie on a line and so are linearly related as

$$y = mx + c.$$

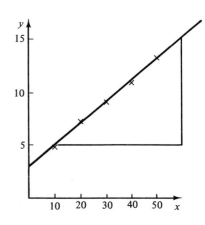

The line is then drawn in 'by eye'. Try to make the sum of all the displacements of the points to the line drawn as small as possible. This may mean that the line does not pass through any of the points plotted.

With the line as drawn we find

$$m \simeq 0.2 \quad \text{and} \quad c \simeq 3.$$

So x and y are linearly related by the approximate equation

$$y = 0.2x + 3$$

Fig. 17.22

When the variables are not linearly related, it is not easy to use the graph to establish the exact form of the relationship. However, many of the relationships which occur can be reduced to linear form as the following example illustrates.

Example 17.11

The variables p and v are believed to be related by a law of the form $p = av^n$, where a and n are constants. Use the following data to show that such a relationship is reasonable, and estimate the values of the constants.

p	1	2	3	4	5
v	3.1	12.6	28.3	50.2	78.5

If the variables are actually related by a law of the form

$$p = av^n$$

then taking logs to base 10 (say) of both sides gives

$$\lg p = n \lg v + \lg a \qquad \text{[see page 15]}$$

Comparing this with

$$Y = mX + c$$

we see that plotting $\lg p$ against $\lg v$ should give rise to a straight line, where $m = n$ and $c = \lg a$.

Now we may readily obtain the following table.

$\lg p$	0	0.30	0.48	0.60	0.70
$\lg v$	0.49	1.10	1.45	1.70	1.89

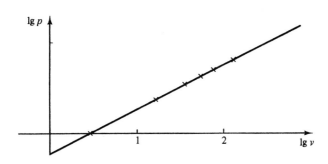

Fig. 17.23

We now draw the graph relating $\lg p$ with $\lg v$ (Fig. 17.23). The straight line suggests that the relationship is reasonable. From the graph we obtain

$$m = n \simeq 0.5$$

and $$c = \lg a \simeq -0.25 \quad \Rightarrow a \simeq 0.56 \qquad \boxed{C}.$$

Thus our estimate of the relationship is

$$p = 0.56\, v^{1/2}.$$

If a law of the form $y = ab^x$ is suspected, where a and b are constants, then this may be reduced to a straight line by again taking logs. We have

$$\lg y = x\lg b + \lg a.$$

This time we plot $\lg y$ against x. If a straight line is obtained then the gradient may be used as an estimate of $\lg b$ and the intercept as an estimate of $\lg a$. Once these are known it is then a simple matter to find estimates for a and b.

It is not always necessary to take logs. For a law of the form

$$\frac{1}{y} = \frac{a}{x} + b$$

we would plot $1/y$ against $1/x$. For $y^2 = a + bx$ we plot y^2 against x.

Exercise 17.4

1. The variables x and y are connected by the relation $y = a + bx^2$. By plotting y against x^2 estimate the values of the constants a and b using the data given.

x	0	2	4	6
y	3.3	8.1	22.5	46.5

2. The variables x and y are related by a law of the form $y = ax^n$, where a and n are integers. Various values of x and y were obtained by experiment. Use these values, given below, to find approximate values of a and n.

x	1	2	3	4	5
y	3.11	1.48	0.92	0.76	0.61

3. The table below gives the values of y for specified values of x as found by direct measurement. Show that the variables appear to be related by an equation of the form

   $$y = a(1 + x)^n$$

   and obtain estimates of a and n.

x	2	4	6	8	10
y	2.75	12.1	34.0	52.1	98.1

4. The variables x and y are connected by a law of the form $y = ax + bx^2$, where a and b are constants. Plot the graph of y/x against x to estimate the values of a and b using the values of x and y given below.

x	1	2	3	4	5
y	4	6	6	4	0

   Find the exact relation using an algebraic method for finding the values of a and b.

5. The table below gives values of the variables x and y which are related by a law of the form

   $$y = ax + \frac{b}{x}.$$

   Find estimates for the constants a and b by plotting xy against x^2 using the data below.

x	1	2	3	4	5
y	7.1	8.2	10.6	13.4	16.3

6. The variables x and y are thought to be connected by a relationship of the form

$$y = \lg(a + bx)$$

where a and b constants. Plot 10^y against x using the table below, and hence obtain estimates for a and b.

x	1	2	3	4	5	6
y	0.86	0.92	0.98	1.10	1.08	1.12

7. Pairs of numerical values (x, y) are collected from an experiment and it is possible that either of the two following equations may be applicable to these data

(a) $ax^2 + by^3 = 1$, where a and b are constants,
(b) $y = cx^d$, where c and d are constants.

   In each case explain carefully how you would use a graph to examine the validity of the equation. Explain how you would estimate the values of the constants if you found the equation to be approximately valid from your graph.

AEB 1983

8. The following corresponding values of x and y are believed to be related by the equation $y = x + ax^b$.

x	2	3	4	5	6	10
y	5.54	7.33	9.00	10.59	12.12	17.90

   Draw a suitable graph to show that this may be so and use your graph to find probable values for a and b.

AEB 1982

9. The following pairs of values of x and y have been found by experiment, and some values of $yx^{3/2}$ have been calculated and tabulated as shown.

x	1.0	2.6	3.2	4.0	6.2
y	5.5	2.0	1.7	1.4	1.0
$yx^{3/2}$	5.50		9.73		15.44

   Copy and complete this table.
   It is believed that x and y are connected by a law of the form

$$yx^{3/2} = ax + b, \quad \text{where } a \text{ and } b \text{ are constants.}$$

   Show graphically that, for these data, this law is approximately valid and use your graph to estimate

(a) suitable values for a and b,
(b) the value of y when $x = 5.1$.

AEB 1982

10. By drawing a suitable graph, show that the corresponding values of x and y in the table are approximately consistent with a law of the form $y = ax^b$.

x	12	15	22	28	35
y	75.9	54.3	30.6	21.3	15.3

From your graph estimate

(a) values for a and b,   (b) the value of x when $y = 42.5$.

AEB 1983

11. In each of the following cases, given experimental values of x and y, explain how straight line graphs may be drawn, using ordinary graph paper only:

(i) $y = ab^{x+1}$,   (ii) $px^2 + qy = x$,

where a, b, p, q are constants.
   In each case express the gradient of the line and its ordinate for $x = 0$ in terms of the constants.

AEB 1982

Revision exercise D

1. Two complex numbers p and q are given respectively by $p = 3 - 2i$ and $q = 1 + 3i$.

   (a) Express the complex number $\dfrac{q}{p}$ in the form $a + ib$.

   (b) Calculate the modulus of p and the modulus of q. Calculate also the argument of pq, giving your answer in radians correct to two significant figures.

   (c) Show on separate Argand diagrams the set of points representing the complex number z in each of the following cases:

   (i) $|q| \leqslant |z| \leqslant |p|$;   (ii) $\arg(z) = \arg(pq)$.

CAMB 1983

2. Solve the differential equation

$$\frac{dy}{dx} = -\frac{(x-1)}{(y-2)},$$

giving the particular solution (in the form $y = f(x)$) that satisfies the condition $y = 5$ when $x = 5$.
   Sketch the solution-curve in this case.

O & C 1982

3. (a) Evaluate $\displaystyle\int_1^2 x^3 \log_e x \, dx$ correct to 3 significant figures.

(b) Evaluate $\displaystyle\int_0^{\pi/3} \sin x \sec^2 x \, dx$.

(c) If $dy/dx = 2x + y$ and $z = 2x + y$, show that $dz/dx = 2 + z$.

Find the equation of the curve which passes through the origin and is such that $dy/dx = 2x + y$ at all points (x, y) on the curve, giving an equation in the form $y = f(x)$.

SU 1981

4. (a) Given that $dy/dx = (\tan x)\sqrt{y}$, $0 \leqslant x < \frac{1}{2}\pi$, and that $y = 1$ when $x = 0$, find an expression for y in terms of x.
    (b) (i) Show that

$$\int_7^{10} \frac{7}{x^2 - 5x - 6} \, dx = \ln\left(\tfrac{32}{11}\right).$$

    (ii) Evaluate $\displaystyle\int_1^e \ln x \, dx$.

CAMB 1983

5. Show that $\displaystyle\int_0^1 e^x \, dx = e - 1$. The integral is approximated by using the trapezium rule, dividing the range into n intervals each of length $h = 1/n$. Show that the result is

$$\frac{(e - 1)(p + 1)}{2n(p - 1)} \qquad \text{where } p = e^h.$$

Hence or otherwise show that

$$\frac{h(e^h + 1)}{2(e^h - 1)}$$

is very close to 1 when h is small.

OX 1983

6. If $y = \sec^2 x$, prove that $\dfrac{dy}{dx} = 2y \tan x$ and $\dfrac{d^2y}{dx^2} = 6y^2 - 4y$. Obtain also an expression for $\dfrac{d^4y}{dx^4}$ in terms of y. Using these results, or otherwise, show that the expansion of $\sec^2 x$ in ascending powers of x up to and including the term in x^4 is $1 + x^2 + \frac{2}{3}x^4$.

Explain briefly why, however far the expansion is continued, you would expect there to be no odd powers of x present.

By drawing rough sketches of the functions $6 \sec^2 x$ and $7 + 4x^4$, show that the equation $6 \sec^2 x = 7 + 4x^4$ has only one root in the range $0 \leqslant x < \frac{1}{2}\pi$. Use the expansion of $\sec^2 x$ to find an approximate value for the root.

O & C 1982

7. Use the binomial series to expand $\left(1 - \dfrac{x^2}{a^2}\right)^{1/2}$ in ascending powers of x^2 up to and including the term in x^6 and simplify each coefficient.

   Use your expansion to find, in ascending powers of x, the first *two* non-zero terms in the expansions of

   (a) $(9 - 4x^2)^{1/2}$,

   (b) $(9 - 4x^2)^{1/2} \sin 6x$.

AEB 1982

8. (a) Differentiate

$$\log_e \left\{ \frac{\sqrt{1 + 2x}}{(1 - x)^3} \right\}$$

   with respect to x, simplifying your answer.

   (b) Find the equation of the normal to the curve given parametrically by

$$x = \sin 2\theta, \qquad y = 2\cos\theta$$

   at the point where $\theta = \pi/3$.

AEB 1983

9. A function f is defined on the set S, where $S = \{x : x \in \mathbb{R}, x \neq 3\}$, by

$$f : x \to \frac{3x + b}{x - 3}, \qquad (b \neq -9).$$

   (a) Show that the inverse of f is f.

   (b) Determine the range of values of b for which there are two invariant values of x and find these values of x when $b = 55$. [x is an invariant value if $x = f(x)$.]

AEB 1982

10. Given that

$P(x) = (x - a)^2 Q(x)$, where $Q(x)$ is a polynomial, show that $P'(a) = 0$.
Hence, factorise the cubic

$$4x^3 + 12x^2 - 15x + 4$$

[*Hint*: substitute each of the roots of $P'(x) = 0$ into the cubic $P(x)$ above].

11. Show, by integration, that the volume of the *larger* segment of a sphere of radius r cut off by a plane distant $tr\,(0 < t < 1)$ from the centre of the sphere is $\frac{1}{3}\pi r^3 (2 + 3t - t^3)$. A flask is made by joining a hollow cylinder of height $\frac{3}{4}r$ onto a hollow part of a sphere as shown in Fig. 17.24 (see opposite page). Show that the total volume of the flask, V, is given by

$$V = \pi r^3 \left(\frac{17}{12} + t - \frac{3}{4}t^2 - \frac{1}{3}t^3 \right).$$

Find the maximum value of V for a given r as t varies.

SU 1980

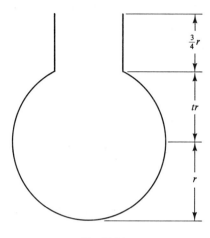

Fig. 17.24

12. The function g is defined by $g(x) = 7\cos^2 x + \sin^2 x - 8\sin x \cos x$.

   (a) Show that $g(x)$ may be expressed in the form $a + b\cos(2x + \alpha)$, where $\tan \alpha = \frac{4}{3}$ and a, b are constants to be determined. Find the greatest and least values of $g(x)$.

   (b) Find the values of x in the range $0 \leqslant x \leqslant \pi$ for which $g(x) = 0$ and sketch the graph of $g(x)$ in this range.

   (c) Assuming that $x_1 = \frac{1}{2}\pi$ is an approximate solution of the equation $3g(x) = 2x$, find a closer approximation by one application of Newton's method.

O & C 1983

13. Given that

$$\frac{x^2 - 2x - 9}{(2x - 1)(x^2 + 3)} \equiv \frac{A}{2x - 1} + \frac{Bx + C}{x^2 + 3},$$

   show that $C = 0$ and obtain the values of A and B.
   Hence evaluate

$$\int_1^2 \frac{x^2 - 2x - 9}{(2x - 1)(x^2 + 3)}\, dx,$$

   giving your answer to two significant figures.
   Given that

$$y = \frac{x^2 - 2x - 9}{(2x - 1)(x^2 + 3)},$$

   find the value of $\dfrac{dy}{dx}$ when $x = -1$.

CAMB 1983

14. (a) Find $\displaystyle\int \frac{1}{(1+x)(2-x)}\,dx.$

(b) Use the substitution $x = \sin^2 \theta$ to show that

$$\int_0^1 \sqrt{\left(\frac{1-x}{x}\right)}\,dx = \int_0^{\pi/2} 2\cos^2\theta\,d\theta,$$

and evaluate either integral.

15. Find the values of the constants a and b for which the expansions, in ascending powers of x, of the two expressions

$$(1 + 2x)^{1/2} \quad \text{and} \quad \frac{1 + ax}{1 + bx},$$

up to and including the term in x^2, are the same.

With these values of a and b, use the result

$$(1 + 2x)^{1/2} \approx \frac{1 + ax}{1 + bx},$$

with $x = \dfrac{1}{100}$, to obtain an approximate value for $\sqrt{2}$ in the form p/q, where p and q are positive integers.

CAMB 1982

16. P is the point $(ap^2, 2ap)$ on the parabola $y^2 = 4ax$ and S is the focus. If the normal at P meets the x axis at N, show that triangle SPN is isosceles.

Hence show that PS and a line through P parallel to the axis of the parabola make equal angles with the normal.

SU 1982

17. Find the co-ordinates of the point on the curve

$$y = \frac{\ln x}{x} \quad (x > 0),$$

at which

$$\frac{dy}{dx} = 0.$$

Determine whether the point found is a maximum point, a minimum point, or a point of inflexion of the curve.

CAMB 1983

18. (a) Differentiate with respect to x:

(i) $e^{\sqrt{x^2+1}}$,   (ii) $x^2 \ln x$,   (iii) $\cos^4 3x$.

(b) Find the equation of the tangent to the curve

$$x^3 + 3xy^2 + y^3 = 5$$

at the point $(1, 1)$.

O & C 1982

19. (a) Solve the equation $2 \tan \theta - 4 \cot \theta = \operatorname{cosec} \theta$, giving answers in radians in the range $-\pi < \theta < \pi$.
    (b) Prove that the area of the minor segment cut off by a chord subtending an angle of θ radians at the centre of a circle of radius r is $\frac{1}{2} r^2 (\theta - \sin \theta)$.

    A chord which subtends an angle α at the centre of a circle divides the area of the circle into 2 segments in the ratio $1:5$.
    Prove that $\sin \alpha = \alpha - \pi/3$.
    Plot the graphs of $y = \sin \alpha$ and $y = \alpha - \pi/3$ for values of α between 0 and π using the same axes. Hence find the value of α.

SU 1981

20. A vertical pole BAO stands with its base O on a horizontal plane, where $BA = c$ and $AO = b$. A point P is situated on the horizontal plane at distance x from O and the angle $APB = \theta$.
    Prove that $\tan \theta = \dfrac{cx}{x^2 + b^2 + bc}$.

    As P takes different positions on the horizontal plane, find the value of x for which θ is greatest.

AEB 1982

21. Show that the mean value of $\sin x \, e^{-x}$ in the interval $0 \leqslant x \leqslant \pi$ is

$$\frac{(e^{-\pi} + 1)}{2\pi}.$$

22. Use the series expansion for $\sin x$ to write down in ascending powers of x the first three non-zero terms in the series expansion of $\dfrac{\sin x}{x}$.

    Write down in ascending powers of y the first three non-zero terms in the series expansion of $\log_e (1 - y)$.

    Use these two series to show that if x is small enough for terms in x^6 and higher powers of x to be neglected, then

$$\log_e \sin x - \log_e x = Ax^2 + Bx^4$$

    for constants A and B whose numerical values are to be found.

AEB 1983

23. (a) An arithmetic series has first term 1000 and common difference -1.4. Calculate the value of the first negative term of the series, and the sum of all the positive terms.

(b) The sum to infinity of a geometric series is equal to fifteen times the sum of the first fifteen terms. Calculate the value of the common ratio, giving your answer correct to 3 decimal places.

CAMB 1983

24. K, P, Q, are the points $\left(ck, \dfrac{c}{k}\right), \left(cp, \dfrac{c}{p}\right), \left(cq, \dfrac{c}{q}\right)$ respectively on the rectangular hyperbola $xy = c^2$. Prove that the chord KP has equation

$$kpy + x = c(k + p)$$

and deduce the equation of the tangent at K. If K is fixed and P, Q vary so that PK is perpendicular to QK,

(a) prove that PQ is perpendicular to the tangent at K;
(b) prove that the mid-point M of PQ lies on a certain straight line through the origin, and that the tangents at P, Q intersect on this line.

O & C 1982

25. The points P, Q on the rectangular hyperbola $xy = c^2$ have co-ordinates $(cp, c/p), (cq, c/q)$ respectively. Find the gradient of the chord PQ.
    The point R also lies on the rectangular hyperbola and $\angle PRQ$ is a right-angle. Draw a rough sketch of this configuration and prove that PQ is perpendicular to the tangent at R.

O & C 1983

26. The normal to the parabola $y^2 = 4ax$ at the point $P(at^2, 2at)$ cuts the x-axis at the point G. The mid-point of PG is M, and O is the origin.

(a) Prove that the equation of the normal is $y + tx = 2at + at^3$.
(b) Prove that M is the point $(at^2 + a, at)$.
(c) Find the Cartesian equation of the locus of the point M as t varies.

AEB 1982

27. Prove that the equation of the tangent to the parabola $y^2 = 4ax$ at the point $P(ap^2, 2ap)$ on the curve is

$$py = x + ap^2.$$

Find the coordinates of the point of intersection, T, of the tangents at P and $Q(aq^2, 2aq)$, simplifying your answers where possible.
Given that S is the pont $(a, 0)$, verify that

$$SP . SQ = ST^2.$$

CAMB 1983

28. (a) Simplify

(i) $20 \times 8^{2n} - 5 \times 4^{3n+1}$,   (ii) $(\log_2 5) \times (\log_5 8)$.

(b) Find x and y given that $e^x + 3e^y = 3$ and $e^{2x} - 9e^{2y} = 6$, expressing each answer as a logarithm to base e.

AEB 1982

29. The parametric equations of a curve are

$$x = 3t^2, \qquad y = t^3 + 3t.$$

Find dy/dx in terms of t and show that

$$\frac{d^2y}{dx^2} = \frac{(t^2 - 1)}{12t^3}.$$

30. (a) If $y = x + \sqrt{1 + x^2}$, find dy/dx and hence prove that

$$\sqrt{1 + x^2}\frac{dy}{dx} = y.$$

(b) If $x = t^2 \sin 3t$ and $y = t^2 \cos 3t$, find dy/dx in terms of t, and show that the curve defined by these parametric equations is parallel to the x-axis at points where $\tan 3t = \dfrac{2}{3t}$.

(c) Differentiate with respect to x: $\dfrac{e^{2x}}{(1 + x)}$.

For what range of values of x is $\dfrac{e^{2x}}{(1 + x)}$ increasing?

Sketch the curve represented by $\dfrac{e^{2x}}{(1 + x)}$ for $-1 < x \leqslant 3$.

SU 1981

31. A curve is given by the parametric equations

$$x = 2\theta - \sin 2\theta, \qquad y = 2\sin^2\theta, \qquad 0 < \theta < \pi.$$

Show that $\dfrac{dy}{dx} = \cot\theta$, and hence find the co-ordinates of the turning point on the curve. Sketch the curve.

The normal to the curve at the point P, where $\theta = \alpha$, meets the x-axis in Q. Show that Q has co-ordinates $(2\alpha, 0)$. The tangent at P meets the line through Q parallel to the y-axis in R. Show that the length of QR is independent of α.

32. Given that α and β are the roots of the equation

$$x^2 - 6x + 1 = 0,$$

state the values of $\alpha + \beta$ and $\alpha\beta$. Evaluate $\alpha^2 + \beta^2$ and $\alpha^3 + \beta^3$.

Find the numerical values of the coefficients of the first four terms in the expansion in ascending powers of t of $e^{\alpha t} + e^{\beta t}$.

JMB 1983

Answers to exercises

Exercise 1.1

1. True: (b), (c), (f)

2. True: (a), (b), (d), (h)
   False: (c) eg. $1 \in \mathbb{N}$ and $2 \in \mathbb{N}$ but $1 - 2 = -1$ and $-1 \notin \mathbb{N}$.
   (e) eg. let $x = y = \sqrt{2}$, then $xy = 2$.
   (f) if $x, y \in \mathbb{R}$ then $x + y \in \mathbb{R}$ and $xy \in \mathbb{R}$
   but $x/y \notin \mathbb{R}$ if $y = 0$.
   (g) $4 \in \mathbb{Q}$, but then $\sqrt{4} = 2$ which is rational.

3. (a) False   (b) True   (c) False—The smallest prime number is 2.
   (d) False—if $x^2 = 2$ then $x = \sqrt{2}$ or $-\sqrt{2}$ and neither belongs to $\mathbb{Q}$.

4. Follow proof that $\sqrt{2}$ is irrational

5. (a) The set contains the two members 1 and 2.
   (b) The set of prime numbers.   (c) The set of even numbers.
   (d) The set of negative reals.   (e) The set of positive reals.
   (f) The set of non negative reals.

7. (a) 0.625   (b) (i) $\frac{1}{4}$ (ii) $\frac{7}{16}$ (iii) $\frac{2}{3}$

Exercise 1.2

1. (a) 11   (b) 300   (c) 15   (d) $\frac{1}{12}$ (e) $6\sqrt{12} - 10\sqrt{6}$ (f) $6(5 + 2\sqrt{6})$

2. (a) $3\sqrt{10}$ (b) 30   (c) $30\sqrt{10}$ (d) $3\sqrt{2}$ (e) $8\sqrt{2}$ (f) $12\sqrt{7}$

3. (a) $\sqrt{18}$ (b) $\sqrt{125}$ (c) $\sqrt{128}$ (d) $\sqrt{\frac{1}{2}}$ (e) $\sqrt{\frac{1}{12}}$ (f) $\sqrt{\frac{5}{12}}$

4. (a) $5\sqrt{2}$ (b) $3\sqrt{2}$ (c) $12\sqrt{3}$ (d) $14\sqrt{2} + 5\sqrt{3}$ (e) $(5 - 2\sqrt{6})$
   (f) $6(11 - 4\sqrt{6})$ (g) $\sqrt{3} + \sqrt{5}$ (h) $2\sqrt{10}$

5. (a) -2 (b) 1   (c) 3

6. (a) $\sqrt{3}/3$ (b) $\sqrt{5}$ (c) $\sqrt{10}/5$ (d) $2 - \sqrt{2}$ (e) $5(\sqrt{3} - \sqrt{2})$
   (f) $(3 + 2\sqrt{15})/17$ (g) $\frac{1}{2}(1 + \sqrt{6} + \sqrt{2} + \sqrt{3})$

6. (h) $\frac{1}{2}(\sqrt{15} + 3 + 2\sqrt{10} + 2\sqrt{6})$ (i) $2\sqrt{2} + 1)$
   (j) $\frac{3}{19}(4 + 3\sqrt{6})$ (k) $2\sqrt{2}$

7. (a) 0.5774 (b) 2.236 (c) 0.6325 (d) 0.5858 (e) 1.589 (f) 0.6321
   (g) 3.298 (h) 9.048 (i) 4.828 (j) 1.792 (k) 2.828

8. (a) $2 - \sqrt{2}$ (b) $2 + \sqrt{3}$ (c) $-(3 + \sqrt{6})$

9. (a) $a = 5$ and $b = 2$, $\sqrt{(7 + 2\sqrt{10})} = (\sqrt{5} + \sqrt{2})$
   (b) $\sqrt{(19 - 8\sqrt{3})} = 4 - \sqrt{3}$

Exercise 1.3

1. (a) 1 (b) 4 (c) $\frac{1}{6}$ (d) 27 (e) $\frac{1}{27}$ (f) $\sqrt{2}/4$ (g) 1 (h) 1000 (i) -2
   (j) $-\frac{1}{3}$ (k) 8 (l) 4 (m) $\frac{1}{8}$ (n) $\frac{3}{5}$ (o) 0.5 (p) 2

2. (a) 3 (b) 2 (c) $\frac{3}{4}$ (d) -2 (e) $\frac{4}{3}$ (f) $\frac{5}{2}$

3. (a) 3^{20} (b) 3^9 (c) 3^2 (d) 3^4 (e) 3^{-1} (f) $3^{7/12}$

4. (a) $8x^2y$ (b) $\dfrac{ab^2}{c^3}$ (c) $2\sqrt{2}xy^{3/2} = 2xy\sqrt{2y}$ (d) $\dfrac{3x}{4(x + 1)^2}$ (e) $2x^{-1/4}$

5. (a) x^2 (b) $6x^5$ (c) $1/3x^2$ (d) $10x^{3/4}$ (e) $4x^2$ (f) $2x^{1/4}$
   (g) $x^{2/3}$ (h) $(1 + x)^2$ (i) $\frac{1}{3}x^{-4}$ (j) $(1 - x)\sqrt{1 + x}$

6. (a) $x^2 + 1$ (b) $x - x^{1/3}$ (c) $\left(\dfrac{1 - x}{x}\right)\sqrt{x}$
   (d) $8x^2 + 2\sqrt{x} - 12x^{3/2} - 3$ (e) $(x - 1) + \dfrac{1}{x(x - 1)}$

Exercise 1.4

1. (a) $\log_3 27 = 3$ (b) $\log_4 16 = 2$ (c) $\log_5 1 = 0$ (d) $\log_4 2 = \frac{1}{2}$
   (e) $\log_3 \frac{1}{3} = -1$ (f) $\log_4 \frac{1}{16} = -2$ (g) $\log_a c = b$ (h) $\log_4 8 = \frac{3}{2}$

2. (a) 7 (b) 3 (c) 2 (d) -1 (e) -2 (f) $\frac{2}{3}$ (g) 1 (h) -2 (i) 2

3. (a) $\log_3 8 + \log_3 4$ (b) $\log_4 3 - \log_4 5$ (c) $3\log_2 5$ (d) $\log_a b + \log_a c$
   (e) $2\log_a b + \log_a c$ (f) $3\log_a b - \log_a c$ (g) $2(\log_a b - \log_a c)$ (h) $\log_a b - 1$
   (i) $3 - 2\log_a b$ (j) $\frac{1}{2}\log_a b + \log_a c - 1$ (k) $\frac{1}{2}(\log_a(x + 1) - \log_a(x - 1)$
   (l) $\log_a(x^2 + 1) - \frac{1}{2}\log_a(x^2 - 1) = \log_a(x^2 + 1) - \frac{1}{2}(\log_a(x + 1) + \log_a(x - 1))$

4. (a) $\log_2 15$ (b) $\log_3 \frac{4}{3}$ (c) $\log_2 125$ (d) $\log_3 \frac{1}{16}$ (e) $\log_5 \frac{1}{4}$ (f) $\log_3 50$
   (g) $\log_2 9$ (h) $\log_2 \frac{1}{3}$ (i) $\log_3 6$ (j) $\log_2 \frac{5}{8}$

5. (a) $\log_a bc$ (b) $\log_c bc$ (c) $\log_c bc$ (d) $\log_a a^2 b$ (e) $\log_a b^2$ (f) $\log_a b^{3/2}$
   (g) 0 (h) $\log_a \left(\dfrac{b^2 c^2}{a^2} \right)$ (i) $\log_a (\sqrt{(x-1)(x-3)})$ (j) $\log_2 \left(\dfrac{x^6}{y^3} \right)$

Exercise 1.5

1. (a) 1.585 (b) 0.631 (c) 1.113 (d) -3.322 (e) -0.631

2. (a) 1.893 (b) 1.683 (c) -3.822 (d) -0.523 (e) -0.339 (f) -1.795

3. (a) 22.637 (b) 2.158 (c) 11.212 (d) 1.035

4. (a) 2.005 (b) 1.260 (c) 3.824

5. (a) $\lg \left[\dfrac{(x+1)^{1/2}}{(x-1)^2} \right]$ (b) $\frac{1}{2} \lg \left(\dfrac{x-1}{x+1} \right)$ (c) 4

6. $x = \frac{1}{2}$ 7. 8.64   8. -0.39 9. 3   10. 6

12. $\pm \sqrt{2}/2$ 13. (a) $\frac{7}{4}$ (b) 6   14. $\frac{1}{2}$

Exercise 2.1

1. (a) 3 (b) 3 (c) 4   2. (a) -8 (b) -29 (c) $-3\frac{5}{8}$

3. (a) $A = 1$, $B = 2$ (b) $A = 3$, $B = 4$ (c) $A = 3$, $B = -1$
   (d) $A = -2$, $B = \frac{5}{3}$, $C = \frac{7}{3}$ (e) $A = \frac{3}{2}$, $B = \frac{3}{2}$, $C = -2$

4. (a) 2 (b) -3 (c) $-\frac{1}{2}$ (d) -2, -5 (e) -1, 1

5. (a) $x^2 + 5x + 1$ (b) $x^2 - x - 2$ (c) $5x^2 + 4x - 3$ (d) $x^3 + 2x^2 - 1$
   (e) $2x^3 + 3x^2 - x - 1$ (f) $5x^2 + 5x$

6. $A = 3$, $B = 1$ 7. $x^2 + x - 1$

8. (a) $A = 3$, $B = -7$ (b) $A = 1$, $B = -2$ (c) $A = \frac{3}{2}$, $B = -\frac{13}{4}$
   (d) $A = -\frac{5}{2}$, $B = -\frac{1}{4}$

9. (a) $A = 1$, $B = -1$ (b) $A = \frac{1}{3}$, $B = 5\frac{2}{3}$ (c) $A = -\frac{1}{2}$, $B = -1\frac{3}{4}$
   (d) $A = -\frac{3}{4}$, $B = 5\frac{1}{8}$

10. (a) $(x + \frac{3}{2})^2 - \frac{1}{4}$, min $= -\frac{1}{4}$ when $x = -\frac{3}{2}$
    (b) $2(x - \frac{1}{4})^2 + 3\frac{7}{8}$, min $= 3\frac{7}{8}$ when $x = \frac{1}{4}$

11. (a) max: (iii), (iv), (vii), (viii) min: (i), (ii), (v), (vi)
    (b) (iii) $3\frac{1}{4}$ when $x = -\frac{3}{2}$ (iv) $1\frac{1}{3}$ when $x = -\frac{1}{3}$
        (vii) $2\frac{1}{4}$ when $x = \frac{1}{2}$ (viii) $1\frac{1}{8}$ when $x = \frac{1}{4}$

Exercise 2.2

1. (a) $x^2 - x + 3$; -5 (b) $x + 3$; $7x + 1$ (c) $x + 1$; 0 (d) 1; $x^2 + x$

3. (a) 20 (b) 1 (c) -6 (d) 35 (e) 6 (f) $2\frac{1}{8}$ (g) $\frac{11}{8}$ (h) $-6\frac{1}{27}$

5. (a) Yes (b) No (c) Yes (d) No (e) Yes (f) Yes (g) No (h) Yes

6. (a) $(x - 1)(x - 2)(x - 3)$ (b) $(x + 1)^3$ (c) $(x - 1)(x^2 + 1)$
   (d) $(x + 1)(x - 1)^2(x + 2)$ (e) $(1 + x)(1 - x + x^2)$
   (f) $(3 - x)(9 + 3x + x^2)$ (g) $(4 + 3x)(16 - 12x + 9x^2)$

7. $-\frac{8}{9}$ 8. -8 10. $(x - 2)(x + 2)(4x - 9)$ $x = 2$, -2 or $\frac{9}{4}$

11. $(x - 2)(2x + 1)(x + 2)$; $x = \pm \lg 2$ 12. $a = 6$

Exercise 2.3

1. (a) $9x^3 + x^2 + 11x - 11$ (b) $8x^4 - 12x^3 + 4x$ (c) $9x^3 + 18x^2 - 10x - 24$
   (d) $4 + 6x - 15x^2 + 8x^3 - 3x^4$ (e) $4 + 4x - 7x^2 + 2x^3$

2. (a) $1 + 6x + 9x^2$ (b) $25 - 20x + 4x^2$ (c) $4 + 12x + 9x^2$
   (d) $1 + 9x + 27x^2 + 27x^3$ (e) $8 - 12x + 6x^2 - x^3$ (f) $8x^3 - 12x^2 + 6x - 1$

3. (a) $1 + 8x + 24x^2 + 32x^3 + 16x^4$ (b) $1 - 4x + 6x^2 - 4x^3 + x^4$
   (c) $32x^5 + 80x^4 + 80x^3 + 40x^2 + 10x + 1$
   (d) $x^7 - 7x^6 + 21x^5 - 35x^4 + 35x^3 - 21x^2 + 7x - 1$
   (e) $1 + 2x + \frac{3}{2}x^2 + \frac{1}{2}x^3 + \frac{1}{16}x^4$ (f) $1 - \dfrac{10x}{3} + \dfrac{40x^2}{9} - \dfrac{80x^3}{27} + \dfrac{80x^4}{81} - \dfrac{32x^5}{243}$
   (g) $32 - 80x + 80x^2 - 40x^3 + 10x^4 - x^5$
   (h) $x^4 + 12x^3 + 54x^2 + 108x + 81$ (i) $16 - 16x + 6x^2 - x^3 + \frac{1}{16}x^4$
   (j) $4x + x^3$ (k) $32 + 80x^2 + 80x^4 + 40x^6 + 10x^8 + x^{10}$

4. (a) -22 (b) 110 (c) 16

5. (a) $401 + 298\sqrt{2}$ (b) $252 + 144\sqrt{3}$ (c) $49 + 20\sqrt{6}$ 6. 1.04060401

7. $a^6 + 6a^5b + 15a^4b^2 + 20a^3b^3 + 15a^2b^4 + 6ab^5 + b^6$

$x^6 + 6x^4 + 15x^2 + 20 + \dfrac{15}{x^2} + \dfrac{6}{x^4} + \dfrac{1}{x^6}$

8. $-\dfrac{1}{x^5} + \dfrac{5}{x^2} - 10x + 10x^4 - 5x^7 + x^{10}$

9. 1.083 10. $2 - 11x + 24x^2$

11. $1 + 3a + 3a^2 + a^3$, $1 + 3x + 9x^2 + 13x^3 + 18x^4 + 12x^5 + 8x^6$

12. $8 + 12x - 6x^2 - 11x^3 + 3x^4 + 3x^5 - x^6$

Exercise 2.4

1. (a) 1 (b) $\dfrac{(x-1)^2}{x+2}$ (c) $\dfrac{x+2}{x+3}$ (d) 1

2. (a) $\dfrac{2}{1-x^2}$ (b) $\dfrac{1}{(1+x)(2+x)}$ (c) $\dfrac{2x^2+x+1}{1-x^2}$ (d) $\dfrac{12+5x}{(2+x)(3+x)(4+x)}$

3. (a) $\dfrac{2}{x+1}$ (b) $\dfrac{2}{x+2} - \dfrac{1}{x-1}$ (c) $\dfrac{1}{2(x-3)} - \dfrac{1}{2(x+3)}$

   (d) $\dfrac{1}{2(x+1)} + \dfrac{1}{2(x-1)}$ (e) $\dfrac{1}{4(x+2)} + \dfrac{3}{4(x-2)}$ (f) $\dfrac{1}{3(x+4)} + \dfrac{2}{3(x+1)}$

4. (a) $\dfrac{1}{2+x} + \dfrac{2}{3+x} - \dfrac{3}{4+x}$ (b) $-\dfrac{2}{x+2} + \dfrac{1}{x-1} + \dfrac{1}{x+1}$

   (c) $-\dfrac{4}{2-x} + \dfrac{8}{(2-x)^2}$ (d) $\dfrac{1}{1+x} - \dfrac{1}{(1+x)^2}$

5. (a) $1 + \dfrac{1}{x-1} - \dfrac{1}{x+1}$ (b) $1 - \dfrac{1}{x+1} - \dfrac{1}{x+2}$

   (c) $2 - \dfrac{4}{x+2} + \dfrac{1}{x+1}$ (d) $x - 2 + \dfrac{4}{x+2} - \dfrac{1}{x+1}$

6. (a) $\dfrac{1}{x+1} - \dfrac{1}{x^2+1}$ (b) $\dfrac{1}{x^2+1} - \dfrac{1}{x^2+2}$

   (c) $\dfrac{x}{x^2+1} + \dfrac{2-x}{x^2+2}$ (d) $\dfrac{x-1}{x^2+9} - \dfrac{1}{x+3}$

7. (a) $\dfrac{1}{x} - \dfrac{1}{x-1} + \dfrac{1}{(x-1)^2}$ (b) $-\dfrac{2(x+2)}{3(x^2+2)} + \dfrac{2}{3(x+1)}$

(c) $1 - \dfrac{4}{x+2} + \dfrac{1}{x+1}$ (d) $\dfrac{1}{x} + \dfrac{2}{x^2} + \dfrac{3x}{x^2+1}$

8. $(x-1)(x-2)(x-3)$; $\dfrac{1}{x-1} - \dfrac{2}{x-2} + \dfrac{1}{x-3}$ 9. $\dfrac{2}{x} - \dfrac{x}{x^2+1}$

Exercise 3.1

1. (a) $x = -1$ or -5 (b) $x = -\frac{5}{2} \pm \sqrt{\frac{13}{4}}$

  (c) $x = -\frac{1}{2}$ or 2 (d) $x = -\frac{5}{4} \pm \sqrt{\frac{33}{16}}$

2. (a) $x = -0.382$ or -2.618 (b) $x = 2.366$ or 0.634
  (c) $x = 1.618$ or -0.618 (d) $x = -2$ or 1
  (e) $x = \frac{1}{2}$ or -2 (f) $x = \pm\sqrt{5}$ (g) $x = 0$ or -4
  (h) $x = 0$ or $-\frac{2}{3}$ (i) $x = \pm\sqrt{\frac{3}{5}}$ (j) $x = \pm\frac{3}{2}\sqrt{2}$

3. (a) No real solutions.   (b) Two real solutions.   (c) Two equal solutions.
  (d) Two real solutions.
  (e) $a < -0.25 \Rightarrow$ two real solutions.
      $a = -0.25 \Rightarrow$ two equal solutions.
      $a > -0.25 \Rightarrow$ two real solutions.

4. (a) $x = -1$ or 3 (b) $x = 2,\ -1$ or -6 (c) $x = -\frac{1}{2}$ or $\frac{5}{3}$
  (d) $x = \frac{1}{2},\ -2$ or $\frac{7}{3}$ (e) $x = -3$ or -4 (f) $x = \frac{1}{2}$ or -2

5. (a) $(x+2)(2x+1)(x-3)$; $x = -2,\ -\frac{1}{2}$ or 3
  (b) $(2x+1)(x-1)$; $x = -\frac{1}{2}$ or 1
  (c) $(3x-2)(2x+1)$; $x = \frac{2}{3}$ or $-\frac{1}{2}$
  (d) $(x^2+1)(x-1)(x+1)$; $x = 1$ or -1

6. (a) Identity   (b) $x = 3$ or -1 (c) Identity   (d) Identity

7. (a) $x = -1$ (b) $x = 1,\ \frac{1}{2}$ or $-\frac{1}{2}$ (c) $x = -1$ (d) $x = -2$

8. $b = \pm 12$ 9. $b^2 = 36a$ 12. $k = -0.072$ or -13.928

Exercise 3.2

1. (a) $x < 2$ (b) $x \leqslant -9$ (c) $x \geqslant \frac{31}{2}$ (d) $x \geqslant \frac{19}{2}$ (e) $x > -\frac{33}{2}$

2. (a) $1 < x < 2$ (b) $3 < x < 4$ (c) $0 < x < 4$ (d) $-1 < x \leqslant 2$

3. (a) $(1,3]$ (b) $[0,5)$ (c) $[2,3)$ (d) $(-\infty,4)$ (e) $[-1,\infty)$ (f) $(-\infty,-1)$
   (g) $(-\infty,\infty)$ (h) ϕ—the empty set

4. (a) 9 (b) 10 (c) 4 (d) 2 (e) 2

5. (a) $x = -1$ or -3 (b) $x = 0$ or 4 (c) No solutions
   (d) $x = \frac{1}{2}$ or $5\frac{1}{2}$

7. (a) $x < -2$ or $x > 6$ (b) $x \leqslant -4$ or $x \geqslant -2$
   (c) $x \leqslant a - N$ or $x \geqslant a + N$

8. (a) $-5 < x < 1$ (b) $-1 \leqslant x \leqslant 1$ (c) $-3 < x < 3$ (d) $-1 \leqslant x \leqslant 7$

Exercise 3.3

1. $x = 0$ or 2   2. $x = 1$ or 2   3. $x = 1$

4. $x = 25$ or 125   5. $x = \frac{1}{3}$ or 9   6. $x = \frac{1}{4}$ or $\sqrt{2}$

7. $x = 1$ or -8 8. $x = \pm\sqrt{2}$ 9. $x = -1$ or 8

10. $x = 2$ ($x = 58$ is inadmissible)   11. $x = 2$ or 6

12. $x = \frac{3}{2}$ ($x = 71.5$ is inadmissible)

13. $n = -\frac{2}{3}t$ or $n = 2t$, $t = -\frac{3}{2}n$ or $t = \frac{1}{2}n$

14. $x = 0$ or -1 15. $n = \frac{1}{2}t$ or $n = -2t$, $t = 2n$ or $t = -\frac{1}{2}n$

16. (a) $x = 0$, $y = 3$; or $x = 1$, $y = 2$
    (b) $x = \frac{3}{2}$, $y = -1$; or $x = 1$, $y = -2$

17. $a = 5$, $b = -3$; or $a = -5$, $b = 3$ 18. $x = 2$

19. $x = 12$, $y = 4$; or $x = \frac{15}{8}$, $y = -5$

20. $a = \frac{3}{2}\sqrt{2}$ and $b = \frac{1}{4}$ 21. $x = 5$, $y = -2$

22. $x = -2$ ($x = 1$ is inadmissible).

23. $x = 2$, $y = -1$; or $x = -1$, $y = 2$; or $x = 0$, $y = 0$;
    or $x = 5$, $y = 5$

24. (a) $x = 1$, $y = -1$ (b) $x = 3$, $y = -4$; or $x = 1\frac{2}{5}$, $y = -4\frac{4}{5}$
    (c) $x = 3\frac{1}{2}$, $y = -\frac{1}{2}$

25. $x = 2$, $y = \frac{1}{2}$; or $x = -1$, $y = -1$ 26. $x = 2$, $y = -\frac{1}{2}$, $z = \frac{5}{2}$

Exercise 3.4

1. (a) -5, -6 (b) -6, 3 (c) -4, 1 (d) $\frac{9}{2}$, $\frac{5}{2}$

2. (a) $x^2 - 9x + 4 = 0$ (b) $3x^2 - 2x - 3 = 0$

3. (a) $\frac{29}{4}$ (b) $-\frac{29}{2}$ (c) -5

4. (a) -14 (b) 4 (c) 8

5. (a) $2x^2 + 2x - 9 = 0$ (b) $5x^2 - 6x - 2 = 0$ (c) $4x^2 - 56x + 25 = 0$

6. (a) $x^2 + 6x - 4 = 0$ (b) $x^2 - 7x + 1 = 0$ (c) $x^2 + 3x - 9 = 0$

7. $-2a, -c^2$; $c^2x^2 - 2ax - 1 = 0$

8. $x^2 - abx + a + b = 0$ $x^2 - 2abx + 4(a + b) = 0$

9. $\frac{3}{2}$ 10. $9mp = 2n^2$ 11. $px^2 + mx + m = 0$ 12. $p = 3$

14. $\dfrac{b^2}{a^2} - \dfrac{2c}{a} - \dfrac{b}{c}$ 15. $16x^2 + 6x + 1 = 0$

16. $\alpha^3 + \beta^3 = (\alpha + \beta)(\alpha^2 - \alpha\beta + \beta^2)$; $\alpha\beta = -4$; $x^2 - 2x - 4 = 0$;
    $1 + \sqrt{5}$ and $1 - \sqrt{5}$

Revision exercise A

1. $x = 6$, $y = 1$; or $x = 2$, $y = 3$; or $x = -3$, $y = -3$

2. $x = -1$ or $x = \pm\sqrt{2}$ 3. $x = 1.53$ (2 d.p.)

4. $x = 3$ or $x = \frac{1}{9}$ 5. (a) $x = 10^{\sqrt{pq}}$ (b) $x = \dfrac{10p + 3q}{2(p + 2q)}$

6. $x = -1$, 2, 1 or $-1 \pm \sqrt{3}$ 7. $x = -2$ ($x = -9$ inadmissible).

8. $x = 4$, $y = 2$ (negative solutions inadmissible)   9. $a = 216$, $n = -\frac{3}{4}$

10. $x = 2$, $y = 4$; or $x = -2$, $y = -4$; and $x = 4$, $y = 2$;
   or $x = -4$, $y = -2$

11. $A = \pm\frac{1}{3}$ and $B = \pm\frac{2}{3}$ 12. $k = \frac{3}{2}$ or $k = \frac{1}{2}$

13. (a) $-\frac{11}{4}$ (b) $4x^2 + 11x + 9 = 0$ 14. $x^2 - (k^2 + 1)x - (k^2 + 2) = 0$

15. $a_1 + a_2 = b_1 + b_2$; $a_1a_2 = b_1b_2 - c$

16. $x = 3$, $y = -2$; or $x = \frac{5}{9}$, $y = 2\frac{8}{9}$ 17. $x = a$ or $x = a^{1/2}$

18. $x = 3^{1/2}$ 19. $p = 6$, $q = 1$; $(x - 1)(2x + 1)(3x - 4)$

Exercise 4.1

1. (a) $\angle C = 50°$, $b = 24.2\,\text{m}\,(1\,\text{d.p.})$, $c = 32.3\,\text{m}\,(1\,\text{d.p.})$
   (b) $\angle A = 73°$, $a = 5.23\,\text{m}\,(2\,\text{d.p.})$, $b = 4.87\,\text{m}\,(2\,\text{d.p.})$
   (c) $\angle B = 66.6°$, $\angle A = 78.4°$, $a = 4.27$; or
   $\angle B = 113.4°$, $\angle A = 31.6°$, $a = 2.28$
   (d) $\angle C = 17.8°$ $\angle B = 112.2°$ $b = 6.0\,(1\,\text{d.p.})$

2. 1554 m

3. (a) $a = 8.63\,\text{cm}$, $\angle B = 85.2°$, $\angle C = 59.8°$
   (b) $b = 1.35\,\text{m}$, $\angle A = 21.0°$, $\angle C = 145°$
   (c) $\angle A = 44.4°$, $\angle B = 34.0°$, $\angle C = 101.5°$
   (d) $\angle A = 12.6°$, $\angle B = 149.3°$, $\angle C = 18.0°$

4. 14.6, 152.4° 5. (a) 27.6 cm² (b) 47.5 cm² (c) 17.9 cm²

6. (a) 4.41 cm (b) 36.70 cm²

8. 7.45 cm; 38.7°; 1.21 cm; 141.3° 9. 33.4°

Exercise 4.2

1. (a) 0 (b) 1 (c) 0 (d) -1 2. (a) 1 (b) 0 (c) 0 (d) 1

3. (i) $n\pi + \pi/2$ for any $n \in \mathbb{Z}$
   (ii) (a) 2nd, 4th (b) 1st, 2nd (c) 1st, 4th (d) 2nd (e) 3rd (f) 1st

4. (a) $-\sin 35°$ (b) $-\cos 50°$ (c) $\sin 26°$ (d) $-\tan 21°$ (e) $\cos 15°$
   (f) $\sin 18°$ (g) $-\tan 72°$ (h) $\cos 5°$ (i) $-\tan 41.5°$

5. (a) $\dfrac{\sqrt{3}}{3}$ (b) $\dfrac{\sqrt{3}}{2}$ (c) 0 (d) 0 (e) $-\sqrt{3}$ (f) $-\sqrt{3}$

5. (g) $-\frac{1}{2}$ (h) $-\frac{1}{2}$ (i) $\dfrac{1}{\sqrt{2}}$ (j) $-\dfrac{1}{\sqrt{2}}$ (k) $\dfrac{1}{\sqrt{2}}$ (l) 0

6. (a) 0.799 (b) -0.891 (c) 0.407 (d) -1.428 (e) 8.144 (f) -0.914

Exercise 4.3

1. (a) 39.7°, 140.3° (b) 139.0° (c) 69.0°

2. (a) 153.4°, 333.4° (b) 40.9°, 220.9° (c) 45°, 135° (d) 225°, 315°
   (e) 60°, 300° (f) 45°, 225°

3. (a) 54.5°, 125.5°, $-234.5°$, $-305.5°$
   (b) 35.5°, 324.5°, $-35.5°$, $-324.5°$
   (c) 234.5°, 305.5°, $-54.5°$, $-125.5°$
   (d) 39°, 219°, $-321°$, $-141°$
   (e) 140.7°, 320.7°, $-39.3°$, $-219.3°$
   (f) 60°, 120°, 240°, 300°, $-60°$, $-120°$, $-240°$, $-300°$
   (g) 60°, 120°, 240°, 300°, $-60°$, $-120°$, $-240°$, $-300°$
   (h) 54.7°, 125.3°, 234.7°, 305.3°, $-54.7°$, $-125.3°$, $-234.7°$, $-305.3°$

4. (a) 0°, 180°, (b) $\pm90°$, $-11.5°$, $-168.5°$ (c) 0°, 180°, $\pm30°$
   (d) 18.4°, $-161.6°$, 116.6°, $-63.4°$ (e) $\pm131.8°$, $\pm60°$
   (f) $-22.5°$, $-157.5°$

5. (a) 226.1°, 313.9° (b) 25.3°, 137.5°, 222.5°, 334.7°
   (c) 63.4°, 166.0°, 243.4°, 346.0°

6. (a) $\frac{3}{5}$ (b) (i) $\frac{5}{13}$ (ii) $-\frac{13}{12}$ (c) (i) $-\frac{5}{13}$ (ii) $-\frac{12}{5}$

7. 108.4°, $-71.6°$, 153.4°, $-26.6°$

9. (a) 13.9, 59.7° (b) 5.8, $-31.0°$ (c) 4.5, $-116.6°$ (d) 5.8, 120.96°

10. 78.7° 11. 48.6°, 131.4°, 210°, 330°

Exercise 4.4

1. (a) 45° (b) 22.5° (c) 112.5° (d) 120° (e) 150° (f) 300° (g) 540°
   (h) 85.9° (i) 72.2° (j) 180.0°

2. (a) $\frac{1}{3}\pi$ (b) $\frac{3}{4}\pi$ (c) $\frac{3}{2}\pi$ (d) $\frac{7}{6}\pi$ (e) $\frac{5}{3}\pi$ (f) $\frac{5}{36}\pi$ (g) $\frac{23}{18}\pi$ (h) $\frac{23}{90}\pi$
   (i) $0.32\pi = \frac{8}{25}\pi$ (j) 0.7π approx.

3. (a) -0.5 (b) $-\dfrac{\sqrt{3}}{2} \simeq -0.866$ (c) 0 (d) 0.174 (e) 0.841 (f) 3.602

4. (a) $\frac{5}{6}\pi$, $\frac{7}{6}\pi$ (b) $\frac{5}{4}\pi$, $\frac{7}{4}\pi$ (c) 1.01, 4.15 (d) $\frac{5}{6}\pi$, $\frac{7}{6}\pi$ (e) $\frac{1}{2}\pi$, $\frac{3}{2}\pi$

5. (a) 0.34, 2.80, -0.41, -2.73 (b) -0.23, 2.91, -0.96, 2.18

6. (a) 16.35 cm², 24.87 cm (b) $\frac{1}{2}r^2(2\theta - \sin 2\theta)$ cm², $2r(\theta + \sin\theta)$ cm

7. (a) 0.6 (b) $\frac{1}{36}\pi$ 8. (a) 0.06, (b) 0.15

9. (a) $\frac{1}{2}$ (b) 2   10. $\frac{1}{2}r^2(2\pi + \sin\theta - \theta)$

13. (a) ± 0.42, ± 2.72, ± 1.21, ± 2.78 (b) 1.11, -2.03, 2.36, -0.79
    (c) 0.62, 2.52 (d) 2.21, -0.93 (e) ± 2.25 (f) 0.76 (g) ± 2.52
    (h) 0.26, -1.83 (i) ± 1.31, ± 1.83 (j) 3.05, -0.09

14. (a) $\dfrac{\sqrt{2}}{4}(1 + \sqrt{3})$ (b) $-\dfrac{\sqrt{2}}{4}(1 + \sqrt{3})$ (c) $-\dfrac{\sqrt{2}}{2}$ (d) $\frac{1}{2}$

    (e) $-\dfrac{\sqrt{3}}{2}$ (f) $-\frac{1}{2}$

15. (a) $\frac{3}{4}$ (b) $\frac{12}{13}$ (c) $\frac{4}{5}$ (d) $-\frac{12}{13}$ (e) $\frac{63}{65}$ (f) $-\frac{16}{65}$ (g) $\frac{56}{65}$ (h) $-\frac{33}{65}$

17. $\sin A \cos B \cos C + \cos A \sin B \cos C + \cos A \cos B \sin C - \sin A \sin B \sin C$

19. (a) 60, 240 (b) 18.4°, 198.4°

Exercise 5.1

1. (a) -4 (b) 5 (c) 0 (d) 0

2. (a) (i) 3 (ii) $\sqrt{2}$ (iii) 0 (b) (i) 14 (ii) 98

3. (a) $[-2, 6]$ or $-2 \leqslant x \leqslant 6$ (b) $[-1, 1]$ or $-1 \leqslant x \leqslant 1$
   (c) $(3, 5]$ or $3 < x \leqslant 5$

4. (a) $[-1, 1]$ (b) $[0, 1]$ (c) $(-1, 1)$

5. (a) $[0, \infty)$ (b) $(-\infty, \infty)$ 6. (a) 2 (b) 1 (c) 2; $[1, \infty)$

7. (a) 15 (b) 6 (c) 15; $[16, \infty)$ 8. $[1, 9]$

9. (i) $(0, \infty)$; $(0, \infty)$ (ii) $(-\infty, 0) \cup (0, \infty)$; $(-\infty, 0) \cup (0, \infty)$
   (iii) $\mathbb{R}$; $[0, \infty)$ (iv) $\mathbb{R}$; $\mathbb{R}$

Exercise 5.2

1.

(i)

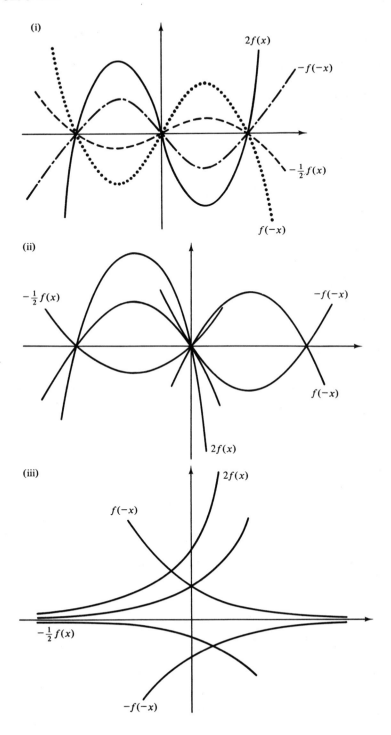

(ii)

(iii)

2.

(a)

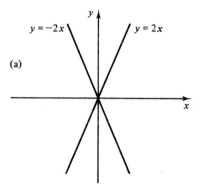

(b)

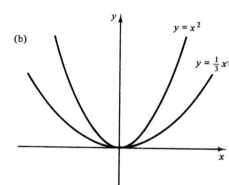

(c)

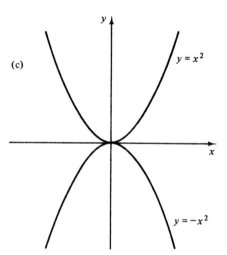

(d)

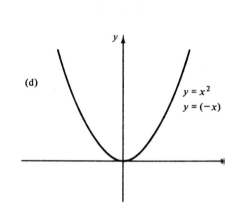

(e)

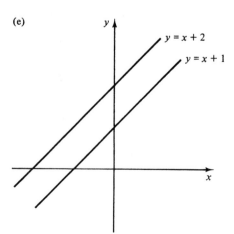

(f)

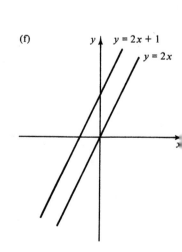

2.

(g)

(h)

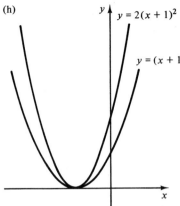

(i)

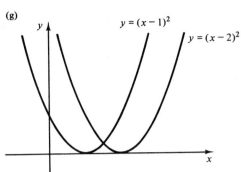

$y = (x+1)^2$

$y = -2(x+1)^2$

3.

(a), (b), (c)

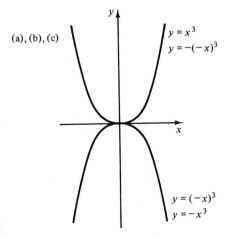

(d)

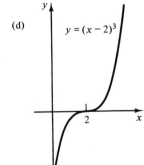

(e)

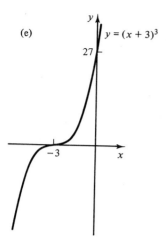

$y = (x + 3)^3$

(f)

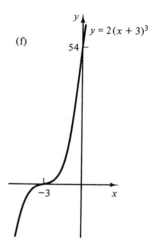

$y = 2(x + 3)^3$

(g)

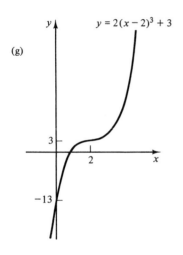

$y = 2(x - 2)^3 + 3$

(h)

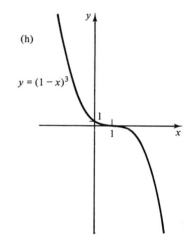

$y = (1 - x)^3$

(i)

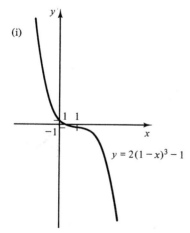

$y = 2(1 - x)^3 - 1$

(j)

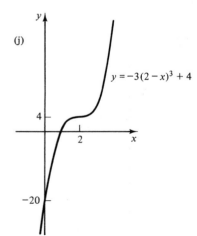

$y = -3(2 - x)^3 + 4$

4.

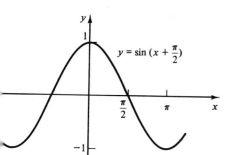

(b)

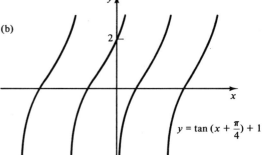

(c)

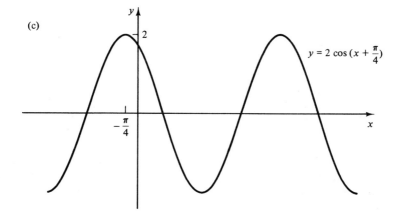

5. (i) odd   (ii) even   (iii) odd, periodic   (iv) odd

6.

(a)

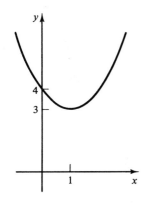

(b)

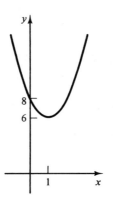

(c)

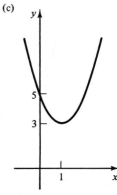

7.

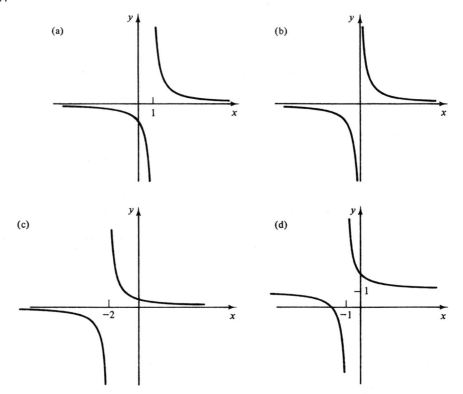

Exercise 5.3

1. (a) $(x-1)^2 - 4$ (b) $(x+2)^2 - 5$ (c) $(x+\frac{3}{2})^2 - \frac{13}{4}$ (d) $2(x+1)^2 - 1$
   (e) $2(x+\frac{3}{4})^2 - \frac{17}{8}$ (f) $-(x+3)^2 + 13$ (g) $-2(x-\frac{3}{4})^2 + \frac{1}{8}$

2. (a) min = 1 (b) min = $3\frac{3}{4}$ (c) max = 8 (d) max = 4

4. (a) 1, -2 (b) -3, -2 (c) -1, -3 (d) -2, -1 (e) 1, -7

5. Cross at: (a) $(-\frac{1}{2}, 0)$, $(0, 1)$ (b) $(3, 0)$, $(0, 3)$ (c) $(1, 0)$, $(0, 1)$
   (d) $(4, 0)$, $(0, 4)$ (e) $(2, 0)$, $(0, 4)$ (f) $(a, 0)$, $(0, b)$

Exercise 5.4

1. (a) $-2 \leqslant x \leqslant -1$ (b) $x < -4$ or $x > 3$ (c) $-1 < x < 2$
   (d) $x \leqslant -\frac{1}{2}$ or $x \geqslant \frac{2}{3}$

2. (a) $x \leqslant -2$ or $x \geqslant 2$ (b) no solutions (c) $-3 \leqslant x \leqslant -2$
   (d) $-\frac{2}{5} \leqslant x \leqslant 1$ (e) $-\frac{1}{2} < x < \frac{1}{4}$

3. $-\frac{1}{2} \leqslant p \leqslant \frac{1}{4}$ 4. $-1 < p < 3$ 5. $-1 < k < 0$

6.

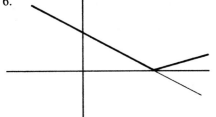 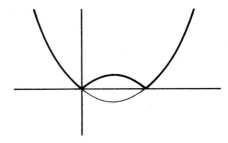

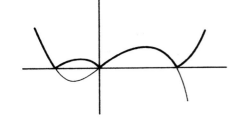 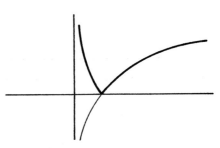

7. $k \leqslant -3 - 2\sqrt{3}$ or $k \geqslant 2\sqrt{3} - 3$

8.

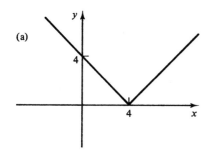 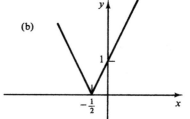

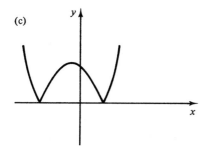 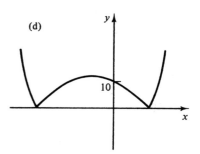

9. (a) $x < 1$ or $x > 3$ (b) $-2\frac{1}{2} \leqslant x \leqslant 1\frac{1}{2}$ (c) $x < \frac{1}{2}$ (d) $-\frac{2}{3} < x < 0$
   (e) $-8 < x < 0$ (f) $x < -4$ or $x > 0$ (g) $-3 \leqslant x \leqslant -\frac{5}{7}$

Exercise 5.5

1. (a) $\cos(1 + x)$, $x \in \mathbb{R}$ (b) $\cos(\cos x)$, $x \in \mathbb{R}$ (c) $1 + \cos x$, $x \in \mathbb{R}$

2. (a) $\sqrt{1 - x^2}$, $-1 \leqslant x \leqslant 1$ (b) $1 - x$, $x \in \mathbb{R}$ (c) $2x^2 - x^4$

3. (a) eg.: $f(x) \equiv \sqrt{x}$, $x \geqslant 0$; $g(x) \equiv 1 + x^2$, $x \in \mathbb{R}$
   (b) eg.:   (i) $f(x) \equiv \cos(x)$, $x \in \mathbb{R}$; $g(x) \equiv 1 + x$, $x \in \mathbb{R}$
         (ii) $f(x) \equiv x^3$, $x \in \mathbb{R}$; $g(x) \equiv 4 + x$, $x \in \mathbb{R}$
         (iii) $f(x) \equiv 5 + x$, $x \in \mathbb{R}$; $g(x) \equiv x^3$, $x \in \mathbb{R}$
         (iv) $f(x) \equiv 1 + \tan x$, $x \neq (n + \tfrac{1}{2})\pi$, $n \in \mathbb{Z}$; $g(x) \equiv x^2$, $x \in \mathbb{R}$

4. (a) $d^{-1}(x) \equiv x - 1$, $x \in \mathbb{R}$ (b) $e^{-1}(x) \equiv \tfrac{1}{2}x$, $x \in \mathbb{R}$

   (c) $f^{-1}(x) \equiv \sqrt{x - 1}$, $x \geqslant 1$ (d) $g^{-1}(x) \equiv 1 + \sqrt{x}$, $x \geqslant 0$

   (e) $h^{-1}(x) \equiv \sqrt{\dfrac{x}{2}}$, $x \geqslant 0$ (f) $i^{-1}(x) \equiv e^x$, $x \in \mathbb{R}$

5. (i) not   (ii) 1–1   (iii) not   (iv) 1–1

6. (a) $f \circ f(x) \equiv x^4 + 2x^2 + 2$, $x \in \mathbb{R}$ (b) $f^{-1}(x) \equiv \sqrt{x - 1}$, $x > 1$

7. $f^{-1}(x) \equiv \sqrt{\dfrac{1 - x}{2x - 1}}$, $\tfrac{1}{2} < x \leqslant 1$ 8. $f^4(x) \equiv x$, $f^7(x) \equiv \dfrac{x - 1}{1 + x}$

9. (a) eg. $h(0) = h(1) = 0$; $A = \tfrac{1}{2}$

   (b) $(0, 1)$, $(0, \infty)$, $(-\infty, 0) \cup \left(\dfrac{e}{e - 1}, \infty \right)$;

   $f^{-1}(x) \equiv \ln\left(\dfrac{1}{x} \right)$, $0 < x < 1$; $g^{-1}(x) \equiv \dfrac{x - 1}{x}$, $x > 0$;

   $(g \circ f)^{-1}(x) \equiv \ln\left(\dfrac{x}{x - 1} \right)$, $x < 0$ or $x > 1$

10. $\mathbb{R}$, $[1, \infty)$; f is one–one; g is not, eg. $g(-1) = g(1)$
   $f \circ g(x) \equiv 3x^2 + 5$, $x \in \mathbb{R}$; $g \circ f(x) \equiv 9x^2 + 12x + 5$, $x \in \mathbb{R}$;

   $x = 0$ or -2; $(f \circ g)^{-1}(x) \equiv \pm\sqrt{\dfrac{x - 5}{3}}$, $x \geqslant 5$, two valued

Exercise 5.6

1. (a) $y = x - 1$ (b) $y = \sqrt[3]{x + 3}$ (c) $y = \log_2 x$ (d) $y = 10^x$

1.

(a)
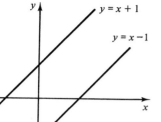

(b)
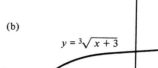

(c)
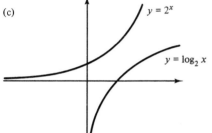

(d)
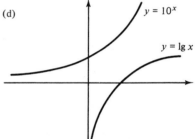

2.

(a)
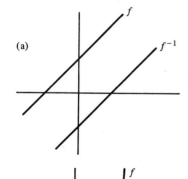

(b)
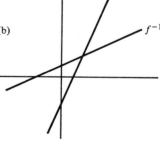

(c)
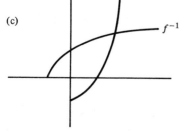

(d)
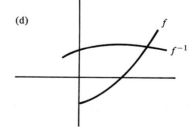

3.

(a)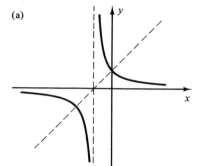

(b)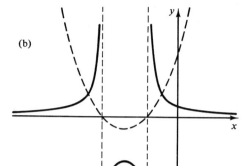

(c)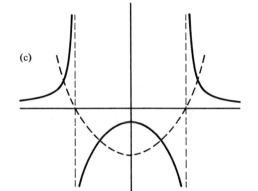

(d)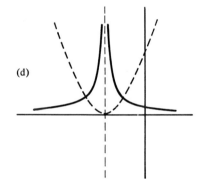

(e)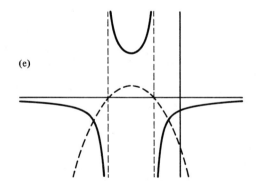

4. $f^{-1}(x) \equiv \dfrac{1 + 2x}{1 - x}, \quad x \neq 1$

Exercise 5.7

(a)
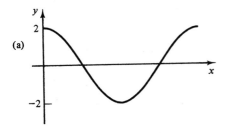

(b)
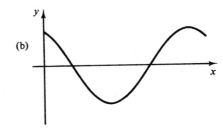

(c)
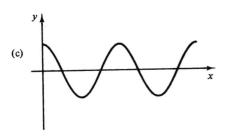

(d)
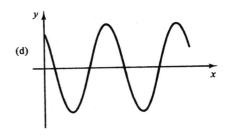

(e)
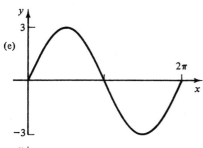

(f)
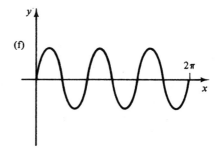

(g)
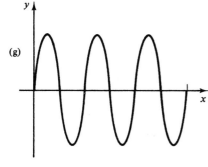

(h)
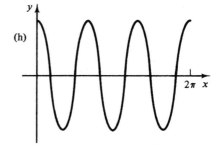

(i)
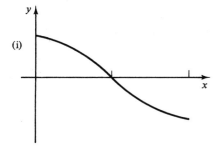

(j)
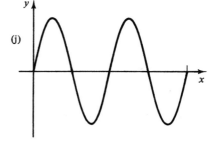

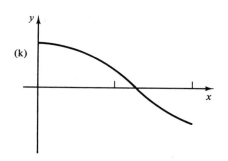

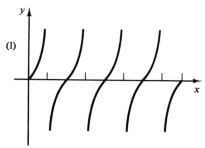

Exercise 5.8

1. (a) 5 (b) $\sqrt{20}$ (c) $5\sqrt{2}$ (d) $\sqrt{74}$ (e) $\sqrt{a^2 + b^2}$ (f) $(x - y)\sqrt{2}$

2. (a) $(-2\frac{1}{2}, 1)$ (b) $(\frac{1}{2}, 7\frac{1}{2})$ (c) $(4, \frac{1}{2})$ (d) $(3, 4)$

3. (a) 3 (b) -3 (c) $\frac{1}{5}$ (d) 0 (e) $\frac{1}{3}$ (f) $\frac{1}{4}$

4. (a) $-\frac{1}{2}$ (b) $\frac{1}{3}$ (c) $-\frac{1}{a}$ (d) $-m$

5. (a) 2 (b) $\frac{1}{2}$ (c) $-\frac{3}{2}$ (d) 1 (e) $\frac{1}{2}$

7. (a) $y = 4x + 2$ (b) $y = x + 1$ (c) $y = mx + b - am$

8. (a) $y = 3x - 5$ (b) $3y + 4x = 10$ (c) $y - y_1 = \left(\dfrac{y_2 - y_1}{x_2 - x_1} \right)(x - x_1)$

9. $3y - 2x + 1 = 0$ 10. $5y - 2x + 5 = 0$ 11. $5y - 2x + 5 = 0$

12. $((a - b), (a + b))$ 13. 78 sq units.   14. (a) $(1, 1)$ (b) $(-\frac{5}{7}, \frac{18}{7})$

15. $(1, 0)$, $\sqrt{2}$ 16. $x = 2$

Exercise 6.1

1. (a) 0 (b) 5 (c) $-\frac{2}{3}$ (d) 3 (e) $\frac{b}{a}$

2. (a) 3 (b) 2 (c) $\frac{1}{3}$ (d) 3 (e) $\frac{1}{5}$

3. (a) 2.72 (b) 0.69 (c) 1.10 (d) 1.00

4. (a) 2 (b) -2 (c) 2 (d) 5 (e) $2b$ (f) b (g) 2

5. (a) $\frac{2}{5}$ (b) 2 (c) $\frac{1}{2}$ (d) 1

Exercise 6.2

1. (a) $6x^5$ (b) $-3x^{-4}$ (c) $\frac{3}{2}x^{1/2}$ (d) $-3/x^4$ (e) $1/x^2$ (f) $6x^2$ (g) $-6x^{-3}$
   (h) $-4/x^3$ (i) $-1/6x^{3/2}$ (j) $-5/x^4$ (k) 0 (l) 9 (m) 3 (n) $-4/x^2$
   (o) $25x^4$ (p) $-1/3x^2$ (q) $-1/x^3$ (r) $-3/4x^{3/2}$ (s) $-5/x^6$ (t) $-6/x^5$

2. (a) $3x^2 - 4x$ (b) $2x - 2$ (c) $3x^2 + 2$ (d) $3x^2 + 4x - 11$

   (e) $3x^2 + 12x + 12$ (f) $\dfrac{1}{2\sqrt{x}} - \dfrac{1}{2\sqrt{x^3}}$ (g) $-\dfrac{1}{x^2} - \dfrac{1}{x^3} - \dfrac{1}{x^4}$

   (h) $12x - 1$ (i) $\frac{3}{2} - \dfrac{1}{2x^2}$ (j) $2x - \dfrac{2}{x^3}$

3. (a) $-\dfrac{1}{u^2} - \dfrac{2}{u^3}$ (b) $36u^3 + 12u$ (c) $-\dfrac{2}{u^2} - \dfrac{2}{u^3}$ (d) 4

4. (a) 2 (b) $\sqrt{2}/4$ (c) 1 (d) -3

5. (a) $(2, 1)$ (b) $(2, 16)$ (c) $(-\frac{1}{2}, \frac{1}{4})$ (d) $(2, -16)$, $(-1, 11)$ (e) $(1, 1)$

6. (a) $(2, -1)$ (b) $(3, 0)$

7. (a) $\pm\sqrt{2}$ (b) $x > \sqrt{2}$ or $x < -\sqrt{2}$ (c) $-\sqrt{2} < x < \sqrt{2}$

8. $(2^{1/3}, 3 . 2^{-2/3})$ 9. $(1, 0)$, $(-3, -4)$

10. $y = 2a(2x - a)$, $a = 8$ or $a = 0$ 11. $(1\frac{1}{2}, 2)$ 12. $\frac{1}{2}$

Exercise 6.3

1. (a) $y = 5x$ (b) $4y + x = 4$ (c) $4y = 8x + 1$

2. $y = x - 2$, $y = -x + 3$ 3. $y = 12x - 16$, $12y + x = 98$

4. $6ay + x - 18a^3 - a = 0$ 5. $t^3y + x - 2t = 0$

6. $x = 1$ and $x = -1$ 7. $y = 2x$, $(\frac{5}{4}, \frac{15}{16})$ 8. $x = \frac{1}{2}$

9. $\sqrt{260}$ 10. meet at $(2, 11)$ each with gradient 12 so the graphs touch

11. -2 12. $2ax + b$ (a) $-\dfrac{b}{2a}$ (b) 0 (c) $-\dfrac{b(2a + 1)}{4a^2}$

13. $(\frac{2}{3}, -4\frac{13}{27})$, $(2, -19)$ 14. ± 1.

Exercise 6.4

1. (a) -131,   212   (b) 1, $1\frac{5}{27}$ (c) 27   (d) -8,   8

2. (a) $(-2, -3)$ a minimum   (b) $(1\frac{1}{2}, 9\frac{1}{2})$ a maximum
   (c) $(-3, -1)$ a maximum, $(-\frac{1}{3}, -10\frac{13}{27})$ a minimum
   (d) $(-1, -3)$ a maximum, $(-\frac{1}{3}, -3\frac{8}{27})$ a minimum

3. (a) $(2, 9)$ (b) $(\frac{1}{2}, -\frac{1}{2})$ (c) $(3, -19)$

4. (a) yes   (b) no   (c) yes   (d) no

5. $(0, 27)$ maximum, $(\sqrt{6}, -9)$ minimum, $(-\sqrt{6}, -9)$ minimum

7. (a) $(-1, 0)$ point of inflexion
   (b) $(0, 0)$ minimum, $(1, 1)$ maximum, $(2, 0)$ minimum

8. $(2, 2\sqrt{2})$ minimum   9. Show that the minimum is positive

10. $\frac{5}{8}$ 11. $-1/4c$ 13. $a = 2$, $b = 3$, $c = 4$ 14. 32

15. $6\sqrt{6}$ 16. $\pi r^2 + \dfrac{1024\pi}{r}$, 192π,   radius 8 cm,   height 8 cm

17. 50   18. $y = 2x + 1 - \frac{4}{x}$ 19. $\frac{3}{2}$, $-\frac{3}{2}$, -3 respectively

21. $y = 2x^3 - 8x^2 + 8x + 1$ 22. Maximum $= -1\frac{22}{27}$,   greatest value $= 0$

23. 64   24. a 25. (b) $\pi\sqrt{\frac{8}{3}}$

Exercise 6.5

1. (a) $-2\,\text{ms}^{-1}$,   0   (b) $1\,\text{ms}^{-1}$, $6\,\text{ms}^{-2}$; $\sqrt{\frac{2}{3}}\,\text{s}$, $6\sqrt{\frac{2}{3}}\,\text{ms}^{-2}$

2. (a) 8 s   (b) (i) $10\,\text{ms}^{-1}$ (ii) 0   (c) $t = 4\,\text{s}$ (d) 80 m

3. (a) $R + \dfrac{16R^3}{k^2}$ (b) $t = \sqrt[3]{\dfrac{-2k}{R}}$ 4. 50 m by 25 m

Exercise 6.6

1. (a) $3(x + 2)^2$ (b) $15(3x + 1)^4$
   (c) $-16(1 - 2x)^3$ (d) $3(2x^2 - 3x + 1)^2(4x - 3)$

1. (e) $\dfrac{1}{2\sqrt{1+x}}$ (f) $\dfrac{x}{\sqrt{1+x^2}}$ (g) $\dfrac{-1}{2\sqrt{2-x}}$ (h) $\dfrac{-1}{(x+1)^2}$ (i) $\dfrac{-6x}{(x^2+2)^2}$

(j) $\dfrac{-16x}{(2x^2-3)^2}$ (k) $\dfrac{-6}{(1+x)^3}$ (l) $\dfrac{-15}{(6+x)^4}$ (m) $\dfrac{-2}{(1+2x)^{3/2}}$ (n) $\dfrac{-3}{(5-x)^{3/2}}$

(o) $\dfrac{-x}{(x^2+1)^{3/2}}$ (p) $\dfrac{-12}{(3x-1)^5}$ (q) $\tfrac{12}{5}(x+1)^{-1/5}$ (r) $36x(2+x^2)^2$

(s) $12(3x+1)^3 - 6(x+1)^2$ (t) $\dfrac{-6}{(2x-1)^4} + \dfrac{1}{\sqrt{2x-1}}$

(u) $\dfrac{-2}{(x+1)^3} + \dfrac{2}{(2x+1)^2}$

2. (a) $\tfrac{1}{3}$ (b) $\dfrac{1}{3(x-1)^2}$ (c) $\tfrac{2}{3}y$ or $\tfrac{2}{3}\sqrt{3x}$ (d) $\tfrac{3}{2}\sqrt{y}$ or $\tfrac{3}{2}\sqrt[3]{x-2}$

3. (a) $\dfrac{1}{4(2y+1)}$ or $\dfrac{1}{4\sqrt{x}}$ (b) $\dfrac{\sqrt{1+y^2}}{y}$ or $\dfrac{x}{\sqrt{1-x^2}}$

Exercise 6.7

1. (a) $3(u+1)^2\dfrac{du}{dx}$ (b) $4u\dfrac{du}{dx}$ (c) $-\dfrac{1}{2\sqrt{1-u}}\dfrac{du}{dx}$ (d) $3(u-1)^2\dfrac{du}{dx}$

(e) $6(2u+1)^2\dfrac{du}{dx}$ (f) $3y^2\dfrac{dy}{dx}$ (g) $4y\dfrac{dy}{dx}$ (h) $4y(y^2+1)\dfrac{dy}{dx}$ (i) $-\dfrac{2}{y^3}\dfrac{dy}{dx}$

(j) $\dfrac{-1}{(y-1)^2}\dfrac{dy}{dx}$ (k) $\dfrac{-\theta}{(1-\theta^2)^{3/2}}\dfrac{d\theta}{dx}$ (l) $\dfrac{-12\theta}{(2\theta^2+1)^2}\dfrac{d\theta}{dx}$

2. $4\pi r^2$. $72\pi\,\text{cm}^3\,\text{s}^{-1}$ 3. $\tfrac{3}{4}\text{cm}^3\,\text{s}^{-1}$

4. (a) $\dfrac{3t}{2}$ (b) $\dfrac{3}{3t}$ (c) $\dfrac{2}{3t^2}$ (d) $-\dfrac{1}{t^2}$ (e) $\dfrac{t^2-1}{t^2+1}$ (f) $\dfrac{-1}{6t(t+1)^2(t^2+1)^2}$

5. $\sqrt{12}$, $44\,\text{s}$ 6. $-\dfrac{1}{t^2}$, (a) $4y+x=4c$ (b) $y+\dfrac{x}{t_1^2}=\dfrac{2c}{t_1}$

7. $y = \dfrac{x}{t_1} + at_1$, $y+t_1x = 2at + at^3$

8. $\dfrac{3t}{2}$, $\dfrac{3}{4t}$ (a) $\tfrac{3}{8}$ (b) ± 3 9. $\pi\text{m}^2\,\text{s}^{-1}$ 10. $2x-4$

12. $\dfrac{t}{3t^2 - 1}$ (a) $(0, 1)$ (b) $(\frac{4}{9}\sqrt{3}, \frac{2}{3})$, $(-\frac{4}{9}\sqrt{3}, \frac{2}{3})$; $y = \pm \frac{x}{2}$ 14. $4\,\text{cm}^2\,\text{s}^{-1}$

15. -4 16. $\dfrac{1}{15\pi}$; $4\sqrt{5}\,\text{cm}\,\text{s}^{-1}$ 17. $-\frac{3}{10}\,\text{radians}\,\text{s}^{-1}$

Exercise 6.8

1. (a) $\dfrac{x^6}{3} + c$ (b) $-\dfrac{1}{x} + c$ (c) $-\dfrac{1}{x^3} + c$ (d) $\dfrac{2x^3}{3} - \dfrac{x^2}{2} + c$

(e) $x - x^2 + \dfrac{x^3}{3} + 3$ (f) $\dfrac{x^4}{16} + \dfrac{x^3}{2} + 6x^2 + 8x + c$ (g) $\dfrac{x^2}{2} - \dfrac{4x^{3/2}}{3} + x + c$

(h) $\dfrac{2x^{3/2}}{3} - 2x^{1/2} + c$ (i) $-\dfrac{1}{x} - \dfrac{1}{x^2} + c$ (j) $2\sqrt{x} + 4x + \dfrac{8x^{3/2}}{3} + c$

(k) $x + x^2 + \dfrac{x^3}{3} + c$ (l) $\dfrac{x^3}{3} + x^2 - 2x + c$ (m) $-\dfrac{3}{4x^2} + c$

(n) $-\dfrac{1}{4x^4} + \dfrac{2}{3x^3} - \dfrac{1}{2x^2} + c$ (o) $\dfrac{2x^{13/2}}{13} + \dfrac{4x^{7/2}}{7} + 2x^{1/2} + c$ (p) $\dfrac{x^3}{3} + \dfrac{x^4}{4} + c$

Exercise 6.9

1. (a) $\frac{1}{4}x^4 + c$ (b) $x^2 + c$ (c) $\frac{1}{3}(x - 1)^3 + c$ (d) $x + c$

(e) $-\dfrac{1}{x} + c$ (f) $-\dfrac{3}{x} + c$ (g) $-\dfrac{1}{x} - \dfrac{1}{2x^2} - \dfrac{1}{3x^3} + c$ (h) $\frac{2}{5}x^{5/2} + c$

(i) $6x^{2/3} + c$ (j) $\frac{1}{8}(2x + 1)^4 + c$ (k) $\frac{2}{5}x^{5/2} + c$ (l) $\frac{1}{2}x^2 - \frac{4}{3}x^{3/2} + x + c$

2. (a) $t^4 + c$ (b) $\frac{1}{3}(u + 3)^3 + c$ (c) $\frac{2}{3}v^{3/2} - 2\sqrt{v} + c$

(d) $-\dfrac{1}{2s^2} + \dfrac{2}{3s^3} - \dfrac{1}{4s^4} + c$ (e) $w^2 - w + c$

3. (a) $\frac{1}{12}(2x + 1)^6 + c$ (b) $-\frac{1}{4}(1 - x)^4 + c$

(c) $\frac{1}{6}(x^2 + 1)^6 + c$ (d) $\frac{1}{4}(x^2 - x)^4 + c$ (e) $-\dfrac{1}{2(x^2 + 1)} + c$

(f) $\frac{2}{9}(x^3 + 4)^{3/2} + c$

(g) $-3(1 - x^2)^{1/2} + c$ (h) $\dfrac{-1}{(1 + x)} + \dfrac{1}{2(1 + x^2)} + c$ (i) $\frac{1}{3}(x^2 + 3x - 2)^3 + c$

(j) $\frac{1}{5}(x^2 + 3)^{5/2} + c$ (k) $\frac{1}{12}(3x - 2)^4 + c$ (l) $-\frac{1}{6}(1 - x^2)^{3/2} + c$

4. (a) $13\,\text{ms}^{-1}$ (b) $0\,\text{ms}^{-1}$ (c) $2g\,\text{ms}^{-1}$, $2g\,\text{m}$ 5. $y = x^3 - x^2 + 3x - 3$

Exercise 7.1

1. (a) $\dfrac{\sqrt{2}}{4}(\sqrt{3}-1)$ (b) $\dfrac{\sqrt{2}}{4}(\sqrt{3}+1)$ (c) 1 (d) $2-\sqrt{3}$

2. (a) $\frac{3}{5}$ (b) $\frac{4}{3}$ (c) $\frac{24}{25}$ (d) $-\frac{7}{25}$ (e) $-\frac{24}{7}$ 3. (a) $-\frac{119}{169}$ (b) $-\frac{120}{119}$

4. (a) $\frac{3}{5}$ (b) $\frac{4}{5}$ (c) $\frac{24}{25}$ 5. $\dfrac{a-1}{1+a}$ 7. (a) 1 (b) $\frac{4}{3}$

8. $2+\sqrt{3}$ 10. (a) $\left|\dfrac{\sqrt{2}}{\tan\theta-1}\right|$ (b) $\frac{1}{2}(1+\tan^2\theta)$ (c) $\dfrac{6\tan\theta+1-\tan^2\theta}{1+\tan^2\theta}$

16. (a) $9.6°$, $90°$, $170.4°$, $270°$ (b) $0°$, $120°$, $180°$, $240°$

17. (a) $\dfrac{1-x}{2\sqrt{x}}$ (b) $-\dfrac{2\sqrt{x}\pm(1+x)}{1-x}$

18. (a) $\frac{3}{5}$ (b) $\frac{1}{5}\sqrt{5}$ (c) $\frac{1}{2}$ (d) $5\frac{1}{2}$ (e) $-\frac{7}{25}$

19. $\frac{5}{2}$; 30, 150, 210, 330   20. $18.4°$, $161.6°$, $198.4°$, $341.6°$

21. $\frac{5}{4}\pi$ 23. $2+\sqrt{3}$ 27. $\frac{7}{25}$ 28. -2 or $-\frac{2}{7}$

Exercise 7.2

1. (a) $\sqrt{34}\cos(\theta-59°)$ (b) $\sqrt{85}\cos(\theta-77.5°)$

2. (a) $\sqrt{29}\sin(\theta-21.8°)$ (b) $\sqrt{34}\sin(\theta-59°)$

3. (a) $128.9°$ or $349.1°$ (b) $126.9°$ or $28.1°$
   (c) $69.8°$ or $153.8°$ (d) $90.0°$ or $208°$

4. (a) $\pm\sqrt{34}$ (b) $1\pm\sqrt{29}$ (c) ±5 ! (d) $\pm\sqrt{17}$

5. (a) $29.6°$ or $-103.3°$ (b) $55.7°$ or $168.0°$
   (c) $129.0°$ or $-10.9°$ (d) $13.8°$ or $-53.6°$

6. 0, -1.1, 2.0, or π

7. $\sqrt{2}\cos(2\theta+225°)$, $1-\sqrt{2}$, 0, 135

8. $5\cos(x+36.9°)$, -5 to 5

9.

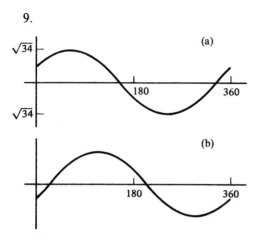

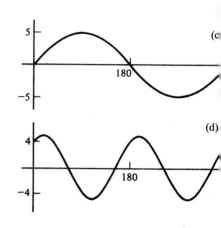

10. $2 \sin(\theta - \frac{1}{3}\pi)$; $\frac{1}{2}\pi$, $\frac{7}{6}\pi$

11. $3\sqrt{2} \sin(\theta + 45)$; (a) 18, 0 (b) 15°, 75°

12. 5, 36.87°, 119.55°, 346.71°

13. 107.59°, 319.79°   14. 18.4°, 30°, 198.4°, 225°

15. (a) 116.6°, 296.6° (b) 90°, 270°; $-\sqrt{5} \leqslant k \leqslant \sqrt{5}$

16. $\dfrac{\pi}{4}$, $\dfrac{\pi}{2}$ 17. (a) 3, $\frac{1}{5}$ (b) $-\frac{4}{5}$, $\frac{12}{13}$, 0.39

18. (a) 1.57, -0.65 (b) 0.48, -0.68

Exercise 7.3

1. (a) $2 \sin 4\theta \cos \theta$ (b) $2 \cos 4\theta \sin \theta$ (c) $2 \cos 5\theta \cos 2\theta$ (d) $-2 \sin 4x \sin x$

2. (a) 0, 22.5°, 67.5°, 112.5°, 157.5° (b) 0°, 90° (c) 45°, 90°, 135°

3. 30°, 90°, 150°, 210°, 270°, 330°

4. $\pm 22.5°$, $\pm 67.5°$, $\pm 112.5°$, $\pm 157.5°$, $\pm 60°$, $\pm 120°$

5. 67.5°, 135°, 157.5°, 247.5°, 315°, 337.5°   6. 0, $\frac{1}{3}\pi$, $\frac{2}{3}\pi$, π

7. $\frac{1}{2}\pi - \alpha$, $\frac{3}{2}\pi - \alpha$ 8. 0°, 90°, 36°, 108°, 180°

9. (a) 0, $\frac{1}{5}\pi$, $\frac{2}{5}\pi$, $\frac{3}{5}\pi$, $\frac{4}{5}\pi$, π (b) $\frac{1}{4}\pi < x < \frac{3}{4}\pi$

10. (a) $\frac{1}{4}\pi$, $\frac{3}{4}\pi$, $\frac{1}{9}\pi$, $\frac{5}{9}\pi$, $\frac{7}{9}\pi$ (b) 0, $\frac{2}{5}\pi$, $\frac{1}{2}\pi$, $\frac{4}{5}\pi$, π

Exercise 7.4

7. $-\frac{1}{2}\pi$, $-\frac{1}{3}\pi$, $\frac{1}{2}\pi$, $\frac{2}{3}\pi$

Exercise 7.5

1. (a) $4\cos^3 A - 2\cos A$ (b) $\sin 2A \cos 2B$ (c) 5

2. $26.1°$, $90°$, $146.7°$, $213.3°$, $270°$, $333.9°$ 4. (c) $\frac{1}{4}(1 + \sqrt{5})$

5. (a) $75.96°$, $135°$, $255.96°$, $315°$
   (b) $30°$, $45°$, $90°$, $135°$, $150°$, $210°$, $270°$, $330°$

6. (a) $16°$, $145°$, $196°$, $325°$ (b) 0, π, $\frac{3}{2}\pi$, 0.8, 2.3

7. $11°$, $27°$, $162°$, $-18°$, $-153°$, $-169°$

8. 1, $-\frac{3}{7}$, 2; $45°$, $156.8°$, $63.4°$

10. $\frac{1}{4}\pi$, $\frac{5}{4}\pi$, 1.33, 4.46 12. (a) $15°$, $-165°$ (b) $39°$, $-100.9°$

13. (a) $7 - 4\sqrt{3}$ (b) $90°$ or $270°$ 14. $c = 1$ or $\frac{1}{4}(1 \pm \sqrt{5})$

15. (a) $a\sin\beta$, $a(\sin\beta + \cos\beta)$ 16. (a) $\frac{1}{8}\pi$, $\frac{3}{8}\pi$, $\frac{9}{8}\pi$, $\frac{11}{8}\pi$ (b) $x = \pm\pi$

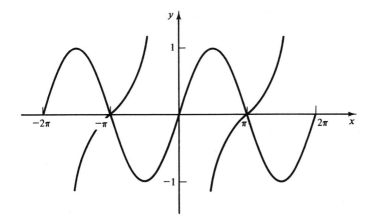

17. (a) $6.45\,\text{cm}^2$ (b) $27°$, $63°$, $207°$, $243°$ 18. $\cos 4\theta + 4\cos 2\theta + 3$

20. (a) $77.13\,\text{cm}$ (b) $45.96\,\text{cm}$ (c) $32.95°$ (d) $59.09\,\text{cm}$ (e) $134.75°$

22. (a) $500\,\text{m}$ (b) $320\,\text{m}$ (c) $208.1\,\text{m}$

23. (a) $\frac{1}{3}\pi$, π, $\frac{5}{3}\pi$ (b) 0, $\frac{1}{6}\pi$, $\frac{5}{6}\pi$, π, $\frac{7}{6}\pi$, $\frac{11}{6}\pi$, 2π (c) $\frac{5}{6}\pi$, $\frac{3}{2}\pi$

Exercise 7.6

1. (a) $3\cos(3x)$　(b) $a\cos ax$　(c) $\cos(x+\pi)$　(d) $a\cos(ax+b)$
   (e) $-5\sin 5x$　(f) $\sin(-x)$　(g) $-\sin(x-\tfrac{1}{2}\pi)$　(h) $-a\sin(ax+b)$
   (i) $a\sec^2(ax+b)$　(j) $2\sin x\cos x$　(k) $3\sin^2 x\cos x$　(l) $9\sin^8 x\cos x$
   (m) $-3\cos^2 x\sin x$　(n) $3\tan^2 x\sec^2 x$　(o) $2x\cos(x^2)$　(p) $2x\sin(x^2+1)$

   (q) $\dfrac{1}{2\sqrt{1+x}}\cos(\sqrt{1+x})$　(r) $\dfrac{-x}{\sqrt{x^2+1}}\sin(\sqrt{x^2+1})$　(s) $-\dfrac{2}{x^3}\sec^2\left(\dfrac{1}{x^2}\right)$

   (t) $-2(1+x)\sin(1+x)^2$

2. (a) $-\operatorname{cosec}x\cot x$　(b) $-2x\operatorname{cosec}^2 x^2$　(c) $-\dfrac{1}{2\sqrt{x}}\operatorname{cosec}^2\sqrt{x}$

   (d) $-\sin x\cos(\cos x)$　(e) $\dfrac{\cos x}{2\sqrt{1+\sin x}}$　(f) $5(1-\cos x)^4\sin x$　(g) $\dfrac{\sec^2 x}{(1-\tan x)}$

   (h) $\dfrac{-\sin x\cos x}{(1-\sin^2 x)^{1/2}}$

3. (a) $-\sin x$　(b) $-\operatorname{cosec}x\cot x$

4. (a) $\dfrac{\pi}{180}\cos\left(\dfrac{\pi x}{180}\right)$　(b) $-\dfrac{\pi}{180}\sin\left(\dfrac{\pi x}{180}\right)$　(c) $\dfrac{\pi}{180}\sec^2\left(\dfrac{\pi x}{180}\right)$

5. (a) $f'(x)\cos(f(x))$　(b) $-f'(x)\sin(f(x))$

6. $\dfrac{dy}{dx}=-\sin x-2\cos x,\quad \dfrac{d^2y}{dx^2}=-\cos 2x+2\sin x=-y$!

7. $\pm 13,\quad \pm 5,\quad \pm 25$

8. (a) $5\sin^4 x\cos x$　(b) $\dfrac{-\sin x}{2\sqrt{\cos x}}$

   (c) $6(1+\sin^2 x)^2\sin x\cos x$　(d) $\dfrac{\cos x-\sin x}{2(\cos x+\sin x)^{1/2}}$

Exercise 7.7

1. (a) $\dfrac{6}{\sqrt{1-36x^2}}$　(b) $\dfrac{1}{\sqrt{1-(x+4)^2}}$　(c) $\dfrac{4}{\sqrt{1-(4x-1)^2}}$　(d) $\dfrac{-2x}{1-x^4}$

   (e) $\dfrac{-1}{2\sqrt{x-x^2}}$　(f) $\dfrac{1}{2x\sqrt{x-1}}$　(g) $-\dfrac{1}{x^2+1}$　(h) $\dfrac{-1}{2(2-x)\sqrt{1-x}}$

(i) $\dfrac{4x(x^2 - 2)}{1 + (x^2 - 2)^4}$ (j) $\dfrac{1}{2\sqrt{\arcsin x}\sqrt{1 - x^2}}$

2. (a) 1 (b) 1

3. (a) $\frac{1}{4}\pi$ (b) $\frac{3}{4}\pi$ (c) $\frac{1}{6}\pi$ (d) $\frac{1}{3}\pi$ (e) $-\frac{1}{6}\pi$ (f) $\frac{1}{2}\pi$ (g) $\frac{1}{6}\pi$ (h) 1.11

4. (a) x (b) $\sqrt{1 - x^2}$ (c) $\dfrac{x + y}{1 - xy}$

5. $\dfrac{-2x}{(1 + x^2)^2}$ 7. (b) $\dfrac{\sqrt{17} - 1}{4}$ 8. 0.28

Exercise 7.8

1. (a) $-2\cos x + c$ (b) $\pi\sin x + c$ (c) $\frac{1}{2}\tan x + c$ (d) $a\sin x - b\cos x + c$
   (e) $\sin 3x + c$ (f) $\frac{a}{b}\tan x + c$ (g) $\frac{1}{2}\tan(2x - 1)$ (h) $\cos(1 - x) + c$

2. (a) $\frac{1}{4}\sin^4 x + c$ (b) $\frac{1}{5}\cos^5 x + c$ (c) $\frac{1}{3}\cos^3 x - \cos x + c$
   (d) $\sin x - \frac{1}{3}\sin^3 x + c$ (e) $\frac{1}{2}x - \frac{1}{4}\sin 2x + c$ (f) $\frac{1}{2}x + \frac{1}{4}\sin 2x + c$
   (g) $\frac{1}{2}x - \frac{1}{12}\sin 6x + c$ (h) $\frac{1}{2}ax + \frac{a}{4b}\sin 2bx + c$ (i) $\frac{1}{4}\tan^4 x + c$
   (j) $\frac{1}{12}\tan^4 3x + c$

3. (a) $-\dfrac{180}{\pi}\cos\left(\dfrac{\pi x}{180}\right) + c$ (b) $\dfrac{180}{\pi}\sin\left(\dfrac{\pi x}{180}\right) + c$ (c) $\dfrac{180}{\pi}\tan\left(\dfrac{\pi x}{180}\right) + c$

4. (a) $\frac{1}{8}(\cos 4x + 2\cos 2x) + c$ (b) $\frac{1}{42}(3\sin 7x + 7\sin 3x) + c$
   (c) $\frac{1}{21}(7\sin 3x - 3\sin 7x) + c$ (d) $\frac{1}{6}(\sin 3x + 3\sin x) + c$

5. (a) $\frac{1}{5}\cos^5 x - \frac{1}{3}\cos^3 x + c$ (b) $\sin x - \frac{2}{3}\sin^3 x + \frac{1}{5}\sin^5 x + c$

Revision exercise B

1. $\dfrac{\lg 2 + 2}{2(1 - \lg 2)} = 1.646$ 2. 4

3. (a) 0.841 (b) $\frac{1}{2}\pi$ 4. 7, 6; -4, -5

5. 7; (a) $x^2 - 4x + 10 = 0$ (b) $x^3 - 9x^2 + 30x - 50 = 0$

6. 1, 1; $5^{-1/3}$, $3.5^{-1/3}$; $10^{-1/3}$, $4.10^{-1/3}$

7. -44; (a) $(x + 2)(2x + 5)(x - 6)$ (b) $x < -2.5$, $-2 < x < 6$

8. 0 max, $\frac{1}{2}\pi$ inflexion; $\frac{3}{2}\pi$ inflexion; π min

9. $(-\infty, 2\frac{1}{4}]$, $(0, 1]$, g possesses an inverse

10. (a) $-\frac{1}{3} \leqslant k \leqslant 1$ (b) 0.892   or -0.320

11. $(x + 2)(2x - 1)(x^2 - x + 2)$ 12. 7.58 cm; 24.8°

13. $1 - 1$ function; $[-1, 1]$, $[-\frac{1}{2}\pi, \frac{1}{2}\pi]$; $\frac{1}{6}\pi$; $\frac{1}{2}$, $\frac{1}{6}\pi$

14.

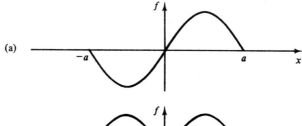
(a)

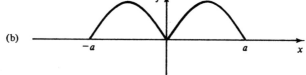
(b)

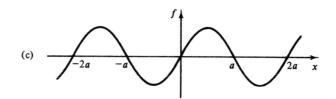
(c)

15. (a) $\frac{2}{3}$, $-\frac{10}{3}$ (b) $-\frac{10}{3} < k < \frac{2}{3}$ 16. 1, 1; $-\frac{3}{5}$, $-\frac{11}{5}$

17. (a) $x = 2y$, $y + 2x = 4$; $\frac{8}{5}$, $\frac{4}{5}$ (b) (i) $1 - 2p$ (ii) $\dfrac{(1 - 2p)}{p}$ (iii) $\dfrac{1}{2p}$

18. even;   even,   periodic;   odd,   periodic

19. $2x^2 - 12x + 30$; $2(x - 3)^3 + 21$;   21; $(7.5, 52.5)$ 20. $-\frac{9}{4}$, $\frac{1}{4}$

21. $-\sqrt{2} \leqslant k < \sqrt{2}$, -5.5 22. $4ab(3a^2 + b^2)(a^2 + 3b^2)$

Exercise 8.1

1. (a) $2e^{2x}$ (b) $-3e^{-3x}$ (c) $4e^{4x} - e^{-x}$ (d) $e^x + e^{-x}$ (e) $\frac{1}{2}e^{x/2}$
   (f) $-\frac{1}{3}e^{-x/3}$ (g) $2xe^{x^2}$ (h) $-xe^{-x^2/2}$ (i) $3e^{3x+4}$ (j) $2e^{2x-3}$
   (k) $-e^{-(x+3)}$ (l) $4xe^{2x^2-1}$ (m) $2(x - 1)e^{(x-1)^2}$ (n) $-4(x - 3)e^{-2(x-3)^2}$
   (o) $-2(x - 2)e^{-(x-2)^2}$ (p) $-\sin x\,e^{\cos x}$ (q) $-2x\sin x^2\,e^{\cos x^2}$
   (r) $2\sin x\cos x\,e^{\sin^2 x}$ (s) $-e^x\sin(e^x)$ (t) $2e^{2x}\sec^2(e^{2x})$ (u) $e^x e^{e^x}$
   (v) $e^x + e^{-x}$ (w) $-2(1 - e^x)e^x$ (x) $-e^{-x} + 2e^x + 3e^{3x}$ (y) e^x

2. (a) 1.10, 0 (b) 1.39   3. $y = \pm \dfrac{x}{\sqrt{2}}$

Exercise 8.2

1. (a) $\dfrac{1}{x}$ (b) $\dfrac{1}{x}$ (c) $\dfrac{2}{2x-1}$ (d) $-\dfrac{1}{1-x} = \dfrac{1}{x-1}$ (e) $\dfrac{3}{x}$ (f) $\dfrac{4}{(1+2x)}$

(g) $-\tan x$ (h) $\dfrac{\sec^2 x}{\tan x}$ (i) $\dfrac{2x}{x^2+4}$ (j) $\dfrac{e^x - e^{-x}}{e^x + e^{-x}}$ (k) $\dfrac{1+6x}{x+3x^2}$ (l) $\dfrac{x}{x^2-1}$

(m) $\dfrac{1}{x+\sqrt{x^2+1}} + \dfrac{x}{x\sqrt{x^2+1} + x^2+1}$ (n) $\dfrac{1}{x(1-x)}$ (o) $\dfrac{1}{x(2x+1)}$

(p) $-\tfrac{1}{2}\tan x$ (q) $\dfrac{x+1}{x}$ (r) $\dfrac{2}{x} + \cot x$ (s) $\dfrac{-2}{x^2-1}$ (t) $\dfrac{2x+3-3x^2}{(x^2+1)(1-3x)}$

(u) $\dfrac{3-2x}{(1-x)(2-x)}$ (v) $\dfrac{1}{x\ln 2}$ (w) $\dfrac{1}{x\ln a}$ (x) $1 - \tan x$

(y) $(\sin^{-1}x)\sqrt{1-x^2}$ (z) $\dfrac{1}{(\tan^{-1}x)(1+x^2)}$

2. $\dfrac{1 + 2\cos x - 2\sin x}{(2 + \cos x)(2 - \sin x)}$

4. (a) $3^x \ln 3$ (b) $a^x \ln a$ (c) $-a^{-x}\ln a$ (d) $10^x \ln 10$ 6. 1

Exercise 8.3

1. (a) $\cos x - x\sin x$ (b) $2x\sin x + x^2 \cos x$ (c) $\dfrac{3x+2}{2\sqrt{x+1}}$ (d) $e^x(1+x)$

(e) $1 + \ln x$ (f) $(1+x)\sec^2 x + \tan x$ (g) $2\cos x \cos 2x - \sin 2x \sin x$

(h) $\dfrac{e^x}{x}(1 + x\ln x)$ (i) $\dfrac{1}{(1+x)^2}$ (j) $\dfrac{-(x\sin x + \cos x)}{x^2}$

(k) $\dfrac{x\cos x - 2\sin x}{x^3}$ (l) $\dfrac{-2}{(1+x)^2}$ (m) $\dfrac{x(2+x)}{(1+x)^2}$ (n) $2x - \dfrac{1}{x^2}$

3. (a) $(1, \tfrac{1}{2})$, $(-1, -\tfrac{1}{2})$ (b) none

4. (a) $\cos^2 y$ (b) $\dfrac{1}{\sec y \tan y}$ (c) $-\sin^2 y$ $\dfrac{1}{1+x^2}$, $\dfrac{1}{x\sqrt{x^2-1}}$, $-\dfrac{1}{x^2(1+x^2)}$

5. (a) $\sqrt{\dfrac{x}{1-x^2}}\left(\dfrac{1}{2x}+\dfrac{x}{1-x^2}\right)$ (b) $\dfrac{x}{\sqrt{1-x^2}}\left(\dfrac{1}{x}+\dfrac{x}{\sqrt{1-x^2}}\right)$ (c) $-\csc^2 x$

(d) $\dfrac{1+\ln x}{2\sqrt{x\ln x}}$ (e) $\dfrac{x\cos x+\sin x}{2\sqrt{x\sin x}}$ (f) $xe^x\cos x+e^x(1+x)\sin x$

6. $\dfrac{9}{8}$ 7. (a) $(x\cos x+\sin x)e^{x\sin x}$ (b) $\left(\sin^{-1}x+\dfrac{x}{\sqrt{1-x^2}}\right)$

10. $y=-4x+22$ 11. $2\dfrac{dy}{dx}-y$ 13. 0 and -0.035

15. (a) $(1,1)$ (b) $(2,0)$, $(-3,0)$, $(-\tfrac{1}{2},\tfrac{625}{16})$ 16. 18, $-\tfrac{14}{3}$

17. (a) $\sec x\tan x$ (b) $\dfrac{\sec^2 x}{2\sqrt{\tan x}}$ 19. $a=3$, $b=5$, $c=2$

20. (a) $g'(x+a)$ (b) $g'(x+g(x))(1+g'(x))$ (c) $g'(xg(x))[g(x)+xg'(x)]$

Exercise 8.4

1. (a) $-\dfrac{y}{x}$ (b) $-\dfrac{x}{y}$ (c) $\dfrac{9x}{4y}$ (d) $-\dfrac{1}{x}$

(e) $-\dfrac{y(1+y\cos x)}{x+2y\sin x}$ (f) $-\dfrac{(\sin y+y\cos x)}{\sin x+x\cos y}$

2. $y+2x=6$

3. (a) $2x_1 y+y_1 x=3x_1 y_1$ (b) $y_1 y+x_1 x=y_1^2+x_1^2$
   (c) $a^2 y_1 y+b^2 x_1 x=a^2 y_1^2+b^2 x_1^2$

4. 0.303 max, -3.303 min   5. -3

6. (a) $\dfrac{\cos(x+y)-y}{x-\cos(x+y)}$ (b) $2e^{-(x+y)}-1$

9. (a) $2^x\ln 2$ (b) $x^x\ln x+x^x$ (c) $\dfrac{2\ln x\, x^{\ln x}}{x}$

10. $\dfrac{2x^2 y-y}{x-2y^2 x}$ 11. $\dfrac{y-x^2}{y^2-x}$

Exercise 8.5

1. (a) $e^x(1 + x)$ (b) $x e^{-x}(2 - x)$ (c) $e^{-x^2/2}(1 - x^2)$

   (d) $6x(x + 4)^3(3x + 4)$ (e) $(1 - x)(1 + x)^4(3 - 7x)$ (f) $\dfrac{x(2 + x)}{(1 + x)^2}$

   (g) $\dfrac{2(1 - 2x^2)}{(1 + x)^2(2x + 1)^2}$ (h) $-\dfrac{(x^2 + 34x + 37)}{(x - 4)^2(x + 5)^2}$ (i) $e^{-x}[\sec^2 x - \tan x]$

2. $\dfrac{u}{v}\left[\dfrac{u'}{u} - \dfrac{v'}{v}\right] \equiv \dfrac{u'v - v'u}{v^2}$ 4. 1

5. (a) $-x^{n-1}[n \ln x + 1]$ (b) $e^{x^2 - x}\dfrac{(2x^3 + x^2 - x + 1)}{(1 + x)^2}$

6. $x = 2 \Rightarrow f(2) = -4$ max; $x = 4 \Rightarrow f(4) = 0$ min   7. 0

Exercise 8.6

1. $t_1 y = t_1^2 x + a$; $t_1 y + x = 2a + \dfrac{a}{t_1^2}$ 2. $\dfrac{\sin \theta}{1 - \cos \theta}$ 4. $\dfrac{3(t^2 - 1)}{4t^3}$

5. (a) $-\cot \theta$ (b) $-\dfrac{1}{a \sin^3 \theta}$ (c) $2ay + 16x = a^2 + 4a\sqrt{3}$

6. $\dfrac{1}{3a} \sec^4 \theta \operatorname{cosec} \theta$ 8. $\dfrac{t(t + 2)}{t + 1}$ 10. $\dfrac{2t}{1 + t^2}$; $\dfrac{(1 - t^2)^3}{(1 + t^2)^3}$

11. $2t e^{t^2}(1 + t)$; $2 e^{t^2}(1 + t)(2t^3 + 2t^2 + 2t + 1)$

13. $\dfrac{x(y + 1)^3}{y(x + 1)^3}$ 14. $-\dfrac{(1 + t^2)^3}{4t^3}$

Exercise 8.7

1. $\dfrac{3\sec^2(3x + 5)}{\tan(3x + 5)}$ 3. $\dfrac{2x(2 - x^2)}{(1 - x^2)^{3/2}}$

4. (a) $8\tan^3 2x \cdot \sec^2 2x$ (b) $\dfrac{-1}{(2x - 3)^2}$ (c) $2x \ln x + x$

5. (a) $-e^{-2x}(3\sin 3x + 2\cos 3x)$ (b) $-\tfrac{3}{4}$ 6. (a) $\dfrac{1 - x^2 - 4x}{(x^2 + 1)^2}$ (b) $\tfrac{5}{3}$

7. (a) $x^2 + 2 - 2x - y$;   (b) 1; 1; 1   8. (a) (i) $\sec x$ (ii) $\dfrac{y+2}{1-x}$ (b) y

9. 2   10. $(0.23, 8.0)$ minimum

11. (a) (i) $\dfrac{\ln 3}{x} e^{\ln x}$ (ii) $\dfrac{-1}{\sqrt{1-x^2}}$ (b) 4.02   12. $x^2 + \dfrac{4V}{x}$; $\tfrac{1}{2}$

13. (a) (i) $\dfrac{3x^2}{(4-x^3)^2}$ (ii) $k\sec^2 kx\, e^{\tan kx}$ (iii) $2x(1 + \ln(x^2 + 1))$

    (b) $\tfrac{2}{3}$ (c) 2; 50

14. $y = \dfrac{x^2}{2} + \dfrac{8}{x^2}$; $(2,4)$, $(-2,4)$;   no point of inflexion

    (there is a discontinuity when $x = 0$)

15. $N\left(\tfrac{1}{7}, \tfrac{15}{7}\right)$

16. (a) (i) $-2(a-x)e^{(a-x)^2}$ (ii) $\dfrac{a^2}{(x^2+a^2)^{3/2}}$ (b) t, $\dfrac{1}{2(1+6t^2)}$

Exercise 9.1

1. 22, 47, $5n - 3$ 2. (a) 26 (b) 82 (c) 18   3. 3, 9

4. (a) $3, 5, 7, 9$ (b) $1, 4, 7, 10$

5. (a) 21, 120   (b) $-\tfrac{8}{5}$, -7 (c) $a + 8d$, $5(2a + 7d)$ (d) $10x$, $50x$
   (e) $10(x + 1)$, $55(x + 1)$

6. 67, 13467   7. 16734   8. 12   9. (a) 21 (b) 241

10. $\tfrac{3}{2}n(n + 1)$ 11. 50   12. (a) $4(7b - 5a)$ (b) $4(12 - 5n)$

13. $2(2n - 1)$, $2n^2$ 15. 646, 18   16. 51, 2

17. 2, 6   18. 2800   20. $\tfrac{3}{5}l$, l, $\tfrac{7}{5}l$

Exercise 9.2

1. 24, 192, ar^{n-1} 2. 2, 6

3. (a) $\tfrac{1}{16}$, $\tfrac{255}{16}$ (b) $-\tfrac{1}{32}$, $\tfrac{85}{32}$

3. (c) x^7, $\dfrac{1 - x^8}{1 - x} = 1 + x + x^2 + \ldots + x^7$ (d) $\dfrac{n^8}{a^7}$, $\dfrac{n(a^8 - n^8)}{a^7(a - n)}$

4. 8   5. (a) 1365 (b) 31.25 (c) $\dfrac{a(r^{n+2} - 1)}{r - 1}$ (d) $a\dfrac{(r^{2n} - 1)}{r - 1}$

6. $113°$ 7. -16 9. $\dfrac{g^n - p^n}{g^{n-2}(g - p)}$ 10. 48.48

11. 16 years (approx.)   12. 3   13. (a) 10 (b) $n + 7$ (c) 1

Exercise 9.3

1. (a) $1 + 4 + 9 + 16$ (b) $0 + 2 + 6$ (c) $3 + 5 + 9 + 17$
   (d) $0 + 1 - 2 + 3 - 4$ (e) $2 + 8 + 18 + 32$

2. (a) $\displaystyle\sum_{s=1}^{6} s$ there are many more possibilities eg. $\displaystyle\sum_{s=0}^{6} s$

   (b) $\displaystyle\sum_{s=1}^{8} (-1)^{s+1} s$ (c) $\displaystyle\sum_{s=1}^{5} (-1)^s s$ (d) $\displaystyle\sum_{s=1}^{n} s^2$

   (e) $\displaystyle\sum_{s=1}^{n} s(s + 2)$ (f) $\displaystyle\sum_{s=1}^{n} sx^3$ (g) $\displaystyle\sum_{s=0}^{10} (-1)^{s+1} x^s$

3. (a) 18 (b) 36

4. (a) $n(2n + 1)$ (b) $\frac{1}{4}t^2(t + 1)^2$ (c) $\frac{1}{6}n(n + 1)(2n + 7)$
   (d) $\frac{1}{4}n(n + 1)(n + 2)(n + 3)$.

5. $n(2n + 3)$ 6. $\frac{1}{2}n(6n^2 + 15n + 11)$

7. $2(2^n - 1) + \frac{3}{2}n(n + 1) - 2n$ 8. $\frac{1}{6}n(2n^2 + 6n + 7)$

10. $\frac{1}{4}n^2(n + 1)^2$, $(n + 1)^2(4n + 1)$

11. (a) $\dfrac{n}{n + 1}$ (b) $\dfrac{3}{2} - \dfrac{1}{n + 1} - \dfrac{1}{n + 2}$ (c) $\dfrac{1}{2} - \dfrac{1}{n + 2}$ (d) $\dfrac{1}{4} - \dfrac{1}{n + 4}$

Exercise 9.4

1. (a) 20 (b) $\frac{9}{4}$ (c) $\dfrac{r}{1 - a}$, $|a| < 1$ (d) $\dfrac{na^2}{a - r}$, $\left|\dfrac{r}{a}\right| < 1$ (e) $\dfrac{1}{1 - 2x}$, $|x| < \frac{1}{2}$

2. (a) $\frac{1}{9}$ (b) $\frac{77}{225}$ (c) $\frac{6}{11}$ (d) $\frac{179}{495}$

3. 4, $\frac{1}{2}$ 4. $\frac{2}{3}$, 7   5. 13   6. $\frac{52}{27}$, 2

7. $\frac{1}{3}$, 2; $\frac{8}{3}$, $\frac{1}{2}$; $\frac{16}{3}$ 8. $\frac{1}{3}$ 9. 22, 688

10. (a) $\frac{100}{9}$ m   (b) Failure to realise that an infinite sum can be finite.

11. $-\sec 2\theta = \dfrac{1}{1 - 2\cos^2\theta}$; $|\cos\theta| < \frac{1}{2}$, so $45 < \theta < 135$ or $225 < \theta < 315$ in
    the range 0 to 360°

12. $\dfrac{1}{1 - \cos\theta}$, $\dfrac{1}{1 - \cos 2\theta}$ 13. $\frac{3}{55}$ 14. $-\sec 2\theta$, $\frac{1}{4}\pi < \theta < \frac{3}{4}\pi$

15. (a) $100n^2$ (b) $2^n - 1$; 15   16. $\frac{1}{2}$; 39.96   17. 20

18. $\dfrac{a}{4}$, $1\frac{1}{2}$ 19. $1\frac{3}{5}$, 9   20. $\frac{3}{4}$, $\frac{3}{16}$, $\frac{3}{64}$; 1

21. (a) $3\frac{1}{3}$, 16000   (b) $\dfrac{a^2}{b}$ 22. $2n(2n + 1)$, 2^{2n}

Exercise 9.5

9. (a) 0, 56   (b) $9^{2n+2} - 5^{2n+2}$

Exercise 9.6

1. (a) $1 - 10x + 40x^2 - 80x^3$ (b) $x^6 + 6x^5y + 15x^4y^2 + 20x^3y^3$
   (c) $1 + \dfrac{7}{x} + \dfrac{21}{x^2} + \dfrac{35}{x^3}$

2. (a) 70   (b) 2835   (c) 6   (d) -5 3. 152   4. $160x^3$

5. -84 6. (a) $2 + 42x^2 + 70x^4 + 14x^6$ (b) $16x^2 + \dfrac{64}{x^2}$

7. (a) $120x^3$ (b) $-22680x^3$ (c) $120x^3$ (d) $292x^3$

8. (a) $1 - 5x + 15x^2 - 30x^3$ (b) $16 - 32x - 40x^2 + 88x^3$

9. (a) 1.0828   (b) 1.1484   (c) 64.19224

10. (a) $1 - 10x + 45x^2$ (b) $256 + \frac{1024}{3}x + \frac{1792}{9}x^2$ (c) $512 + 2816x + 6912x^2$
    (d) $1 + 3x - 6x^2$ (e) $64 - 96x - 48x^2$

11. 2, $\frac{9}{4}$

Exercise 9.7

1. (a) $1 - 2x + 4x^2 - 8x^3 + 16x^4$, $|x| < \frac{1}{2}$
   (b) $1 + 2x + 3x^2 + 4x^3 + 5x^4$, $|x| < 1$

   (c) $1 + \dfrac{x}{2} - \dfrac{x^2}{8} + \dfrac{x^3}{16} - \dfrac{5x^4}{128}$, $|x| < 1$

   (d) $2^{1/3}\left(1 - \dfrac{x}{6} - \dfrac{x^2}{36} - \dfrac{5x^3}{648} - \dfrac{x^4}{7776}\right)$, $|x| < 2$

   (e) $\frac{1}{3} - \dfrac{x}{9} + \dfrac{x^2}{27} - \dfrac{x^3}{81} + \dfrac{x^4}{243}$, $|x| < 3$

   (f) $\dfrac{1}{\sqrt{3}}\left(1 + \dfrac{x}{6} + \dfrac{x^2}{24} + \dfrac{5x^3}{432} + \dfrac{35x^4}{10368}\right)$, $|x| < 3$

3. (a) $1 - 2x + 2x^2 - 2x^3$ (b) $1 + 3x + 6x^2 + 12x^3$

5. $10 + \dfrac{x}{20} - \dfrac{x^2}{8000} + \dfrac{x^3}{1600000}$, $|x| < 100$;   10.050   6. 10.003

7. (a) $5 + 8x + 14x^2$, $|x| < \frac{1}{2}$ (b) $-\frac{1}{6} + \frac{17}{36} + \frac{11}{216}x^2$, $|x| < \frac{1}{3}$

9. -4, $-\frac{1}{2}$ 10. $1 - x + \dfrac{x^2}{2} - \dfrac{x^3}{2}$

12. $\pm\sqrt{\frac{5}{2}}$ 13. $-\dfrac{2^5 \cdot 20!}{15!5!} = -496\,128$

14. $x^4 + 4x^3y + 6x^2y^2 + 4xy^3 + y^4$; $p^4 - 4qp^2 - 14q^2$

15. (a) $1 + 12x + 58x^2$ (b) $-\dfrac{2^5 \cdot 15!}{10!5!} = -96096$

16. $1 - \dfrac{x}{2} + \dfrac{3x^2}{8} - \dfrac{5x^3}{16}$, 0.099504   18. 15; $1 - \dfrac{x}{2} - \dfrac{x^2}{8}$, $\dfrac{1351}{780}$

19. $1 + apx + \frac{1}{2}a^2p(p - 1)x^2 + \frac{1}{6}a^3p(p - 1)(p - 2)x^3$; -9, $-\frac{2}{9}$; $|x| < \frac{1}{9}$

20. $1 + 2ny + 2n(n - 1)y^2$ 21. $1 + x + \frac{5}{4}x^2$

22. $A = 4$, $B = 1$, $C = 2$; $5 + 9x + \frac{57}{2}x^2 + \frac{287}{4}x^3$; $|x| < \frac{1}{2}$

23. Need to divide first! $a = 4$; $b = -1$; $\frac{1}{2}x^2 + \frac{3}{4}x^3$; $|x| < 1$

24. $\dfrac{2}{1-x} + \dfrac{1}{(1-x)^2} + \dfrac{2}{2-x}$; $4 + \frac{17}{4}x + \frac{21}{4}x^2$

25. $\frac{3}{8}$ 26. $2,\ \frac{3}{4};\ 1,\ \frac{3}{2}$

Exercise 10.1

1. (a) 15 (b) 243 (c) 60 (d) 2 (e) 0 (f) 0 (g) $\frac{74}{3}$ (h) $\frac{1}{2}$
   (i) $\frac{1}{3}(8 - 3\sqrt{3})$ (j) $\frac{1}{2}\pi$ (k) $\frac{1}{2}\pi$ (l) 0 (m) $8 + \ln 9$ (n) 1

2. (a) 63 (b) $1\frac{1}{3}$ (c) $4\frac{1}{2}$! (d) 1 (e) 12   3. $\frac{1}{6}$ 4. $4\frac{2}{3}$

5. $\frac{1}{2}$ 6. 1, 3   7. 2.52   8. $2\frac{2}{3}$ 9. 8

Exercise 10.2

3. (a) 0 (b) π (c) 0 (d) 0   4. 0, 1.5

Exercise 10.3

1. $10\frac{2}{3},\ 34.13\pi$ 2. π^2 3. $\sqrt{3}\pi$ 4. $\pi r^2 h/3$ 5. $\pi r^2 h$

6. $11\frac{13}{15}\pi$ 7. (a) 7 (b) $\dfrac{2}{\pi}$ (c) 0 (d) $\frac{1}{2}$ (e) $\frac{1}{2}$ 8. $\dfrac{3}{2} + \dfrac{4}{\pi}$

10. (a) $1\frac{5}{9}$ (b) 2 (c) $2\pi \ln 2$ 11. (a) $2 + \dfrac{2}{\pi}$ (b) $\dfrac{\pi}{2}(9\pi + 16)$

12. $\frac{1}{2}\pi(\pi + 2)$ 13. (a) $(-1, \frac{1}{3})$ (b) 2.15   14. $\frac{1}{3}\pi a^3$

16. $y = 6x + 15$,   (b) $\frac{48}{5}\pi$

17. 0, π, 2π; $(\frac{1}{3}\pi, \frac{3}{4}\sqrt{3})$, $(\pi, 0)$, $(\frac{5}{3}\pi, -\frac{3}{4}\sqrt{3})$

   2, $\dfrac{2}{\pi}$

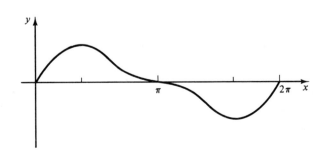

19. $\dfrac{5a^3\pi}{96}, \dfrac{a^3\pi}{3}$

20. Max 2 when $x = \frac{1}{6}\pi$; min -2 when $x = \frac{2}{3}\pi$; $\frac{5}{12}\pi$, $\frac{11}{12}\pi$

$\frac{1}{4}\pi(2\pi + 5\sqrt{3})$

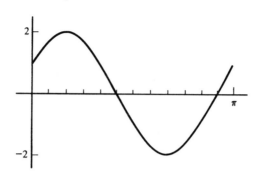

Exercise 10.4

1. (a) $\frac{1}{3}\tan^{-1}(3x)$ (b) $\frac{1}{2}\sin^{-1}(2x)$ (c) $\frac{1}{4}\sin^{-1}(4x)$ (d) $8\sin^{-1}(\frac{1}{4}x)$

(e) $\frac{1}{2}\tan^{-1}(\frac{1}{2}x)$ (f) $2\sin^{-1}(\frac{1}{2}x)$ (g) $\sin^{-1}(\frac{1}{4}x)^2$ (h) $\frac{1}{3}\sin^{-1}(\frac{3}{2}x)$

(i) $\frac{1}{5}\sin^{-1}(\frac{5}{4}x)$ (j) $\frac{1}{3}\tan^{-1}(\frac{1}{3}x)$

Exercise 10.5

1. (a) $\frac{1}{8}(2x+1)^4 + c$ (b) $\frac{2}{9}(3x+2)^{3/2} + c$ (c) $\frac{1}{6}(1+x)^6 - \frac{1}{5}(1+x)^5 + c$
(d) $\frac{1}{8}(1+x^2)^4 + c$ (e) $-\frac{1}{3}(1-x^2)^{3/2} + c$
(f) $\frac{9}{2}\sin^{-1}(\frac{1}{3}x) + \frac{9}{4}\sin(2\sin^{-1}\frac{1}{3}x) + c$ or $\frac{9}{2}\sin^{-1}(\frac{1}{3}x) + \frac{1}{2}x(9-x^2)^{1/2} + c$
(g) $\tan^{-1}(\frac{1}{3}x) + c$ (h) $\frac{1}{6}\tan^{-1}(\frac{2}{3}x) + c$

2. (a) $\frac{1}{12}(3x-1)^4 + c$ (b) $-\frac{1}{3}(1-\frac{1}{2}x)^6 + c$ (c) $-\frac{1}{30}(1-5x)^3 + c$

(d) $\dfrac{2}{3a}(ax+b)^{3/2} + c$ (e) $\frac{1}{3}(x^2+1)^{3/2} + c$ (f) $\frac{2}{5}(x+1)^{5/2} - \frac{2}{3}(x+1)^{3/2} + c$

(g) $\frac{1}{20}(2x+1)^5 - \frac{5}{16}(2x+1)^4 + c$ (h) $\frac{1}{2}x(1-\frac{1}{2}x^2)^{1/2} + \sin^{-1}(\frac{1}{2}x) + c$

(i) $\frac{1}{2}\tan^{-1}(2x) + c$ (j) $\sqrt{3}\tan^{-1}\sqrt{2}x + c$ (k) $-\tan^{-1}(1-x) + c$

(l) $\frac{1}{3}\tan^{-1}\left(\dfrac{x+2}{3}\right) + c$ (m) $\dfrac{\sqrt{2}}{2}\tan^{-1}\left(\dfrac{x+1}{\sqrt{2}}\right) + c$

(n) $2\sqrt{2}\tan^{-1}\left(\dfrac{x+2}{\sqrt{2}}\right) + c$ (o) $\tan^{-1}\left(\dfrac{x-1}{2}\right) + c$ (p) $2\sin^{-1}\left(\dfrac{x}{2}\right) + c$

(q) $\frac{3}{2}\sin^{-1}(2x) + c$

Exercise 10.6

1. (a) $-\frac{1}{2}e^{-2x} + c$ (b) $2e^{x/2} + c$ (c) $3e^x + c$ (d) $e^{5x} + c$ (e) $e^{x^2} + c$
   (f) $e^{\tan x} + c$ (j) $x + 2e^x + \frac{1}{2}e^{2x} + c$ (k) $e^{x^3 + x^2} + c$
   (l) $\frac{1}{2}e^{2x} - \frac{1}{2}e^{-2x} - 2x + c$ (m) $-\frac{1}{2}e^{-2x} - 3e^{-x} + c$ (n) e

   (o) $\dfrac{1}{\ln a}(a^2 - 1)$ (p) 1 (q) $\frac{1}{2}e^{x^2} + c$

2. $\dfrac{a^x}{\ln a} + c$

Exercise 10.7

1. (a) $\frac{1}{3}\ln|x| + c$ (b) $3\ln|x| + c$ (c) $\ln(x + 3) + c$ (d) $\frac{1}{3}\ln|3x + 1| + c$
   (e) $-\ln|1 - 3x| + c$ (f) $\frac{1}{2}\ln|x^2 + 1| + c$ (g) $\frac{1}{3}\ln|x^3 - 1| + c$
   (h) $\frac{2}{9}\ln|3x^3 - 2| + c$ (i) $\ln|\sin x| + c$ (j) $\ln(1 + \sin x) + c$
   (k) $-\frac{1}{2}\ln|1 - 2\sin x| + c$ (l) $\ln(1 + \sin^2 x) + c$ (m) $\ln|\ln x| + c$

2. (a) $\frac{1}{2}\ln 5$ (b) $\ln\sqrt{2}$ (c) $\frac{1}{3}\ln\frac{2}{9}$ (d) $3\ln\frac{4}{3}$ (e) $\frac{1}{2}\ln(x^2 + 1) + \tan^{-1}x$

   (f) $\ln(x^2 + 4) - \frac{1}{2}\tan^{-1}\left(\dfrac{x}{2}\right) + c$ (g) $\ln\left(\dfrac{x + 1}{x + 2}\right) + c$

Exercise 11.1

1.

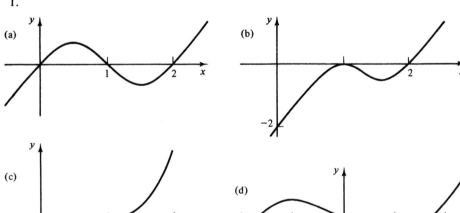

(e)
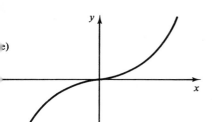

(f)
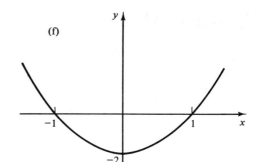

(g)
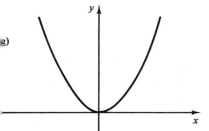

$(-1,0),\quad (2,9);\quad 6.75$

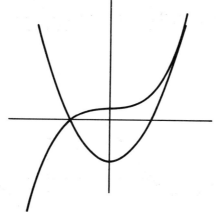

$0,\quad -\frac{9}{4}$

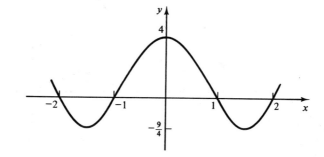

4. $x = \frac{1}{2}$ min; $x = -\frac{3}{2}$ max

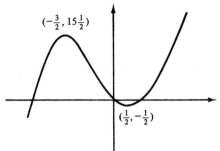

5. $(0, 0)$ min; $(1, 5)$ max, $(3, -27)$ min; $-27 < k < 0$

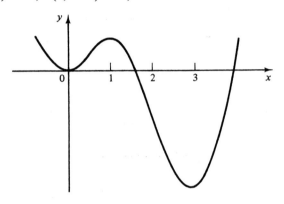

6. $9\frac{1}{3}$ square units

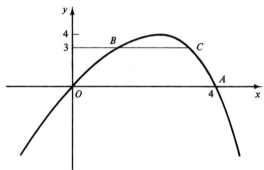

7. -3, -9, 9

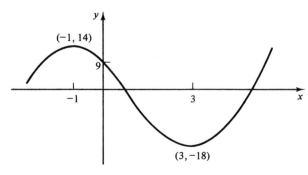

8. $x = \sqrt{30}$ minimum;

$x = 0$ point of inflexion;

$x = -\sqrt{30}$ minimum

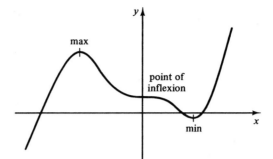

9. $f(x) \equiv -2(x - 1)(x - 3)^2$, positive, $k > 18$

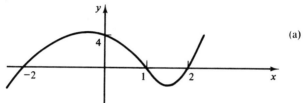

(a)

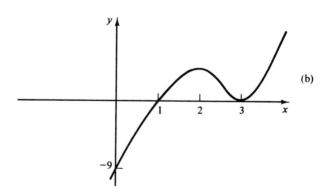

(b)

10. (a) -1, 5, -7, 3   (c) $\frac{7}{3}$

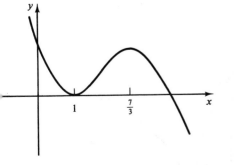

11. (a) 0, -48 (c) -1

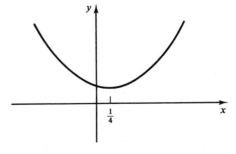

Exercise 11.2

6. (a) $0 \leqslant y \leqslant 12$ (b) $y < -6$ or $y > -2$ (c) $\frac{1}{2}(1 - \sqrt{2}) < y < \frac{1}{2}(1 + \sqrt{2})$

9. $\dfrac{3}{\sqrt{5}}$

10. -1, $\frac{5}{3}$

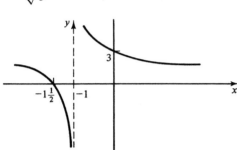

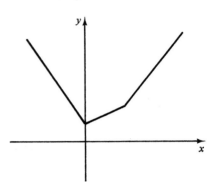

12. (a) (i) 1   (ii) 1   (c) $\frac{2}{3} < y < 2$ (d) -1,   1

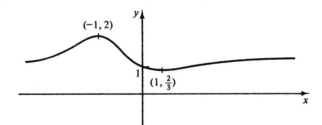

13.

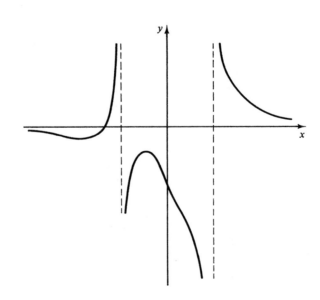

14. min at $(1, 0)$, max at $(5, \frac{2}{27})$

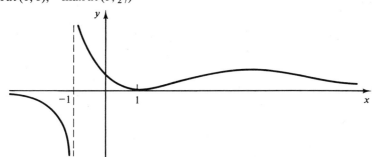

15.

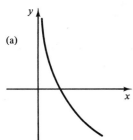

16. max at $(2, -4)$,
min at $(4, 0)$

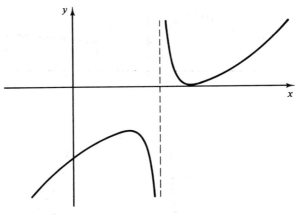

17.

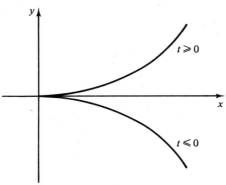

18. $\ln(x + 4)$, $x > -4$,
    $(-\infty, \infty)$; $e^x - 4$

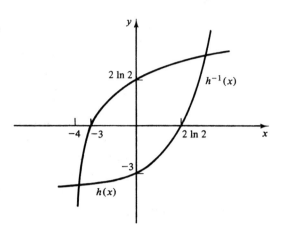

19. 1.10

(a)

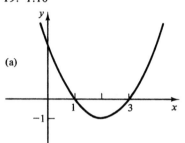

(b)

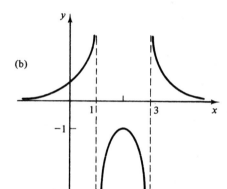

20.

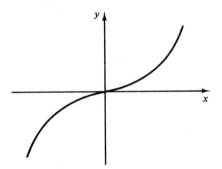

21.

(a)

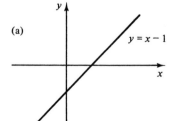

(b)

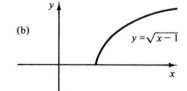

21.

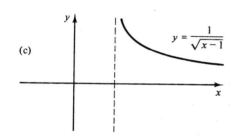

(c)

22. $\dfrac{4\sqrt{3}}{45}a^2$

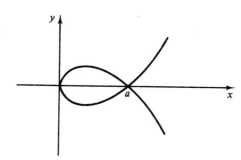

Exercise 11.3

1. (a) $x < -3$ or $-2 < x < 1$ (b) $x \leqslant -2$ (c) $x \geqslant 1$
   (d) $x < \frac{1}{2}$ or $1 < x < 2$ (e) $\frac{1}{2} < x < 2$ (f) $-4 < x < 1$ (g) all x
   (h) $-\frac{1}{4} \leqslant x < \frac{3}{2}$

2. $x < -1$ or $3 < x < 5$ 3. $x < 1$ or $x > \frac{5}{2}$

4. $2 < x < \frac{5}{2}$ or $x > 3$ 5. $\frac{3}{4} < k < 6$

6. $x = -3$, $y = 0$;
   $y = 0$, $x = 1$

   $x < -3$ or
   $-1 < x < 1$

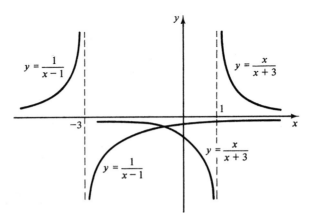

7. (a) $x < -3$ or $x > 1$ (b) $-1 < x < \frac{1}{5}$

8. (a) $x < -1$ or $x > 4$ (b) $-2 < x < 0$ or $x > 3$

9. -2, $1 \pm \sqrt{5}$; $x < -2$ or $1 - \sqrt{5} < x < 1 + \sqrt{5}$

Exercise 11.4

1.

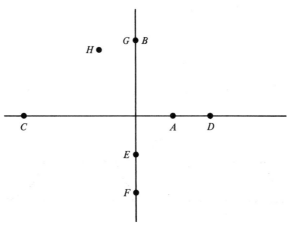

2. (a) (i) $\sqrt{5}$ (ii) $2\sqrt{2 + \sqrt{2}}$ (iii) $\sqrt{5 + 2\sqrt{2}}$
   (b) (i) 1 (ii) $\sqrt{2}$ (iii) $(3 + \sqrt{2})/\sqrt{2}$

3. (a) $(\frac{3}{2}, \frac{3}{2}\sqrt{3})$ (b) $(-\sqrt{2}, \sqrt{2})$ (c) $(\sqrt{2}, -\sqrt{2})$ (d) $(-2\sqrt{2}, -2\sqrt{2})$

4. (a) $(2, \frac{1}{3}\pi)$ (b) $(4, 0)$ (c) $(3, \frac{1}{2}\pi)$ (d) $(-\sqrt{2}, \frac{1}{4}\pi)$

5. (a) $r^2 \sin 2\theta = 8$ (b) $r^2 = 9$ or $r = 3$ (c) $r^2(a^2 \cos^2\theta + b^2 \sin^2\theta) = a^2 b^2$
   (d) $r^2 - 2r \sin\theta - 3 = 0$

6. (a) $x^2 + y^2 = 4$ (b) $y = -\sqrt{3}x$ (c) $x^2 + y^2 = y$ (d) $x = 4$
   (e) $a^2 y = (x^2 + y^2)^{3/2}$ (f) $(x^2 + y^2)^{3/2} = (x^2 - y^2)$ (g) $x^2 - y^2 = 4$
   (h) $x^2 + y^2 = \sqrt{x^2 + y^2} - y$

Revision exercise C

1. $x^2 - 78x + 225 = 0$;   2. (a) 3π (b) $\frac{1}{6}$ (c) $\frac{1}{2}\pi(e^2 - 4e + 5)$
   75, 3

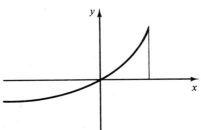

3. 3   4. (b) $((c - 1)^2 + b^2)x^2 - b(c + 1)x + c = 0$

5. (a) $3^{3/2}$, $3^{-3/2}$ (b) 21

6. $\dfrac{3\tan A - \tan^3 A}{1 - 3\tan^2 A}$; $0°$, $40.2°$, $180°$

7. $4.3\,\text{km}$, $256.7°$ 8. $x > 2$ or $x < 0$; $2x - 1$

9. (a) 101 (b) 69   11. (b) $2ay = x^2 + a^2$ 12. (a) $-\dfrac{1}{x^2}\sec^2\!\left(\dfrac{1}{x}\right)$

13.

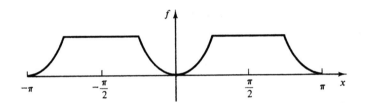

14. $13\cos(\theta + 26.6°)$;   (a) 13 when $\theta = 333.4°$; -13 when $\theta = 153.4°$
    (c) $225.5°$ or $81.3°$

15. $e^{x-2} - 2$; $(-\infty, \infty)$, $(-2, \infty)$ 16. (a) $(4, 2)$ 17. $2\sqrt{2}$

18. (a) (i) $-\cos x/(1 + \sin x)^2$ (ii) $2\tan 2x$ (b) 0.607
    (c) $(2t - 2)/(2t + 2)$, $1/(t + 1)^3$; 3, -1, min

21. (a) $\dfrac{2t}{3\,e^{3t}}$

(c)

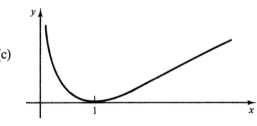

22. max $6e^{-3}$, min $-2e$;

$7e^{-2} - 14e^{-3}$

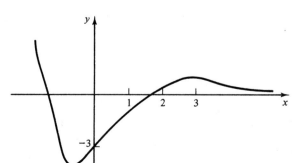

23. (a) 2 (b) -6 24. (a) $3^{2p+\frac{1}{4}q}$, $3^{4p-\frac{1}{4}q}$ (b) $\ln 2$

25. $\dfrac{32\pi h r^2}{27}$ 27. $\dfrac{1}{x+2} - \dfrac{1}{x+3}$, $\frac{1}{6}\ln\frac{4}{3}$ 28. 16, 6, 4, 16

Exercise 12.1

1. (a) $x\tan^{-1}x - \frac{1}{2}\ln(x^2+1) + c$ (b) $x\sin^{-1}x + \sqrt{(1-x^2)} + c$
   (c) $x\sin x + \cos x + c$ (d) $\frac{1}{2}x^2\ln x - \frac{1}{4}x^2 + c$ (e) $-e^{-x}(x^2+2x+2) + c$
   (f) $\frac{1}{2}(\ln x)^2 + c$ (g) $\frac{1}{2}e^x(\sin x - \cos x) + c$ (h) $\frac{1}{4}(e^2+1)$ (i) $\frac{1}{5}(e^{2\pi}+1)$
   (j) $\frac{1}{2}(x^2+1)\tan^{-1}x - \frac{1}{2}x + c$ (k) $x\tan x - \ln|\sec x| + c$
   (l) $\pi - 2$ (m) $\frac{1}{2}e^{-x}(\sin x - \cos x)$ (n) 1

2. $\frac{1}{9}\left(1 - \dfrac{4}{e^3}\right)$

Exercise 12.2

1. (a) $-2\ln|1-x| + c$ (b) $\frac{3}{2}\ln|2x-3| + c$

   (c) $\dfrac{2}{(1-x)} + c$ (d) $\dfrac{3}{2(2-x)} + c$ (e) $-\dfrac{1}{2(1+3x)^2} + c$

   (f) $-4x - 4\ln|1-x| + c$ (g) $\frac{2}{3}x + \frac{4}{9}\ln|3x-2| + c$ (h) $x + \ln|2x-1| + c$

   (i) $2x - 3\ln|x+2| + c$ (j) $\ln\left|\dfrac{1+2x}{1+x}\right| + c$ (k) $\tan^{-1}(x+1) + c$

   (l) $\ln\left|\dfrac{x}{1-x}\right| + \dfrac{1}{1-x} + c$ (m) $\ln\left|\dfrac{1+x}{x}\right| - \dfrac{1}{x} + c$ (n) $\tan^{-1}(x+2) + c$

   (o) $\frac{1}{3}\tan^{-1}\left(\dfrac{x+1}{3}\right) + c$ (p) $\frac{3}{2}\tan^{-1}(2x+1) + c$ (q) $\frac{1}{4}\tan^{-1}\left(\dfrac{2x-1}{2}\right) + c$

   (r) $\frac{1}{4}\tan^{-1}\left(\dfrac{3x-4}{4}\right) + c$ (s) $\frac{1}{2}\ln|x^2+2x-2| + c$

   (t) $\frac{1}{4}\ln|(x-1)^5(x+3)^7| + c$ (u) $2\ln|x^2+6x+18| - \frac{5}{3}\tan^{-1}\left(\dfrac{x+3}{3}\right) + c$

   (v) $x + \ln\left|\dfrac{x-1}{x+1}\right| + c$ (w) $\frac{1}{2}x^2 - x + 2\ln|x+1| + c$ (x) $\frac{1}{3}x^3 + \frac{1}{2}x^2 + x + c$

Exercise 12.3

1. (a) $2\sqrt{x-4} + c$ (b) $\ln 2$ (c) $4 + 8\ln 2$

2. $\frac{2}{3}$ 4. 38   5. (a) $1 - \frac{2}{e}$ (b) 14.3

8. (a) $\ln\frac{3}{2}$ (b) $\frac{1}{8}\pi + \frac{1}{4}\ln 2$ (c) $\frac{1}{2}$ (d) $\ln\frac{5}{4}$ 9. (a) $\frac{1}{2}\pi$ (b) $\frac{3}{4}$

10. $\ln\left|\dfrac{(x+4)^2}{x-1}\right| + c$ 11. (b) $2 - \frac{1}{2}\pi$ 12. $\frac{1}{20}$

Exercise 12.4

1. 3   2. $-\cos x - \frac{1}{3}\cos^3 x + c$ 3. $\frac{7}{9}$ 4. $\dfrac{4\pi}{3} - \dfrac{\sqrt{3}}{2}$

6. $\dfrac{1}{\sqrt{2}}\tan^{-1}\left(\dfrac{1}{\sqrt{2}}\tan\left(\dfrac{x}{2}\right)\right) + c$ 7. $\dfrac{\sqrt{3}}{8} + \dfrac{\pi}{12}$ 8. 0.59

10. $-\frac{8}{5}$ 11. $\frac{1}{2}xe^{2x} + \frac{1}{4}e^{2x} + c$ 12. $-e^{-x}(x+1) + c$

13. $\frac{1}{2}x^2\ln x - \frac{1}{4}x^2 + c$ 14. (a) 3.45

15. $(x-1)(x+1)(x^2+1)$; $\frac{1}{4}\ln\left|\dfrac{(x-1)^3}{(x+1)(x^2+1)}\right| - \tan^{-1}x + c$

16. $-\dfrac{1}{x} + 2\tan^{-1}x + c$ 17. $\frac{1}{2}\ln\dfrac{x+1}{\sqrt{x^2+1}} - \dfrac{1}{2(x+1)} + c$

18. (a) $-\frac{1}{72}$ (b) $\frac{1}{8}\pi \times \frac{1}{12}$ 19. (a) $\frac{1}{4}e^{2x}(2x-1) + c$ (b) $-\frac{1}{2}\cos x^2 + c$

20. (a) $-2\cot x - \operatorname{cosec} x + c$ (b) 0.18   21. (a) $\ln\frac{9}{4}$ (b) $\frac{4}{3} - \frac{1}{2}\sqrt{3}$

22. (a) $\frac{1}{4}\ln\frac{5}{3}$ (b) $\frac{2}{3}$ 23. (a) $\frac{1}{24}(x^4-3)^6 + c$ (b) $\ln|\sec x| + c$ (c) $\frac{1}{2}e^{x^2} + c$

24. (a) (i) 0.105 (ii) 0.283 (b) $\dfrac{e^{3x}}{27}[9x^2 - 6x + 2]$ (c) $-\dfrac{1}{a}\sin^{-1}\left(\dfrac{a}{x}\right) + c$

25. (a) $x - \frac{1}{4}\cos 4x + c$ (b) 0.571 (c) 0.380

26. $\dfrac{2}{2x-1} - \dfrac{x}{x^2+1}$; $\ln\left|\dfrac{2x-1}{\sqrt{x^2+1}}\right| + c$ 27. (a) 0.347 (b) $2 - \frac{1}{2}\pi$

28. $\sin^{-1}mx + \dfrac{mx}{\sqrt{1-m^2x^2}}$; $x\sin^{-1}mx + \dfrac{1}{m}\sqrt{1-m^2x^2}$

29. $(\frac{1}{2}\pi, 0)$, $(\pi, 0)$

30. $\frac{1}{6}$; $\frac{5}{6}\pi$; $\frac{1}{2}\pi^2$; -2; $\frac{1}{2}\pi$;
$\frac{1}{3}\pi^4 - 4\pi + 2\pi^2$

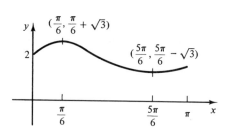

Exercise 13.1

1. (a) $(2, -2)$, $(0, 2)$ (b) $(-1, 6)$, $(-2, 7)$ (c) $(2, -\frac{1}{3})$, $(3, \frac{4}{3})$
(d) $(\frac{1}{3}(2a + c), \frac{1}{3}(2b + d))$, $(\frac{1}{3}(a + 2c), \frac{1}{3}(b + 2d))$

2. (a) $(3, 5)$ (b) $(6, 1)$ (c) $(3a, b)$ 3. (a) $(\frac{2}{3}, 2)$ (b) $(\frac{11}{3}, \frac{5}{3})$

4. (a) $\frac{1}{2}\pi$ (which is not acute!) (b) $\frac{1}{4}\pi$ (c) 0.72 (d) 1.00

5. $y = x - 2$ 6. (a) $\frac{4}{5}$ (b) $\frac{3}{5}$ (c) $\frac{1}{2}\sqrt{2}$ 7. (a) $\frac{25}{13}$ (b) $\frac{6}{17}\sqrt{34}$

9. $c^2 - m^2 = 1$ 10. 13   11. $(2, -3)$; $\frac{4}{5}\sqrt{5}$

12. $(x - y)^2 - 8(x + y) + 16$ 13. $5(x - y)^2 = 2(2x - y)^2$

15. $(2, -4)$ 16. (a) $(10, 9)$ (b) $\frac{1}{10}\sqrt{2}$ (c) 6   17. $28°$

Exercise 13.2

1. (a) $(2, 5)$; 4 (b) $(-1, 2)$; 3 (c) $(-\frac{1}{2}, \frac{3}{2})$; 5 (d) $(-1, -2)$; $\sqrt{5}$
(e) $(-f, -g)$; $\sqrt{c + f^2 + g^2}$ $c + f^2 + g^2 > 0$

2. (a) 0 (b) 0 (c) undefined; $(2 + \sqrt{2}, 3 + \sqrt{2}), (2 - \sqrt{2}, 3 - \sqrt{2})$

3. $y = -\frac{1}{2}\sqrt{5}x + \frac{1}{2}(\sqrt{5} + 5)$ or $y = +\frac{1}{2}\sqrt{5}x - \frac{1}{2}(\sqrt{5} - 5)$

4. $(-3, 2)$; 4   5. $(x - 2)^2 + (y - 1)^2 = 5$ 6. $\sqrt{29}$

7. $(x - 2)^2 + (y + 1)^2 = 10$ 8. $\sqrt{5}y = 2\sqrt{5}x \pm 9$

9. $5x^2 + 5y^2 = 4$ 10. $(2, 3)$; 3; $\sqrt{3}y + x = 3\sqrt{3} + 8$

12. $(2x - 7)^2 + (2y - 7)^2 = 10$ 14. $(-5, 3)$; 6   15. $(-2a, 0)$; $2a$

16. $(2, 3)$; 5; $(2, 8)$, $(-2, 0)$; $y + 2x - 2 = 0$ 17. $(5, 7)$, $(-1, 13)$

18. $x^2 + y^2 - 6x - 2y + 9 = 0$, $x^2 + y^2 + 2x - 10y + 1 = 0$

20. $y = 7x - 37$;   The tangent is perpendicular to the diameter with one end the point of contact of the tangent with the circle; $(6, 5)$

21. 10,   26

22. $x^2 + y^2 + (x - 1)^2 + y^2 + (x - 1)^2 + (y - 1)^2 + x^2 + (y - 1)^2$;   (b) $(\frac{1}{2}, \frac{1}{2})$
    (c) $(\frac{1}{2}, \frac{1}{2})$;   1   (d) 2

23. 8 sq units   24. 22 cm^2

Exercise 13.3

1.

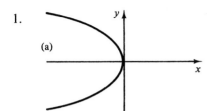

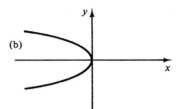

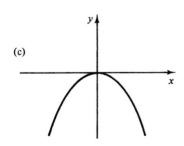

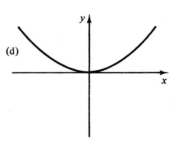

2. (a) $(3, 0)$, $(4, 0)$ (b) $(1, -2)$, $(\frac{3}{2}, -2)$ (c) $(-1, 1)$, $(1, 1)$
   (d) $(-9, 3)$, $(-\frac{35}{4}, 3)$

3. (a) $y^2 = 12x$ (b) $y^2 = 4(x - 3)$ (c) $x^2 = 4(y - 1)$ 4. $x = 2$

5. $2y + 4x = 19$ 6. $16x^2 - 24xy - 156x + 9y^2 - 208y + 624 = 0$

7. $x + y = \pm\sqrt{8}$ 8. $y^2 = 8x$ 9. (a) $y^2 = 4x$ (b) $y^2 = 4ax$

10. $(y - 2)^2 + (x - 2)^2 = 1$ 11. $\dfrac{x^2}{a^2} + \dfrac{y^2}{b^2} = 1$ 12. $2y = x$

13. $2y + 2x = 7$ 14. $yt = x + at^2$; $(2at^2, 2at)$

15. $y + tx = 2at + at^3$; $(\frac{81}{4}a, -9a)$

Exercise 13.4

1. (a) 4, $2\sqrt{2}$ (b) 6, $2\sqrt{3}$ 2. (a) $y + x = 6$ (b) $4y + x + 12 = 0$

5. 10, 5   6. (a) $\frac{35}{2}$, $(7, -9)$, $163.7°$, $16.3°$

7. (a) $m^2 + 30m - 15 = 0$; -30, -15 (b) $y = 12x \pm 16$

9. $m^2 + 24m - 6 = 0$, $\frac{12}{13}$

Exercise 13.5

1. $(25, -10)$ 2. $y = x + a$ 3. $ty = x + at^2$, $\dfrac{|at^2|}{\sqrt{1 + t^2}}$

4. $y + tx = t^3 + 2t$, $(1, 2)$ 5. $y^2 = 4x - 8$

6. $y = 2x + 2$, $y + x + 4 = 0$; $(-2, -2)$

7. $a\sin\theta y + b\cos\theta x = ab$ 8. $ty = t^3x + c - ct^4$, $4xy = c^2$

Exercise 13.6

1. $-\frac{1}{4} \leqslant m \leqslant \frac{1}{4}$, $y = \pm\frac{1}{4}x$ 2. $y = 2(\alpha - 2)x - 2(\alpha - 2)\alpha + \beta$

3. $y = \pm 4x - 2$ 4. $y = 2x^2 + 2$

5. $y - 2at_1 = \dfrac{2}{t_1 + t_2}(x - at_1^2)$ (could interchange t_1 with t_2)

   $t_1y = x + at_1^2$, $t_2y = x + at_2^2$; $(at_1t_2, a(t_1 + t_2))$; $y = 2x + 2a$

6. $2ax = y^2 - y$ 7. $x = -b$ 9. $pq = -3$; ± 1, $8\sqrt{2}$

10. $(a, 0)$, $(8a, 0)$

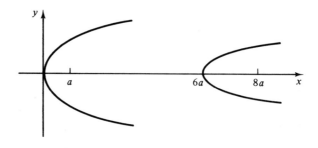

11. $3ty + 2x = 3t^4 + 2t^2$, $\frac{1}{8}$

Exercise 13.7

1. $y = tx - at^3$, $(6a, 4a\sqrt{2})$ or $(6a, -4a\sqrt{2})$ 2. $y^2 - x^2 = 4$

3. $\frac{3}{4}(x + 2)^2 + 1 = y$ 4. $\left(\dfrac{2 + m^2}{m^2}, \dfrac{2}{m}\right)$, $\sqrt{2}y = x + 2$

5. $x = -2$, $y = 1$;
   0,   2

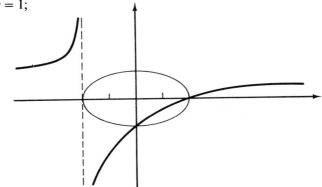

6. $\left(\dfrac{a}{2}(p^2 - 1), \dfrac{3a}{2p}(p^2 - 1)\right)$

7. $(0, 8)$, $4\sqrt{2}$; $(9, -1)$, $5\sqrt{2}$; $(4, 4)$; $167°$ 8. (a) $py = x + ap^2$

9. $y^2 = 2a(x + 2a)$

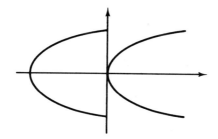

10. $(x - 5)^2 + (y + 4)^2 = 25$; $(x - 1)^2 + y^2 = 1$; $(1, -1)$

11. (a) $(0, 3)$ (b) $(0, \frac{7}{4})$; $x^2 + (y - 3)^2 = 9$ 13. (a) $72°$

14. $(x - 6a)^2 + y^2 = 18a^2$, $(6a, 0)$; $3a\sqrt{2}$;
     $A(3a(2 - \sqrt{2}), 0)$, $B(3a(2 + \sqrt{2}), 0)$; $\frac{1}{2}\pi$, $y = x$, $y = -x$; $18a^2(3 - 2\sqrt{2})$

15. (a) $y + tx = 2at + at^3$ (b) $2y^2 = a(x - a)$

16. $t^2y + x = t + \dfrac{1}{t}$;   (a) $x = 2y$ (b) 4   17. (a) $x = -\frac{1}{4}$ (b) $8y^2 = 4x - 1$

Exercise 14.1

1. (a) $1 + 2x + 2x^2 + \frac{4}{3}x^3$ (b) $1 - x + \frac{1}{2}x^2 - \frac{1}{6}x^3$ (c) $1 + \frac{1}{2}x + \frac{1}{8}x^2 + \frac{1}{48}x^3$
    (d) $1 - x^2 + \frac{1}{2}x^4 - \frac{1}{6}x^6$ (e) $1 + 2x + \frac{3}{2}x^2 + \frac{2}{3}x^3$ (f) $1 - 2x + \frac{3}{2}x^2 - \frac{2}{3}x^3$
    (g) $1 + 2x + x^2 - \frac{2}{3}x^3$ (h) $\frac{8}{3} - \frac{5}{2}x + x^2 - \frac{1}{6}x^3$

2. (a) 2.718   (b) 1.649   (c)1.105

4. (a) $2x - 2x^2 + \frac{8}{3}x^3 - 4x^4 + \ldots;$ $|x| < \frac{1}{2}$
    (b) $x^2 - \frac{1}{2}x^4 + \frac{1}{3}x^6 - \frac{1}{4}x^8 + \ldots;$ $|x| < 1$
    (c) $\ln 10 [x - \frac{1}{2}x^2 + \frac{1}{3}x^3 - \frac{1}{4}x^4 + \ldots;$ $|x| < 1$

5. $x + x^2 + \frac{1}{3}x^3 + \ldots$ 6. $\frac{1}{2}, \frac{1}{12}$

7. $x + \frac{1}{4}x^2 + \frac{1}{8}x^3 + \frac{5}{64}x^4 + \ldots$ 8. $1, \ 0; \ -1, \ -2$

9. (a) $1 + x + \frac{1}{2}x^2 + \ldots$ (b) $1 + x \ln a + \frac{1}{2}x^2(\ln a)^2 + \ldots$

11. $x - \frac{1}{3}x^3 + \frac{1}{5}x^5 - \frac{1}{7}x^7 + \ldots$ 12. $1 - x + \frac{3}{2}x^2$

13. (a) and (b) $x + \dfrac{x^3}{3!} + \dfrac{x^5}{5!} + \ldots$ 14. $-3x - \frac{3}{2}x^2 - 3x^3$

15. $1 - x - \frac{1}{2}x^2 + \frac{5}{6}x^3 - \frac{1}{6}x^4$

16. $2x - x^2 + \frac{17}{6}x^3;$ $0.0396, \ 0.0000226, \ 0.0396$ 17. $-\frac{1}{8}$

18. (a) $1 + x + \dfrac{x^2}{2!} + \dfrac{x^3}{3!} + \ldots + \dfrac{x^n}{n!} + \ldots$ (b) $1 - \dfrac{3x}{2} + \dfrac{11x^2}{8} + \ldots$

19. $3x + \frac{3}{2}x^2 - 5x^3 + \frac{15}{4}x^4 + \frac{33}{5}x^5 - \frac{43}{2}x^6 + \ldots;$ $|x| < \frac{1}{2}$

20. $\frac{1}{2}x - \frac{3}{4}x^2 - \frac{1}{8}x^3 \ldots;$ $|x| < 1$

21. $-8x - 7x^2 - \dfrac{62}{3}x^3 - \ldots;$ $\dfrac{2}{n}x^n[(-1)^n 2^{n-1} - 3^n];$ $|x| < \frac{1}{3}$

22. $1 - \dfrac{1}{x} + \dfrac{\frac{3}{2}}{x^2} - \dfrac{\frac{5}{2}}{x^3} + \ldots;$ 2.9704

24. $\sin x + \frac{1}{3}\sin^3 x + \frac{1}{5}\sin^5 x + \ldots;$ $1 + \sin^2 x + \sin^4 x + \ldots;$
   all values other than $\frac{1}{2}\pi, \frac{3}{2}\pi$

25. (a) $-6x - \frac{3}{2}x^2 - \frac{29}{2}x^3 + \ldots;$ (b) $-6x - \frac{3}{2}x^2 - x^3 - \ldots;$
    (c) $1 - 6x + \frac{33}{2}x^2 + \ldots;$ $(-1)^n \left(\dfrac{3^n}{n!} - \left(\dfrac{n+2}{2} \right) \right); \ -\dfrac{3}{n!}$

26. 10,   4

Exercise 14.2

1. (a) $y^2 + x^2 = c$ (b) $3y^2 = 6x + 2x^3 + c$ (c) $\ln y = e^x + c$

   (d) $\dfrac{1}{1 + y} = e^{-x} + c$ (e) $y = \sin^{-1}(cx)$ (f) $y^2 = \ln(1 + x^2) + c$

   (g) $\tfrac{1}{2}y^2 + \ln y = -e^x + c$ (h) $\ln y = \ln x + \tfrac{1}{2}x^2 + c$

   (i) $y^2 = 2\sin x + c$ (j) $y = -\ln(\cos x + c)$

2. (a) $y = \sin^{-1} x - \tfrac{1}{2}\pi$ (b) $y = \ln|\sin x| + 1 - \ln(\tfrac{1}{2}\sqrt{3})$

3. $\theta = \tfrac{1}{2}(1 + (2\pi - 1)e^{2t})$ 4. $\dfrac{-\pi}{\pi + 6}$ 6. $y^2 = \ln(x^2) - x^2 + 5$

7. $y = \dfrac{2}{1 + e^{-2x}}$ 8. $y = \ln(x^2 + x + 1)^{1/2}$ 9. $y = x\ln x + 1$

10. $y = \dfrac{2e^{x-1}}{x^2}$ 11. $y = \tfrac{1}{2}(1 + e^{1-x^2})$

12. $y = 3 \Big/ \left(\ln\left(\dfrac{x + 1}{x - 2}\right) + c\right)$; $y = 3 \Big/ \left(\ln\left(\dfrac{x + 1}{2(x - 2)}\right) + 3\right)$

13. (a) $y + 2x = 12$ (b) $y = \dfrac{8x}{x + 1}$

Exercise 14.3

1. 900;   1 hr 38 mins   2. $\ln\tfrac{5}{6}$ 3. $v^2 = u^2(1 - e^{-2kx})$

4. (a) $\theta = \theta_0(1 + e^{-kt})$ (b) $\theta = \tfrac{1}{2}\theta_0(2 - e^{-kt})$ 5. 2.04

6. $\dfrac{3}{1 + 3x} - \dfrac{1}{1 + x}$; $y = \dfrac{x - 1}{x + 1}$ 7. 9.5

8. $y = 20e^{-t^2/(1 + t^2)}$; $\dfrac{20}{e}$ (Hint: let $t \to \infty$)   9. $y = 2x(x^2 + 1)$

10. $\dfrac{dy}{dx} = 3\left(\dfrac{1 - y}{2 - x}\right)$; $y = 1 + (x - 2)^3$

   translate 2 unit along the positive x axis then 1 unit along the positive y axis.

Exercise 15.1

1. (a) $\mathbf{a} + \mathbf{b}$ (b) $\mathbf{a} - \mathbf{b}$ (c) $\mathbf{a} + \mathbf{b}$

2. $\mathbf{b} - \mathbf{a}$, $2(\mathbf{b} - \mathbf{a}) = 2\mathbf{b} - 2\mathbf{a}$ hence parallel

3. (a) $\mathbf{c}$ (b) $\mathbf{b} + \mathbf{c}$ (c) $\mathbf{a}$ (d) $\mathbf{a} + \mathbf{b} + \mathbf{c}$ (e) $\mathbf{b} + \mathbf{c}$ (f) $-\mathbf{a} + \mathbf{b} + \mathbf{c}$
   (g) $\mathbf{a} + \mathbf{b} - \mathbf{c}$ (h) $\mathbf{a} - \mathbf{b} + \mathbf{c}$

4. (a) scalar   (b) scalar   (c) scalar   (d) vector   (e) vector   _

5. AB is parallel to CD and is λ times longer

6. (a) $\mathbf{b}$ (b) $\mathbf{a}$ (c) $2\mathbf{a}$ Hint: show that $\vec{AB} = \mathbf{a}$

7. (a) $\mathbf{a}$ (b) $2\mathbf{a}$ (c) $\mathbf{a} - \mathbf{b}$ (d) $\mathbf{a} - \mathbf{b}$ (e) $2\mathbf{b} - \mathbf{a}$

Exercise 15.2

1. (a) 5   (b) 13   (c) $\sqrt{2}$

2. (a) $\frac{1}{13}(-5\mathbf{i} + 12\mathbf{j})$ (b) $\frac{1}{\sqrt{2}}(\mathbf{i} - \mathbf{j})$ (c) $\frac{1}{\sqrt{13}}(3\mathbf{i} - 2\mathbf{j})$

3. (a) $5\mathbf{i}$ (b) $4\mathbf{i} - 7\mathbf{j}$ (c) $\binom{6}{4}$ (d) $\mathbf{i} - \mathbf{j}$ (e) $-\mathbf{i} + 4\mathbf{j}$ (f) $-6\mathbf{i} + 5\mathbf{j}$

4. (a) $\mathbf{a} + \mathbf{b}$ (b) $2\mathbf{a} + \mathbf{b}$ (c) $\mathbf{a} - 2\mathbf{b}$ (d) $2\mathbf{a} + 3\mathbf{b}$

5. (a) $6\mathbf{i} - 2\mathbf{j} + 3\mathbf{k}$ (b) $-2\mathbf{i} - 4\mathbf{j} - \mathbf{k}$ (c) $4\mathbf{i} - 6\mathbf{j} + 2\mathbf{k}$ (d) $-7\mathbf{j}$
   (e) $-2\mathbf{i} - 11\mathbf{j} - \mathbf{k}$

6. (a) $\sqrt{x^2 + y^2 + 1}$ (b) $\sqrt{10}$ (c) $\sqrt{82}$

7. (a) $\frac{1}{\sqrt{14}}(2\mathbf{i} + \mathbf{j} - 3\mathbf{k})$ (b) $\frac{1}{\sqrt{21}}(4\mathbf{i} - \mathbf{j} + 2\mathbf{k})$

8. $2\mathbf{a} - \mathbf{b}$ 10. $\mathbf{b} = 2\mathbf{a}$; $\frac{3}{2}$

Exercise 15.3.

2. $\frac{11}{5}$ 3. (a) $-\frac{\sqrt{42}}{7}$ (b) $-\frac{\sqrt{10}}{10}$ (c) $\frac{5\sqrt{7}}{21}$

4. (a) $85.8°$ (b) $60.5°$

6. (a) $\frac{2}{7}$, $\frac{3}{7}$, $\frac{6}{7}$ (b) $\frac{2}{7}$, $-\frac{3}{7}$, $-\frac{6}{7}$ (c) $\dfrac{1}{\sqrt{3}}$, $\dfrac{1}{\sqrt{3}}$, $\dfrac{1}{\sqrt{3}}$

7. $\mathbf{i}+\mathbf{j}+\mathbf{k}$, $-\mathbf{i}+\mathbf{j}+\mathbf{k}$, $\mathbf{i}+\mathbf{j}-\mathbf{k}$, $\mathbf{i}-\mathbf{j}+\mathbf{k}$ or the negative of any of these

8. (a) $\mathbf{a}-\mathbf{c}$ (b) $-\mathbf{a}-\mathbf{b}+\mathbf{c}$ (c) $\frac{1}{2}(\mathbf{a}+\mathbf{c})$

9. (a) $-\mathbf{a}=-(2\mathbf{i}+\mathbf{j}-\mathbf{k})$ (b) 7 (c) 0 (d) 0   10. -11, -5

12. (a) 7 (b) 4

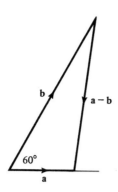

13. $\frac{9}{2}$

Exercise 15.4

3. (a) $\frac{1}{2}(\mathbf{a}+\mathbf{b})$ (b) $\frac{1}{3}(2\mathbf{a}+\mathbf{b})$ (c) $\frac{1}{5}(3\mathbf{a}+2\mathbf{b})$

4. (a) $\mathbf{b}-\mathbf{a}$ (b) $\frac{1}{3}(\mathbf{b}-\mathbf{a})$ (c) $\mathbf{c}-\mathbf{a}$ (d) $\frac{1}{3}(\mathbf{c}+\mathbf{b}-2\mathbf{a})$ (e) $\frac{1}{3}(\mathbf{a}+\mathbf{b}+\mathbf{c})$

5. $\frac{1}{2}(2\mathbf{i}-\mathbf{j}-\mathbf{k})$ 6. (a) $\mathbf{i}+\mathbf{j}-\mathbf{k}$ (b) $\pm\dfrac{1}{\sqrt{3}}(\mathbf{i}+\mathbf{j}-\mathbf{k})$

7. $\pm\dfrac{1}{\sqrt{18}}(4\mathbf{i}-\mathbf{j}-\mathbf{k})$

8. (a) For some $\lambda\in\mathbb{R}$, $\mathbf{a}=\lambda\mathbf{b}$ (b) $\mathbf{a}\cdot\mathbf{b}=0$; $-2\mathbf{j}$: not a rhombus

9. $45°$

10. (a) $\mathbf{r}=(1+t)\mathbf{i}+(2t-1)\mathbf{j}+\mathbf{k}$ (b) $\mathbf{r}=(1+2t)\mathbf{i}-(1+t)\mathbf{j}+(1+t)\mathbf{k}$
    (c) $\mathbf{r}=t(\mathbf{i}+2\mathbf{j}-\mathbf{k})$ (d) $\mathbf{r}=3(1+t)\mathbf{i}+(1+t)\mathbf{j}-(1+t)\mathbf{k}$

11. Many possibilities
    (a) $\mathbf{r}=3\mathbf{i}-\mathbf{j}+t(\mathbf{i}-5\mathbf{j})=(3+t)\mathbf{i}-(1+5t)\mathbf{j}$
       or $\mathbf{r}=2\mathbf{i}+4\mathbf{j}+t(-\mathbf{i}+5\mathbf{j})=(2-t)\mathbf{i}+(4+5t)\mathbf{j}$

11. (b) $\mathbf{r} = 2\mathbf{i} + \mathbf{j} - \mathbf{k} + t(\mathbf{i} - \mathbf{j} + 2\mathbf{k})$
$= (2 + t)\mathbf{i} + (1 - t)\mathbf{j} + (2t - 1)\mathbf{k}$ with similar variations
(c) $\mathbf{r} = t(\mathbf{i} + \mathbf{j} + \mathbf{k})$

12. $3.2°$; (a) L_1 (b) L_2; $\mathbf{k}$!!

13. (a) $\mathbf{r} = \mathbf{c} - \mathbf{b} + t\mathbf{b} = \mathbf{c} + (t - 1)\mathbf{b}$
(b) $\mathbf{r} = \mathbf{b} + t(\mathbf{c} - 2\mathbf{b}) = t\mathbf{c} + (1 - 2t)\mathbf{b}$; $\mathbf{c} - \mathbf{b}$
Hint: equate $\mathbf{c} + (t - 1)\mathbf{b}$ with $s\mathbf{c} + (1 - 2s)\mathbf{b}$

14. (a) 109 (b) $\mathbf{r} = \begin{pmatrix} 3 + 4t \\ 1 + 3t \\ 7 + 5t \end{pmatrix}$ (c) $\begin{pmatrix} -1 \\ -2 \\ 2 \end{pmatrix}$

15. $3 + 9\lambda$; $\frac{1}{3}(\mathbf{i} + \mathbf{j} + 4\mathbf{k})$; $\sqrt{2}$ 16. $4\mathbf{i} + 7\mathbf{j} - 4\mathbf{k}$; $(8, 14, -8)$

17. (a) $\frac{1}{6}(5\mathbf{a} + \mathbf{b})$ (b) $\frac{1}{3}(4\mathbf{b} - \mathbf{a})$ (c) $\frac{1}{4}(\mathbf{a} + 3\mathbf{b})$; $3:1$

18. 6; 5 19. $\frac{1}{18}\sqrt{6}(2\mathbf{i} + \mathbf{j} - 7\mathbf{k})$

20. (a) $\frac{1}{5}\sqrt{5}(\mathbf{i} + 2\mathbf{k})$ (b) $\sqrt{\frac{3}{5}}$ (c) $\mathbf{i} - 2\mathbf{j} + \mathbf{k}$ (d) $\mathbf{i} - 4\mathbf{j}$

21. $2:1$ 22. (a) $71°$ (b) $-\frac{1}{2}$, $\frac{4}{7}$, $-\frac{1}{14}$

23. $\dfrac{1}{r - q}(r(1 - q)\mathbf{c} + q(r - 1)\mathbf{b})$; $\dfrac{1}{(r - p)}(r(1 - p)\mathbf{c} + p(r - 1)\mathbf{a})$ (b) $71.6°$

24. $\frac{2}{5}(5\mathbf{i} - 4\mathbf{j} + 3\mathbf{k})$; $\frac{1}{5}(5\mathbf{i} - 21\mathbf{j} - 3\mathbf{k})$

25. $\mathbf{r} = (4 + 3t)\mathbf{i} + (6t - 8)\mathbf{j} + 3t\mathbf{k}$
or $\mathbf{r} = (4 + s)\mathbf{i} + (2s - 8) + s\mathbf{k}$ or other variant; $6\mathbf{i} - 4\mathbf{j} + 2\mathbf{k}$

26. (a) $2:3$ (b) $\frac{1}{8}(3\mathbf{a} + 5\mathbf{b})$

27. $\frac{1}{5}\sqrt{5}(2\mathbf{i} + \mathbf{k})$; $\mathbf{r} = (a + 2\lambda)\mathbf{i} + (\lambda - 2a)\mathbf{k}$ with other variants

28. $\mathbf{r} = (4 + t)\mathbf{i} + (2 + 2t)\mathbf{j} + (6 + t)\mathbf{k}$; $(5, 4, 7)$,
$\mathbf{r} = (4 + 5t)\mathbf{i} + (2 - 2t)\mathbf{j} + (6 + 5t)\mathbf{k}$

29. $\frac{63}{2}\sqrt{1 - (\frac{59}{63})^2} \simeq 11.05$ 30. $5\mathbf{i} - \mathbf{k}$; 1

Exercise 15.5

1. $2x - 3y + 6z = 41$ 3. (a) $\pm\frac{1}{3}(\mathbf{i} + 2\mathbf{j} + 2\mathbf{k})$ (b) $\pm\dfrac{1}{\sqrt{26}}(3\mathbf{i} - \mathbf{j} + 4\mathbf{k})$

4. (a) $(0, 2, 1)$ (b) $(2, 4, 6)$ (c) $(5, 0, -5)$ 5. (a) 1 (b) $\frac{5}{7}$ (c) 2

6. $\frac{2}{7}$ 7. $76.9°$ 8. (a)$18.7°$ (b) $54.7°$

9. $\mathbf{i} + 2\mathbf{j} + 5\mathbf{k}$; $\mathbf{i} + 2\mathbf{j} + 5\mathbf{k}$ or multiple

0. (a) $3x + 2y + 3z = 9$ (b) $x + y - 3z = 8$

1. (b) $(5, 8, -6)$ (d) $46.9°$; $43°$

2. $\mathbf{r} = (2 + 3t)\mathbf{i} + (6t - 9)\mathbf{j} + (11 - 2t)\mathbf{k}$; $3x + 6y - 2z = 28$.
   Point A, B in the line are equidistant to the plane

Exercise 16.1

1. (a) $3 + 3i$ (b) $4 - 5i$ (c) $1 - \sqrt{3}i$

2. (a) 6 (b) $1 - 2i$ (c) $8 - i$ (d) $6 + i$

3. (a) 2, -5 (b) 6, 5 (c) 3, 2   4. $-i$, 1, i, -1, $-i$, 1, i

5. (a) $2 + 3i$ (b) $-4 + 2i$ (c) $14 + 5i$ (d) $17 + 7i$ (e) 25 (f) $8 + 6i$
   (g) $-1 + 2\sqrt{2}i$ (h) $-2 + 2i$ (i) $-3 + i$ (j) 29

6. (a) $\frac{1}{2} - \frac{1}{2}i$ (b) $-\frac{2}{5} + \frac{1}{5}i$ (c) $\frac{4}{5} + \frac{3}{5}i$ (d) $\frac{13}{5} - \frac{9}{5}i$ (e) $-\frac{1}{2}i$ (f) $-\frac{1}{4} - \frac{1}{4}i$
   (g) 1 (h) $-\frac{2}{5} - \frac{4}{5}i$

7. (a) $3 + 2i$ (b) $4 - 5i$ (c) $3 - i$ (d) 6 (e) 0

8. (a) 2 (b) 5 (c) $-\frac{3}{5} - \frac{4}{5}i$ (d) $-3 + 4i$ (e) $-3 + 4i$ (f)$\frac{1}{5}$

11. (a) $7 - 7i$ (b) $7 + 7i$ (c) $8 - 6i$ (d) $25i$ (e) $\frac{1}{25}(24 - 7i)$ (f) $\frac{1}{25}(24 + 7i)$

13. $x = 3$ and $y = 3$ or -1

14. (a) $\frac{1}{25}(7 + i)$ (b) $x^4 - 6x^2 + 1 + i4x(x^2 - 1)$

15. (a) 1 (b) $\cos 2\theta + i \sin 2\theta$ (c) $\cos \theta - i \sin \theta$ 16. $2 \cos \theta$, $2i \sin \theta$

Exercise 16.2

1. (a) $-1 \pm i$ (b) $\frac{1}{2}(-1 \pm \sqrt{7}i)$ (c) $\frac{1}{2}(-3 \pm \sqrt{3}i)$ (d) $\frac{1}{4}(-5 \pm \sqrt{7}i)$
   (e) $\frac{1}{2}(-3 \pm \sqrt{7}i)$

2. (a) $z^2 - 2z + 2 = 0$ (b) $z^2 - 4z + 29 = 0$ (c) $z^2 - 6z + 25 = 0$
   (d) $z^2 + (3i - 5)z + 8 - 15i = 0$ (e) $z^2 - 5z + 7 - i = 0$

3. (a) 3 (b) $2 - i$ 4. $\frac{1}{2}(1 - i)$

5. Let $w = -\frac{1}{2}(1 + \sqrt{3}i)$ and $w^2 = -\frac{1}{2}(1 - \sqrt{3}i)$ (a) -1, $-w$, $-w^2$
   (b) 2, $2w$, $2w^2$ (c) 4, $2(1 + w)$, $2(1 + w^2)$ or 4, $-2w^2$, $-2w$
   (d) 2, $1 + w$, $1 + w^2$ or 2, $-w^2$, $-w$

6. (a) 0 (b) 1 (c) $4w^2$ (d) -1 (e) $-w$ (f) -1 (g) $\frac{1}{3}(2 + w)$ (h) $-\frac{1}{3}w$

7. (a) $\pm(2 + i)$ (b) $\pm(1 - i)$ 8. 4, -2 10. $\frac{7}{3}$, 1

11. $6\alpha\beta$ 12. Normally 0, but check for $n = 0$

13. $2 + i$, $2 - i$, $-2 + i$, $-2 - i$ 14. $\frac{1}{3}$, $\frac{1}{3}w$, $\frac{1}{3}w^2$

15. (a) $-\frac{1}{2}(1 + \sqrt{3}i)$ (b) $x^2 + 6x + 25 = 0$ 16. (a) $2 - 11i$ (b) $-(2 + 3i)$

17. $-2, -4$ 18. 1, -2; $\frac{1}{10}(3 + 21i)$

Exercise 16.3

1. (a) 2, $\frac{1}{2}\pi$ (b) 3, π (c) 5, $\tan^{-1}(-\frac{4}{3}) \simeq 2.21$ radians
   (d) 13, $-\tan^{-1}\frac{12}{5} \simeq -1.97$ radians
   (e) $\sqrt{2}$, $\frac{1}{4}\pi$ (f) 5, $\tan^{-1}\frac{4}{3} \simeq 0.93$ radians

2. $\sqrt{13}$; $-\tan^{-1}(\frac{2}{3}) \simeq -0.59$ radians; $\sqrt{13}$; 0.59 radians

3. (a) 13, $\pi - \tan^{-1}\frac{12}{5} \simeq 1.97$ radians (b) 13, $\tan^{-1}\frac{12}{5} \simeq 1.18$ radians
   (c) $\sqrt{2}$, $\frac{1}{4}\pi$ (d) $\sqrt{290}$, $\tan^{-1}\frac{1}{17} \simeq 0.06$ radians
   (e) 3, $-\frac{1}{4}\pi$ (f) $\sqrt{34}$, $-\tan^{-1}\frac{5}{3} \simeq -1.03$

4. (a) 3, $\frac{1}{3}\pi$ (b) 5, $-160° \simeq -2.79$ radians (c) 1, $-\frac{3}{4}\pi$ (d) 1,
   $-\theta$ (e) 1, $\theta - \pi$

5. (a) $\pm(\frac{1}{2}\sqrt{2} + \sqrt{\frac{3}{2}}i)$ (b) $\pm(1 - i)$ (c) $\pm(2 + i)$

10. -10, (a) $b > -4$ (b) 2, $-1 \pm 2i$ 11. (a) $18 - i$, $\frac{1}{25}(6 + 17i)$

12. (a) 1, θ; $2\cos\frac{1}{2}\theta$, $\frac{1}{2}\theta$; $2\sin\frac{1}{2}\theta$, $\frac{1}{2}\pi + \frac{1}{2}\theta$

13. $\sqrt{2}$, $-\frac{1}{4}\pi$; $2\sqrt{2}$, $-\frac{3}{4}\pi$; 1, $-\frac{1}{3}\pi$

14. (a) $\sqrt{2}$, $-\frac{1}{4}\pi$ (b) 2, $\frac{1}{3}\pi$ (c) $2\sqrt{2}$, $\frac{1}{12}\pi$ (d) $\frac{1}{2}\sqrt{2}$, $-\frac{7}{12}\pi$

15. $(x + 1)(x^2 - x + 1)$; -1. $\frac{1}{2}(1 \pm \sqrt{3}i)$; $-(1 + i)$, $\frac{1}{2}(1 - \sqrt{3} + (1 + \sqrt{3})i)$,
$\frac{1}{2}(1 + \sqrt{3} + (1 - \sqrt{3})i)$;   1

17. $\pm(1 + 2i)$ 18. $4(\cos\frac{2}{3}\pi + i\sin\frac{2}{3}\pi)$;   64; $\frac{1}{256}$, $-\frac{2}{3}\pi$

Exercise 16.4

1.

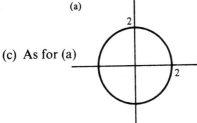

(a)   (b)

(c) As for (a)

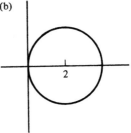

2. (a) Circle centre $(\frac{1}{2}, \frac{1}{2})$ with radius $\frac{1}{2}\sqrt{2}$ (b) Straight line with equation $x + y = 1$

3. z follows the line $y = 2$, w a circle radius 1 centre $(-2, 0)$ (a) 1   (b) $2 + 2i$

4.

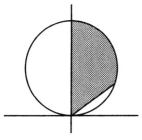

6. (a) $\frac{5}{13} - \frac{14}{13}i$ (b) $\sqrt{34}$, -0.54;   a circle centre the origin with radius $\sqrt{34}$

7. (b) $z^2 - 4z + 5 = 0$ (c) locus is $y = x - 1$—parallel to the line $y = x$

8. (a) $2 + 3i$, $5 - 4i$ (b) $a = 2$, $b = +3$ or $a = -2$, $b = -3$; $\pm(2 + 3i)$
   (c) (i) a circle centre $(2, 0)$ with radius 3   (ii) $x = \frac{5}{2}$

9. (a) $-3\sqrt{3} + 5i$; $\sqrt{52}$, $2.38 \equiv \pi - \tan^{-1}\left(\dfrac{5}{3\sqrt{3}}\right)$ (b) 5, -20

10. (a) $\frac{2}{5}(3 + 4i)$;   2,   0.927   (b) $(x + \frac{5}{3})^2 + y^2 = \frac{16}{9}$,
    a circle with centre $(-\frac{5}{3}, 0)$ and radius $\frac{4}{3}$

11. (a) $2i$,   63°;   a circle with centre $(5, 1)$ and radius 12   (b) $1 - 2i$

12. 1, $\frac{1}{3}\pi$; 1, $\frac{1}{2}\pi$

Exercise 17.1

1. (a) $1.5 - 1.6$ (b) $1.3 - 1.4$ 2. 1.07

3. $1.9 \simeq \frac{3}{5}\pi$;   describe Newton's approximation   4. 3.43

5. 1.11; the Newton's Raphson method   6. 0.35 7. 3.20 (2 d.p.)

8. -1.21 9. -1.6, 0.6 10. 0.766 11. 1.8 12. 0.44

13. $2.0147 \simeq 2.01$;   subsequent approximations will not affect the first two decimal places;   show that $f(x_1) \div f'(x_1)$ is small

14. 0.53 15. 3.53; 0.278

Exercise 17.2

1. 0.7077 2. 0.5008 3. $3.77\,\text{cm}^2$; $3.90\,\text{cm}^2$ 4. $11.31\,\text{cm}^3$

5. 0.025; 2.025 6. $22.6\,\text{cm}^3$ 7. $-1/\pi$

Exercise 17.3

1. (a) 0.987 (b) 0.873 (c) 0.696 (d) 1.405

2. (a) 1.000; 1 (b) 4.014 3. 2.005; 0.005 approx.

4. 0.693; 0.693 5. (a) 0.836 (b) 4.041; 4.047

6. 0.18 7. (a) 0.721 (b) 0.719

8. (a) 10.86 (b) 11.96

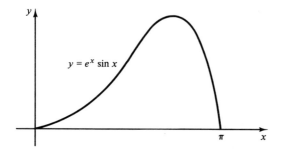

Little curvature from 0 to $\frac{1}{2}\pi$. From $\frac{1}{2}\pi$ to π the curvature is high compared with the size of the intervals. So Simpson's rule will give a more accurate answer because it attempts to approximate with quadratic curves rather than straight line segments.

9.

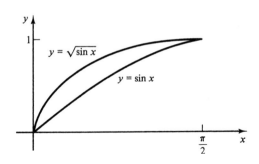

I corresponds to an area which can be enclosed with a rectangle of area $\frac{1}{2}\pi$;   1.14

10. 4.69   11. 2.76   12. -0.026

13. $\frac{2}{3}\pi(e^{-9} + 4e^{-4} + 2e^{-1} + 2)$

Exercise 17.4

1. 3.3,   1.2   2. 3, -1.0 3. 0.25,   2.5

4. 5, -1 5. 3.1,   4   6. 6.2,   1.2

7. (a) Plot y^3 against x^2, find gradient and intercept;
   $b = (\text{intercept})^{-1}$, $a = -b \times$ gradient
   (b) Plot $\ln y$ against $\ln x$, find gradient and intercept; $c = e^{(\text{intercept})}$,
   $d =$ gradient

8. 2.5,   0.5   9. 8.38,   11.2   (a) 1.9,   3.6   (b) 1.15

10. (a) 3150, -1.5 (b) 17.6

11. (i) Take logs and plot $\ln y$ against x; $\ln b$, $\ln(ab)$

    (ii) Plot $\dfrac{y}{x}$ against x; $-\dfrac{p}{q}$, $\dfrac{1}{q}$

Revision exercise D

1. (a) $-\dfrac{3 + 11i}{13}$ (b) $\sqrt{13}$, $\sqrt{10}$,   0.66

(c)

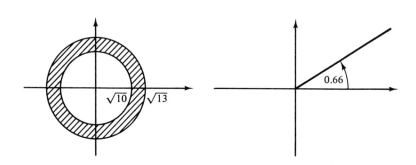

2. $y = 2 + \sqrt{25 - (x - 1)^2}$ [Hint: Think of a circle]

3. (a) $4\ln 2 - \frac{15}{16} = 1.83$ (b) 1   (c) $y = 2(e^x - 1 - x)$

4. (a) $y = (1 - \ln\sqrt{\cos x})^2$ (b) (ii) 1

6. $\left(\dfrac{d^3y}{dx^3} = 4(3y - 1)\dfrac{dy}{dx}\right),\quad \dfrac{d^4y}{dx^4} = 8y(15y^6 - 15y + 2);$

   $\sec^2 x$ is an even function; $\frac{1}{6}\sqrt{6}$

7. $1 - \dfrac{x^2}{2a^2} - \dfrac{x^4}{8a^4} - \dfrac{x^6}{16^6}$ (a) $3 - \frac{2}{3}x^2$ (b) $18x - 112x^3$

8. (a) $\dfrac{4 + 5x}{(1 + 2x)(1 - x)}$ (b) $2\sqrt{3}y + 2x = 3\sqrt{3}$ 9. (b) $b > -9$;   5,   11

10. $4(x - \frac{1}{2})^2(x + 4)$ 11. $\frac{27}{16}\pi r^3$ when $t = \frac{1}{2}$

12. (a) 4,   5;   9, -1 (b) 0.79,   1.43   (c) 1.577

13. -3,   2; -1.1, $\frac{11}{12}$ 14. (a) $\frac{1}{3}\ln\left|\dfrac{1 + x}{2 - x}\right|$ (b) $\dfrac{\pi}{2}$

15. $\frac{3}{2}$, $\frac{1}{2}$;   1970,   1393   17. $(e, 1/e)$ maximum

18. (a) (i) $\dfrac{x}{\sqrt{x^2 + 1}}\,e^{\sqrt{x^2+1}}$ (ii) $x(\ln(x^2) + 1)$ (iii) $-12\cos^3 3x \sin 3x$

    (b) $3y + 2x = 5$

19. (a) ± 1.05, ± 2.30 (b) 1.97   20. $\sqrt{b(b + c)}$

22. $1 - \dfrac{x^2}{3!} + \dfrac{x^4}{5!}$; $\quad -y - \dfrac{y^2}{2} - \dfrac{y^3}{3}$; $\quad -\frac{1}{6}, \quad -\frac{1}{180}$

23. (a) -1, $357\,643$ (b) 0.995 24. $k^2 y + x = 2kc$

25. $-1/pq$ 26. (c) $y^2 = a(x - a)$ 27. $(apq, a(p + q))$

28. (a) (i) 0 (ii) 3 (b) $\ln \frac{5}{2}$, $-\ln 6$ 29. $(t^2 + 1)/2t$

30. (a) $1 + \dfrac{x}{\sqrt{1 + x^2}}$ (b) $\dfrac{2 \cos 3t - 3t \sin 3t}{2 \sin 3t + 3t \cos 3t}$ (c) $\dfrac{1 + 2x}{(1 + x)^2} e^{2x}$; $\quad x > -\frac{1}{2}$

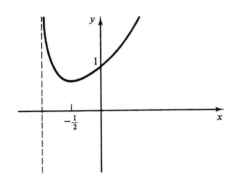

31. $(\pi, 2)$

32. $6, \quad 1, \quad 34, \quad 198;\quad 2, \quad 6, \quad 17, \quad 33$

Index